The Biology of Sole

Editors

José A. Muñoz-Cueto

Department of Biology
Faculty of Marine and Environmental Sciences
Marine Research Institute (INMAR)
Marine Campus of International Excellence (CEI·MAR)
Agri-food Campus of International Excellence (ceiA3)
University of Cádiz
Campus Río San Pedro
Puerto Real (Cádiz), Spain

Evaristo L. Mañanós Sánchez

Fish Reproduction and Diversification Group
Institute of Aquaculture Torre de la Sal (IATS)
Spanish National Research Council (CSIC)
Ribera de Cabanes, Castellón, Spain

F. Javier Sánchez Vázquez

Department of Physiology, Faculty of Biology
Regional Campus of International Excellence "Campus Mare Nostrum"
University of Murcia
Murcia, Spain

CRC Press
Taylor & Francis Group
Boca Raton London New York

CRC Press is an imprint of the
Taylor & Francis Group, an **informa** business

A SCIENCE PUBLISHERS BOOK

Cover illustrations reproduced by kind courtesy of Evaristo L. Mañanós Sanchez y David Cordero Otero

CRC Press
Taylor & Francis Group
6000 Broken Sound Parkway NW, Suite 300
Boca Raton, FL 33487-2742

© 2019 by Taylor & Francis Group, LLC
CRC Press is an imprint of Taylor & Francis Group, an Informa business

No claim to original U.S. Government works

Printed on acid-free paper
Version Date: 20190111

International Standard Book Number-13: 978-1-498-72783-9 (Hardback)

Library of Congress Cataloging-in-Publication Data

Names: Muñoz-Cueto, José A. (José Antonio), editor. | Mañanós-Sánchez, Evaristo L. (Evaristo Luis), editor. |Sánchez-Vázquez, F. Javier (Francisco Javier), editor.
Title: The biology of sole / editors, José A. Muñoz-Cueto, Department of Biology, Faculty of Marine and Environmental Sciences, Marine Research Institute (INMAR), Marine Campus of International Excellence (CEI-MAR), Agrifood Campus of International Excellence (CeiA3), University of Cádiz, Campus Río San Pedro, Puerto Real (Cádiz), Spain; Evaristo L. Mañanós-Sánchez, Fish reproduction and diversification Group, Institute of Aquaculture of Torre la Sal, CSIC, Ribera de Cabanes, Castellón, Spain; F. Javier Sánchez-Vázquez, Department of Physiology, Faculty of Biology, Regional Campus of International Excellence Campus Mare Nostrum, University of Murcia, Murcia, Spain.
Description: Boca Raton, FL: CRC Press, Taylor & Francis Group, [2019] | "A science publishers book." | Includes bibliographical references and index.
Identifiers: LCCN 2018057447 | ISBN 9781498727839 (hardback)
Subjects: LCSH: Flatfishes.
Classification: LCC QL637.9.P5 B56 2018 | DDC 591.5/18--dc23
LC record available at https://lccn.loc.gov/2018057447

Visit the Taylor & Francis Web site at
http://www.taylorandfrancis.com

and the CRC Press Web site at
http://www.crcpress.com

Preface

Senegalese sole, *Solea senegalensis*, is a flatfish highly appreciated in Spain and Portugal. Its name in Spanish (*"lenguado"*) and Portuguese (*"linguado"*) derives from the latin *"linguatus"*, meaning tongue-like shape. Actually, solea has a flat oval shape with numerous skin dots and shades that mimic the surrounding environment. This fish species is widely distributed from eastern Atlantic (southern Great Britain and Ireland) to western African (Angola-Senegal, herein its name), inhabiting coastal and estuarine sandy seabed. It is also found in western Mediterranean as far east as Tunisia, as a migration phenomenon through the Straits of Gibraltar.

Senegalese sole and common sole, *Solea solea*, are very similar in appearance, although the latter one has a compact black spot near the margin of the pectoral fin of the eyed-side. Historically, there is evidence for *S. solea* aquaculture back in the late XIX century in France, although its production never succeeded really despite research and production efforts made in France and the UK in the late XX century. *Solea senegalensis*, however, offered better prospects as this species proved stronger and better adapted to warm waters, assessing a great success when reproduction in captivity was achieved in the 1990s from wild-caught broodstocks. With sustained spawning and larvae supply, juvenile on-growing and mass production was finally possible, although major bottlenecks remained regarding (1) reproduction control and larval mortality, (2) optimum feeding protocols and diets, and (3) disease control and malformations.

Aquaculture of *S. senegalensis* started in Portugal and Spain in the late 1970s mid 1980s with very few tonnes per year. This species was introduced as a promising model for fish farming diversification in southern Europe. Sole farming has been expanding exponentially since then, reaching in 2017 over 1.600 tonnes, which were produced mainly in four countries: Spain (830 tonnes), Portugal, France and Iceland. Outside Europe, S. sole is being cultured in China, although there is no reliable production data (300–500 tonnes unofficial rough estimates).

In the last decade, devoted research programmes provided substantial advance in our scientific knowledge of Senegalese sole and helped solving problems and improving sole farming prospects. Actually, over seven hundred scientific papers have been published on *S. senegalensis*, involving researchers mainly from Spain (68%), Portugal (39%), France (7%) and UK (7%). This book brings together the work of the world's most relevant researchers working on different and complementary aspects of the general Biology, Physiology, Behaviour and Pathology of Senegalese sole.

The book is divided into two Sections: A (General Biological and Engineering overviews) and B (Specific Physiological functions). Section A begins with an introductory chapter (A-1) entitled *"An Overview of Soleid (Pleuronectiformes) Fisheries and Aquaculture"*, written by the most respected senior scientists working in sole: Bari Rhys Howell and Maria Teresa Dinis. They will stress the commercial interest of this flatfish species, discussing the current fisheries situation with declining catches and the increasing interest for fish farming to supply the market. The chapter introduces *S. senegalensis* as the best candidate because of its faster growth and greater tolerance to captivity conditions. The second chapter (*A-2*) deals with the *"Engineering of Sole Culture Facilities: Hydrodynamic Features"*, providing technical advice for the design of aquaculture systems adapted to sole. This chapter highlights important particularities of flatfish and bottom feeding, and the need to optimize space introducing the concept of multilevel layer, flow-throw recirculating systems for intensive on growing facilities. Special attention is paid to tank geometry and water inlet configurations and velocities to safeguard the welfare of fish with bottom and sedentary habits.

Section B begins with 5 chapters devoted to Reproduction (B-1). The first one (*B-1.1*) is entitled *"Sensing the Environment: The Pineal Organ of the Senegalese Sole"*, and focusses on phototransduction mechanisms and the key role of the pineal, which is a neural photosensory structure that encodes environmental information into rhythmic neural projections and neuroendocrine (melatonin) signals. This chapter put together current knowledge on the functional neuroanatomy of the pineal organ, the characterisation of its photoreceptor and melatonin-synthesizing cells, and the tract-tracing studies revealing the pinealofugal and pinealopetal projections. The second chapter (*B-1.2*) reviews the *"Neuroendocrine Systems in Senegalese Sole"*, and reports as the brain, and particularly the forebrain, plays a central role in the neuroendocrine control of many physiological processes in fish, such as reproduction, feeding, metabolism, growth and stress. This chapter describes how external and internal signals are perceived by fish through specific sensory systems and are integrated into a cascade of neurohormones transported via neurosecretory fibres that enter through the pituitary stalk and are released from axon terminals directly into the surroundings area of the target adenohypophyseal cells. In the third chapter (*B-1.3*), the *"Reproductive Physiology and Broodstock Management of Soles"* are addressed. Different issues are discussed such as the anatomy, morphology and maturation of the gonads, the annual reproductive cycles, the endocrinology of reproduction, sex differentiation and the hormone-based therapies developed to stimulate sole spermiation and spawning. The fourth chapter (*B-1.4*) is devoted to *"Sperm Physiology and Artificial Fertilization"*, as most breeders are still wild-caught and gamete management is required for *in vitro* fertilization. Issues regarding factors leading to poor sperm quality will be reviewed and the use of new techniques (apoptosis, DNA damage and transcript analysis) will be discussed. The last chapter (*B-1.5*) focus on the *"Mating Behaviour"* of sole as their reproduction failure in captivity is most likely linked to unpaired behavioural responses of the broodstocks. Courtship patterns and dysfunctions of sole raised in captivity are discussed based on behavioural analysis of different aspects such as dominance, mating systems, genetics, reproductive development, chemical communication and learning.

The next section of the book is related to Early Development (B-2) and comprises 4 chapters. The first one (*B-2.1*) introduces *"The Biological Clock of Sole: From Early Stages to Adults"*, starting with the characterization of the molecular clock in Senegalese sole and its development during embryonic and larval stages. The synchronizing effect of light during these early stages is discussed, highlighting its role in the onset of the molecular clock mechanism and gene expression. The second chapter (*B-2.2*) is entitled *"Embryonic and Larval Ontogeny of the Senegalese sole: Normal Patterns and Pathological Alterations"* and considers the main events during embryogenesis, larval development and metamorphosis. The complex flatfish metamorphosis and the changes in the brain, neuroendocrine, bone, digestive, epidermal, cardio-respiratory, hematopoietic, excretory and immune cell-tissues, as well as pigmentary disorders and skeletal abnormalities, will be reviewed.

The third chapter (*B-2.3*) is focused on the *"Effects of Light and Temperature Cycles during Early Development"*, as these two main environmental factors synchronizes the clock that triggers hatching rhythms. The underwater photo-environment and the effect of light wavelengths on larval performance are also discussed, as well as the long-lasting effects of daily thermocycles on reproduction rhythms and sex ratio. Finally, the last chapter (*B-2.4*) reviews the *"Larval Production Techniques"*, introducing key issues of the production techniques used for Senegalese sole larvae and post-larvae, focusing on rearing systems, feeding protocols and early nutrition.

The following section deals with Nutrition (B-3) and contains one chapter (*B-3.1*) devoted to *"Macronutrient Nutrition and Diet Formulation"*. Here, the nutritional requirements of Senegalese sole are discussed in order to design high quality, cost-effective feeds. This chapter reviews major achievements in research on macronutrient requirements and recommended levels to include in diets for juveniles. The feasibility of using new feedstuffs to replace the marine sources by more sustainable aqua feeds is also addressed, with particular focus on plant sources.

Welfare issues will be dealt in this section (B-4) along three chapters. The first one (*B-4.1*) is entitled *"Welfare, stress and immune system"* and looks at current research on the S. sole stress response and the crosstalk with the immune system. This chapter further explores circadian rhythms in the hypothalamus-pituitary-interenal (HPI) stress axis and its connection with the thyroid axis. The second chapter (*B-4.2*) is focused in *"Ecotoxicology"* and reviews the bioaccumulation of chemicals and their effects in S. sole as early-warning biomarkers to monitor water pollution and ultimately human health. The last chapter (*B-4.3*) is devoted to *"Pathology and Diseases Control"*, which reviews the main diseases caused by bacteria, viruses and parasites in both farmed and wild sole. Special attention is paid to the particularities of sole and its susceptibility in rear conditions in captivity, suggesting guidelines for disease prevention management.

The osmoregulatory capabilities of S. sole to cope with changes in environmental salinity is discussed in the Section B-5 in one chapter (*B-5.1*) entitled *"Osmoregulation"*. This chapter summarizes current knowledge on osmoregulatory tissues (gills, intestine and kidney), ion transport strategies and metabolic changes in sole inhabiting coastal waters and riverine estuaries. The endocrine control (cortisol, growth hormone, insulin/insulin-like growth factors, thyroid and renin/angiotensin system) is also discussed.

The last section is dedicated to Genetics (B-6), which is reviewed in the chapter *B-6.1*, *"Genetic and Genomic Characterization of Soles"*. The book ends summarizing the advances made recently in genetic and genomic resources in *Solea*, including main transcriptome assemblies, genetic maps and existing molecular markers for genetic studies. This information is further used to research on epigenetics and the implementation of genetic breeding programs for the sole aquaculture industry.

This book represents a dissemination activity of the research projects CRONOSOLEA (AGL2010-22139-C03), SOLEMBRYO (AGL2013-49027-C3) and BLUESOLE (AGL2017-82582-C3), developed by the editors of this book and funded by the Spanish Ministry of Economy and Competitiveness (MINECO).

Contents

Section A
Fisheries, Aquaculture and Engineering

A-1.1

An Overview of Soleid (Pleuronectiformes) Fisheries and Aquaculture

Bari Rhys Howell[1,*] and *Maria Teresa Dinis*[2]

1. Introduction

Flatfish occur throughout the world and numerous species are caught and highly valued as a source of flavorsome and nutritious food. A high demand, often combined with diminishing resources, results in high prices that make the fish attractive to fishermen and farmers alike. In some parts of the world species of the family Soleidae are among the most favoured. This chapter reviews the status of catch fisheries for these species and the progress made towards increasing supply through the development and application of aquaculture techniques.

Reviews of flatfish catch fisheries are included in a recent publication on the biology and exploitation of flatfish (Gibson 2005). Information on soleid fisheries has largely been sourced from three chapters of that book that comprehensively review Atlantic (Millner et al. 2005), Pacific (Wilderbuer et al. 2005) and tropical (Munroe 2005a) flatfish fisheries. Data on landings have been updated from the most recent FAO fisheries statistics (FAO 2014).

Research on the development of aquaculture techniques for soles has a long history dating back to the beginning of the 20th century. The somewhat spasmodic progress is presented in roughly chronological order together with an explanation of the changing context that motivated the prioritization of the research. This historic account provides the basis for recent advances that have enabled, and will undoubtedly continue to underpin,

---
[1] Cardiff, UK.
[2] Centre of Marine Sciences, University of Algarve, Portugal.
* Corresponding author: bari-howell@virginmedia.com

long-awaited commercial developments. These most recent scientific advances are the principal motivation for this book and are described in detail in the following chapters.

2. Fisheries

2.1 Commercial importance of flatfish

Important industrial fisheries for flatfish have become established in areas where larger species are sufficiently abundant. Invariably, national or international management measures have been designed to maintain a maximum sustainable yield. Despite their popularity as food, the contribution of flatfish to global commercial fisheries is relatively small. In 2012, for example, the global capture production of fish, crustaceans, molluscs and other invertebrates was about 92 million tonnes (FAO 2014). Apart from minor fluctuations, this amount has changed little over the past 10 years. Of this total, 12 million tonnes derived from inland waters with the remaining 80 million tonnes coming from the marine fishing areas defined by FAO (Fig. 1). Catches of marine fishes accounted for most (66 million tonnes) of this. The contribution of all flatfish was just under 1 million tonnes, i.e., about 1.5% of all marine fish landed. This was slightly less than the contribution of cod (*Gadus morhua* L.)!

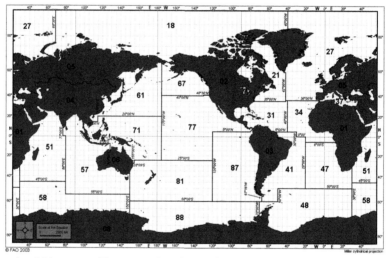

Fig. 1. FAO marine fishing areas. The areas referred to in the text are: 1 (Egypt, Inland Waters), 21 (Atlantic, Northwest), 27 (Atlantic, Northeast), 34 (Atlantic, Eastern Central), 37 (Mediterranean and Black Sea), 41 (Atlantic, Southwest), 47 (Atlantic, Southeast), 51 (Indian Ocean, Western), 57 (Indian Ocean, Eastern), 61 (Pacific, Northwest), 67 (Pacific, Northeast), 71 (Pacific, Western Central).

The significance of flatfish catches expressed in terms of weight somewhat understates their worth because in many instances the value of flatfish per unit weight is considerably greater than that of many other exploited species. Their economic contribution is consequently greater than would appear from weight data only.

2.1.1 Commercial importance of the Soleidae

A brief overview of dominant flatfish families by FAO fishing areas reveals the importance of the Pleuronectidae in world flatfish fisheries. This family is dominant in the north Atlantic and throughout the Pacific. Exceptions include the south-west Atlantic, where

Table 1. Recorded landings of soleid species in 2012 (data from FAO, 2014).

Common name	Species	FAO Fishing Areas	Landings (t)
Common sole	*Solea solea* (L.)	1, 27, 34, 37	32,746
Sand sole	*Solea lascaris* (Risso, 1810)	27, 34	239
Senegal sole	*S. senegalensis* (Kaup, 1858)	27	60
Wedge sole	*Dicologlossa cuneata* (Moreau, 1881)	27, 34, 37	1,198
Thick-back soles	*Microchirus* spp.	27	202
West coast sole	*Austroglossus microlepis* (Bleeker, 1863)	47	1,561
Agulhas or mud sole	*A. pectoralis* (Kaup, 1858)	47, 51	338

paralichthyids dominate, and the south-east Atlantic where soleids, bothids and cynoglossids are the most dominant. However, the most important soleid fisheries occur in the north-east Atlantic alongside pleuronectids and scophthalmids. This apparent restriction of the commercially important soleids to the eastern side of the Atlantic is illustrated in the 2012 catch data for soleid species shown in Table 1. The common sole, *Solea solea* (L.), is the only species for which there is a substantial targeted fishery, which in 2012 yielded about 11% of the total flatfish catch of the northeast Atlantic. The landings of other soleid species in this fishing area were relatively small and were largely by-catches of demersal fisheries targeting other species. This is also the case for the *Austroglossus* spp. fished in the south-east Atlantic which are mainly a by-catch of the productive hake (*Merluccius capensis*, Casteinan) fishery.

This overview of commercial soleid fisheries does not reflect the much wider global distribution of soleids. Although soleids are curiously absent from the west-Atlantic and only occur rarely in the eastern Pacific, the family is widely distributed throughout the Indo-Pacific with maximum diversity in the Indo-Malayan archipelago and north Australian waters (Munroe 2005b). Although the contribution of soleids in these and other areas is largely unquantifiable they undoubtedly have substantial local importance (see below).

2.1.2 Soleids in the northeast and east-central Atlantic, and Mediterranean

Common sole. The common sole is by far the most important commercial soleid in the world and the only one for which there is a targeted fishery. The species is widely distributed from the Trondheim Fjord in Norway southward, around the whole of the British Isles and most of the Mediterranean, to Senegal on the west coast of Africa. Annual landings from 2003 to 2012 ranged from 33 to 42 thousand tonnes (Fig. 2), showing a gradual decline that continued the trend of the 1990s that Millner et al. (2005) described and attributed to over-exploitation.

In 2012, the total landings were just less than 33 thousand tonnes (Fig. 2), the majority of which (73%) was caught in the northeast Atlantic. The remainder was caught in the Mediterranean (16%), the east-central Atlantic (7%) and in Egypt's inland waters (4%). The latter seemingly paradoxical contribution is presumed to have originated from Lake Qarun into which the local sole (named *S. vulgaris* at the time) was introduced in the 1930s and 1940s following salinization of the lake after its disconnection from the freshwater of the Nile (Ishak 1980). Within the productive northeast Atlantic area, the greatest landings were from the south and central North Sea but with significant landings from the eastern English Channel, the Bay of Biscay and the Irish Sea. Although significant, the size of the catch is a poor reflection of the value of the fishery. In 1998, Millner et al. (2005) calculated that, although the total catch of sole was only 10% of the total flatfish catch by weight, it represented 34% of its value.

Beam trawling is the usual form of capture, although there is an important gill net fishery for sole (as well as other species) along the Dutch coast. The seasonal spawning concentrations of sole close inshore is exploited by inshore fishing vessels that use trawls and fixed tangle nets, especially in the English Channel. However, it is the large fleets of trawlers from the Netherlands, Belgium and England that generate the greatest part of the catch through their exploitation of the stocks of the North Sea and the English Channel. The development of intensive targeted fisheries for sole began in the 1950s. Vessels from the Netherlands began using heavy chains to increase the effectiveness of the beam trawls in catching sole and this was accompanied by a rapid increase in the size and power of vessels. By the 1980s these developments had spread to all fleets around the British Isles (Millner et al. 2005).

The decline in stocks since the late 1980s emphasized the need for effective management. This is primarily achieved by regulations governing the gear used and the imposition of quotas that aim to maintain the spawning stock biomass (SSB) above the threshold at which recruitment is impaired, while maintaining fishing mortality below the level that would drive the SSB to that threshold. In 2013, nearly all stocks of sole in the northeast Atlantic were assessed as being fished inside safe biological limits. Only the Irish Sea stock was deemed to be overfished and operating outside safe biological limits. These assessments are made by the International Council for the Exploration of the Sea (ICES) that provides independent scientific advice to countries bordering the North Atlantic, including the European Union (Seafish 2013). In the Mediterranean the abundance of sole is too low to justify a targeted fishery and so its capture is part of a multi-species fishery in which vessels rely on the capture of a large number of species for their income.

Other Soleidae. French, Portuguese and Spanish boats, fishing in the northeast Atlantic, eastern-central Atlantic and Mediterranean land small quantities of four other soleid species which are caught as by-catches rather than targeted fisheries. Of these species the Senegal sole, *Solea senegalensis* Kaup, is of particular interest because it is almost indiscernible from the common sole and the two species may not always be differentiated in areas where their distribution overlaps. This includes the Bay of Biscay, northwest Africa and the western

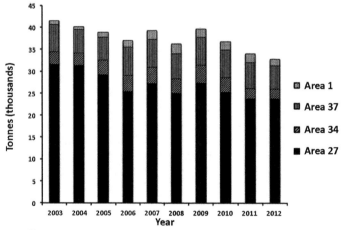

Fig. 2. Landings of common sole, Solea solea, by FAO fishing areas from 2003 to 2013.

Mediterranean. The catch data for these two species in these areas may not therefore be very reliable. In 2012, for example, the common sole catch in the eastern-central Atlantic was reported to be 2,223 tonnes with a further 8,274 tonnes of 'unidentified' soleids, about 70% of which was landed by the Nigerian fleet. Surprisingly, no catches of Senegal sole were reported from this area, the only catch of that species recorded (FAO 2014) was 60 tonnes from French vessels in the Bay of Biscay. It seems possible that the reported catch of *S. senegalensis* has been underestimated either by being grouped with the common sole or by being included in the unidentified Soleidae category.

2.1.3 Soleids in the southeast Atlantic

According to FAO statistics (FAO 2014), *Austroglossus* spp. are the only commercially significant soleids in the southeast Atlantic. The two commercially important species, the west coast sole (*A. microlepis* Bleeker) and the Agulhas or mud sole (*A. pectoralis* Kaup), represent only a small proportion of the total catch as a by-catch of the hake fishery. However, although hake accounts for up to 70% of the catch by weight the soles are by far the most important finfish species because of their higher value per unit weight. They are consequently the main target of small trawlers particularly off the south coast (Diaz de Astarloa 2002).

The west coast sole is distributed from northern Namibia to Cape Town, South Africa while the Agulhas sole is mainly found off the south coast of south and east coast of South Africa. During the 1970s and 1980s annual catches of west coast sole off South Africa were variable but showed peaks of around 2000 tonnes. Landings diminished to virtually zero during the 1990s (Millner et al. 2005) and have not recovered since. This reduction in catches by South African fishermen coincides with Namibia becoming independent in 1990 with the consequence that South African boats were limited to the area south of the Orange river, which marks the border between South Africa and Namibia (Diaz de Astarloa 2002). Landings of this species since that period are now limited to Angola & Namibia but with a significant proportion being caught by Korean Republic fishermen (Fig. 3).

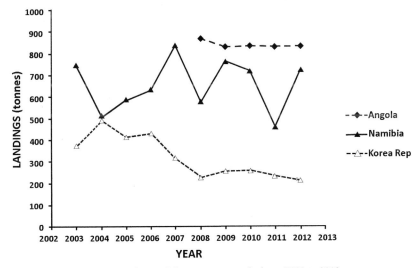

Fig. 3. Landings of the west coast sole from 2003 to 2012.

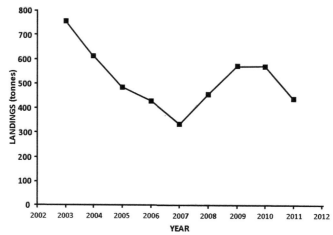

Fig. 4. Landings of the mud sole from 2003 to 2012.

The distribution of the Agulhas sole does not extend to the African west coast and is mainly caught off the south coast of South Africa. During the 1980s and 1990s the annual landings averaged about 850 tonnes (Millner et al. 2005) but from 2003–2012 landings had decreased to an average of about 500 tonnes (Fig. 4). They remain, however, one of South Africa's most valued fish.

2.1.4 Soleids in the tropics

Despite the absence of quantitative data on the commercial value of soleids in tropical waters, Munroe's (2005a) review of flatfish in these areas provides some quantitative evidence of their value.

Flatfish are common in suitable soft-bottom habitats but most are too small and too sparsely distributed to be directly targeted as a commercial fishery. Larger fish caught as a by-catch can be marketed for human consumption and provide an important source of nutrition in areas where they are abundant. In the coastal fisheries of Indo-west Pacific, several soleid genera are prominent including *Solea* spp., *Synaptura* spp., *Brachirus*, *Zebrias* spp. and *Pardachirus* spp. (Table 2). In the estuarine waters of the Indo-West Pacific, the ovate sole, *Solea ovata* Richardson, also has some prominence as does the black sole, *Achlyopa nigra* Macleay, whose distribution extends to the Northern Australian estuaries.

The type of fisheries ranges from traditional subsistence and artisanal fisheries to large-scale industrial fisheries. Subsistence and artisanal fisheries play a major role in providing food for local communities as well as an occupation for a high proportion of the population.

The fishing gear used in the tropics is similar to that used elsewhere for flatfish. Beam trawls and other benthic trawls are used in deep waters and mini-trawls, barrier or stake nets, seines, traps and many other gears are used in shallow water. In African waters, artisanal canoes are the most important part of the fishing fleet with the fishermen landing more fish (50–90% of the total landings) than the industrial fisheries in their region. The industrial trawl fisheries use large boats with modern equipment targeting specific or groupings of fish. Unlike their more traditional subsistence and artisanal counterparts, the majority of the by-catch is invariably discarded or processed into fishmeal or other products.

Table 2. Examples of soleid species marketed from by-catch in the Indo-west Pacific (compiled from Munroe 2005).

FAO fishery area	Species
Oriental sole	*Brachirus orientalis* (Bloch & Schneider 1801)
Elongate sole	*Solea elongate* (Day, 1877)
Convict zebra sole	*Zebrias captivus* (Randall, 1995)
Indian zebra sole	*Zebrias synapturoides* (Jenkins, 1910)
Finless sole	*Pardachirus marmoratus* (Lacepède, 1802)
Stanaland's sole	*Solea stanalandi* (Randall & McCarthy, 1989)
Commerson's sole	*Synaptura commersonii* (Lacepède, 1802)

In these tropical areas landings of soleids, as for most flatfish, cannot be estimated with any degree of accuracy mainly because only a small proportion of catches is identified to species level. Even the proportion identified to family level is relatively small. Landings of all flatfish, however, though variable, have increased during the last decade. In the Indo-Pacific region, for example, catches increased from about 45,000 tonnes in 2003 to 65,000 tonnes in 2012. The financial contribution of soleids is impossible to estimate, but in terms of the provision of nutritious food is likely to be significant in local areas.

3. Aquaculture

3.1 Early pioneers

The development of farming techniques for flatfish has had a long and sporadic history since its origins in the latter half of the 19th century. The 'early hatchery movement', as it was known and described in some detail by Shelbourne (1964), aimed to arrest the decline of natural stocks by the release into the sea of countless millions of eggs and yolk-sac larvae. By the 1930s this approach to fish stock management had fallen into disrepute having failed to demonstrate the efficacy of such a practice, but the demonstration that marine fish eggs could be fertilized successfully and hatched in extremely large numbers did, however, lay the foundations and inspiration for subsequent work on the rearing of marine fish larvae, including those of soles. The first reported success was that of Fabre-Domergue and Biétrix (1905), working at the Concarneau Laboratory in France. They reared common sole from eggs through to the completion of metamorphosis in 50-litre barrel-shaped glass containers. The larvae were fed first on a flagellate culture followed by plankton collected from local rock pools.

Further attempts to rear any marine fish beyond the yolk-sac stage generally foundered on an inability to provide the larvae with adequate nourishment. The discovery by the Norwegian biologist Rollefsen (1939) that the nauplii of the brine shrimp, *Artemia salina* (L.), were a suitable food for plaice larvae was therefore of paramount importance to the subsequent development of rearing methodologies. It was not until the 1960s, however, that the impact of this discovery was fully realized when Shelbourne (1964) demonstrated that larval plaice, *Pleuronectes platessa* L. could be mass-reared using brine shrimp nauplii as food. In addition to meeting the nutritional requirements of the larvae he attributed his success to the use of good quality eggs from stocks acclimated to captive conditions, the control of bacterial infestation of the eggs with antibiotics, the provision of high light levels and black tanks to promote a good feeding response, and the maintenance of high standards of hygiene.

The methodology developed for plaice was readily adapted to the common sole with appropriate adjustments to water temperatures (Shelbourne 1975). In small-scale experiments using 35 l tanks up to 80% of eggs survived to the completion of metamorphosis with up to 90% normal pigmentation. Encouraged by these results Shelbourne undertook larger scale trials in 6 m^{-3} tanks and achieved a survival rate of just over 40%, normal pigmentation of more than 95% and a final survivor density of four to five thousand juveniles per m^2 at an average length of just under 2 cm.

At that time these unequaled achievements were regarded with considerable global interest and were the stimulus for work on a wide range of species. There is certainly no case for detracting from Shelbourne's achievements but, in retrospect, he seems to have been somewhat fortunate in his choice of species, i.e., the plaice and common sole. Firstly, the larvae of these species are relatively large-mouthed and, in contrast to many other species of commercial interest, such as the turbot, *Scophthalmus maximus* (L.), and lemon sole, *Microstomus kitt* (Walbaum), are able to ingest brine shrimp nauplii as a first food. It was sometime later that the rotifer, *Brachionus plicatilis* Müller, was subsequently identified as a suitable first food for most smaller-mouthed species and was also shown to be a suitable alternative for *Artemia* nauplii for sole (Howell 1973). Secondly, and perhaps most importantly, neither of these two organisms form part of the natural diet of marine fish larvae and it was only after it was appreciated that these prey organisms were deficient in certain essential fatty acids (EFAs) that the rearing of some of the more valuable flatfish species, such as the turbot and the Japanese flounder, *Paralichthys olivaceous* Temminck & Schlegel, was accomplished. Success with these species depended on enriching the EFA content of the prey organisms offered by first feeding them on algae rich in these substances (Watanabe et al. 1978, Howell 1979). Fortunately for Shelbourne, the requirement of the sole (and of the plaice) for these EFAs appears to be less stringent than for many other species, high survival through the larval stages being achievable without any such enrichment of the prey organisms. However, enriching the EFA content of the live food fed to sole larvae has subsequently been shown to have other important beneficial consequences (see below).

The vast majority of subsequent studies on soleids remained focused on the common sole although in the last decade or two the emphasis has strongly shifted to its more southerly counterpart, the Senegal sole. Some studies have been undertaken in South Africa on the white margined sole, *Dagetichthys marginatus* (Boulenger), but as yet there is no record of any commercial developments (Ende and Hecht 2011).

3.1.1 Further development of rearing techniques for the common sole

The ability to mass-produce juvenile fish not only met the pre-requisite for stock enhancement trials as envisaged by Shelbourne (1975), but stimulated a considerable interest in the possibility of on-growing the juveniles to market size in land-based systems. In colder countries, such as the UK, the possibly of promoting faster growth rates by using a continuous supply of warm water generated by nuclear power stations was also considered. The common sole was one of many target species identified because of their high market value and depressed availability through over fishing.

Further research on the common sole was greatly facilitated by the ability to provide a regular supply of naturally spawned eggs from captive stocks of 'wild' fish fed natural feeds (Baynes et al. 1993). Larvae were reared using techniques largely based on those developed by Shelbourne (1975) with the exception that eggs were generally incubated without antibiotics but in containers gently agitated by aeration and continuously irrigated with a continuous flow of fresh seawater. Research focused on the principal requirements of an intensive farming approach, namely, formulated feeds for weaning and on-growing, good growth rates under crowded conditions and effective protocols for disease prevention and control.

Weaning and on-growing. Initially, juvenile plaice and sole of about 1 g in weight were successfully weaned from brine shrimp nauplii on to another live food, the oligochaete worm *Lumbricillus rivalis* (Levinsen) (Kirk and Howell 1972). These were abundant in rotting seaweed in the littoral zone but, although ample quantities could be gathered to support laboratory experiments, securing the quantities needed for commercial-scale production was not feasible. The use of formulated feeds was therefore the only realistic option but this proved to be much more challenging for sole than for other species such as turbot. Results of weaning trials were highly variable, usually yielding relatively low rates of survival and growth (Imsland 2003). This apparent dislike of fish-based feeds probably reflected the fact that the natural diet of sole is comprised of invertebrates and, as a nocturnal species, olfaction is a likely to be a dominant factor in prey selection. This was experimentally demonstrated by Mackie et al. (1980), who identified the precise chemical attractants (glycine betaine and certain L-amino acids) in mussel flesh that induced a strong feeding response. Subsequently, Cadena Roa et al. (1982) evaluated the use of these attractants in formulated feeds and found they elicited a greater response with corresponding higher survivals and growth rates than supplements of ground molluscs, polychaetes or brine shrimp. Significant progress had been made, but further advances were needed to secure the prospects of commercial farming (see below).

The effects of crowding on growth. Tolerance to crowding is an essential requirement for economic viability of intensive systems, the productivity per unit area being a function of both mean growth rate and the density at which the fish are stocked. High stocking densities can have adverse effects on growth through reduced water quality but in common sole it has been shown that even when such conditions are negated growth rate may still significantly diminish with increasing stocking density (Fig. 5). Such a decline in growth rate was not observed in a similar experiment with turbot (Howell 1998).

These results were interpreted as reflecting the different natural behaviour of the two species. Sole take food from the bottom, rather than the water column, and their intestine is adapted to eat small but frequent meals, rather than just one or two per day. These behavioural differences, coupled with the tendency of the oligochaete worms on which they were fed to form clumps, would create a situation in which a proportion of the stock could dominate the food supply either by active aggression or passive inhibition. This would suggest that any tendency for negative interactions within communal populations might be overcome by adopting appropriate feeding strategies. More recent studies (Schram et al. 2006, Lund et al. 2013) also found a negative correlation of stocking density with growth rate. However, their results differed in some important respects. Schram et al. (2006) obtained a positive correlation between size variation and stocking density implying that larger fish were growing faster than smaller fish. In contrast, Lund et al. (2013) found no such correlation, fish of all sizes growing at the same rate at each density. They concluded that at increased densities growth was depressed primarily due to a lower feed intake of all individuals, but also to less efficient use of the feed. The latter may have arisen from increased activity from social interactions and feed browsing, and possibly even from chronic stress.

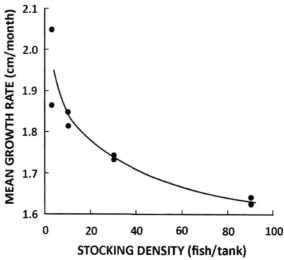

Fig. 5. The relationship between stocking density and growth rate of common sole on an ad libitum diet of the oligochaete worm Lumbricillus rivalis (from Howell 1998).

It seems clear that stocking density in common sole is an important determinant of growth rate although there is still some scope for conjecture regarding the behavioural mechanisms involved and hence the feasibility of minimizing these effects by adjusting rearing procedures.

Disease prevention and control. Disease rapidly becomes an issue with all species subjected to intensive rearing conditions. This proved to be the case in large-scale trials undertaken by the British White Fish Authority (now the Seafish Authority) at their site in Hunterston, Scotland. Juvenile sole were being weaned from brine shrimp nauplii on to either oligochaete worms or formulated feeds with some success until they experienced heavy mortalities caused by a disease known as Black Patch Necrosis (BPN). The condition was described by McVicar and White (1979) and caused by *Flexibacter maritimus* (Bernadet et al. 1990), now known as *Tenacibaculum maritimum*. Outbreaks devasted stocks and the only effective means of control appeared to be the provision of a sand substrate in which the fish could bury (McVicar and White 1982). This experience encouraged the view that sole was susceptible to disease unless cultured on a sand substrate, a requirement that would be a major impediment to the effective cleaning of tanks and the maintenance of hygienic conditions. Subsequent work (Baynes and Howell 1993) promoted a more optimistic view. In a small-scale experiment to assess the relative value of cooked and stored mussel with fresh mussel as food for juvenile sole, an outbreak of BPN appeared only among groups that received less than 2 feeds of fresh mussel per week (Fig. 6). The disease didn't spread to other groups despite the proximity of the tanks and the lack of stringent precautions to contain it. This experience suggested that a combination of a nutritious diet and hygienic conditions may be sufficient to avoid the occurrence of BPN without the provision a sand substrate. Common sole are, of course, susceptible to other diseases, for example vibriosis, but there is no evidence that the species is more vulnerable to disease than other species when subjected to appropriate rearing protocols.

New directions. Despite some positive outcomes, studies in these key areas did little to strengthen the belief that the intensive farming of sole was an economically viable option,

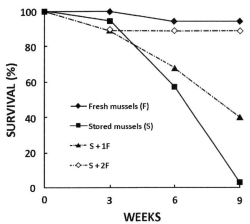

Fig. 6. The survival of groups of sole fed fresh mussel (F), cooked mussel (S) and cooked mussel with one (S + 1) or two (S + 2) feeds per week of fresh mussel (Redrawn from Baynes and Howell 1993).

other species, such as the turbot (*S. maximus*), appearing to be a more attractive candidate. By the onset of the 1980s the challenging problems of rearing turbot through its larval stages had largely been overcome and a dependable supply of juveniles could be assured. In contrast to sole, high juvenile growth rates of turbot could be achieved at high stocking densities on fish-based formulated feeds. The focus of commercial and research effort consequently shifted to turbot.

In the UK there was a further change of direction towards the end of the 1980s. Renewed concern about the continued decline of natural fish stocks rekindled interest in the possibility of arresting these trends by enhancing natural recruitment with hatchery-produced fish. This notion was encouraged by the considerable effort the Japanese had invested in this subject with an encouraging degree of success (see, for example, Howell and Yamashita 2005). In the UK, the sole was seen to be a suitable target because of the high value of the fishery and the relative ease of rearing the larvae beyond metamorphosis in large numbers.

Japanese and other studies suggested that, because of diminishing predatory pressures as fish grow, reared fish would need to be grown to a length of at least 5 cm before release in order to achieve a favourable return. It had also been established that the post-release survival of reared fish was significantly less than their wild counterparts (see reviews by Howell 1994, Olla et al. 1994). New research programmes were therefore focused on (1) developing methods for growing large numbers of metamorphosed juveniles to a length of at least 5 cm, and (2) assessing the suitability of reared fish for survival in the sea.

Further weaning and on-growing developments. Weaning sole onto formulated feeds under laboratory conditions had proved relatively successful but success on a commercial scale had not been demonstrated (Day et al. 1997). Apart from the importance of chemo-reception in stimulating a feeding response, the digestive system of the sole significantly differs from that of other candidates for farming. In particular, pepsin activity is not detected before day 200 and this is likely to impair the efficient digestion of certain feed components. For this reason, they assessed the effect of incorporating an enzymatically-hydrolysed fish meal into a weaning diet for 3 cm-long juvenile sole that had been reared on brine shrimp nauplii alone. The results were quite striking. The survival of the juveniles was directly dependent on the content of hydrolysed fish protein content (HFPC). Despite the inclusion of betaine-glycine to improve the attractiveness of the diets, increasing the HFPC content from 0 to 80% resulted in a progressive increase in survival from just over 40% to more than 75%, a survival rate approaching that of the control diet that consisted of high levels of invertebrate tissue.

Day et al. (1997) postulated that, although the hydrolysate may have increased the attractiveness of the diets over and above that of the added attractants, it was also possible that the more readily digested HFPC resulted in higher assimilation rates at a time when ingestion rates were relatively low. Such effects had been reported in European seabass, *Dicentrarchus labrax* (L.), by Cahu and Zambonino Infante (1995).

These trials were followed by weaning trials to evaluate a commercial larval feed produced by a process of agglomeration by the Norwegian Herring Oil and Meal Industry Research Institute (SSF), Fyllingsdalen, Norway (Day et al. 1999). The incorporation of water-soluble components into small-diameter feeds presented problems of particle stability and water pollution due to excessive leaching. SSF provided a technical solution that allowed high levels of soluble protein (30% of total protein) to be incorporated into small diameter feeds (100–2000 μm) while maintaining particle stability. This diet was customized for sole by the inclusion of a feeding attractant (betaine) at 6% of dry weight and by increasing the water-soluble protein to about 30% of the total protein. This allowed newly-metamorphosed sole weighing about 30mg to be weaned with a survival rate over 90% and with growth rates exceeding that on live foods. Weaned juveniles were successfully transferred from the agglomerated feed to a more conventional pellet with a 100% survival and without loss of growth rate. This positive result provided an effective means of growing juveniles to an appropriate size for stock enhancement purposes as well as greatly enhancing the prospect of developing viable intensive farming methods.

Fitness of hatchery-reared fish for survival in the sea. Morphological, such as abnormal pigmentation, and behavioural impairments, particularly those related to feeding and predator avoidance, had been shown to be among the major determinants of the post-release survival (Howell 1994, Olla et al. 1994) of hatchery-reared fish. Less predictable was the discovery in the common sole that temperature tolerance of juveniles was significantly influenced by nutritional factors during the larval stages. This was discovered when trials designed to determine the tolerance of reared juveniles to a winter temperature profile that simulated that in areas where they may be liberated produced widely differing results in consecutive years (Howell 1994). These conflicting results were subsequently experimentally demonstrated to be attributable to the poly-unsaturated fatty acid content of the live food fed to the larvae rather than that of the subsequent on-growing feed (Fig. 7) (Howell et al. 1996).

Further work on other environmental factors, including salinity, high temperature and hypoxia, concluded that susceptibility to environmental stress was responsive to dietary n-3 manipulation, possibly due to altered tissue development or the overproduction of eicosanoids (Logue et al. 2000).

The practical implication of these studies is that survival during the larval stages is not necessarily a good indicator of subsequent 'hardiness'. This would have important implications if the fish were destined to be released into the sea but might also have a detrimental impact in more controlled intensive farming systems in which fish are exposed to other stressors such as handling and crowding.

Other studies generated further evidence of the importance of morphological and behavioural factors in determining the survival of reared fish released into a natural environment (Ellis et al. 1997). Post-release survival largely depends on evasion of

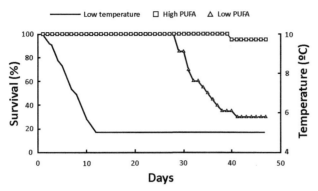

Fig. 7. The effect of low temperature (continuous line) on the survival of 8–10 cm long sole which had been reared to a length of 3 cm on Artemia nauplii of high (squares) and low (triangles) PUFA content (from Howell et al. 1996).

predators either by avoiding discovery or by escape. Sole, like many other flatfish, inhabit soft substrata largely devoid of vegetation. They adapt their skin colour to that of their surroundings and bury in the sediment to further reduce detection. The hatchery in which the fish are raised provides a very different environment with no sand substrate in which they can bury and tank colour and lighting that combine to induce a totally different skin colour to that of their wild counterparts. Their early life in hatcheries is also predator free and the fish will have become habituated to handling and presence of humans. Ellis et al. (1997) demonstrated that naïve (reared) sole are significantly disadvantaged with regard to factors such as burying ability and colour. The differences did, however, diminish when exposed to more natural conditions indicating that exposure to suitably designed pre-conditioning regimes for up to two weeks would enhance post-release survival.

3.1.2 Development of rearing methods for Senegal sole

Interest in the laboratory rearing of *S. senegalensis* began in the 1980s in Spain and Portugal where techniques for rearing other species of marine fish, particularly seabass, *Dicentrarchus labrax* (L.), and gilthead seabream, *Sparus aurata* L., were already quite advanced and showing considerable commercial promise. Work on the early life stages revealed considerable similarity to the common sole, although, as would be expected from their geographical distribution, the Senegal sole was adapted to higher temperatures regime. As with the common sole, a reliable supply of naturally-spawned fertilised eggs could be obtained from captive stocks fed on 'natural' food (squid and polychaetes in this case) and survival rates of larvae fed on brine shrimp nauplii and metanauplii to the completion of metamorphosis were similarly high (Dinis et al. 1999). Rotifers offered for the first few days of feeding were consumed but it was shown this did not improve survival or growth (Magalhães and Dinis 1996).

The reluctance of juveniles to accept formulated foods also reflected that of the common sole. The problem was to a large extent relieved by employing the same methods used for common sole but further advances have been made more recently by Engrola et al. (2009), who found that offering an inert diet at first-feeding together with brine shrimp nauplii promoted growth and better quality juveniles.

Progress was also made in the development of on-growing diets. For example, Coutteau (2001) (cited by Imsland et al. 2003) importantly demonstrated that formulations developed for other species were not necessarily optimal for sole. In terms of growth and feed conversion efficiency, sole performed better on a specially formulated feed with a crude protein to crude fat ration of 55/16 than on a standard commercial turbot feed with a protein/fat ratio of 52/20, though other factors, such as feed attractability, may have been involved.

Disease is a common feature of all cultured fish and the Senegal sole is no exception. Like the common sole, the Senegal sole is susceptible to tenacibaculosis (also know as fin-rot or Black Patch Necrosis) and vibriosis. Senegal sole is also susceptible to photobacteriosis, caused by a particularly virulent pathogen, *Photobacterium damsela* ssp. *piscicida*, which causes high losses in many cultured species, such as gilthead seabream (Morais et al. 2014). Several sole farms in the south of Spain have suffered high mortalities from this pathogen. Isolation from other fish species that are vulnerable to this disease is helpful as is the selection of sites or systems where temperatures do not rise above 22°C (Cañavate 2005).

Important differences between the species include that of their response to stocking density. Morais et al. (2014) observed that most studies reported that growth in Senegal sole is not affected by high stocking density and frequent handling stress, citing the work of Aragão et al. (2008), Costas et al. (2008, 2012, 2013) and Salas-Leiton et al. (2010). This would be a favourable trait in an intensive farming environment and is in contrast to the studies on the common sole cited above.

Growth rate is a key determinant of economic viability, and in this regard the Senegal sole would seem to have a significant advantage over common sole. Growth data to market size is sparse in the literature for both species. Howell (1997) used published experimental data to estimate the time required for common sole to reach the minimum market size of 24 cm (125 g) at a near optimum temperature of 19°C. The data indicated it would take 300 days, though different growth rates of the two sets of data used suggested that could have been an over-estimate. Nevertheless, that is a considerably slower growth rate than that of Senegal sole grown in an earthen pond at a naturally fluctuating temperature ranging from 15 to 24°C (Dinis et al. 1999). The fish were fed pellets but benthic invertebrates, notably polychaetes, were probably also exploited. After one year the fish had grown to a mean length of 35 cm (456 g). This evidence does suggest that Senegal sole may have a considerably faster growth rate than the common sole.

3.1.3 Further advances

By the end of the 20th century, the developments in feeding technology that facilitated ready transfer from live to formulated feeds with consistently good survival and growth rates encouraged industry to believe that the main obstacle to the farming of soles could be overcome. This coincided with saturated markets for established farmed marine species, such as turbot, seabass and gilthead seabream, and the desire of industry for further diversification. In an attempt to avoid further 'false dawns' in 2002 a workshop was organized at the CEFAS Laboratory, Weymouth, UK to provide an opportunity for researchers and commercial operators who had experience of working with soles to form a considered view of the potential for farming these species and to identify the obstacles that would still need to be overcome. The group reached a positive view and, because of the perceived usefulness of such a forum, four more workshops were held over the next 10 years, each one reassessing the status of the industry, its problems and the most recent research aimed at alleviating those problems (Howell et al. 2006, 2009, 2011). This period saw a significant shift in emphasis in research from the common sole to the Senegal sole as it became clear that the latter species offered the greater prospect of commercial viability. Comprehensive reviews of that research are presented elsewhere in this book.

3.1.4 Application of rearing technology

Stock enhancement. Shelbourne's (1964) main aim in developing mass-production techniques for flatfish was to enhance natural recruitment to depleted stocks with hatchery-reared juveniles. There had been no convincing evidence of the benefits of releasing eggs or yolk-sac larvae but the success of transplantation experiments using larger wild fish encouraged the belief that releasing older life stages when larger and less vulnerable to predation and starvation would be more productive (Shelbourne 1975). Shelbourne did not undertake any releases of sole but in 1964, large numbers of juvenile plaice were released in two sites off the west coast of Scotland. All the fish were lost, either through predation or other reasons (Blaxter 2000). Blaxter (2000) implied better results may have been obtained had there been some consideration of carrying capacity, predator abundance, optimal conditions for release and the fitness of the reared fish for release into the wild. There are no recorded releases of sole at that time or since, despite the vast amount of information now available to support the construction of appropriate protocols. Although catches of sole in the north-east Atlantic in 2012 were 44% less than in 1967, the preferred remedy remained the traditional management practice of limiting fishing mortality principally through quotas and gear restrictions. To attempt to make good such a deficit by the release of reared fish would certainly be a formidable challenge and highly expensive.

Extensive and semi-intensive cultivation. It has been recently estimated that there are over 92,000 ha of coastal wetlands in southern Europe that are already being used for extensive and semi-intensive production (Anras et al. 2010). Total fish production from these is far from trivial, Portugal, Spain and Italy collectively producing a total of about 13,000 tonnes per annum. About 77% of this production comes from semi-intensive systems. Detailed statistics for individual species are not available but sole is one of the main species listed for both Portugal and Spain (Anras et al. 2010).

Statistics for aquaculture production of soles (FEAP 2015) for the period 2005 to 2014 shows that production for Portugal and Spain was relatively constant from 2005 to 2008 (Table 3). Thereafter, Spanish and French production dramatically increased reflecting the onset and growth of intensive production. It may be deduced, therefore, that the 2005–2008 production is an approximation of production of soles from extensive and semi-intensive systems, i.e., 70–90 tonnes per annum.

Spain and Portugal have a long tradition of extensive cultivation in the now largely deserted salt ponds ('esteros' in Spain) of their southern coast. Yúfera and Arias (2010) describe the annual cycle that begins after the harvest at the end of autumn. The pond-monks are left open for several months so that the tidal flow of water will populate the pond with larval fish and other organisms. After this period of natural recruitment the pond-monks are kept closed except for periodic exchanges of water undertaken with netting installed in the monks to prevent escapes. The fish trapped in the pond feed entirely on natural food production within the pond until they are captured at the end of autumn. The catches mainly comprise mullets, seabass, gilthead seabream and sole with a small amount of eels and crustacean (crabs and shrimps). In a study in the Bay of Cádiz, the production after one year was 320 kg/ha of which only 5.1% was crustacea and 0.2% fish of no commercial value, a reflection of the high productivity of the area (Yúfera and Arias 2010).

Table 3. Annual aquaculture production of soles (tonnes) for the period 2005 to 2014 (FEAP 2015).

	2005	2006	2007	2008	2009	2010	2011	2012	2013	2014
France	0	0	0	0	0	142	200	220	223	261
Portugal	11	9	8	13	14	14	50	100	35	60
Spain	60	80	60	55	180	204	110	194	313	786
Italy	0	0	0	19	14	14	10	0	0	0
TOTAL	71	89	68	87	208	232	170	294	348	846

Productivity is, of course, a key issue with farmers and the profitability of totally extensive aquaculture is highly marginal. Following successful trials in the mid-1990s many farmers converted to semi-intensive cultivation systems that deliver greater productivity and profitability (Anras et al. 2010). The main target species were seabass and gilthead seabream. Senegal sole occur in these ponds either having been carried in naturally or released after purchase from local hatcheries. They feed on the natural benthos and on any formulated feed left by the target species and may reach a size of about 400 g after 16 months (Morais et al. 2014). The use of reared sole is not, however, always successful. Failures have been attributed to stocking fish of too small a size (< 0.2 g) before they were weaned coupled with a failure to rid the pond of predators prior to release. Releasing the fish at a larger size and sun-drying the pond before use could avoid this problem (Imsland et al. 2003). As previously indicated, a degree of conditioning of the fish for release may also prove beneficial.

Intensive production. Extensive and semi-intensive systems have considerable value in terms of exploiting a natural resource in a sustainable way and providing a living for coastal communities. However, productivity is relatively low and outputs can be uncertain, being dependent on the vagaries of nature, so that the demands of the market cannot be fully met. To a large extent intensive cultivation may overcome these problems because the environment in which the fish are reared can be almost completely controlled. As the obstacles of rearing soles through the whole of their life history have been overcome, industry has progressively shown interest in adapting and developing systems that would permit the intensive production of the species.

Intensive farming is normally undertaken in fibre-glass, concrete tanks or shallow raceways, the latter of which one would naturally presume to be appropriate for flatfish, tank area having more importance than volume (Imsland et al. 2003). Producers have become attracted to these systems, often in conjunction with recirculation systems that, with the advantages of modern technology, enable a high degree of environmental control. There is no doubt that the use of these systems has brought about a dramatic improvement in disease control by both eliminating contact with other fish species and providing the required environmental control (Morais et al. 2014). Over the last 5 years the annual production of farmed soles has steadily increased and by 2014 had almost reached 1000 tonnes. Problems still have to be overcome but the sound scientific support that has been established and the experience and expertise of the industry will undoubtedly combine to secure the future of this relatively novel industry.

4. Markets

The demand for soles in Europe is widespread, especially in its coastal countries. The common sole is the most abundant species but, as pointed out above, other species, notably the Senegal sole, are also highly regarded and often not distinguished from the common sole in the market place.

The natural distribution of sole does not coincide with the pattern of its demand. In 2012, for example, the Netherlands and France caught about 50% of the total landings of the common sole with Belgium and the UK landing a further 15%. Spain and Italy, two countries with a high demand for soles, captured only 1 and 6% of the total catch respectively (FAO 2014) and consequently had to import much of their requirement from more northern countries. This follows a general trend for demersal fish as a whole in this region, many of which are competitors with sole in the market place (Bjørndal and Guillen 2014). In this case the principal 'producers' are Norway, Denmark and the Netherlands with the first two countries being the only countries that export significantly more fish than they import. The reverse is true for all other European coastal countries, especially Spain and Italy where the demand is high. In the case of Italy, 85% of their total supply of demersal fish is imported (data from FAO 2014). These are of course general trends and the situation for different species may vary. The UK, for example, is a net importer of demersal fish as a whole but a net exporter of the common sole (UKFS 2014), although the quantities are relatively small. In general, there is a significant shortfall in the supply of demersal fish, including soles, particularly in southern European countries and so there is almost certainly a ready market for any aquaculture production given the constraints of price and quality.

The role of aquaculture as a source of supply has become an attractive proposition in Spain as it has a favorable climate for the preferred species (Senegal sole), a consumer demand that greatly exceeds the country's natural resources, and a substantial body of research and industrial expertise that has become established through the development of its existing aquaculture enterprises. The over 40% reduction in sole catches from 1995 to the present has added to the pressure for, and commercial viability of, such innovations. An important consequence of the decline in catches is the accompanying decrease in average size of the fish. The restaurant sector favours large fish (400–600 g) and this provides a good market opportunity for the aquaculture industry if it can economically produce fish of that size, or larger (Bjørndal and Guillen 2014).

MercaMadrid, one of the largest fish markets in Europe, categorizes sole as small (< 0.5 kg), medium or large (> 1 kg). Over the period 2002–2013 the price of these size categories have ranged from 7–12, 13–18 and 18–25 €/kg respectively showing that large fish attract a price up to double that of small fish. The value of frozen fish can be up to 5 €/kg less than that of small fish. Farmed fish have been on the market for only a short time but the price was closer to that of small fish than that of medium fish, possibly because the fish were less than 0.5 kg in weight (Bjørndal and Guillen 2014).

Farmed fish bring the benefits that they are fresh, devoid of damage due to the rigours of capture, and their availability and size can, to a large extent, be reliably tailored to meet the demands of the market. Despite these advantages there is a tendency for the consumer to prefer 'wild' fish, being less comfortable with a 'novel' fish that may differ in appearance, such as its shape or colour, to its wild counterpart. Experience with other farmed species suggests that such prejudices do diminish with time, providing the product being presented has the necessary eating and safety qualities and is produced by ethical methods.

References

Anras, L., C. Boglione, S. Cataudella, M.T. Dinis, S. Livi, P. Makridis et al. 2010. The current status of extensive and semi-intensive aquaculture practices in Southern Europe. Aquaculture Europe 35(2).

Aragão, C., J. Corte-Real, B. Costas, M.T. Dinis and L.E.C. Conceição. 2008. Stress response and changes in amino acid requirements in Senegalese sole, *Solea senegalensis* Kaup 1758. Amino Acids 34: 143–148.

Baynes, S.M. and B.R. Howell. 1993. Observations on the growth, survival and disease resistance of juvenile common sole, *Solea solea* (L.), fed *Mytilus edulis* L. Aquacult. Fish. Manage. 24: 95–100.

Baynes, S.M., B.R. Howell and T.W. Beard. 1993. A review of egg production by captive sole, *Solea solea* (L.). Aquacult. Fish. Manage. 24: 171–180.

Bernadet, J.F., A.C. Campbell and J.A. Buswell. 1990. *Flexibacter maritimus* is the agent of 'black patch necrosis' in Dover sole in Scotland. Dis. Aquat. Org. 8: 233–237.

Blaxter, J.H.S. 2000. The enhancement of marine fish stocks. Adv. Mar. Biol. 38: 1–54.

Bjørndal, T. and J. Guillen. 2014. The future of sole farming in Europe: Cost of production and markets. Aquaculture Europe 39(2): 5–12.

Cadena-Roa, M., C. Huelvan, Y. le Borgne and R. Metailler. 1982. Use of rehydratable extruded pellets and attractive substances for the weaning of sole (*Solea vulgaris*). J. World Maric. Soc. 13: 246–253.

Cahu, C.L. and J.L. Zambonino Infante. 1995. Effect of the molecular form of dietary nitrogen supply in seabass larvae: response of pancreatic enzymes and intestinal peptidases. Fish Physiol. Biochem. 14: 209–214.

Cañavate, J.P. 2005. Opciones del lenguado senegal_es *Solea senegalensis* Kaup, 1858 para diversificar la acuicultura marina. Bol. Inst. Esp. Oceanogr. 21(1–4): 147–154.

Costas, B., C. Aragão, J.M. Mancera, M.T. Dinis and L.E.C. Conceição. 2008. High stocking density induces crowding stress and affects amino acid metabolism in Senegalese sole, *Solea senegalensis* (Kaup 1858) juveniles. Aquacult. Res. 39: 1–9.

Costas, B., C. Aragão, J.L. Soengas, J.M. Míguez, P. Rema, J. Dias et al. 2012. Effects of dietary amino acids and repeated handling on stress response and brain monoaminergic neurotransmitters in Senegalese sole (Solea senegalensis Kaup, 1858) juveniles. Comparative Biochem. Physiol. Part A 161: 18–26.

Costas B., C. Aragão, J. Dias, A. Afonso and L.E.C. Conceição. 2013. Interactive effects of a high quality protein diet and high stocking density on the stress response and some innate immune parameters of Senegalese sole, *Solea senegalensis*. Fish Physiol. Biochem. 39: 1141–1151.

Coutteau, P., R. Robles and W. Spruyt. 2001. On-growing feed for Senegal sole (*Solea senegalensis* Kaup). *In*: Abstracts of Contribution Presented at the International Conference Aquaculture Europe 2001. Special Publication No. 29: 58–59. European Aquaculture Society.

Day, O.J., B.R. Howell and D.A. Jones. 1997. The effect of dietary hydrolysed fish protein concentrate on the survival and growth of juvenile Dover sole, *Solea solea* (L.), during and after weaning. Aquacult. Res. 28: 911–921.

Day, O.J., B.R. Howell, A. Aksnes and E. Nygard. 1999. Recent advances in the weaning of sole, *Solea solea* (L.). Aquaculture Europe 99, Trondheim, Norway, August 7–10. European Aquaculture Society Special Publication 27: 40–41.

Díaz de Astarloa, J.M. 2002. A review of the flatfish fisheries of the south Atlantic Ocean. Revista de Biologia Marina y Oceanografia 37(2): 113–125.

Dinis, M.T., L. Ribeiro, F. Soares and C. Sarasquete. 1999. A review of the cultivation of *Solea senegalensis* in Spain and Portugal. Aquaculture 176: 27–38.

Ellis, T., B.R. Howell and R.N. Hughes. 1997. The cryptic responses of hatchery-reared sole to a natural sand substratum. J. Fish Biol. 51: 389–401.

Ende, S.S.W. and T. Hecht. 2011. Ontogeny of the feeding apparatus of hatchery-reared white margined sole, *Dagetichthys marginatus* (Soleidae): implications for cultivation. J. Appl. Ichthyol. 27(1): 112–117.

Engrola, S., L. Figueira, L.E.C. Conceição, P.J. Gavaia, L. Ribeiro and M.T. Dinis. 2009. Co-feeding in Senegalese sole larvae with inert diet from mouth opening promotes growth at weaning. Aquaculture 288: 264–272.

Fabre-Domergue, P. and E. Biétrix. 1905. Développement de la Sole (*Solea vulgaris*). Travail du Laboratoire de Zoologie Maritime de Concarneau. Vuibert et Nony, Paris.

FAO. 2014. Yearbook of Fishery and Aquaculture Statistics, 2012. FAO, Rome.

FEAP. 2015. Federation of European Aquaculture Producers, European Aquaculture Production Report 2005–2014.

Gibson, R.N. 2005. Flatfishes: Biology and Exploitation. Blackwell Science, Oxford, UK.

Howell, B., P. Cañavate, R. Prickett and L. Conceição. 2006. Farming soles – a reality at last? Aquaculture Europe. *In*: World Aquaculture 37(3): 1–6.

Howell, B., L. Conceição, R. Prickett, P. Cañavate and E. Mañanos. 2009. Sole farming: nearly there but not quite?! Aquaculture Europe 34(1): 24–27.

Howell, B., R. Prickett, P. Cañavate, E. Mañanos, M. Dinis, L. Conceição and L. Valente. 2011. Sole farming: there or thereabouts! Aquaculture Europe 36(3): 42–45.

Howell, B.R. 1979. Experiments on the rearing of larval turbot, *Scophthalmus maximus* L. Aquaculture 18: 215–225.

Howell, B.R. 1994. Fitness of hatchery-reared fish for survival in the sea. Aquacult. Fish. Manage. Supplement 1: 3–17.

Howell, B.R., T.W. Beard and J.D. Hallam. 1996. Larval diet quality as a determinant of juvenile characteristics. pp. 293–296. *In*: Chatain, B., M. Saroglia, J. Sweetman and P. Lavens [eds.]. Seabass and Seabream Culture: Problems and Prospects. Handbook of Contributions and Short Communications Presented at the International Workshop on "Seabass and Seabream Culture: Problems and Prospects" Verona, Italy, October 16–18, 1996. European Aquaculture Society: Oostende. 388 pp.

Howell, B.R. 1997. A re-appraisal of the potential of the sole, *Solea solea* (L.), for commercial cultivation. Aquaculture 155: 359–369.

Howell, B.R. 1998. The effect of stocking density on growth and size variation in cultured turbot, *Scophthalmus maximus*, and sole, *Solea solea*. ICES CM 1998/L: 10.

Howell, B.R. and Yamashita, Y. 2005. Aquaculture and stock enhancement. pp. 345–371. *In:* Gibson, R.N. [ed.]. Flatfishes: Biology and Exploitation. Blackwell Science, Oxford, UK.

Imsland, A.K., A. Foss, L.E.C. Conceição, M.T. Dinis, D. Delbare, E. Schram, A. Kamstra, P. Rema and P. White. 2003. A review of the culture potential of *Solea solea* and *S. senegalensis*. Rev. Fish Biol. Fish. 13: 379–407.

Ishak, M.M. 1980. The Fisheries of Lake Qarun, Egypt. ICLARM Newsletter (3)1: 14–15.

Kirk, R.G. and B.R. Howell. 1972. Growth rates and food conversion in young plaice (*Pleuronectes platessa* L.) fed on natural and artificial diets. Aquaculture 1: 29–34.

Logue, J.A., B.R. Howell, J.C. Bell and A.R. Cossins. 2000. Dietary n-3 long-chain polyunsaturated acid deprivation, tissue lipid composition, ex vivo prostaglandin production and stress tolerance in juvenile Dover sole (*Solea solea* L.). Lipids 35(7): 745–755.

Lund, I., S.J. Steenfeldt, B. Herrmann and P.B. Pedersen. 2013. Feed intake as explanation for density related growth differences of common sole *Solea solea*. Aquaculture 44(3): 367–377.

Mackie, A.M., J.W. Adron and P.T. Grant. 1980. Chemical nature of feeding stimulants for the juveniles Dover sole, *Solea solea* (L.). J. Fish Biol. 16: 701–708.

McVicar, A.H. and P.G. White. 1979. Fin and skin necrosis of cultivated Dove sole, *Solea solea* (L.). J. Fish. Dis. 2: 557–562.

McVicar, A.H. and P.G. White. 1982. The prevention and cure of an infectious disease in cultivated juvenile Dover sole, *Solea solea* (L). Aquaculture 26: 213–222.

Magalhães, N. and Dinis, M.T. 1996. The effect of starvation and feeding regimes on the RNA, DNA and protein content of *Solea senegalensis* larvae. Book of Abstracts, World Aquaculture '96, Bangkok, p. 242.

Millner, R., S.J. Walsh and J.M. Diaz de Astarloa. 2005. Atlantic flatfish fisheries. pp. 240–271. *In*: Gibson, R.N. [ed.]. Flatfishes: Biology and Exploitation. Blackwell Science, Oxford, UK.

Morais, S., C. Aragão, E. Cabrita, L.E.C. Conceição, M. Constenla, B. Costas et al. 2014. New developments and biological insights into the farming of *Solea senegalensis* reinforcing its aquaculture potential. Review in Aquaculture 6: 1–37.

Munroe, T.A. 2005a. Tropical flatfish fisheries. pp. 292–318. *In*: Gibson, R.N. [ed.]. Flatfishes: Biology and Exploitation. Blackwell Science, Oxford, UK.

Munroe, T.A. 2005b. Distibutions and biogeography. pp. 42–67. *In*: Gibson, R.N. [ed.]. Flatfishes: Biology and Exploitation. Blackwell Science, Oxford, UK.

Olla, B.L., M.W. Davis and C.H. Ryer. 1994. Behavioural deficits in hatchery-reared fish: Potential effects on survival following release. Aquacult. Fish. Manage, Supplement 1: 19–34.

Rollefsen, R.F. 1939. Artificial rearing of fry of sea water fish. Preliminary communication. Rapp. Cons. Explor. Mer. 109: 133.

Salas-Leiton E., V. Anguis, B. Martín-Antonio, D. Crespo, J.V. Planas, C. Infante et al. 2010. Effects of stocking density and feed ration on growth and gene expression in the Senegalese sole (*Solea senegalensis*): potential effects on the immune response. Fish Shellfish Immunol. 28: 296–302.

Schram, E., J.W. Van der Heul, A. Kamstra and M.C.J. Verdegem. 2006. Stocking density-dependent growth of Dover sole (*Solea solea*). Aquaculture 252: 339–347.

Seafish. 2013. Responsible Sourcing Guide, Version: Dover Sole. Version 7–May 2013. 9 pp.

Shelbourne, J.E. 1964. The artificial propagation of marine fish. Adv. Mar. Biol. 2: 1–83.

Shelbourne, J.E. 1975. Pioneering studies on the culture of the larvae of the plaice (*Pleuronectes platessa* L.) and sole (*Solea solea* L.). Fisheries Investigation Series II, 27: 29pp.

UKFS. 2014. UK Fisheries Statistics 2013. Marine Management Organisation, Newport, South Wales, UK.

Watanabe, T., C. Kitajima, T. Arakawa, K. Fukusho and S. Fulita. 1978. Nutritional quality of the rotifer, *Brachionus plicatilis*, as a living feed from the viewpoint of essential fatty acids for fish. Bull. Jpn. Soc. Sci. Fish. 44(10): 1109–1114.

Wilderbuer, T., B. Leaman, Chang Ik Zhang, J. Fargo and L. Paul. 2005. Pacific flatfish fisheries. pp. 272–291. *In*: Gibson, R.N. [ed.]. Flatfishes: Biology and Exploitation. Blackwell Science, Oxford, UK.

Yúfera, M. and A.M. Arias. 2010. Traditional polyculture in "Esteros" in the Bay of Cádiz (Spain). Aquaculture Europe 35(3).

A-1.2

Engineering of Sole Culture Facilities
Hydrodynamic Features

Joan Oca* and Ingrid Masaló

1. Specific constraints in the design of sole culture facilities

In intensive land-based aquaculture, tank design should be adapted to the way species behave and swim while also reducing stress levels and improving fish welfare, which in turn contributes to enhancing its growth (Palstra and Planas 2011).

Sole has been described as a fish with low growth rates (Mas-Muñoz et al. 2011), presenting a high size dispersal when cultured. The low growth observed with sole and the high cost of the land surface makes it necessary to culture them in intensive systems, where growing technologies have to be adapted to their specific needs and behavior. Being one of the most sedentary flatfish species, sole rests on the bottom for long periods and shows a natural tendency to grouping when cultivated in a tank. Moreover, it feeds strictly from the bottom (De Groot 1971) and is adapted to eating small but frequent meals. Sole activity is limited to moving slowly over short distances across the bottom when they detect food; thus there is a time lag between when they detect and eat the food. The low swimming activity of sole has allowed development of specific techniques for determining sole biomass in tanks with laser scanning (Almansa et al. 2015) and for compiling a sole activity index with image analysis techniques (Duarte et al. 2009).

This chapter presents an analysis of growing technologies, tank geometry and hydrodynamic conditions in order to provide some specific guidelines for the design and management of sole facilities.

---

[1] C/Esteve Terrades 8. UPC BarcelonaTECH. Barcelona School of Agricultural Engineering. 08860 Castelldefels (Spain).
* Corresponding author: joan.oca@upc.edu

2. Growing technologies

Solea senegalensis used to be cultivated in salt marshes that were associated with salt works in the south of Spain and Portugal. They were kept in polyculture with mugilids, sea bass and sea bream. With the decline of the salt industry over the 20th century, many ponds were partially adapted to extensive fish farming systems (Yúfera and Arias 2010).

Recent advances have been made in the knowledge of the reproductive biology and behavior of Senegalese sole in captivity as well as in their specific nutritional requirements. This has allowed the regular production of fry and the formulation and commercialization of improved species-specific diets.

These improvements have led to production systems undergoing major changes as they shifted from predominantly earth ponds or salt marshes with water renovation provided by tidal energy to intensive flow-through systems in fiberglass or concrete tanks that use a pumping system to supply a continuous water flow passing a single time through the facility before being discharged to the sea. Flow-through systems have enabled improvements with disease issues on sole farms by eliminating contact with other fish species, increasing the stocking densities and enabling greater control over environmental parameters.

The high cost of the farm's surface along the coast areas has led to adopting strategies that tend to optimize the use of land area and increase the potential area for locating the farm around the source of the water supply. This is done by reducing the required water flow rate.

Multilevel tank configurations have allowed taking advantage of the sole's morphology by breeding it in shallow tanks (around 20 cm depth), which are vertically stacked in order to reduce the farm's surface as well as vulnerability to disease, predators and natural disasters (Øiestad 1999, Labatut and Olivares 2004).

Kamstra et al. (2012) analyzed the economic rationale behind a multilevel shallow raceway system by comparing the capital costs for the rearing space as a function of productivity and the number of rearing levels. Their calculations were based on representative costs for building and land in The Netherlands. The analysis was based on a system with 7 levels that were 50 m long and in 4 rows, which was built at a price of 50 €/m² of rearing area. In this way, a 600 m² floor area supported 2000 m² of rearing area. The results showed that, assuming a productivity of around 20 kg/m²/year, the capital costs per kg produced for a 7-level system was around 1.0 €/kg. Reducing the number of layers to 4, 2 and 1, the capital cost increased to about 1.5, 2.5 and 5.0 €/kg, respectively.

2.1 Flow-through versus recirculating systems

Introducing recirculating aquaculture technologies in single or multilevel facilities has been the key to reduce make-up water needs. This allows promoting versatility in terms of farm location and therefore reduces the costs associated to water transport. It has also made it feasible to control temperatures, which promotes consistent growth rates to market size throughout the production cycle and also contributes to a decline in many disease outbreaks that intensify when temperatures rise above 20–22°C (Morais et al. 2014). Furthermore, environmental and regulatory constraints are pushing fish farmers to move towards Recirculation Aquaculture Systems (RAS). In this type of system, fish are reared at high stocking densities and the make-up water needs are divided by 10 to 100 in comparison to flow-through systems (Martins et al. 2010). Consequently, wastewater flow rates decrease proportionally and waste concentration increases. Intensive or "fully-recirculating" RAS are typically defined as systems with water replacement ratios of less than 10% per day.

The use of water recirculating technologies has been increasingly adopted by fish hatcheries. Flow-through hatcheries that are supplied by marine water are subject to large fluctuations in water quality that are difficult to control. However, RAS technologies provide a rearing medium that is constant and adjustable, showing only slight and slow variations. Moreover, there is minimal heat loss in recirculated water systems, which normally operate above ambient water temperature. Despite the higher initial investment relative to flow-through systems, RAS technologies reduce production costs mainly because much less energy is required for heating, and the survival rate of the fingerlings is much higher (Blancheton 2000).

In comparing recirculating vs. flow-through systems for grow-out stages, the economic factors are usually determinant,in addition to the above mentioned aspects regarding the farm's surface, the water availability and the ability to control environmental parameters. The higher investment required by RAS technologies will condition the design and operation of the growing system, which will require continuous and intensive production, large production units and adequate farm size to minimize production costs. Also, when choosing flow-through or RAS technologies, it is necessary to consider not only the costs associated with moving water from the culture tanks to the different unit processes that restore used water quality, but also the costs linked to oxygen consumption by biological filters as well as those related to preventing catastrophic failures of the system (Timmons and Ebeling 2010).

Kamstra et al. (2001) (in Imsland et al. 2003) analyzed the prospects for growing *Solea solea* using a recirculation system based on a bio-economic model. The authors collected data from the literature and their own experimental research and projected these data into the infrastructure of a Dutch recirculation system. A case study was performed for a farm that produced 50 tons/year. The relative importance of the most important items in the total cost per kg of final product were estimated to be: fingerlings (5 g) 18.7%; feed 17.9%; electricity 9.2%; oxygen, gas and water 5.9%. Interest and depreciation were estimated to be 11.1% and 9.1%, respectively. Labor costs were estimated at around 26.4%.

2.2 Tank geometries and flow patterns

A proper tank design must combine hydrodynamics and the biological requirements of the species. It has to promote uniformity of rearing conditions, fast elimination of biosolids (non-ingested feed and faeces) and uniform distribution of fish throughout the tank (Tvinnereim 1988, Cripps and Poxton 1992, Timmons et al. 1998). Moreover, tanks should facilitate daily labors like feeding, removing dead fish and other routines like fish harvesting and grading.

Tank geometry, inlet/outlet characteristics and water flow rate determine the flow pattern and water velocities (Oca et al. 2004). Also the presence of fish and their stocking density have an effect on tank hydrodynamics by increasing the turbulence (Masaló et al. 2008) and modifying the tank bottom shape in flat fish facilities.

2.2.1 Raceway versus circular tank

Raceways and circular tanks are the most commonly used geometries in aquaculture. The advantages and disadvantages of raceways and circular tanks must be analyzed by considering: (a) the flow patterns and their repercussion on the environmental conditions into the tank, (b) the efficient use of the available land area and ease of fish handling, and (c) self-cleaning capacity. In sole growing facilities, raceways are the most commonly used tanks nowadays. This is due to their advantages in terms of point (b), despite the fact that circular tanks are more efficient in terms of points (a) and (c), as will be shown below.

Flow pattern and environmental conditions: A raceway is a rectangular tank with a length/ width ratio of about 10 and a depth of less than 1.0 m (Summerfelt et al. 2000b). Water flows through the raceway in a plug-flow manner with minimal back mixing, generating gradients of environmental conditions which often promote a heterogeneous fish distribution. Sole's tendency to remain in groups will lead to weak individuals being displaced to zones with low water quality, enhancing hierarchies that are the major cause for growth heterogeneity (Salas-Leiton et al. 2010). On the other hand, the average water velocity in a raceway cross section can only be adjusted by modifying the flow rate per unit width or modifying the water depth. Both parameters are not easy to modify in commercial scale facilities, and the velocities achieved are frequently insufficient for effectively removing settled solids from the rearing area.

In contrast, water in circular tanks is usually injected tangentially to the wall, and the outlet is located at the bottom center of the tank, which creates a rotating flow that provides highly uniform water quality conditions (Westers and Pratt 1977, Ross et al. 1995), due to the effective mixing achieved (Ross and Watten 1998, Timmons et al. 1998). The rotating velocity in these kinds of tanks can be increased not only by increasing the flow rate of inlet water (Q), but also by reducing the size of the water inlet orifices, in order to increase the impulse produced by the water inlet jets. The capability of modifying the impulse force in rotating flow tanks without changing the flow rate of the incoming water makes it much easier to control average velocities, as will be explained in more detail below when we analyze the hydrodynamic conditions in flatfish tanks.

Besides the average velocity, the distribution of velocities was also analyzed by Oca and Masaló (2013), who proposed a model to determine the distribution of velocities in circular tanks. This was later improved upon by Masaló and Oca (2016) in order to introduce the influence of fish swimming in the water column.

Use of land area and fish handling: Circular tanks show some advantages from a hydraulic point of view. Nevertheless, the choice of tank geometry is also determined by the cost of floor space (Timmons et al. 1998). Furthermore, it is necessary to consider their ease of handling, for example, the ability to sort fish and perform routine tasks like cleaning, removing dead fish, etc. In sole facilities, fish have to be on the farm for long periods in order to reach a commercial size, due to their low growth rates. Therefore, two key points to consider are optimization of the culturing area and the ease of tank and fish handling.

The tank size used in the facilities depends on the species, growing stage, and economic considerations. Large tanks are used in grow-out stages, as they provide capital and labor cost savings because tank maintenance is rather independent of tank size (Timmons et al. 1998). Nevertheless, larger tanks are more difficult to handle, especially circular tanks. Summerfelt et al. (2009) proposed different technologies to improve harvesting and grading in large circular tanks; but their handling disadvantages increase with large diameters, while the length can be increased in rectangular tanks without increasing the difficulty in fish handling.

Also, the rectangular geometry allows better use of the available area, since the percentage of land not occupied by tanks is reduced when compared with circular tanks. Moreover, low water depths can be used in the culture of some flatfish species (especially sole), and this facilitates the stacking of shallow raceways on various floors (multilevel tanks), which in turn increases the ratio between rearing area and land area by a factor equivalent to the floors used.

Finally, another feature to be considered in analyzing the land area use is the stocking strategy adopted. Using a continuous stocking strategy rather than a batch strategy, the total system production increases (Watten 1992). Batch stocking will lead to poor space profitability, because maximum biomass will only be reached when fish are near the end of the production stage or close to commercial size. Nevertheless, continuous stocking of fish in large tanks is difficult, because regular grading and harvesting are required. Raceways allow farming different sized fish in one tank by delimiting different areas in the tank with movable partition nets, which is much more difficult to do in circular tanks.

Self-cleaning capacity: Sole faeces are semi-liquid, dissolve very fast in water, and are difficult to collect (Dias et al. 2010). Consequently, it is desirable to concentrate and eliminate biosolids from the rearing area as fast as possible by transporting them the shortest distance from their point of origin to the water outlet.

In raceways, solids removal from the tank is time- and labor-intensive and generally inefficient. To achieve self-cleaning properties in raceways, high velocities are needed for directing biosolids to the outlet faster, which means high flow rates. That, in turn, represents greater power requirements. It is common, and recommended (IDEQ 1998), to leave the end area of the raceway free of fish in order to concentrate biosolids and allow them to settle undisturbed. Such areas are known as quiescent zones.

Larger raceways that are used in intensive production contain more fish, which means more waste particles. Waste particle production is nearly homogeneous in all tank areas (when fish are evenly distributed), but their concentration increases with the distance from the inlet (Brinker and Rösch 2005). Thus, it may be necessary to place quiescent zones along the lengths of raceways in longer tanks.

Fish swimming can aid in resuspending particles by directing them to the outlet. Nevertheless, flatfish species rest on the bottom, and sedimentation is therefore enhanced. Merino et al. (2007b) studied the settling characteristics of solids in California halibut (*Paralichthys californicus*) tanks, and they showed that, in raceways that were stocked over 150% PCA (Percentage of Covered Area), the settled solids were resuspended by fish activity and swept out of the culture area of the tank.

In contrast to raceways, circular tanks show good self-cleaning properties. In circular tanks, the water injected tangentially to the wall creates a primary rotating flow parallel to the tank wall. The primary flow generates a secondary rotation in a thin layer next to the tank bottom and that flows radially inward, carrying settleable solids towards the bottom-center drain (this phenomenon is called the "tea-cup" effect) (Paul et al. 1991). Therefore, the distance travelled by solids is shorter in circular tanks, thus avoiding high concentrations on the tank floor as well as leaching.

Another advantage of circular tanks is the possibility of concentrating a significant percentage of solids in a small volume of water. The use of two drains in the tank center allows having a stream with high solids concentrations (secondary or concentrated flow) and a stream with low concentration (primary or clarified flow). These types of systems have been applied in aquaculture tanks and are known as Dual-drain systems (Fig. 1).

In dual-drain systems, only 5–20% of recirculating water is drained from the bottom centre of the tank (secondary or concentrated flow), but 80–90% of suspended solids are removed (Lunde et al. 1997, Van Toever 1997, Schei and Skybakmoen 1998, Summerfelt et al. 2000a, Davidson and Summerfelt 2004). More recently, triple-drains have been

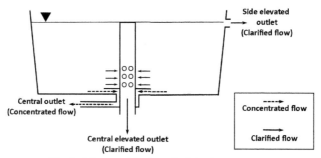

Fig. 1. Scheme of dual- and triple-drain systems.

designed (e.g., Wright et al. 2012), where three flows are differentiated (two clarified and one concentrated flow).

Dual-drains are in-tank systems that concentrate and separate biosolids in a small fraction of the water flow, allowing an increase in treatment efficiencies (Cripps and Bergheim 2000).

2.2.2 Raceway tanks with rotating flow cells

The ease of fish management and better land-use efficiency when employing raceways is in contrast to the advantages of circular tanks, specifically their self-cleaning and velocity control. Some authors have proposed tank designs that combine the hydrodynamic advantages of circular tanks with the handling and land-use advantages of rectangular tanks, creating rotating flow cells in rectangular tanks (Watten et al. 2000, Ebeling et al. 2005, Oca and Masaló 2007, Labatut et al. 2007a, 2007b). In these types of tanks, inlets are placed tangential to the cells and outlets in the center of each rotating flow cell.

Watten et al. (2000) designed the Mixed-Cell Raceway (MCR; Fig. 2), converting linear raceways (14.5 m long) into a series of hydraulically separated cells (each 2.4 m wide by 2.4 m long), with three inlets per cell.

Another similar but simpler design is the multivortex tank (Masaló and Oca 2014), where 4 rotating flow cells of 1 m diameter are created in a rectangular tank (4 × 1 m with 18 cm water depth) by injecting the water tangentially to the cells with only one inlet per cell (Fig. 3A). Baffles can be added between two consecutive water inlets to reduce the dissipation of energy due to the frontal collision of two entering plumes of water. The use of baffles increases the average velocities obtained without baffles by about 30%.

In tanks with rotating flow cells, the flow patterns obtained are very similar to those observed in circular tanks (Masaló and Oca 2014) (Fig. 3B), and no dead volumes or short-circuiting are observed (Watten et al. 2000), which indicates an adequate degree of mixing (Labatut et al. 2007a). In addition, average velocities are ten times higher than those which could be obtained in the same tank working as a linear raceway (Oca et al. 2004). Furthermore, they are proportional to impulse force (Masaló and Oca 2014), which is what happens in circular tanks (Oca and Masaló 2007) and thus allows for controlling velocities, as will be shown in the next section.

As in circular tanks, the distance travelled in these types of tanks by biosolids from tank bottom to outlet is shorter than in classic raceways, which allows for faster elimination from the tank bottom and a reduction in leaching.

The effect of flatfish on flow pattern in tanks with rotating flow cells was studied by Masaló and Oca (2013). These authors compared velocities obtained in the multivortex tank (18 cm water depth with 32% and 53% PCA), and they found an average velocity

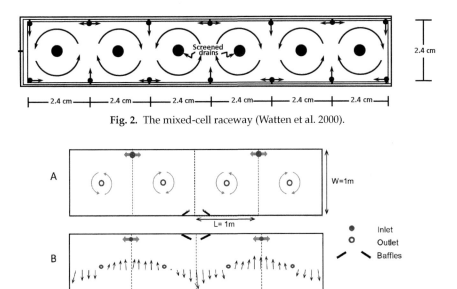

Fig. 2. The mixed-cell raceway (Watten et al. 2000).

Fig. 3. (A) The multivortex tank proposed by Masaló and Oca (2014); (B) Velocity maps obtained in the longitudinal axis of the multivortex tank.

diminution of about 36% and 50% in respect to the tank without fish. Flatfish at the tank bottom increased the resistance to water flow and, consequently, the water velocities were lower.

Rectangular tanks with rotating flow cells combine the superior hydrodynamic characteristics of circular tanks (high velocities and self-cleaning properties) with the easier fish management and better land use of the raceways, which could turn them into a good alternative for sole culture. This would be especially true if they could be vertically stacked in a multilevel system to achieve a much higher rearing area per land area unit, which is a critical factor in a slow-growth species like sole. Moreover, they can include dual-drain systems (Ebeling et al. 2005) to concentrate biosolids and reduce the effluent treatment cost.

3. Hydrodynamic conditions in flatfish growing tanks

3.1 Adjustment of water velocities in flatfish tanks

Water velocities in flatfish tanks must be adjusted to reach self-cleaning conditions and to optimize fish growth and welfare. Before analyzing the influence of water velocity in flatfish growth, it is important to identify the kind of strategies that can be used for fitting the required velocity in flatfish tanks. These strategies must always consider the flow rate required for maintaining water quality levels, but they will be very different in raceways than in circular or rotating flow tanks.

The minimal flow rate required in any fish tank will be related mainly to the amount of dissolved oxygen (DO) that must be transported by water to supply the needs of fish. This amount will be roughly proportional to the amount of feed consumed by fish. The water flow required per unit of biomass (q_0) can be calculated from Eq. 1, where M is the oxygen consumption per unit of biomass, and DO_{in} and DO_{out} are the oxygen concentrations in the inlet water and at the discharge, respectively.

$$q_0 = \frac{M}{DO_{in} - DO_{out}}$$

Eq. 1

For instance, if we estimate a feed consumption of around 0.01 kg/day per kg of biomass, an oxygen consumption of around 250 g DO per kg of feed and a decrease in DO concentration ($DO_{in} - DO_{out}$) of 3.5 gm^{-3}, the oxygen consumption per unit of biomass (M) would be 2.5 g/day per kg of biomass, and the specific water flow requirement (q_0) would be 0.714 m^3/day per kg of biomass.

In a raceway, the average water velocity in the cross section (V) will be determined by the flow rate (Q), water depth (h) and tank width (w) (Eq. 2). This means that, when the tank width is fixed, water velocity can only be modified by adjusting the flow rate per unit width (Q/w) or the water depth (h).

$$V = \frac{Q}{w \cdot h}$$

Eq. 2

With flatfish, the maximal fish biomass is not constrained by the water tank volume, but by the surface available ($L \cdot w$). A maximal stocking fish density per unit surface (SDS) is set. For this reason, water depths used in flatfish raceways are much smaller than in those for round fish. In grow-out stages, the depths of sole tanks range from 14 to 20 cm, giving a relatively low capacity for velocity control.

For a specific flatfish raceway, with known L, w and h, the required flow rate for oxygen supply (Q_0) and the corresponding water velocity (V_0) can be calculated using Eqs. 3 and 4.

$$Q_0 = q_0 \, SDS \, L \, w$$

Eq. 3

$$V_0 = \frac{q_0 \, SDS \, L}{h}$$

Eq. 4

Therefore, supposing a raceway with q_0 = 0.714 m^3/day per kg of biomass, and a maximal stocking density per unit surface (SDS) of 30 kg/m^2, the water velocity needed to deliver the required flow rate for oxygen supply (V_0) can be calculated using Eq. 4 and will be proportional to the ratio length/depth (L/h); giving V_0 = 21.43 · L/h (m/day) or V_0 = 0.025 · L/h (cm/s).

In the design of a flatfish raceway, we can optimize the cost of pumping by choosing the suitable tank length so that the flow rate imposed by oxygen needs matches the optimal water velocity required to reach the optimal growth and self-cleaning conditions in the tank. In the above mentioned case, with a water depth of 0.2 m and a tank width of 1 m, if the water velocity required to reach the optimal growth were 10 cm/s, the water flow rate required in each tank should be 20 L/s (Eq. 2) and the optimal length of the raceway, calculated from Eq. 4, would be 80 m. With this length, the required flow rate calculated from Eq. 2 for achieving a velocity of 10 cm/s coincides with those calculated from Eq. 3, based on fish oxygen needs. If we used a shorter tank, for example 40 m, we would need to double the raceways in order to maintain the same amount of biomass, and the flow rate in each raceway should also be 20 L/s if the optimal growing velocity of 10 cm/s is to be achieved (Eq. 2). Therefore, the total flow rate required for the same amount of biomass will also be twofold, with the corresponding increase in pumping costs.

The difference must be highlighted between the design criteria in raceways for flatfish and for round fish. For the latter, the maximal stocking biomass is not determined by the available surface, but by the water volume, as the tank design is constrained by the maximal stocking density per unit volume (*SDV*). Therefore, Eq. 4 should be replaced by Eq. 5, in which the velocity, calculated for oxygen requirements, is independent from the water depth. In practice, raceways for round fish are managed much closer to their design requirement for oxygen supply than for cleaning requirements or recommended velocities for fish conditioning, which are usually much higher (Timmons and Ebeling 2010).

$$V_0 = L \, SDV \, q_0 \qquad \qquad \text{Eq. 5}$$

The control of water velocity in rotating flow tanks, including circular and raceways with rotating flow cells, is totally different. The average circulating velocity (V_{avg}) is controlled by the water inlet impulse force, *Fi*, which can by calculated by Eq. 6.

$$Fi = \rho Q \, (V_{in} - V_{avg}) \qquad \qquad \text{Eq. 6}$$

where ρ is the water density, and V_{in} is the water inlet velocity.

Considering that the water inlet velocity (V_{in}) is commonly much higher than V_{avg}, Eq. 6 can usually be simplified as Eq. 7a, or as Eq. 7b if we substitute V_{in} with the flow rate (*Q*) divided by the total area of the water inlet orifices (A_O).

$$Fi = \rho Q V_{in} \qquad \qquad \text{Eq. 7a}$$

$$Fi = \frac{\rho Q^2}{A_O} \qquad \qquad \text{Eq. 7b}$$

Equation 7b shows that, with the same flow rate, *Fi* can be increased by reducing A_O.

Oca and Masaló (2007) defined a non-dimensional tank resistance coefficient (C_t) (Eq. 8), which is suitable for characterizing the resistance to water circulation offered by a specific rotating flow tank (circular or rectangular with rotating flow cells).

$$C_t = \frac{2QV_{in}}{A_w V_{avg}^2} \qquad \qquad \text{Eq. 8}$$

where A_w is the tank wet area of the circular tank or rotating flow cell.

Once C_t is experimentally determined, the desired average velocity can be obtained by using Eq. 9.

$$V_{avg} = \left(\frac{2QV_{in}}{C_t A_w} \right)^{1/2} = \left(\frac{2}{A_w \rho C_t} \right)^{1/2} Fi^{(1/2)} \qquad \qquad \text{Eq. 9}$$

Assuming A_w, ρ and *Ct* to be constant, it can be observed that, in a rotating flow tank, the average velocity is proportional to the square root of the impulse force *Fi*, and the constant of proportionality can be experimentally determined (Oca and Masaló 2007). Therefore, the control of velocities will be much easier than in raceways, because independently from the required flow rate for oxygen supply to fish, we have the capacity to modify the size of the water inlet orifices in order to achieve optimal water velocities. For a constant flow rate, reducing the size of water inlet area leads to an increase in *Fi* and therefore in the average velocity, as shown in Eqs. 7b and 9.

3.2 Optimal water velocities in flatfish growing tanks

Water flow velocities can be related to fish size through the estimation of a relative swimming velocity (*RSV*) that is expressed as body lengths per second (bl/s) (Hammer 1995).

Among flatfish species, *RSV* effects on growth have been quantified under farm-like conditions only for Japanese flounder (*Paralichthys olivaceous*) (Ogata and Oku 2000), Summer flounder (*Paralichthys dentatus*) (Bengtson et al. 2004) and California halibut (*Paralichthys californicus*) (Merino et al. 2007a).

Ogata and Oku (2000) analyzed the growth performance of juvenile Japanese flounder (initial mean body weight and length 5.7 g and 9.1 cm), reared at *RSV* 0.3, 0.9 and 2.1 bl/s. Weight gain and final length of the group reared at 2.1 bl/s were significantly lower than those of the group reared at 0.9 bl/s, and they suggested that the optimum water velocity in Japanese flounder occurred at about 1.0 bl/s. The feed efficiency at 2.1 bl/s was significantly lower than those at 0.3 and 0.9 bl/s. These findings are similar to those obtained by Merino et al. (2007a) for California halibut juveniles (initial mean body weight and length 1.53 g and 5.4 cm), which proved to grow faster and made more efficient use of the feed provided at *RSV* of 0.5 and 1.0 bl/s than at 1.5 bl/s.

Working with bigger sizes, Bengtson et al. (2004) found that summer flounder (124 g, 257 g and 387 g) grew best when reared at 0.5 bl/s, in comparison to those grown at an *RSV* of less than 0.3 bl/s or greater than 1.3 bl/s.

Almansa et al. (2011) evaluated the fish distribution in turbot raceways (*Scophthalmus maximus*) with higher sizes (mean weight 234.02 g, mean length 22.4 cm) and different water velocities. They observed that velocities between 0.33 and 0.46 bl/s promoted a highly homogenous distribution of turbot. Nevertheless, when they reduced the width in the middle of the unit and gave the fish the opportunity to choose between the central narrow area with higher velocities (trial 1: 0.58 bl/s and trial 2: 0.98 bl/s) or the upstream and downstream areas (maintained at 0.29 bl/s in both trials), the turbot avoided swimming against the 0.58 bl/s stream and, therefore, less fish biomass was observed in the area upstream than downstream from the narrowing. When the water velocity in the narrow area increased to almost 1 bl/s, the turbot were not able to maintain their position and were distressed, resulting in reduced feed ingestion.

3.3 Vertical oxygen stratification in flatfish tanks

A reduction in specific growth rates was reported by Ogata and Oku (2000), Bengtson et al. (2004) and Merino et al. (2007a) for flatfish species grown in still water. One possible explanation for this reduction has been proposed by Reig et al. (2007), who found that dissolved oxygen is depressed significantly in the immediate vicinity of California halibut at very low water velocities.

Almansa et al. (2014) analyzed the hydrodynamic conditions that determine the oxygen gradient that occurs in the layer of water adjacent to flatfish grown in tanks.

Experiments were conducted with *Solea senegalensis* tanks with two geometries and different water flow rates. Results showed that a vertical oxygen gradient is created in certain hydrodynamic conditions, as a consequence of oxygen consumption by fish resting on the tank bottom.

The vertical profile of oxygen in flatfish tanks is influenced by the fish oxygen consumption and by the presence of the boundary layer. Flatfish extract oxygen from the water layer immediately above them. This water layer can be a semi-stagnant zone with a depressed oxygen concentration due to fish consumption, causing an oxygen gradient from the fish layer surface to the water bulk. The boundary layer adjacent to the wall in a turbulent flow is defined as the distance from the wall to a point where the flow is undisturbed by the wall. One must differentiate between the turbulent boundary layer, where fluid particles show irregular motions and erratic paths, and the viscous sublayer, which occurs immediately next to the wall. In the viscous sublayer, the velocity gradient is very steep, and the fluid particles move in parallel straight lines. The poor movement of fluid particles between adjacent layers can result in large solute concentration gradients. There is a transition zone where a gradual change between both behaviors (viscous and turbulent) can be observed. The Reynolds number (*Re*) expresses the ratio of inertial forces to viscous forces. Low Reynolds numbers indicate that viscous forces are dominant and will lead to an increase in the thickness of the viscous sublayer and, therefore, in the oxygen gradient from the fish layer to the bulk of the tank water.

The Reynolds number can be calculated by Eq. 10, where *v* is the kinematic viscosity of water (around 10^6 m²/s), *V* the average velocity in a tank cross section and R_h the hydraulic radius, which can be calculated by Eqs. 11 and 12 for raceways and circular tanks, respectively, with *w* being the raceway width, *R* the radius and *h* the water depth.

$$R_e = \frac{4R_h \, V}{v} \qquad \text{Eq. 10}$$

$$R_{h \, (Raceway)} = \frac{w \, h}{w + 2h} \qquad \text{Eq. 11}$$

$$R_{h \, (Circular)} = \frac{R \, h}{R + 2h} \qquad \text{Eq. 12}$$

Almansa et al. (2014) found that with Reynolds numbers under 6000, significant decreases in *DO* concentrations were observed in both kinds of tanks at a distance of 2.5 cm from the tank bottom, which is more than two times the average thickness of the fish layer (1.2 cm). In a raceway, the average velocity and the flow rate required to achieve Re = 6000 ($V_{(R;Re=6000}$ and $Q_{R;Re=6000})$ for a tank with width *w* and water depth *h* can be calculated with Eqs. 13 and 14.

$$V_{R; \, Re=6000} = 1500 \, v \frac{w + 2h}{w \, h} \qquad \text{Eq. 13}$$

$$Q_{R; \, Re=6000} = 1500 \, v \, (w + 2h) \qquad \text{Eq. 14}$$

In circular tanks, the required average velocity to achieve Re = 6000 ($V_{C; \, Re=6000})$ can be calculated in a similar way as in raceways, by using Eq. 15. Nevertheless, as explained above (Eqs. 7b and 9), the actual velocity will depend on the impulse force produced by the water entering the tank. Therefore, using Eqs. 7b and 9, it is easy to adjust the values *Q* and A_0 in order to achieve an average velocity higher than $V_{C; \, Re=6000}$ and, consequently, Re > 6000.

$$V_{C; \, Re=6000} = 1500 \, v \frac{R + 2h}{Rh} \qquad \text{Eq. 15}$$

In circular tanks it is possible to achieve the desired velocities, even with small water flow rates, by adjusting the water inlet orifices area, as has been shown previously (Eq. 7 and 9).

References

Almansa, C., L. Reig and J. Oca. 2011. Use of laser scanning to evaluate turbot (*Scophthalmus maximus*) distribution in raceways with different water velocities. Aquacult. Eng. 51: 7–14.

Almansa, C., I. Masaló, L. Reig, R. Piedrahita and J. Oca. 2014. Influence of tank hydrodynamics on vertical oxygen stratification in flatfish tanks. Aquacultural Engineering 63: 1–8.

Almansa, C., L. Reig and J. Oca. 2015. The laser scanner is a reliable method to estimate the biomass of a Senegalese sole (*Solea senegalensis*) population in a tank. Aquacult. Eng. 69: 78–83.

Bengtson, D., S. Willey, E. McCaffrey and D. Alves. 2004. Effects of water velocity on conditioning of summer flounder, *Paralichthys dentatus*, for net pens. J. Appl. Aquacult. 14: 133–142.

Blancheton, J.P. 2000. Developments in recirculation systems for Mediterranean fish species. Aquacult. Eng. 22: 17–31.

Brinker, A. and R. Rösch. 2005. Factors determining the size of suspended solids in a flow-through fish farm. Aquacult. Eng. 33: 1–19.

Cripps, S.J. and M.G. Poxton. 1992. A review of the design and perform of tanks relevant to flatfish culture. Aquacult. Eng. 11: 71–91.

Cripps, S.J. and A. Bergheim. 2000. Solids management and removal for intensive land-based aquaculture production systems. Aquacult. Eng. 22: 33–56.

Davidson, J. and S. Summerfelt. 2004. Solids flushing, mixing, and water velocity profiles within large (10 and 150 m³) circular "Cornell-type" dual-drain tanks. Aquacult. Eng. 32: 245–271.

De Groot, S.J. 1971. On the interrelationships between morphology of the alimentary tract, food and feeding behaviour in flatfishes (Pisces: Pleuronectiformes). Neth. J. Sea Res. 5: 121–196.

Dias, J., M. Yúfera, L.M.P. Valente and P. Rema. 2010. Feed transit and apparent protein, phosphorus and energy digestibility of practical feed ingredients by Senegalese sole (*Solea senegalensis*). Aquacult 302: 94–99.

Duarte, S., L. Reig and J. Oca. 2009. Measurement of sole activity by digital image analysis. Aquacult. Eng. 41: 22–27.

Ebeling, J.M., M.B. Timmons, J.A. Joiner and R.A. Labatut. 2005. Mixed-Cell Raceway: Engineering design criteria, construction, and hydraulic characterization. N. A. J. Aquacult. 67: 193–201.

Hammer, C. 1995. Fatigue and exercise test with fish. Comp. Biochem. Physiol. 112A: 1–20.

Idaho Division of Environmental Quality (IDEQ). 1998. Idaho Waste Management: Guidelines for Aquaculture Operations. Idaho Department of Health and Wel-fare, Division of Environmental Quality, Twin Falls.

Imsland, A.K., A. Foss, L.E.C. Conceiçao, M.T. Dinis, D. Delbare, E. Schram et al. 2003. A review of the culture potential of *Solea solea* and *S. senegalensis*. Rev. Fish Biol. Fisher 13: 379–407.

Kamstra, A., V. van den Briel, J. van der Vorst and J. Wilde. 2001. Farming of sole (*Solea solea*) in recirculation systems; prospects and constraints. European Aquaculture Society, Special Publication no. 29: 127–128.

Kamstra, A., E. Blom and R. Blonk. 2012. Design and economic performance of an innovative raceway system for sole (*Solea solea*). AQUA 2012 Abstracts (Prague-Czech Republic): 538.

Labatut, R.A. and J.F. Olivares. 2004. Culture of turbot (*Scophthalmus maximus*) juveniles using shallow raceways tanks antd recirculation. Aquacult. Eng. 32: 113–127.

Labatut, R.A., J.M. Ebeling, R. Bhaskaran and M.B. Timmons. 2007a. Hydrodynamics of a Large-Scale Mixed-Cell Raceway (MCR): Experimental studies. Aquacult. Eng. 37: 132–143.

Labatut, R.A., J.M. Ebeling, R. Bhaskaran and M.B. Timmons. 2007b. Effects of inlet and outlet flow characteristics on Mixed-Cell Raceway (MCR) hydrodynamics. Aquacult. Eng. 37: 158–170.

Lunde, T., S. Skybakmoen and I. Schei. 1997. Particle Trap. US Patent # 5,636,595.

Martins, C.I.M., E.H. Eding, M.C.J. Verdegem, L.T.N. Heinsbroek, O. Schneider, J.P. Blancheton et al. 2010. New developments in recirculating aquaculture systems in Europe: A perspective on environmental sustainability. Aquacult. Eng. 43: 83–93.

Masaló, I., L. Reig and J. Oca. 2008. Study of fish swimming activity using acoustical Doppler velocimetry (ADV) techniques. Aquacult. Eng. 38: 43–51.

Masaló, I. and J. Oca. 2013. Influence of baffles and fish presence in the velocities distribution of a multivortex tank. Abstracts Celebrating 40 years of Aquaculture. Elsevier, Las Palmas de Gran Canaria-Spain: P3.036.

Masaló, I. and J. Oca. 2014. Hydrodynamics in a multivortex aquaculture tank: Effect of baffles and water inlet characteristics. Aquacult. Eng. 58: 69–76.

Masaló, I. and J. Oca. 2016. Influence of fish swimming on the flow pattern of circular tanks. Aquacult. Eng. 74: 84–95.

Mas-Muñoz J., H. Komen, O. Schneider, S.W. Visch and J.W. Schrama. 2011. Feeding Behaviour, Swimming Activity and Boldness Explain Variation in Feed Intake and Growth of Sole (*Solea solea*) Reared in Captivity. PLoS ONE 6(6): e21393. doi: 10.1371/journal.pone.0021393.

Merino, G.E., R.H. Piedrahita and D.E. Conklin. 2007a. Effect of water velocity on the growth of California halibut (*Paralichthys californicus*) juveniles. Aquaculture 271: 206–215.

Merino, G.E., R.H. Piedrahita and D.E. Conklin. 2007b. Settling characteristics of solids settled in a reciruclating system for California halibut (*Paralichthys californicus*) culture. Aquacult. Eng. 37: 79–88.

Morais, S., C. Aragão, E. Cabrita, L.E.C. Conceição, M. Constenla, B. Costas et al. 2014. New developments and biological insights into the farming of *Solea senegalensis* reinforcing its aquaculture potential. Rev. Aquacult. doi: 10.1111/raq.12091.

Oca, J., I. Masaló and L. Reig. 2004. Comparative analysis of flow patterns in aquaculture rectangular tanks with different water inlet characteristics. Aquacult. Eng. 31: 221–236.

Oca, J. and I. Masaló. 2007. Design criteria for rotating flow cells in rectangular aquaculture tanks. Aquacult. Eng. 36: 36–44.

Oca, J. and I. Masaló. 2013. Flow pattern in aquaculture circular tanks: influence of flow rate, water depth and water inlet and outlet features. Aquacult. Eng. 52: 65–72.

Ogata, H. and H. Oku. 2000. Effects of water velocity on growth performance of juvenile Japanese flounder Paralichthys olivaceus. J. World Aquacult. Soc. 31: 225–231.

Øiestad, V. 1999. Shallow raceways as a compact, resource-maximizing farming procedure for marine fish species. Aquac. Res. 30: 831–840.

Palstra, A.P. and J.V. Planas. 2011. Fish under exercise. Fish Physiol. Biochem. 37: 259–272.

Paul, T.C., S.K. Sayal, V.S. Sakhuja and G.S. Dhillon. 1991. Vortex-settling basin design considerations. J. Hydraul. Eng. 11: 172–189.

Reig, L., R.H. Piedrahita and D.E. Conklin. 2007. Influence of California halibut (Paralichthys californicus) on the vertical distribution of dissolved oxygen in a raceway and a circular tank at two depths. Aquacult. Eng. 36: 261–271.

Ross, R.M., B.J. Watten, W.F. Krise and M.N. DiLauro. 1995. Influence of tank design and hydraulic loading on the behaviour, growth, and metabolism of rainbow trout (*Oncorhynchus mykiss*). Aquacult. Eng. 14: 29–47.

Ross, R.M. and B.J. Watten. 1998. Importance of rearing-unit design and stocking density to the behaviour, growth and metabolism of lake trout (*Salvelinus namaycush*). Aquacult Eng. 19: 41 56.

Salas-Leiton, E., V. Anguís, B. Martín-Antonio, D. Crespo, J.V. Planas, C. Infante, J.P. et al. 2010. Effects of stocking density and feed ration on growth and gene expression in the Senegalese sole (*Solea senegalensis*): Potential effects on the immune response. Fish Shellfish Immun. 28: 296–302.

Schei, I. and S. Skybakmoen. 1998. Control of water quality and growth performance by solids removal and hydraulic control in rearing tanks. Fisheco'98, First International Symposium on Fisheries & Ecology, Trabzon, Turkey.

Summerfelt, S.T., J.W. Davisdon and M.B. Timmons. 2000a. Hydrodynamics in the "Cornell-Type" Dual-Drain tank. Third International Conference of Recirculating Aquaculture, July 19–21, 2000 Roanoke, VA.

Summerfelt, S.T., M.B. Timmons and B.J. Watten. 2000b. Tank and Raceway Culture. pp. 921–928. *In*: Stickney, R.R. [ed.]. Encyclopedia of Aquaculture. Wiley, New York, USA.

Summerfetl, S.T., J. Davidson, G. Wilson and T. Waldrop. 2009. Advances in fish harvest technologies for circular tanks. Aquacult. Eng. 40: 62–71.

Timmons, M.B., S.T. Summerfelt and B.J. Vinci. 1998. Review of circular tank technology and managment. Aquacult. Eng. 18: 51–69.

Timmons, M.B. and J.M. Ebeling. 2010. Recirculating aquaculture. Second ed. Cayuga Aqua Ventures, Ithaca, New York.

Tvinnereim, K. 1988. Design of water inlets for closed fish farms. *In*: Proceedings of the Conference: Aquaculture Engineering: Technologies for the Future. Sterling Scotland. IChemE Symposium Series No111, EFCE Publications Series No 66: 241–249. Rugby, UK.

Van Toever, E. 1997. Water Treatment System Particularly for use in Aquaculture. US Patent # 5,593,574.

Watten, B.J. 1992. Modeling the effects of sequential rearing on the potential production of controlled environment fish-culture systems. Aquacult. Eng. 11: 33–46.

Watten, B.J., D.C. Honeyfield and M.F. Schwartz. 2000. Hydraulic characteristics of a rectangular mixed-cell rearing unit. Aquacult. Eng. 24: 59–73.

Westers, H. and K.M. Pratt. 1977. Rational design of hatcheries for intensive salmonid culture, based on metabolic characteristics. Progress. Fish-Cultur. 39: 157–165.

Wright, D., A. Desbarats and C. Doucette. 2012. Triple Drain Apparatus for an Aquaculture Recirculation System. US Patent # 13,322,736.

Yúfera, M. and A.M. Arias. 2010. Traditional polyculture in "esteros" in the Bay of Cádiz (Spain). Hopes and expectancies of a unique activity in Europe. AE magazine 35: 22–25.

Section B

Reproduction, Chronobiology, Development, Nutrition, Welfare, Ecotoxicology, Pathology, Osmoregulation and Genetics

B-1.1

Sensing the Environment
The Pineal Organ of the Senegalese Sole

José A. Muñoz-Cueto,[1,*] *Francesca Confente,*[1] *Águeda Jimena Martín-Robles,*[2] *María del Carmen Rendón*[1] and *Patricia Herrera-Pérez*[1]

1. Introduction

Fish, as other living organisms, have evolved in a cyclic environment and have developed timekeeping mechanisms (molecular oscillators or biological clocks) to anticipate the forthcoming of highly recurrent events and increase survival probabilities. As a result, many behavioural and physiological functions, including locomotor activity, shoaling behaviour, feeding, reproduction, metabolism, skin pigmentation, sedation, oxygen consumption, thermoregulation, hormonal secretion, etc., show a rhythmic pattern (Falcón et al. 2010). Some environmental factors are unpredictable (rain, cloudiness, earthquakes) but other factors are associated to cyclic geophysical events (Earth's rotation and revolution around the Sun, Moon phases, tidal cycles) and are repetitive, recurrent and reliable, determining biological rhythms with particular periods. According to their periodicity we can classify these biological rhythms in circadian (~ 24 h), ultradian (< 2 0 h, e.g., tidal rhythms) and infradian (> 28 h, e.g., lunar and seasonal rhythms) rhythms. The origin of these rhythms is

[1] Department of Biology, Faculty of Marine and Environmental Sciences and Institute of Marine Research (INMAR). Marine Campus of International Excellence (CEIMAR) and Agrifood Campus of International Excellence (ceiA3), University of Cádiz, 11510 Puerto Real, Cádiz, Spain.
[2] Biochemistry and Molecular Biology, Department of Biomedicine, Biotechnology and Public Health. Faculty of Sciences, Marine Campus of International Excellence (CEIMAR) and Agrifood Campus of International Excellence (ceiA3), University of Cádiz, 11510 Puerto Real, Cádiz, Spain.
* Corresponding author: munoz.cueto@uca.es

endogenous, i.e., they persist in absence of the environmental factors that entrained them, and they are generated by the own organism through the activity of a central oscillator or pacemaker. In mammals, this central pacemaker is located in the suprachiasmatic nucleus of the hypothalamus. Nevertheless, such a hypothalamic pacemaker involved in the generation and synchronisation of rhythms has not yet been identified in fish. In contrast to mammals, fish possess endogenous oscillators in other central areas, like the pineal organ, which is a key component of the circadian system (Falcón et al. 2010). In addition, the pineal organ of fish is a direct photosensitive organ that possesses true photoreceptive pinealocytes, whereas in mammals, the perception of light resides exclusively in the retina (Ekström and Meissl 2003).

The fish pineal organ has not only photoreceptor and oscillatory functions but also effector properties because it codes this light information into nerve (neural projections and neurotransmitters) and neuroendocrine (rhythmic melatonin secretion) signals (Falcón et al. 2010). Up to date numerous rhythms of melatonin production have been described in both freshwater and marine fish species, including Senegalese sole (Zachmann et al. 1992, Sánchez-Vázquez et al. 1997, Bromage et al. 2001, Bayarri et al. 2004a, Vera et al. 2007). In all of them, plasma melatonin levels rise during the night and fall during the day. These cyclical daily changes are the result of the expression and rhythmic activity of the enzyme arylalkylamine N-acetyltransferase (AANAT), the key enzyme in the biosynthesis of melatonin (Ekström and Meissl 1997, Falcón et al. 2007b, 2010). But these melatonin rhythms not only provide information about the time of the day. In fish and other poikilotherms, melatonin secretion also undergoes seasonal variations both in duration and amplitude, which are related to the duration of the night and the temperature of the water, respectively. The goldfish exhibits higher nocturnal values of melatonin at high ambient temperatures than at low temperatures (Iigo and Aida 1995) and similar evidence has been obtained in salmon and sea bass, both of them showing higher nocturnal melatonin levels in the warm months (Randall et al. 1995, García-Allegue et al. 2001). In this way, melatonin can represent both a clock and a calendar for these species. Moreover, melatonin rhythms can also transduce the Moon phases (higher and lower melatonin production in new moon and full moon, respectively), synchronising different lunar behavioural and physiological rhythms in fish (Rahman et al. 2004).

In this chapter, we will review some works developed in our laboratory focused on the functional neuroanatomy of the pineal organ, in the characterisation of its photoreceptor and melatonin-synthesizing cells, and in tract-tracing studies revealing the pinealofugal and pinealopetal projections of this photoneuroendocrine organ to better understand how environmental information is integrated in the central nervous system of the Senegalese sole, *Solea senegalensis*.

2. Anatomical, Histological and Immunohistochemical Analysis of the Pineal Organ of Sole

The pineal anatomy, histology, ultrastructure and function have been studied in many fish orders (Ekström and Meissl 1997), but much less is known on the characteristics of the pineal complex in pleuronectiform species, being most of the studies focused in developing pineal organ (Forsell et al. 1997, 2001, 2002, Flamarique 2002, Vuilleumier et al. 2007). In our laboratory, we performed an anatomical and histological study of the pineal complex of Senegalese sole and characterized the presence of rod opsin-like and cone opsin-like photoreceptors and melatonin-producing cells by immunohistochemistry (Fig. 1, Confente et al. 2008a,b,c, Confente 2009).

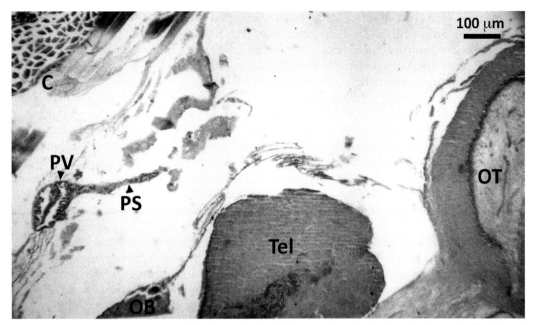

Fig. 1. Sagittal histological section of the brain of sole (*Solea senegalensis*) showing the long pineal stalk and a prominent pineal vesicle in close proximity to the cartilage and skull bone. Abbreviations: C, cartilage; OB, olfactory bulb; OT, optic tectum; ps, pineal stalk; pv, pineal vesicle; Tel, telencephalon. Bar scale: 100 µm.

The pineal complex of sole consisted of the pineal and parapineal organs, and a highly lobulated dorsal sac. In turn, the pineal organ of sole was composed of a large and expanded pineal vesicle, and a narrow and very long pineal stalk (Fig. 1), both of them exhibiting a conspicuous internal lumen that was interconnected with the cerebral third ventricle (Confente et al. 2008a). The anterior margin of the proximal pineal stalk entered the brain at the level of the habenular commissure (Fig. 3D) while the posterior tract of the proximal stalk was smaller and joined the brain at the rostral pole of the posterior commissure. Whereas in most teleosts the pineal vesicle is emitted dorsally to reach the roof of the cranium, between the optic tectum and the telencephalic hemispheres (Ekström and Meissl 1997), the pineal vesicle of adult sole extended beyond the rostral forebrain, occupying in its anterior pole a dorsal position in relation to the olfactory bulbs (Fig. 1, Confente et al. 2008a). The pineal vesicle of sole is hypertrophied, probably reflecting an adaptation to enhance light sensitivity in low-light benthonic environments (Fig. 1). Benthonic fish exhibit larger pineal organs and/or photoreceptor outer segments compared to pelagic species, reflecting a "deeper = larger" adaptive mechanism for increasing the efficiency in light perception (McNulty and Nafpaktitis 1976, 1977, Wagner and Mattheus 2002). Pineal sensitivity to light and light intensity thresholds appears species-dependent, depending on their daily behavioural patterns, habitats and light transmission through the cranial bone (Migaud et al. 2006, 2007, Ziv et al. 2007). In this respect, although a pineal window was not evident in sole, the skull exhibited thinner bone and cartilage and a prominent hole at the level where the pineal vesicle is attached (Fig. 1). These characteristics can improve the passage of light toward the photoreceptive pineal cells and seem to represent additional adaptive mechanisms to optimise light perception in benthonic habitats.

The metamorphic process that occurs during early developmental stages of sole induced an asymmetry of the cranium, of the rostral forebrain areas and in the position of the eyes, both of them being placed in the upper-right pigmented side in post-metamorphic animals (Rodríguez-Gómez et al. 2000a, Fernández-Díaz et al. 2001). Moreover, left and right eyes of sole exhibited significant differences in weight and melatonin content (Bayarri et al. 2004a). This asymmetry was also extended to the position of the pineal organ, which leaved the midline position characteristic of other teleosts and shifted its photosensitive pineal vesicle towards the upper-right pigmented side, where both eyes were also placed (Confente et al. 2008a). Both readjustments represent amazing adaptive mechanisms permitting a better access of light to photosensitive structures in this benthonic species. The pineal complex of sole also exhibited a small parapineal organ, an unpaired and asymmetric structure positioned in the left hemisphere of the brain (Confente et al. 2008a). The parapineal organ of sole was connected to the left habenula by a prominent nerve tract containing both parapinealopetal and parapinealofugal fibres (van Veen et al. 1980, Yáñez et al. 1996, 1999, Confente et al. 2008a). A photoreceptive function has been proposed for the parapineal organ of fish (Rüdeberg 1969, Vigh-Teichmann et al. 1983, Ekström et al. 1987, García-Fernández et al. 1997), which is also responsible of the asymmetry of the epithalamic habenula (Concha and Wilson 2001, Gamse et al. 2005). Although the parapineal is present in many teleost families, including flatfishes, it is absent in several others as medaka, in which it has been demonstrated that the supposed lack of a distinct parapineal is due to ontogenetic incorporation into the habenula (Ishikawa et al. 2015).

Histologically, we recognized three kinds of cells scattered in the pineal parenchyma: large periventricular cells that seemed to represent photoreceptor pinealocytes, because they exhibited long apical cytoplasmic processes (outer segments) oriented towards the pineal lumen; smaller, rounded to ovoid cells that adopted a more interne position in the pineal parenchyma and represented, probably, interstitial or glial-like supporting cells; and cells exhibiting a large blue/grey stained nucleus that may represent pineal neurons (Confente et al. 2008a).

Immunohistochemical analysis using rod opsin and cone opsin antibodies revealed that large periventricular cells with long apical cytoplasmic processes present in both pineal stalk and vesicle of sole represented photoreceptor cells (Fig. 2A, Confente et al. 2008a), which were much more abundant in sole compared to a pelagic species as sea bass (Herrera-Pérez et al. 2011). Both rod opsin and cone opsin antisera provided a very similar pattern of immunostaining suggesting the existence of an overlapping of the two photoreceptor cell types in the pineal organ of sole. The presence of rod-like and cone-like photopigments has been extensively reported in the pineal organ of numerous fish species using different rod and cone opsin specific antibodies (Vigh-Teichmann et al. 1983, Ekström et al. 1987, García-Fernández et al. 1997, Vigh et al. 2002, Foster and Hankins 2002, Herrera-Pérez et al. 2011, Rincón-Camacho et al. 2016). Nevertheless, the pineal complex of fish expresses a much larger set of opsins, such as vertebrate ancient-opsin, pinopsin, parietopsin, parapinopsin, melanopsin and a subset of visual opsins (Fejér et al. 1997, Philp et al. 2000a,b, Forsell et al. 2001, 2002, Koyanagi et al. 2004, Su et al. 2006, Kawano-Yamashita et al. 2007, Eilertsen et al. 2014). High levels of novel non-visual photopigments (tmt3a, opn4m2, opn4x2, opn6a) have also been revealed recently in the pineal of zebrafish (Davies et al. 2015). Many opsin genes have been identified in the SoleaDB bioinformatic platform (http://www.juntadeandalucia.es/agriculturaypesca/ifapa/soleadb_ifapa/, Benzekri et al. 2014) but their presence and expression in the pineal complex of sole remains to be investigated.

Melatonin-secreting cells were also identified in the pineal organ of sole using antisera against its precursor serotonin and against hydroxyindole-*O*-methyltransferase (Fig. 2B), the enzyme responsible of the last step in melatonin synthesis (Confente et al. 2008a).

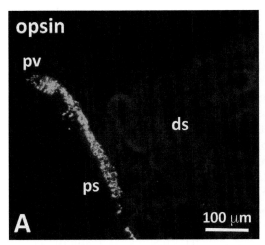
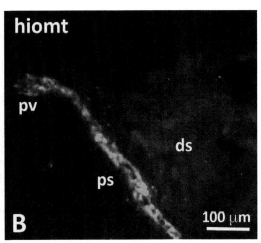

Fig. 2. Immunohistochemical identification of photoreceptor and melatonin-synthesizing cells in the pineal organ of sole (*Solea senegalensis*). A. Cone opsin-immunoreactive photoreceptor cells in the pineal organ of sole. B. HIOMT-immunoreactive cells in the pineal vesicle and distal pineal stalk of sole. Abbreviations: ds, dorsal sac; ps, pineal stalk; pv, pineal vesicle. Bar scale represents 100 μm.

These cells adopted a similar distribution in the pineal stalk and vesicle and had a similar appearance than those labelled with cone opsin- and rhodopsin-antisera (Fig. 2A,B), suggesting that, in addition to produce melatonin, they represent direct photoreceptive cells, as reported in other teleosts (Ekström and Meissl 1997, Falcón et al. 2010).

3. Afferent and efferent connections of the pineal organ of sole

The pineal organ of fish is a photosensitive structure that is involved in the transduction of light and photoperiodic information to the central nervous system and to the organism (Meissl and Dodt 1981, Ekström and Meissl 1997, Falcón et al. 2010). Such information is transmitted through two rhythmic output signals, an endocrine signal and a neural signal (Ekström and Meissl 2003, Falcón et al. 2010). The endocrine signal exiting from the pineal organ is melatonin, an indoleamine that is rhythmically synthesized and released in the cerebrospinal fluid and in the vascular blood system (Fenwick 1970, McNulty 1984, Ekström and Meissl 2003, Falcón et al. 2007a). This neurohormonal message reaches the central and peripheral tissues, allowing the numerous rhythmic processes of the individual to be synchronised with the environment (Falcón et al. 1984, 1994, 2010, Bolliet et al. 1996, Iigo et al. 1991, 2004).

The second pathway exiting from the pineal is represented by the neural circuitry that connects the pineal complex to the brain targets (Ueck 1979, Ekström and Meissl 1997, Falcón et al. 2010). Most of the tract-tracing studies performed to reveal these connections used neural tracers such as the horseradish-peroxidase (Ekström and van Veen 1983, 1984, Ekström 1984, Puzdrowski and Northcutt 1989, Jiménez et al. 1995) or fluorescent lipophilic carbocyanine dyes such as the 1,1'-dioctadecyl 3,3,3',3'-tetramethylindocarbocyanine perchlorate (DiI) (Yáñez et al. 1993, 1996, Yáñez and Anadón 1998, Pombal et al. 1999, Mandado et al. 2001, Yáñez et al. 2009). Although direct photoreceptor axonal projections have been proved to reach the brain in rainbow trout (*Salmo gairdneri*) and lampreys (*Lampetra japonica, Lampetra fluviatilis, Petromyzon marinus*, Ekström 1987, Samejima et al. 1989, Pombal et al. 1999), most of the pineal efferents seem to represent the axons of second order neurons, which are postsynaptic to photoreceptor cells. These pinealofugal axons extend their processes to innervate central areas as the habenula, the preoptic area, the thalamus, the periventricular hypothalamus, the pretectal area, the optic tectum, the torus semicircularis and the rostral tegmentum (Ekström 1984, 1985, Ekström and van Veen 1984, Ekström and Meissl 1997, Falcón et al. 2010). Otherwise, the pineal organ is the target of afferent projections arising from distinct brain areas as occur in lamprey (Yáñez et al. 1993), elasmobranchs (*Raja montagui* and *Scyliorhinus canicula*) (Mandado et al. 2001) or teleost species (*Dicentrarchus labrax*) (Servili et al. 2011). Some of these regions encompass an extensive overlap of both retinofugal and pinealofugal innervations becoming from the retina and the pineal organ, respectively, suggesting their relevant role in the integration of photic inputs (Ekström et al. 1994, Yáñez et al. 1993, Holmqvist et al. 1994).

In a previous study, we have analysed the pinealofugal and pinealopetal projections of the pineal complex in the Senegalese sole, *Solea senegalensis*, by using a carbocyanine dye tract-tracing method (Figs. 3 and 4) (Confente et al. 2008b,c, Confente 2009). The pinealofugal projections of the sole pineal complex, which corresponded to anterogradely labelled tracts and terminal fields, entered the brain through the habenular and posterior commissures (Fig. 3D), and reached the preoptic area (Figs. 3B, C, 4A, B), habenula (Fig. 3D), ventral and dorsal thalamus (Figs. 3D–F, 4C, D), fasciculus retroflexus (Figs. 3E–G), optic tectum (Figs. 3D, E), periventricular pretectum (Figs. 3D–F, 4D), central pretectal area (Figs. 3E–G), superficial pretectum (Figs. 3D, E), posterior tuberculum (Figs. 3D–F, 4D), dorsal synencephalon and rostral tegmentum (Figs. 3G) (Confente et al. 2008b,c). The main target areas of sole pineal organ did not differ too much from those reported in lamprey (Puzdrowski and Northcutt 1989, Yáñez et al. 1993, Pombal et al. 1999), sturgeon (Yáñez and Anadón 1998), elasmobranchs (Mandado et al. 2001) or other teleosts (Hafeez and Zerihun 1974, Ekström and van Veen 1983, 1984, Yáñez et al. 2009, Servili et al. 2011), but some interspecific differences were observed.

The most rostral pinealofugal terminal fields were found in the ventrocaudal telencephalon and preoptic region (Figs. 3B, C, 4A), which exhibited few labelled fibres descending until the proximity of the optic chiasm, as described in sturgeon (Yáñez and Anadón 1998). At the rostral pole of the preoptic recess, the lateral part of the parvocellular preoptic nucleus received pineal projections in sole (Fig. 3B). Moreover, in the caudal preoptic area networks of labelled axonal processes also reached the suprachiasmatic nucleus and the posterior periventricular nucleus (Figs. 3C, 4B), where abundant beaded terminals become evident around the wall of the third ventricle (Confente et al. 2008a,b). As in sole, the preoptic suprachiasmatic nucleus receives a pineal innervation in Atlantic salmon *Salmo salar* (Holmqvist et al. 1994), zebrafish *Danio rerio* (Yáñez et al. 2009), sturgeon *Acipenser baeri* (Yáñez and Anadón 1998) and ray (Mandado et al. 2001) but not in dogfish (Mandado et al. 2001) or the European sea bass *Dicentrarchus labrax* (Servili et al. 2011). Moreover, the suprachiasmatic nucleus expressed melatonin receptors in sea bass (Herrera-Pérez et al. 2010) and the presence of abundant melatonin binding sites was reported in the preoptic area of different teleost species (Martinoli et al. 1991, Ekström and Vaněcek 1992, Mazurais et al. 1999, Herrera-Pérez et al. 2010). This nucleus is also known as a retinorecipient area in fish (Fernald 1982, Presson et al. 1985, Repérant et al. 1986, Medina et al. 1990, Northcutt and Butler 1993). In Senegalese sole, retinofugal projections and retinopetal neurons were also found at the level of the suprachiasmatic area (Confente 2009), suggesting that this nucleus could play a relevant role in the transduction of light information from both photoreceptor centres. Although the suprachiasmatic nucleus represents the central pacemaker controlling mammalian circadian rhythms, the presence

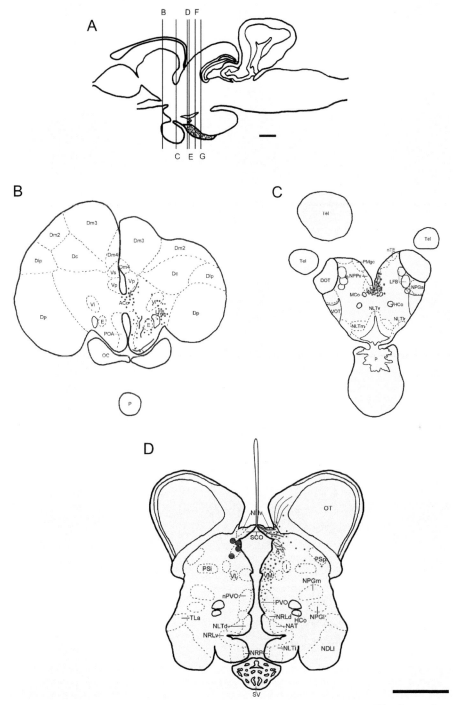

Fig. 3 contd. ...

...*Fig. 3 contd.*

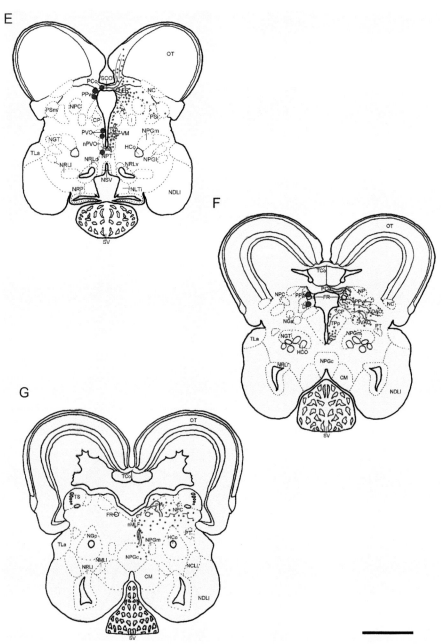

Fig. 3. (A). Lateral view of the brain of *Solea senegalensis*. Lettered lines indicate the levels of the transverse sections shown in B–G. (B–G). Series of transverse sections through the brain of *Solea senegalensis* from rostral (B) to caudal (G), showing the efferent and afferent connections of the pineal organ of sole. Crystals of the DiI tracer were applied to the pineal stalk of sole. Although the distribution of pinealofugal fibers (red dashes and red points) and pinealopetal neurons (red circles) are bilateral and symmetrical, they are represented in the right and left side of the sections, respectively. Scale bars: 1 mm. For abbreviations see Table 1.

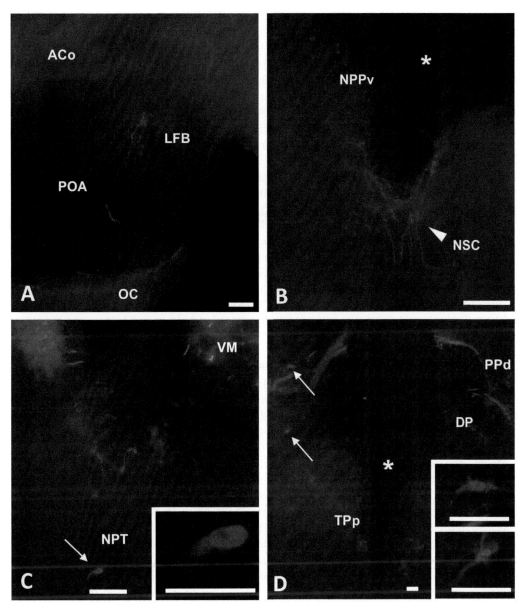

Fig. 4. Anterogradely-labeled pinealofugal fibres and retrogradely-labelled pinealopetal neurons following DiI application into the pineal stalk. A. Scarce pineal fibres reaching the ventral telencephalic/rostral preoptic region (POA) (level of Fig. 3B). B. Profuse varicose pinealofugal fibres within the suprachiasmatic nucleus (NSC), just below the third ventricle (asterisk). Note the sparse innervation in the posterior periventricular nucleus (NPPv) (level of Fig. 3C). C. Conspicuous terminal fields were found surrounding the walls of the third ventricle (asterisk), in the thalamic ventromedial nucleus (VM) and dorsally to the posterior tuberal nucleus (NPT) (level of section from Fig. 3E). Arrow indicates the presence of a pinealopetal neuron magnified in the inset. D. Pinealofugal fibres stained in the dorsal periventricular pretectal nucleus (PPd), in the periventricular nucleus of the posterior tuberculum (TPp), and in the dorsal posterior thalamic nucleus (DP) (Level of Fig. 3F). Arrows mark two pinealopetal neurons present in the caudal part of PPd and in the DP and magnified in the insets. Other abbreviations: ACo, anterior commissure; LFB, lateral forebrain bundle; OC, optic chiasm. Scale bars: 50 μm.

of an endogenous clock has not been yet determined in teleost brain (Ekström 1984, Rusak and Zucker 1979, Cassone et al. 1986, 1987, Klein et al. 1991, Falcón et al. 2010). However, a recent study carried out in flounder (*Paralichthys olivaeus*) embryos has reported a rhythmic per2 expression in the suprachiasmatic nucleus under 24 h dark conditions, suggesting that a molecular clock could exist in this cell mass of flounder (Mogi et al. 2015).

Except for lamprey, in which pineal innervations do not appear to enter through the habenula (Puzdrowski and Northcutt 1989, Yáñez et al. 1993), in most fish species studied up to date the most rostral pineal efferents entered at the level of the habenular commissure and coursed along the left and right habenula to expand bilaterally into the thalamic and pretectal regions (Hafeez and Zerihun 1974, Ekström and van Veen 1984, Yáñez and Anadón 1998, Mandado et al. 2001, Yáñez et al. 2009, Servili et al. 2011). The presence of pinealofugal terminal fields containing beaded fibres has been reported in the habenula of rainbow trout (Hafeez and Zerihun 1974, Yáñez and Anadón 1996), sturgeon (Yáñez and Anadón 1998), three-spined stickleback (Ekström and van Veen 1983, Ekström 1984), goldfish (Jiménez et al. 1995), zebrafish (Yáñez et al. 2009) and sea bass (Servili et al. 2011). An important labelling was also observed in the habenular commissure and habenula of sole (Fig. 3D) but it corresponded to fibres in passage without forming local terminal fields (Confente et al. 2008b,c). Pinealofugal fibres also run through the habenula of crucian carp, European eel and skate but, as in sole, these fibres do not form terminal fields in the habenula (Ekström and van Veen 1984, Mandado et al. 2001).

In sole, the posterior commissure exhibited a dense fluorescent labelling and represented the main entrance of the nerve tracts descending from the pineal stalk and spreading bilaterally to the diencephalic (Figs. 3E, F, 4D) and mesencephalic pineal targets (Fig. 3G) (Confente et al. 2008b,c). The rich pineal innervation of the thalamus and pretectum is a shared feature of cyclostomes (Puzdrowski and Northcutt 1989, Yáñez et al. 1993), chondrosteans (Yáñez and Anadón 1998), chondrichthyans (Mandado et al. 2001) and teleosteans (Hafeez and Zerihun 1974, Ekström and van Veen 1984, Ekström et al. 1994, Jiménez et al. 1995, Yáñez et al. 2009, Servili et al. 2011). The tract-tracing study performed in sole confirmed a comparable richness in these visually-related brain areas. A huge amount of fibres coursed ventrally through the ventral and dorsal thalamic nuclei (Figs. 3D–F, 4C, D), laterally to reach the pretectal area (Figs. 3E–G) and dorsally towards the rostral optic tectum (Fig. 3D, E) (Confente et al. 2008b,c). This pinealotectal innervation was also evident in lamprey (Puzdrowski and Northcutt 1989, Yáñez et al. 1993), sturgeon (Yáñez and Anadón 1998), goldfish (Jiménez et al. 1995) and zebrafish (Yáñez et al. 2009). In contrast, pineal projections to the optic tectum appeared completely absent in elasmobranchs (Mandado et al. 2001) and most teleosts studied (Hafeez and Zerihun 1974, Ekström and van Veen 1984, Servili et al. 2011). It should be noted that retinofugal and pinealofugal axons overlapped at the level of the optic tectum, caudal preoptic area, ventral thalamus and several pretectal cell masses from sole (Confente et al. 2008b,c, Confente 2009). In lampreys and bony fishes, these pretectal and diencephalic pinealorecipient areas also received retinal projections (Ekström and van Veen 1984, Yáñez et al. 1993, Yáñez and Anadón 1998). This overlapping was also reported in the optic tectum of lamprey (Puzdrowski and Northcutt 1989, de Miguel et al. 1990, Yáñez et al. 1993), sturgeon (Yáñez and Anadón 1998, Ito et al. 1999), goldfish (Springer and Gaffney, 1981, Jiménez et al. 1995) and zebrafish (Yáñez et al. 2009) but is lacking in most advanced teleost fish studied. Taking into account that the Senegalese sole is a member of Pleuronectiform order and belongs to one of the most modern and evolved teleost groups (Percomorpha), the idea that retinal and pineal innervations in the optic tectum is not a characteristic of advanced bony fishes should be reconsidered. Furthermore, the presence of pinealofugal terminal fields in brain areas containing melatonin binding sites and/or melatonin receptor transcripts in teleosts (Martinoli et al. 1991, Mazurais et al. 1999, Bayarri et al. 2004b, Oliveira et al. 2008, Herrera-Pérez et al. 2010, Confente et al. 2010), such as the preoptic area, thalamus, pretectal area and optic tectum remarks their important role in the integration of neural and neurohormonal messages conveying photic signals to the brain.

The posterior tuberculum and the rostral mesencephalic tegmentum were also revealed as target areas for pineal projections in sole (Figs. 3E, F, 4D) (Confente et al. 2008b,c). The presence of pineal efferents in these brain areas seems to be a conserved characteristic between cyclostomes (Puzdrowski and Northcutt 1989, Yáñez et al. 1993), chondrostei (Yáñez and Anadón 1998), elasmobranchs (Mandado et al. 2001) and teleosts (Hafeez and Zerihun 1974, Ekström and van Veen 1984, Ekström et al. 1994, Jiménez et al. 1995, Yáñez et al. 2009, Servili et al. 2011). In several species, extensive pineal projections progress caudally along the midline of the brain to reach the mesencephalic reticular area, the oculomotor region and, in some cases, the interpeduncular nucleus and the region of the Edinger-Westphal nucleus (Puzdrowski and Northcutt 1989, Yáñez et al. 1993, Mandado et al. 2001, Yáñez and Anadón 1998). In sole, the most caudal pinealofugal fibres were observed at the transitional area between the nucleus of the medial longitudinal fascicle and the oculomotor nucleus (Fig. 3G) (Confente et al. 2008b,c). As in all vertebrates, this tegmental area is known for containing gonadotrophin-releasing hormone (GnRH)-2 cells in sole (Rodríguez-Gómez et al. 1999, Zohar et al. 2010). Our results in sole and previous data obtained in elasmobranchs (Mandado et al. 2001) showing the close topographical relationship of midbrain pinealofugal fibres and GnRH cells suggest the existence of interactions between the pineal organ and this neuroendocrine system. In the European sea bass, we have provided neuroanatomical, molecular and physiological evidences of direct links between the GnRH-2 cells and the pineal organ, showing that both structures exhibited bidirectional connections (González-Martínez et al. 2004, Servili et al. 2010). Moreover, these studies permitted us to assign to GnRH-2 a functional role in the modulation of the activity of this photoreceptive organ by stimulating pineal melatonin secretion (Servili et al. 2010). Interestingly, this tegmental GnRH-2 system appears highly conserved in vertebrates (Zohar et al. 2010) and does not project to the pituitary (González-Martínez et al. 2002). Additional experiments are required to clarify if GnRH-2 from sole has similar functions as intermediary system mediating in the integration/modulation of photoreceptor information perceived in the pineal organ.

It is noteworthy mentioning that some of the cell masses that received pinealofugal projections and/or are the source of pineal afferents in sole also represented neuroendocrine centres containing serotonin-, galanin-, tyrosine hydroxylase-or neuropeptide Y-immunoreactive cells (Rodríguez-Gómez et al. 2000b, 2000c, 2000d, 2001). Many of these brain nuclei also contain hypophysiotrophic cells in teleost brain (Anglade et al. 1993, Holmqvist and Ekström 1995) and, probably, in sole because serotonin-, galanin-, catecholamine-, and neuropeptide Y-immunoreactive fibres were present in the pituitary of this species (Rodríguez-Gómez et al. 2000b, 2000c, 2000d, 2001). These aminergic and neuropeptidergic systems have been involved in the modulation of metabolism, feeding behaviour, food intake and reproduction in fish (Lin et al. 2000, Zohar et al. 2010, Volkoff 2016, Delgado et al. 2017), all these processes exhibiting circadian and/or circannual rhythms. Therefore, pineal connections observed in these cell masses could participate in the transduction of environmental information between the pineal organ and these neuroendocrine cells, synchronizing such processes in a daily and/or annual basis.

In turn, we have identified the presence of central cells projecting to the sole pineal complex in the periventricular pretectum, the ventral and dorsal thalamus, and the posterior tuberculum (Fig. 3, 4C) (Confente et al. 2008b,c). DiI application in the pineal stalk of *Solea senegalensis* revealed retrogradely-labelled neurons in the periventricular pretectum (Figs. 3D–F, 4D), the ventral and dorsal thalamus (Figs. 3D–F), and the posterior tuberculum (Figs. 3E, 4C). Although we found abundant neurons projecting to the pineal organ in the brain of sea bass (Servili et al. 2011), no pinealopetal cells were found previously in the brain of other teleosts as the rainbow trout, three-spined stickleback, crucian carp, European eel and goldfish (Hafeez and Zerihun 1974, Ekström and van Veen 1983, 1984, Jiménez et al. 1995). Besides, comparable pinealopetal cell masses to those found in sole were identified in the brain of *Scyliorhinus canicula* and *Raja montagui* (Mandado et al. 2001). Subhabenular/thalamic neurons projecting to the pineal organ were also present in the brain of larval lamprey (Yáñez et al. 1993) and sturgeon (Yáñez and Anadón 1998). The presence of putative pinealopetal neurons was also reported in the rostral mesencephalic tegmentum of dogfish, skate, lamprey, sturgeon, zebrafish and sea bass (Yáñez et al. 1993, Yáñez and Anadón 1998, Mandado et al. 2001, Yáñez et al. 2009, Servili et al. 2011), but not in sole nor in other teleost species studied (Hafeez and Zerihun 1974, Ekström and van Veen 1983, 1984, Jiménez et al. 1995, Confente et al. 2008b,c). Species-specific differences between teleosts could be determining the presence or absence of central neurons projecting to the pineal organ. However, it is also possible that methodological constraints of the neural tracers used in different studies (i.e., in the ability to progress retrogradely) or plasticity in central innervation of the pineal organ associated with the physiological stage of animals, sex, day time or season in which the experiments were performed, were conditioning this identification. In fact, GnRH fibres in the pineal of the Indian major carp were evident only during particular reproductive periods (Sakharkar et al. 2005). Moreover, GnRH (Ekström and Meissl 1997, Sakharkar et al. 2005, Servili et al. 2010), NPY (Subhedar et al. 1996) and FMRFamide (Ekström et al. 1988) fibres have been previously detected in the pineal organ of different teleosts, suggesting that the existence of central pinealopetal neurons is the rule and not the exception.

Table 1. Abbreviations used for cell masses in the Senegalese sole brain.

A	anterior thalamic nucleus
ACo	anterior commissura
CM	corpus mamillare
CP	central posterior thalamic nucleus
DAO	dorsal accessory optic nucleus
Dc	central part of the dorsal telencephalon
Dlp	lateral posterior part of the dorsal telencephalon
Dm2	subdivision 2 of the medial part of dorsal telencephalon
Dm3	subdivision 3 of the medial part of dorsal telencephalon
Dm4	subdivision 4 of the medial part of dorsal telencephalon
DOT	dorsal optic tract
Dp	posterior part of the dorsal telencephalon
DP	dorsal posterior thalamic nucleus
DT	dorsal tegmental nucleus
E	entopeduncular nucleus
FR	fasciculus retroflexus
HCo	horizontal commissure
I	intermediate thalamic nucleus
LFB	lateral forebrain bundle
MCo	minor commissure
MLF	medial longitudinal fascicle

NAT	anterior tuberal nucleus
NC	cortical nucleus
NCLI	central nucleus of the inferior lobe
NDLI	diffuse nucleus of the inferior lobe
NGa	anterior part of the glomerular nucleus
NGp	posterior part of the glomerular nucleus
NGT	tertiary gustatory nucleus
NHv	ventral part of the habenular nucleus
NLTd	dorsal part of the lateral tuberal nucleus
NLTi	inferior part of the lateral tuberal nucleus
NLTl	lateral part of the lateral tuberal nucleus
NLTm	medial part of the lateral tuberal nucleus
NLTv	ventral part of the lateral tuberal nucleus
NMLI	medial nucleus of the inferior lobe
NP	paracommissural nucleus
NPC	central pretectal nucleus
NPGa	anterior preglomerular nucleus
NPGl	lateral preglomerular nucleus
NPGc	commissural preglomerular nucleus
NPGm	medial preglomerular nucleus
NPPv	posterior periventricular nucleus
NPT	posterior tuberal nucleus
nPVO	nucleus of the paraventricular organ
NRLd	dorsal part of the nucleus of the lateral recess
NRLl	lateral part of the nucleus of the lateral recess
NRLv	ventral part of the nucleus of the lateral recess
NRP	nucleus of the posterior recess
NSC	suprachiasmatic nucleus
NSV	nucleus of the vascular sac
nTE	nucleus of the thalamic eminentia
OC	optic chiasm
OT	optic tectum
P	pituitary
PCo	posterior commissure
PMgc	gigantocellular part of the magnocellular preoptic nucleus
POA	preoptic area
PPd	dorsal part of the periventricular pretectal nucleus
PPv	ventral part of the periventricular pretectal nucleus
PSi	intermediate superficial pretectal nucleus
PSm	magnocellular superficial pretectal nucleus
PSp	parvocellular superficial pretectal nucleus
PT	posterior thalamic nucleus
PVO	paraventricular organ
SCO	subcommissural organ
SV	vascular sac
TCo	tectal commissure
Tel	telencephalon
TLa	nucleus of the lateral torus
TPp	periventricular nucleus of the posterior tuberculum
TS	semicircular torus
VAO	ventral accessory optic nucleus
Vi	part intermedia of the ventral telencephalon
VL	ventrolateral thalamic nucleus
VM	ventromedial thalamic nucleus
VOT	ventral optic tract
Vp	postcommissural part of the ventral telencephalon
Vs	supracommissural part of the ventral telencephalon

Acknowledgements

The works presented in this chapter were supported by grants from the Spanish Ministry of Science and Innovation (MICINN, AGL2004-07984-C02-02, AGL2007-66507-C02-01).

References

Anglade, I., H.A. Zandbergen and O. Kah. 1993. Origin of the pituitary innervation in the goldfish. Cell Tissue Res. 273: 345–355.

Bayarri, M.J., J.A. Muñoz-Cueto, J.F. López-Olmeda, L.M. Vera, Rol de M.A. Lama, J.A. Madrid et al. 2004a. Daily locomotor activity and melatonin rhythms in Senegal sole (*Solea senegalensis*). Physiol. Behav. 81: 577–583.

Bayarri, M.J., R. García-Allegue, J.A. Muñoz-Cueto, J.A. Madrid, M. Tabata, F.J. Sánchez-Vázquez et al. 2004b. Melatonin binding sites in the brain of European sea bass (*Dicentrarchus labrax*). Zoolog. Sci. 21: 427–434.

Benzekri, H., P. Armesto, X. Cousin, M. Rovira, D. Crespo, M.A. Merlo et al. 2014. De novo assembly, characterization and functional annotation of Senegalese sole (*Solea senegalensis*) and common sole (*Solea solea*) transcriptomes: integration in a database and design of a microarray. BMC Genomics 15: 952.

Bolliet, V., M.A. Ali, F.J. Lapointe and J. Falcón. 1996. Rhythmic melatonin secretion in different teleost species: an *in vitro* study. J. Comp. Physiol. B 165: 677–683.

Bromage, N., M., Porter and C. Randall. 2001. The environmental regulation of maturation in farmed finfish with special reference to the role of photoperiod and melatonin. Aquaculture 197: 63–98.

Cassone, V.M., M.J. Chesworth and S.M. Armstrong. 1986. Entrainment of rat circadian rhythms by daily injection of melatonin depends upon the hypothalamic suprachiasmatic nuclei. Physiol. Behav. 36: 1111–1121.

Cassone, V.M., M.H. Roberts and R.Y. Moore. 1987. Melatonin inhibits metabolic activity in the rat suprachiasmatic nuclei. Neurosci. Lett. 81: 29–34.

Concha, M.L. and S.W. Wilson. 2001. Asymmetry in the epithalamus of vertebrates. J. Anat. 199: 63–84.

Confente, F., El M' Rabet, A., Ouarour, A., Voisin, P., De Grip, W.J., Rendón, M.C., Muñoz-Cueto, J.A. 2008a. The pineal complex of Senegalese sole (*Solea senegalensis*): Anatomical, histological and immunohistochemical study. Aquaculture 285: 207–215.

Confente, F., A. El M'Rabet, M.C. Rendón, A. Ouarour, L. Besseau, J. Falcón et al. 2008b. The pineal organ of *Solea senegalensis*: central integration of its neural and neurohormonal messages. pp. 115–120. *In*: J.A. Muñoz-Cueto, J.M. Mancera, G. Martínez [eds.]. Avances en Endocrinología Comparada. Vol. IV. Servicio de Publicaciones de la Universidad de Cádiz. Cádiz.

Confente, F., M.C. Rendón, J.A. Muñoz-Cueto, A. El M'Rabet, A. Ouarour, L. Besseau et al. 2008c. Integration of photoperiod signals in the brain–pituitary–gonadal (BPG) axis in *Solea senegalensis*. Comp. Biochem Physiol. Part A Mol. Int. Physiol. 151: S13.

Confente, F. 2009. Study of the pineal organ and melatonin receptors along the brain-pituitary-gonad axis of sole, *Solea Senegalensis*. Ph.D. Thesis, University of Cádiz, Cádiz, Spain.

Confente, F., M. Rendón, L. Besseau, J. Falcón and J.A. Muñoz-Cueto. 2010. Melatonin receptors in a pleuronectiform species, *Solea senegalensis*: Cloning, tissue expression, day-night and seasonal variations. Gen. Comp. Endocrinol. 167: 202–214.

Davies, W.I., T.K. Tamai, L. Zheng, J.K. Fu, J. Rihel, R.G. Foster et al. 2015. An extended family of novel vertebrate photopigments is widely expressed and displays a diversity of function. Genome Res. 25: 1666–1679.

Delgado, M.J., J.M. Cerdá-Reverter and J.L. Soengas. 2017. Hypothalamic integration of metabolic, endocrine, and circadian signals in fish: involvement in the control of food intake. Front. Neurosci. 11: 354.

de Miguel, E., M.C. Rodicio and R. Anadon. 1990. Organization of the visual system in larval lampreys: an HRP study. J. Comp. Neurol. 302: 529–542.

Eilertsen, M., O. Drivenes, R.B. Edvardsen, C.A. Bradley, L.O. Ebbesson and J.V. Helvik. 2014. Exorhodopsin and melanopsin systems in the pineal complex and brain at early developmental stages of Atlantic halibut (*Hippoglossus hippoglossus*). J. Comp. Neurol. 522: 4003–4022.

Ekström, P. and T. van Veen. 1983. Central connections of the pineal organ in the three-spined stickleback, *Gasterosteus aculeatus* L (Teleostei). Cell Tissue Res. 232: 141–155.

Ekström, P. 1984. Central neural connections of the pineal organ and retina in the teleost *Gasterosteus aculeatus* L. J. Comp. Neurol. 226: 321–335.

Ekström, P. and T. van Veen. 1984. Pineal neural connections with the brain in two teleosts, the crucian carp and the European eel. J. Pineal Res. 1: 245–261.

Ekström, P. 1985. Anterograde and retrograde filling of central neuronal systems with horseradish peroxidase under *in vitro* conditions. J. Neurosci. Methods 15: 21–35.

Ekström, P. 1987. Photoreceptors and CSF-contacting neurons in the pineal organ of a teleost fish have direct axonal connections with the brain: an HRP-electron-microscopic study. J. Neurosci. 7: 987–995.

Ekström, P., R.G. Foster, H.-W. Korf and J.J. Schalcken. 1987. Antibodies against retinal photoreceptor-specific proteins reveal axonal projections from the photosensory pineal organ in teleosts. J. Comp. Neurol. 265: 25–33.

Ekström, P. and J. Vaněcek. 1992. Localization of 2-[125I]iodomelatonin binding sites in the brain of the Atlantic salmon, *Salmo salar* L. Neuroendocrinology. 55: 529–537.

Ekström, P., T. Östholm and B.I. Holmqvist. 1994. Primary visual projections and pineal neural connections in fishes, amphibians and reptiles. pp. 1–18. *In:* M. Moller and P. Pévet [eds.]. Advances in Pineal Research: 8. John Libbey, London, UK.

Ekström, P. and H. Meissl. 1997. The pineal organ of teleost fishes. Rev. Fish. Biol. Fish 7: 199–284.

Ekström, P., T. Honkanen and S.O. Ebbesson. 1988. FMRFamide-like immunoreactive neurons of the nervus terminalis of teleosts innervate both retina and pineal organ. Brain Res. 460: 68–75.

Ekström, P. and H. Meissl. 2003. Evolution of photosensory pineal organs in new light: the fate of neuroendocrine photoreceptors. Phil. Trans. R. Soc. Lond. B. 358: 1679–1700.

Falcón, J., L. Besseau and G. Bœuf. 2007a. Molecular and cellular regulation of pineal organ responses. pp. 203–406. *In:* Hara, T. and B. Zielinski [eds.]. Sensory Systems Neuroscience. Fish Physiology. Academic Press Elsevier, Amsterdam, The Netherlands.

Falcón, J., L. Besseau, S. Sauzet and G. Boeuf. 2007b. Melatonin effects on the hypothalamo-pituitary axis in fish. Trends Endocrinol. Metab. 18: 81–88.

Falcón, J., V. Bolliet, J.P. Ravault, D. Chesneau, M.A. Ali and J.P. Collin. 1994. Rhythmic secretion of melatonin by the superfused pike pineal organ: thermo-and photoperiod interaction. Neuroendocrinology 60: 535–543.

Falcón, J., M. Geffard, M.T. Juillard, H.W. Steinbusch, P. Seguela and J.P. Collin. 1984. Immunocytochemical localization and circadian variations of serotonin and N-acetylserotonin in photoreceptor cells. Light and electron microscopic study in the teleost pineal organ. J. Histochem. Cytochem. 32: 486–492.

Falcón, J., H. Migaud, J.A. Muñoz-Cueto and M. Carrillo. 2010. Current knowledge on the melatonin system in teleost fish. Gen. Comp. Endocrinol. 165: 469–482.

Fejér, Z., A. Szél, P. Röhlich, T. Görcs, M.J. Manzano e Silva and B. Vigh. 1997. Immunoreactive pinopsin in pineal and retinal photoreceptors of various vertebrates. Acta Biol. Hung. 48: 463–471.

Fenwick, J.C. 1970. Demonstration and effect of melatonin in fish. Gen. Comp. Endocrinol. 14: 86–97.

Fernald, R.D. 1982. Retinal projections in the African cichlid fish, *Haplochromis burtoni*. J. Comp. Neurol. 206: 379–389.

Fernández-Díaz, C., M. Yúfera, J.P. Cañavate, F.J. Moyano, F.J. Alarcón and M. Díaz. 2001. Growth and physiological changes during metamorphosis of Senegal sole reared in the laboratory. J. Fish Biol. 58: 1086–1097.

Flamarique, I.N. 2002. A novel function for the pineal organ in the control of swim depth in the Atlantic halibut larva. Naturwissenschaften 89: 163–166.

Forsell, J., B. Holmqvist, J.V. Helvik and P. Ekström. 1997. Role of the pineal organ in the photoregulated hatching of the Atlantic halibut. Int. J. Dev. Biol. 41: 591–595.

Forsell, J., P. Ekström, I. Novales Flamarique and B. Holmqvist. 2001. Expression of pineal ultraviolet- and green-like opsins in the pineal organ and retina of teleosts. J. Exp. Biol. 204: 2517–2525.

Forsell, J., B. Holmqvist and P. Ekström. 2002. Molecular identification and developmental expression of UV and green opsin mRNAs in the pineal organ of the Atlantic halibut. Brain Res. Dev. Brain. Res. 136: 51–62.

Foster, R.G. and M.W. Hankins. 2002. Non-rod, non-cone photoreception in the vertebrates. Prog. Retin. Eye Res. 21: 507–527.

Gamse, J.T., Y.S. Kuan, M. Macurak, C. Brosamle, B. Thisse, C. Thisse et al. 2005. Directional asymmetry of the zebrafish epithalamus guides dorsoventral innervation of the midbrain target. Development 132: 4869–4881.

García-Allegue, R., J.A. Madrid and F.J. Sánchez-Vázquez. 2001. Melatonin rhythms in European sea bass plasma and eye: influence of seasonal photoperiod and water temperature. J. Pineal. Res. 31: 68–75.

García-Fernández, J.M., A.J. Jiménez, B. González, M.A. Pombal and RG. Foster. 1997. An immunocytochemical study of encephalic photoreceptors in three species of lamprey. Cell Tissue Res. 288: 267–278.

González-Martínez, D., N. Zmora, S. Zanuy, C. Sarasquete, A. Elizur, O. Kah et al. 2002. Developmental expression of three different prepro-GnRH (gonadotrophin-releasing hormone) messengers in the brain of the European sea bass (Dicentrarchus labrax). J. Chem. Neuroanat. 23: 255–267.

González-Martínez, D., T. Madigou, E. Mañanos, J.M. Cerdá-Reverter, S. Zanuy, O. Kah et al. 2004. Cloning and expression of gonadotropin-releasing hormone receptor in the brain and pituitary of the European sea bass: an *in situ* hybridization study. Biol Reprod. 70: 1380–1391.

Hafeez, M.A. and L. Zerihun. 1974. Studies on central projections of the pineal nerve tract in rainbow trout, *Salmo gairdneri* Richardson, using cobalt chloride iontophoresis. Cell Tissue Res. 154: 485–510.

Herrera-Pérez P., M.C. Rendón, L. Besseau S. Sauzet, J. Falcón and J.A. Muñoz-Cueto. 2010. Melatonin receptors in the brain of the European sea bass: An in situ hybridization and autoradiographic study. J. Comp. Neurol. 518: 3495–3511.

Herrera-Pérez, P., A. Servili, M.C. Rendón, F.J. Sánchez-Vázquez, J. Falcón and J.A. Muñoz-Cueto. 2011. The pineal complex of the European sea bass (*Dicentrarchus labrax*): I. Histological, immunohistochemical and qPCR study. J. Chem. Neuroanat. 41: 170–180.

Holmqvist, B.I., T. Östholm and P. Ekström. 1994. Neuroanatomical analysis of the visual and hypophysiotropic systems in Atlantic salmon (*Salmo salar*) with emphasis on possible mediators of photoperiodic cues during parr-smolt transformation. Aquaculture 121: 1–12.

Holmqvist, B.I. and P. Ekström. 1995. Hypophysiotrophic systems in the brain of the Atlantic salmon. Neuronal innervation of the pituitary and the origin of pituitary dopamine and nonapeptides identifies by means of combined carbocyanine tract tracing and immunocytochemistry. J. Chem. Neuroanat. 8: 125–145.

Iigo, M., H. Kezuka, K. Aida and I. Hanyu. 1991. Circadian rhythms of melatonin secretion from superfused goldfish (*Carassius auratus*) pineal glands *in vitro*. Gen. Comp. Endocrinol. 83: 152–158.

Iigo, M. and K. Aida. 1995. Effects of season, temperature, and photoperiod on plasma melatonin rhythms in the goldfish, *Carassius auratus*. J. Pineal Res. 18: 62–68.

Iigo, M., Y. Fujimoto, M. Gunji-Suzuki, M. Yokosuka, M. Hara, R. Ohtani-Kaneko et al. 2004. Circadian rhythm of melatonin release from the photoreceptive pineal organ of a teleost, ayu (*Plecoglossus altivelis*) in flow-through culture. J. Neuroendocrinol. 16: 45–51.

Ishikawa, Y., K. Inohaya, N. Yamamoto, K. Maruyama, M. Yoshimoto, M. Iigo et al. 2015. The parapineal is incorporated into the habenula during ontogenesis in the medaka fish. Brain. Behav. Evol. 85: 257–270.

Ito, H., M. Yoshimoto, J.S. Albert, N. Yamamoto and N. Sawai 1999. Retinal projections and retinal ganglion cell distribution patterns in a sturgeon (*Acipenser transmontanus*), a non-teleost actinopterygian fish. Brain Behav. Evol. 53: 127–141.

Jiménez, A.J., P. Fernández-Llebrez and J.M. Pérez-Fígares. 1995. Central projections from the goldfish pineal organ traced by HRP-immunocytochemistry. Histol. Histopathol. 10: 847–852.

Kawano-Yamashita, E., A. Terakita, M. Koyanagi, Y. Shichida, T. Oishi and S. Tamotsu. 2007. Immunohistochemical characterization of a parapinopsin-containing photoreceptor cell involved in the ultraviolet/green discrimination in the pineal organ of the river lamprey *Lethenteron japonicum*. J. Exp. Biol. 210: 3821–3829.

Klein, D.C., R.Y. Moore and S.M. Reppert. 1991. Suprachiasmatic Nucleus: the Mind's Clock. Oxford Press, New York, USA.

Koyanagi, M., E. Kawano, Y. Kinugawa, T. Oishi, Y. Shichida, S. Tamotsu et al. 2004. Bistable UV pigment in the lamprey pineal. Proc. Natl. Acad. Sci. USA 101: 6687–6691.

Lin, X., H. Volkoff, Y. Narnaware, N.J. Bernier, P. Peyon and R.E. Peter. 2000. Brain regulation of feeding behavior and food intake in fish. Comp. Biochem. Physiol. A 126: 415–434.

Mandado, M., P. Molist, R. Anadón and J. Yáñez. 2001. A DiI-tracing study of the neural connections of the pineal organ in two elasmobranchs (*Scyliorhinus canicula* and *Raja montagui*) suggests a pineal projection to the midbrain GnRH-immunoreactive nucleus. Cell Tissue Res. 303: 391–401.

Martinoli, M.G., L.M. Williams, O. Kah, L.T. Titchener and G. Pelletier. 1991. Distribution of central melatonin binding sites in the goldfish (*Carassius auratus*). Mol. Cell. Neurosci. 2: 78–85.

Mazurais, D., I. Brierley, I. Anglade, J. Drew, C. Randall, N. Bromage et al. 1999. Central melatonin receptors in the rainbow trout: comparative distribution of ligand binding and gene expression. J. Comp. Neurol. 409: 313–324.

McNulty, J.A. and B.G. Nafpaktitis. 1976. The structure and development of the pineal complex in the lanternfish *Triphoturus mexicanus* (family Mycotphidae). J. Morphol. 150: 579–605.

McNulty, J.A. and B.G. Nafpaktitis. 1977. Morphology of the pineal complex in seven species of lanternfishes (Pisces: Myctophidae). Am. J. Anat. 150: 509–529.

McNulty, J.A. 1984. Organ culture of the goldfish pineal body. An ultrastructural and biochemical study. Cell Tissue Res. 238: 35–75.

Medina, M., N. Le Belle, J. Repérant, J.P. Rio and R. Ward. 1990. An experimental study of the retinal projections of the European eel (*Anguilla anguilla*), carried out at the catadromic migratory silver stage. J. Hirnforsch. 31: 467–480.

Meissl, H. and E. Dodt. 1981. Comparative physiology of pineal photoreceptor organs. pp. 61–80. *In:* A. Oksche and P. Pévet [eds.]. The Pineal Organ: Photobiology-Biochronometry-Endocrinology. Elsevier/North-Holland, Amsterdam, The Netherlands.

Migaud, H., J.F. Taylor, G.L. Taranger, A. Davie, J.M. Cerdá-Reverter, M. Carrillo et al. 2006. A comparative *ex vivo* and *in vivo* study of day and night perception in teleosts species using the melatonin rhythm. J. Pineal Res. 41: 42–52.

Migaud, H., A. Davie, C.C. Martínez-Chávez and S. Al-Khamees. 2007. Evidence for differential photic regulation of pineal melatonin synthesis in teleosts. J. Pineal Res. 43: 327–335.

Mogi, M., S. Uji, H. Yokoi and T. Suzuki. 2015. Early development of circadian rhythmicity in the suprachiamatic nuclei and pineal gland of teleost, flounder (*Paralichthys olivaeus*), embryos. Dev. Growth Differ. 57: 444–4452.

Northcutt, R.G. and A.B. Butler. 1993. The diencephalon of the Pacific herring, *Clupea harengus*: retinofugal projections to the diencephalon and optic tectum. J. Comp. Neurol. 328: 547–561.

Oliveira, C., J.F. López-Olmeda, M.J. Delgado, A.L. Alonso-Gómez and F.J. Sánchez-Vázquez. 2008. Melatonin binding sites in senegal sole: day/night changes in density and location in different regions of the brain. Chronobiol. Int. 25: 645–652.

Philp, A.R., J. Bellingham, J.M. García-Fernández and R.G. Foster. 2000a. A novel rod-like opsin isolated from the extra-retinal photoreceptors of teleost fish. FEBS Lett. 468: 181–188.

Philp, A.R., J.M. García-Fernández, B.G. Soni, R.J. Lucas, J. Bellingham and R.G. Foster. 2000b. Vertebrate ancient (VA) opsin and extraretinal photoreception in the Atlantic salmon (*Salmo salar*). J. Exp. Biol. 203: 1925–1936.

Pombal, M.A., J. Yáñez, O. Marín, A. González and R. Anadón. 1999. Cholinergic and GABAergic neuronal elements in the pineal organ of lampreys, and tract-tracing observations of differential connections of pinealofugal neurons. Cell Tissue Res. 295: 215–223.

Presson, J., R.D. Fernald and M. Marx. 1985. The organization of retinal projections to the diencephalon and pretectum in the cichlid fish, *Haplochromis burtoni*. J. Comp. Neurol. 235: 360–374.

Puzdrowski, R.L. and R.G. Northcutt. 1989. Central projections of the pineal complex in the silver lamprey *Ichthyomyzon unicuspis*. Cell Tissue Res. 255: 269–274.

Rahman, M.S., B.H. Kim, A. Takemura, C.B. Park and Y.D. Lee. 2004. Effects of moonlight exposure on plasma melatonin rhythms in the seagrass rabbitfish, *Siganus canaliculatus*. J. Biol. Rhythms. 19: 325–334.

Randall, C.F., N.R. Bromage, J.E. Thorpe, M.S. Miles and J.S. Muir. 1995. Melatonin rhythms in Atlantic salmon (*Salmo salar*) maintained under natural and out-of-phase photoperiods. Gen. Comp. Endocrinol. 98: 73–86.

Repérant, J., D. Miceli, J.P. Rio, J. Peyrichoux, J. Pierre and E. Kirpitchnikova. 1986. The anatomical organization of retinal projections in the shark *Scyliorhinus canicula* with special reference to the evolution of the selachian primary visual system. Brain Res. 396: 227–248.

Rincón-Camacho, L., L. Morandini, A. Birba, L. Cavallino, F. Alonso, F.L. LoNostro et al. 2016. The pineal complex: a morphological and immunohistochemical comparison between a tropical (*Paracheirodon axelrodi*) and a subtropical (*Aphyocharax anisitsi*) characid species. J. Morphol. 277: 1355–1367.

Rodríguez-Gómez, F.J., M.C. Rendón, C. Sarasquete and J.A. Muñoz-Cueto. 1999. Distribution of gonadotropin-releasing hormone immunoreactive systems in the brain of the Senegalese sole, *Solea senegalensis*. Histochem. J. 31: 695–703.

Rodríguez-Gómez, F.J., C. Sarasquete and J.A. Muñoz-Cueto. 2000a. A morphological study of the brain of *Solea senegalensis*. I. The telencephalon. Histol. Histopathol. 15: 355–364.

Rodríguez-Gómez, F.J., M.C. Rendón-Unceta, C. Sarasquete and J.A. Muñoz-Cueto. 2000b. Distribution of serotonin in the brain of the Senegalese sole, *Solea senegalensis*: an immunohistochemical study. J. Chem. Neuroanat. 18: 103–115.

Rodríguez-Gómez, F.J., C. Rendón-Unceta, M.C. Sarasquete and J.A. Muñoz-Cueto. 2000c. Localization of galanin-like immunoreactive structures in the brain of the Senegalese sole, *Solea senegalensis*. Histochem. J. 32: 123–132.

Rodríguez-Gómez, F.J., M.C. Rendón-Unceta, C. Sarasquete and J.A. Muñoz-Cueto. 2000d. Localization of tyrosine hydroxylase-immunoreactivity in the brain of the Senegalese sole, *Solea senegalensis*. J. Chem. Neuroanat. 19: 17–32.

Rodríguez-Gómez, F.J., C. Rendón-Unceta, C. Sarasquete and J.A. Muñoz-Cueto. 2001. Distribution of neuropeptide Y-like immunoreactivity in the brain of the Senegalese sole (*Solea senegalensis*). Anat. Rec. 262: 227–237.

Rüdeberg, C. 1969. Structure of the parapineal organ of the adult rainbow trout, *Salmo gairdneri* Richardson. Z. Zellforsch. 93: 282–304.

Rusak, B. and I. Zucker. 1979. Neural regulation of circadian rhythms. Physiol. Rev. 59: 449–526.

Sakharkar, A., P. Singru and N. Subhedar. 2005. Reproduction phase-related variations in the GnRH immunoreactive fibers in the pineal of the Indian major carp *Cirrhinus mrigala* (Ham.). Fish Physiol. Biochem. 31: 163–166.

Samejima, M., S. Tamotsu., K. Watanabe and Y. Morita. 1989. Photoreceptor cells and neural elements with long axonal processes in the pineal organ of the lamprey, *Lampetra japonica*, identified by use of the horseradish peroxidase method. Cell Tissue Res. 258: 219–224.

Sánchez-Vázquez, F.J., M. Iigo, J.A. Madrid, S. Zamora and M. Tabata. 1997. Daily cycles in plasma and ocular melatonin in demand-fed sea bass, Dicentrarchus labrax, L. J. Comp. Physiol. B 167: 409–415.

Servili, A., C. Lethimonier, J.J. Lareyre, J.F. López-Olmeda, F.J. Sánchez-Vázquez, O. Kah et al. 2010. The highly conserved gonadotropin-releasing hormone-2 form acts as a melatonin-releasing factor in the pineal of a teleost fish, the european sea bass *Dicentrarchus labrax*. Endocrinology 151: 2265–2275.

Servili, A., P. Herrera-Pérez, J. Yáñez, J.A. Muñoz-Cueto. 2011. Afferent and efferent connections of the pineal organ in the European sea bass *Dicentrarchus labrax*: a carbocyanine dye tract-tracing study. Brain Behav. Evol. 78: 272–285.

Springer, A.D. and J.S. Gaffney. 1981. Retinal projections in the goldfish: a study using cobaltous-lysine. J Comp Neurol. 203: 401–424.

Su, C.Y., D.G. Luo, A. Terakita, Y. Shichida, H.W. Liao, M.A. Kazmi et al. 2006. Parietal-eye phototransduction components and their potential evolutionary implications. Science 311: 1617–1621.

Subhedar, N., J. Cerdá and R.A. Wallace. 1996. Neuropeptide Y in the forebrain and retina of the killifish, *Fundulus heteroclitus*. Cell Tissue Res. 283: 313–323.

Ueck, M. 1979. Innervation of the vertebrate pineal. Prog. Brain Res. 52: 45–88.

van Veen, T., P. Ekström, B. Borg and M. Møller. 1980. The pineal complex of the three-spined stickleback, *Gasterosteus aculeatus* L. (Teleostei). Cell Tissue Res. 209: 11–28.

Vera, L.M., C. De Oliveira, J.F. López-Olmeda, J. Ramos, E. Mañanós, J.A. Madrid et al. 2007. Seasonal and daily plasma melatonin rhythms and reproduction in Senegal sole kept under natural photoperiod and natural or controlled water temperature. J. Pineal Res. 43: 50–55.

Vigh, B., M.J. Manzano, A. Zadori, C.L. Frank, A. Lukats, P. Rohlich Szel et al. 2002. Nonvisual photoreceptors of the deep brain, pineal organs and retina. Histol. Histopathol. 17: 555–590.

Vigh-Teichmann, I., H.-W. Korf, F. Nürnberger, A. Oksche, B. Vigh and R. Olson. 1983. Opsin immunoreactive outer segments in the pineal and parapineal organs of the lamprey (*Lampetra fluviatilis*), the eel (*Anguilla anguilla*), and the rainbow trout (*Salmo gairdneri*). Cell Tissue Res. 230: 289–307.

Volkoff, H. 2016. The neuroendocrine regulation of food intake in fish: a review of current knowledge. Front. Neurosci. 10: 540.

Vuilleumier, R., G. Boeuf, M. Fuentes, W.J. Gehring and J. Falcón. 2007. Cloning and early expression pattern of two melatonin biosynthesis enzymes in the turbot (*Scophthalmus maximus*). Eur. J. Neurosci. 25: 3047–3057.

Wagner, H.J. and U. Mattheus. 2002. Pineal organs in deep demersal fish. Cell Tissue Res. 307: 115–127.

Yáñez, J., R. Anadón, B.I. Holmqvist and P. Ekström. 1993. Neural projections of the pineal organ in the larval sea lamprey (*Petromyzon marinus* L.) revealed by indocarbocyanine dye tracing. Neurosci. Lett. 164: 213–216.

Yáñez, J. and R. Anadón. 1996. Afferent and efferent connections of the habenula in the rainbow trout (*Oncorhynchus mykiss*): an indocarbocyanine dye (DiI) study. J. Comp. Neurol. 372: 529–543.

Yáñez, J. and R. Anadón. 1998. Neural connections of the pineal organ in the primitive bony fish *Acipenser baeri*: a carbocyanine dye tract-tracing study. J. Comp. Neurol. 398: 151–161.

Yáñez, J., M.A. Pombal and R. Anadón. 1999. Afferent and efferent connections of the parapineal organ in lampreys: a tract tracing and immunocytochemical study. J. Comp. Neurol. 403: 171–189.

Yáñez, J., J. Busch, R. Anadón and H. Meissl. 2009. Pineal projections in the zebrafish (*Danio rerio*): overlap with retinal and cerebellar projections. Neuroscience 164: 1712–1720.

Yáñez, Y., H. Meissl and R. Anadón. 1996. Central projections of the parapineal organ of the adult rainbow trout (*Oncorhynchus mykiss*). Cell Tissue Res. 285: 69–74.

Zachmann, A., M.A. Ali and J. Falcón. 1992. Melatonin and its effects in fishes: an overview. pp. 149–165. *In*: Ali, M.A. [ed.]. Rhythms in Fishes. Plenum Press, New York, USA.

Ziv, L., A. Tovin, D. Strasser and Y. Gothilf. 2007. Spectral sensitivity of melatonin suppression in the zebrafish pineal gland. Exp. Eye Res. 84: 92–99.

Zohar, Y., J.A. Munoz-Cueto, A. Elizur and O. Kah. 2010. Neuroendocrinology of reproduction in teleost fish. Gen. Comp. Endocrinol. 165: 438–455.

B-1.2

Neuroendocrine Systems in Senegalese Sole

Ángel García-López,[1] *María Aliaga-Guerrero,*[1]
José A. Paullada-Salmerón,[1] *María del Carmen Rendón-Unceta,*[1] *Carmen Sarasquete,*[2] *Evaristo Mañanós*[3] *and José A. Muñoz-Cueto*[1,*]

1. Introduction

It has been clearly established that the brain, and especially the forebrain, plays an important role in the neuroendocrine control of many physiological processes in fish, acting as a receptor, integrative and effector organ (Flik et al. 2006, Canosa et al. 2007, Zohar et al. 2010, Nardocci et al. 2014, Volkoff 2016, Delgado et al. 2017). Fish perceive external and internal factors into the brain using specific sensory systems and, then, integrate and transduce this information into a cascade of neurohormones and hormones involving the brain, pituitary and peripheral glands. Therefore, the elucidation of the nature and origin of these neurohormones can provide very significant information on how the brain controls the adenohypophyseal function and these physiological processes. Teleost fishes lack the median eminence and a system to put the hypothalamus in vascular contact with the pituitary, like that existing in tetrapods. In this way, the neurohormones are transported via neurosecretory fibres that enter through the pituitary stalk and are released from axon terminals directly into the surroundings area of the target adenohypophyseal cells (Batten and Ingleton 1987). This direct innervation of the pituitary of teleost fishes, together with the identification of a specific neurosecretory terminal in the vicinity of a

[1] Department of Biology. Faculty of Marine and Environmental Sciences. INMAR. University of Cádiz. CEIMAR. Campus Rio San Pedro. E11510, Puerto Real. Spain.
[2] Institute of Marine Sciences of Andalucía. CSIC. Campus Universitario Río San Pedro, E11510, Puerto Real, Cádiz, Spain.
[3] Institute of Aquaculture of Torre de la Sal. CSIC. Ribera de Cabanes. E12595, Castellón. Spain.
* Corresponding author: munoz.cueto@uca.es

particular adenohypophyseal cell type, has enabled researchers to characterise those brain factors that may be involved in controlling the secretion of the pituitary hormones and the functions related to them.

In the present chapter, we will review the available information on the distribution and bioactivity of the main neuroendocrine systems involved in the control of key physiological processes such as reproduction, feeding, metabolism, growth and stress in the Senegalese sole, *Solea senegalensis*.

2. Gonadotropin-releasing hormone (GnRH)

It is widely accepted that the decapeptide gonadotropin-releasing hormone (GnRH) is the central neuroendocrine system regulating reproduction in vertebrates, due to its prominent role in the stimulation of both luteinising hormone (LH) and follicle-stimulating hormone (FSH) release from the pituitary. In addition to its role in releasing gonadotropins, GnRH seems to be implicated in other functions such as nesting and sexual behaviour (Muske and Moore 1994, Yamamoto et al. 1995, 1997). The first fish GnRH peptide was identified in chum salmon *Oncorhynchus keta*, and named as sGnRH (Sherwood et al. 1983). To date, eighteen different forms of GnRH have been identified in vertebrates (Weltzien et al. 2004, Zohar et al. 2010, Shahjahan et al. 2014, Chang and Pemberton 2018). Among them, twelve GnRH sequences have been reported and characterised in fish species: salmon (sGnRH), gilthead seabream *Sparus aurata* L. (sbGnRH; Powell et al. 1994), lake whitefish *Coregonus clupeaformis* (Mitchill, 1818) (wfGnRH; Adams et al. 2002), medaka *Oryzias latipes* (Temminck & Schlegel, 1846) (mdGnRH; Okubo et al. 2000), North African catfish *Clarias gariepinus* (Burchell, 1822) (cfGnRH; Bogerd et al. 1992), Pacific herring *Clupea harengus pallasii* Valenciennes, 1847 (hrGnRH; Carolsfeld et al. 2000), three-spined stickleback, *Gasterosteus aculeatus* (stGnRH; Chang and Pemberton 2018), dogfish *Squalus acanthias* L. (dfGnRH; Lovejoy et al. 1992), elephant shark, *Callorhinchus millii* (esGnRH; Nock et al. 2011, Chang and Pemberton 2018), and sea lamprey *Petromyzon marinus* L. (which has three specific types named lGnRH-I, lGnRH-II and lGnRH-III; Sower et al. 1993, Sower 2018). Moreover, mammalian GnRH (mGnRH) and chicken-II GnRH (cII-GnRH) have also been described in fish species (Okubo et al. 2001, Zohar et al. 2010). Since a given GnRH form originally identified in one species can be also present in other species, a new classification of the GnRH variants on the basis of phylogenetic analysis of known sequences and their respective sites of expression was proposed to avoid nomenclature confusions (Fernald and White 1999, White et al. 1995). This phylogenetic analysis shows the existence of three main GnRH branches. One branch contains hypophysiotropic variants, mainly expressed in the hypothalamus of amphibians and mammals but also a number of fish hypophysiotropic variants (e.g., mammalian GnRH, guinea pig GnRH, chicken-I GnRH, frog GnRH, seabream GnRH, pejerrey/medaka GnRH, herring GnRH, catfish GnRH, whitefish GnRH, stickleback GnRH, coelacanth GnRH, elephant shark GnRH), which were named as GnRH-1. Another branch clusters all GnRH forms consistently expressed in the synencephalon/mesencephalon of vertebrates, from fish to mammals, and is referred to as GnRH-2 (chicken GnRH-II and lamprey-II GnRH). A third GnRH branch includes the salmon GnRH isoform, mainly expressed in the rostral forebrain, the dogfish GnRH and lamprey-I and lamprey-III GnRHs, and is named GnRH-3 (Chang and Pemberton 2018). Most vertebrates possess two forms of GnRH in the brain, although a number of teleost fish species have three different types (Zohar et al. 2010).

In the case of Senegalese sole, three different GnRH variants have been identified, namely sbGnRH as GnRH-1, chicken-II GnRH as GnRH-2, and sGnRH as GnRH-3 (Aliaga-

Guerrero et al. 2018). The different fish GnRH forms are all initially synthesized as a pre-propeptide, which includes a signal peptide (21–23 residues), the specific GnRH decapeptide followed by a three amino acids proteolytic cleavage site (GKR), and a C-terminal GnRH-associated peptide (GAP, 40–60 amino acids) (King and Millar 1997, White and Fernald 1998). The GAP, which seems to have prolactin regulating properties both in mammals (Nikolics et al. 1985) and teleosts (Planas et al. 1990), use to be highly divergent among the different GnRH forms within one species (as well as across distinct species) being very useful for antibody generation for the specific detection of GnRH variants (e.g., Gonzalez-Martínez et al. 2002).

The distribution of GnRH-immunoreactive (ir) cells in the brain and pituitary of Senegalese sole has been described by Rodríguez-Gómez et al. (1999) by immunohistochemistry using two antibodies against chicken-II GnRH (GnRH-2) and sGnRH (GnRH-3) forms (Fig. 1). It is important to note that these antisera against the two GnRH forms exhibited cross-reactivity on the same cell masses and did not allow cell populations expressing different GnRH forms to be discriminated clearly (see below). GnRH-ir cell bodies were observed at the junction between the olfactory bulbs and the telencephalon (terminal nerve ganglion cells), in the ventral telencephalon, in the preoptic parvocellular nucleus, and in the synencephalic nucleus of the medial longitudinal fascicle (Fig. 1A, B, C, I). GnRH-ir fibres were extensively found throughout the brain, located in the telencephalon, preoptic area, hypothalamus, preoptic-hypophyseal tract, neuro- and adenohypophysis, optic tectum, midbrain and rhombencephalon. Despite the cross-reactivity of the antibodies used, the we noted that the anti-cIIGnRH serum showed a higher immunoreactivity on synencephalic cells of the medial longitudinal fascicle (Fig. 1I), which corresponded with GnRH-2 cells, whereas the anti-sGnRH serum immunostained much more intensely the GnRH cells and fibres of the terminal nerve ganglion (Fig. 1A), the ventral telencephalon (Fig. 1B) and the preoptic area (Fig. 1C), the former cell mass corresponding to GnRH-3 cells and the two later brain areas probably containing GnRH-1 cells (Rodríguez-Gómez et al. 1999). These patterns are, thus, in accordance with the main expression regions of the GnRH-1, GnRH-2 and GnRH-3 variants found in vertebrates (Zohar et al. 2010). Rodríguez-Gómez et al. (1999) also observed abundant GnRH fibres in the proximal *pars distalis* of the adenohypophysis, where gonadotrophs, somatotrophs and thyrotrophs are located in Senegalese sole (Rendón et al. 1997). Despite the presence of GnRH-2 and/ or GnRH-3 have been previously reported in the pituitary of other fish species (Chang and Jobin 1994, Montero et al. 1994, Parhar and Iwata 1994, Kim et al. 1995, Parhar 1997, Rodríguez et al. 2000, Andersson et al. 2001, González-Martínez et al. 2002, Amano et al. 2004, Pham et al. 2006), it is generally accepted that the main hypophysiotropic GnRH form in vertebrates is GnRH-1 (Zohar et al. 2010). Accordingly, and due to the cross-reactivity of the antibodies used, it is also possible that at least some of the immunoreactivity detected in the adenohypophysis of Senegalese sole by Rodríguez-Gómez (1999) could be attributed to the GnRH-1 form. This has been recently supported by experimental data from Guzmán et al. (2009a) showing the detection of the three GnRH variants in Senegalese sole pituitary extracts by ELISA using highly specific antibodies for each GnRH type. This study also indicated that the GnRH-2 form is relatively more abundant than the GnRH-1 variant in the pituitary of Senegalese sole at spawning although, as pointed out by the authors, this situation could be transitory as the abundance of the different GnRH types is subjected to variations along the reproductive cycle (Gothilf et al. 1996, Holland et al. 1998a, 2001, Pham et al. 2006, Rodríguez et al. 2000).

Regarding the specific functionality of GnRH in Senegalese sole, multiple studies have shown the ability of synthetic GnRH analogues (GnRHa), such as [D-Ala^6, Pro^9, NEt],

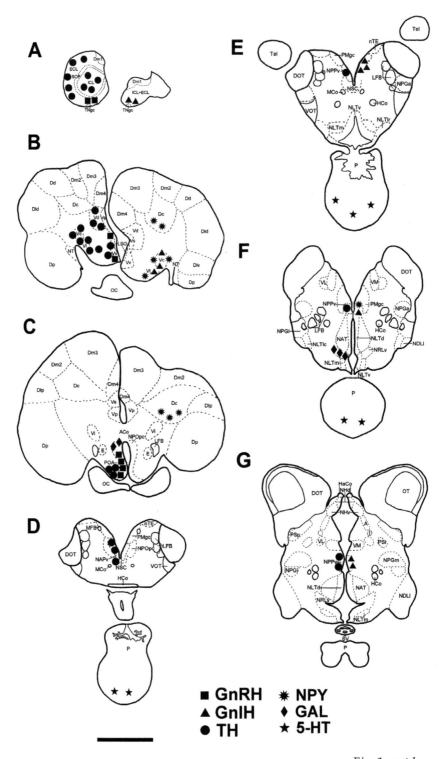

Fig. 1 contd. ...

...*Fig. 1 contd.*

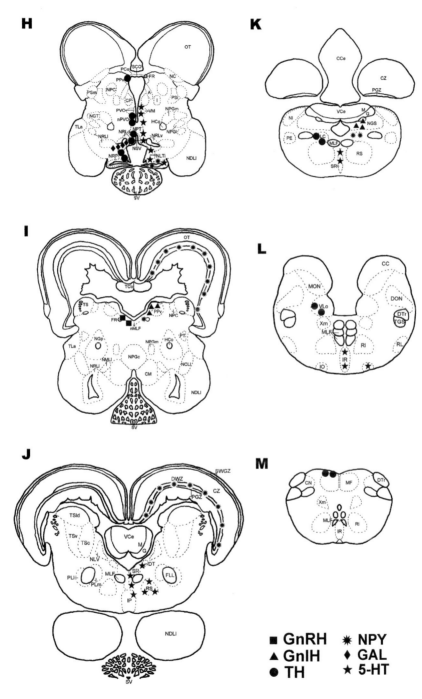

Fig. 1. Series of transverse sections through the brain of *Solea senegalensis* from rostral (A) to caudal (M) showing the distribution of different neuroendocrine cells. Black squares: GnRH cells; black asterisks: NPY cells; black triangles: GnIH cells; black diamond: galanin cells; black circles: catecholaminergic (TH) cells; black stars: serotoninergic (5-HT) cells. Bar scale 1 mm. For abbreviations see Table 1.

to stimulate oocyte maturation and spawning and, in some cases, to enhance also testis maturation and sperm quality (Agulleiro et al. 2006, 2007, Guzmán et al. 2009b, 2011a,b, Cabrita et al. 2011) (see Chapters B-1.3 and B-1.4). These effects are probably mediated by a stimulation of gonadotropin release from the pituitary although, at present, only elevated LHβ and GPα mRNA levels in the pituitary of females and increased GPα expression in the pituitary of males have been demonstrated in the species (Guzmán et al. 2011a). Additionally, the treatments with GnRHa consistently induced significant elevations of plasma steroid levels both in females (17-β estradiol and testosterone) and males (testosterone and 11-ketotestosterone) (see references above), although these alterations did not show any apparent feedback effect on the pituitary content of the endogenous GnRH forms of Senegalese sole (Guzmán et al. 2011a). Finally, Guzmán et al. (2009a) have also analysed the levels of each GnRH variant in the pituitary of Senegalese sole from wild and cultured origins at the time of spawning, in order to determine the relationship in the amounts of different GnRH forms with the reproductive failure found in cultured broodstocks (Agulleiro et al. 2006, García-López et al. 2007, Guzmán et al. 2009a). This study showed that pituitary peptide levels of the three GnRH forms were slightly higher, but not statistically significant, in wild than in cultured females, whereas in males, no differences were observed in GnRH-1 or GnRH-2 pituitary content between wild and cultured breeders. No statistical analysis was possible for GnRH-3, as cultured males had undetectable pituitary levels of this variant (Guzmán et al. 2009a).

3. Catecholamines: dopamine

Catecholamines are released both as neurotransmitters in the hypothalamic-sympathetic nervous system and as classical neurohormones from the adrenal medulla or chromaffin cells of the kidney being involved in a rapid major neuroendocrine pathway integrating the primary response to stress in fish (Wendelaar Bonga 1997, Pankhurst 2011). However, central catecholamines can also act as neurohormones that reach the pituitary and modulate the release of adenohypophyseal hormones (Zohar et al. 2010). Rapid increases in plasma levels of catecholamines occur mainly due to acute stress, particularly if that stress is accompanied by, or involves significant reductions in blood oxygen content. This rapid response is primarily mediated by cholinergic nerve fibres that directly innervate the chromaffin tissue; however, there is also evidence for involvement of serotonin, adrenocorticotrophic hormone (ACTH), angiotensin II and non-cholinergic innervation of chromaffin tissue in modulation of catecholamine release (Perry and Bernier 1999). Effects of increased plasma levels of catecholamines include increased haemoglobin oxygen affinity, in some species increased arterial blood pressure (Perry and Bernier 1999) and in most species examined, release of stored hepatic glycogen to the plasma as glucose (Wendelaar Bonga 1997). In mammals, in addition to their roles in regulating stress responses, the catecholamines acts to modulate early steps of sensory perception in the olfactory bulbs and the retina, motor programming, learning, and memory, affective and motivational processes in the forebrain, control of body temperature, food intake, and several other hypothalamic functions as well as chemosensitivity (Yamamoto and Vernier 2011). Some of these functions are conserved also in fish, while others are not, and this is reflected in the anatomical distribution of catecholaminergic systems in the forebrain and midbrain (Yamamoto and Vernier 2011).

The localization of catecholaminergic neurons and fibres in the brain of the Senegalese sole has been determined by Rodríguez-Gómez et al. (2000d) using antibodies against tyrosine hydroxylase (TH), the rate limiting enzyme for catecholamine synthesis

(Fig. 1). Although the general pattern of distribution of catecholamines in Senegalese sole brain is consistent with that reported in other teleosts, some remarkable differences are observed. The most rostral TH immunoreactive cells were identified in the olfactory bulbs (Fig. 1A), in which a clear asymmetry in the number and location of TH-immunoreactive perikarya and fibres is observed. The number of TH-immunoreactive cells and fibres was manifestly higher in the right olfactory bulb, the one located in the ocular pigmented side of the fish, which is also larger than the left bulb (Rodríguez-Gómez et al. 2000d). This asymmetry in the number and distribution of rostral catecholaminergic cells in Senegalese sole could be a consequence of developmental mechanisms, being possibly related to the establishment of differences in the connections of the right and left bulbs in this species. Other TH-immunoreactive cell masses were identified in the ventral telencephalon (Fig. 1B), preoptic area (Fig. 1C-G), caudoventral hypothalamus (Fig. 1H), posterior tuberculum (Fig. 1H), synencephalon (Fig. 1H), isthmic region (Fig. 1K) and rhombencephalon (Fig. 1L, M). Surprisingly, no TH-immunoreactive cell bodies were found in the ventromedial thalamic nucleus, which exhibits a large number of TH-immunoreactive cells in other teleosts (Meek 1994). The presence of TH-immunoreactive fibres in the brain of Senegalese sole was particularly evident within and around the nuclei in which immunoreactive cells were found. However, other zones such as the dorsal telencephalon, posterior commissure, optic tectum, torus semicircularis, reticular formation or inferior olive also displayed TH-immunoreactive fibres (Rodríguez-Gómez et al. 2000d). TH-immunoreactive axons also entered the infundibulum, reaching the proximal *pars distalis* of the adenohypophysis, where gonadotropin- and growth hormone-producing cells are found in this species (Rendón et al. 1997).

Different studies in fish indicate that catecholamines can regulate the release of these pituitary hormones, although in the case of gonadotropins, it seems that this regulation is particularly relevant in cyprinid and salmonid species but not in marine fishes (Dufour et al. 2005, Zohar et al. 2010). This is especially true for dopamine which exerts inhibitory actions on both the brain and pituitary through the reduction of GnRH synthesis and release, a down regulation of GnRH receptors and the interference with the GnRH-signal transduction pathways, thus affecting gonadotropin secretion from the pituitary (Dufour et al. 2005, Popesku et al. 2008, Zohar et al. 2010). In Senegalese sole, a study performed by Guzmán et al. (2011a) evaluated the integrated effects of the dopamine inhibitor pimozine on the functionality of the reproductive axis in both males and females. This study showed that the blockage of endogenous dopaminergic signalling stimulates spermatogenesis, testicular maturation and sperm production in mature male Senegalese sole, although no effects were observed in females. The stimulation observed in males was associated with increased transcription of pituitary gonadotropin α subunit, although the mRNA levels of LHβ and FSHβ and the amounts of GnRH-1 and GnRH-2 peptides in the pituitary were not modified. Based on these results, Guzmán et al. (2011a) concluded that a dopamine-mediated inhibition of the reproductive axis could be present in Senegalese sole males, being absent or weakly manifested in females.

The involvement of the dopaminergic signalling in response to challenging stress situations has also been well characterized in Senegalese sole juveniles. In this regard, several studies making use of different chronic (e.g., high stocking density, low water replacement or repeated handling) and acute (e.g., exposure to exogenous ammonia, air exposure, net handling, or chasing) stress situations in Senegalese sole juveniles, have evidenced a significant increase in the cerebral dopaminergic activity that can be partially or totally counteracted by the presence of melatonin in the water (Costas et al. 2012, Weber et al. 2012, 2015, López-Patiño et al. 2013, Gesto et al. 2016).

Complete or partial cDNA sequences predicting for Senegalese sole dopamine and adrenergic receptors and tyrosine hydroxylase are now available in the SoleaDB bioinformatic platform (e.g., Unigene name solea_v4.1_unigene6194, Unigene name solea_v4.1_unigene2162, and Unigene name solea_v4.1_unigene21080, respectively; http://www.juntadeandalucia.es/agriculturaypesca/ifapa/soleadb_ifapa/; Benzekri et al. 2014). Full-length molecular characterization of these sequences will permit the specific functional assays required to decipher the precise actions of catecholaminergic signalling systems in the neuroendocrine regulation of different physiological processes in this species.

4. Gonadotropin-inhibitory hormone (GnIH)

As it has been indicated in the previous section, dopamine constitutes the main inhibitor of gonadotropins in many fish species (Trudeau 1997, Yaron et al. 2003, Levavi-Sivan et al. 2003, Dufour et al. 2005). However, highly evolved teleosts as perciforms do not appear to exhibit such a dopaminergic inhibition (Zohar et al. 1995, Holland et al. 1998b, Prat et al. 2001) and studies approaching the role of dopamine in the reproduction of pleuronectiforms, including the sole, have provided inconclusive results (Guzmán et al. 2011a). Recently, a new hypothalamic dodecapeptide that suppressed the synthesis and release of gonadotropins has emerged as an important regulator of reproduction in tetrapod vertebrates (Tsutsui et al. 2000, Ukena et al. 2002). Besides, this RFamide neuropeptide, termed as gonadotropin-inhibitory hormone (GnIH), controls the synthesis and/or release of GnRH and kisspeptins, but also functions in the transduction of photoperiod and melatonin effects, in the mediation of stress response, in the modulation of reproductive and social behaviour and in the control of food intake (Ubuka et al. 2012, Tsutsui et al. 2015). These actions are mediated via G protein-coupled GnIH, of which two different subtypes, GPR147 and GPR74, have been described to date, although GPR147 appears to represent the functional receptor for GnIH (Ogawa and Parhar 2014). Following pioneer studies performed in birds, it was shown that GnIH orthologues are present in all vertebrate groups, from lampreys to humans (Tustsui 2009, Osugi et al. 2012). However, the study of GnIH system and its physiological actions on reproduction and other physiological processes has only been addressed in a few teleost species, being results obtained rather conflicting.

In flatfishes, the GnIH system has only been studied in the Senegalese sole (Aliaga-Guerrero et al. 2018) and half-smooth tongue sole, *Cynoglossus semilaevis* (Wang et al. 2018a,b). In Senegalese sole, we cloned a full-length sequence for the GnIH precursor that encodes for three putative GnIH peptides (Aliaga-Guerrero et al. 2018). This sole GnIH precursor sequence exhibited highest identity (53%–62%) with pleunonectiforms (Wang et al. 2018a) and perciforms (Paullada-Salmerón et al. 2016), moderate identity with tetraodontiforms (49%–51%) (Ikemoto and Park 2005, Shahjahan et al. 2011) and lower identity with cypriniforms (39%) (Sawada et al. 2002) and tetrapods (29%–36%) GnIH orthologues (Muñoz-Cueto et al. 2017).

We also generated specific antibodies against sole GnIH peptides, which were used to identify the distribution of GnIH perikarya (Fig. 1) and GnIH innervation in the brain and pituitary (Aliaga-Guerrero et al. 2018). The GnIH-immunoreactive (ir) cell bodies were located in the olfactory bulbs (Fig. 1A), ventral telencephalon (Fig. 1B), caudal preoptic area (Fig. 1F, G), dorsal tegmentum (Fig. 1I) and rostral rhombencephalon (Fig. 1K), in agreement with quantitative real time PCR studies that showed GnIH mRNA expression in the same areas. A similar brain distribution of GnIH cells was reported in the European sea bass (Paullada-Salmerón et al. 2016) but positive GnIH cell masses were less widespread in the brain of other teleost species (see Muñoz-Cueto et al. 2017 for a revision). We also found a profuse GnIH innervation in the brain and pituitary of Senegalese sole (Aliaga-Guerrero et al. 2018). The presence of GnIH-ir perikarya and abundant GnIH-ir projections in neuroendocrine areas known for projecting to the pituitary, such as the ventral telencephalon and the preoptic area, suggests that these forebrain GnIH cell populations have a role in the modulation of hormonal secretion in the adenohypophysis of sole. In sole, these GnIH-ir axons course through the hypothalamus up to the pituitary stalk, and ended in the proximal *pars distalis* and *pars intermedia*, both areas containing gonadotropic cells (Rendón et al. 1997). The presence of GnIH-ir projections in the rostral *pars distalis* of the sole pituitary suggests that GnIH effects are pleiotropic and can be exerted on other adenohypophyseal cells (e.g., lactotropic or adrenocorticotropic cells), as was proposed in tilapia and sea bass (Ogawa et al. 2016, Paullada-Salmerón et al. 2016). We also detected GnIH fibres entering into melatoninergic/photoperiod sensor systems such as the pineal organ and the vascular sac of sole (Aliaga-Guerrero et al. 2018), which could imply a role of GnIH in phototransduction and in mediating the effects of photoperiod and melatonin in this species. Moreover, the pattern of distribution of GnIH cells and fibres in the brain of sole could suggest the existence of interactions of GnIH system with other neuroendocrine systems secreting neuropeptide Y (Rodríguez-Gómez et al. 2001), cathecolamines (Rodríguez-Gómez et al. 2000d) and/or GnRH (Rodriguez-Gómez et al. 1999).

In addition, we have determined in our laboratory the physiological effects of GnIHs in Senegalese sole (Aliaga-Guerrero et al. 2018). Intramuscular injection of GnIH-3 was efficient in decreasing transcript levels of brain GnRH-3 and pituitary LH-ß, whereas GnIH-2 injection did not modify the expression of the main reproductive genes analysed (Aliaga-Guerrero et al. 2018). These results are supported by morphological studies performed in sole and other fish species, which evidenced the localization of GnIH-ir cells and/or fibres in the terminal nerve area of the olfactory bulbs, known for containing GnRH-3 cells (Rendón et al. 1997, González-Martínez et al. 2002), and in the proximal *pars distalis* of the pituitary in the vicinity of gonadotropic cells (Rendón et al. 1997). However, the effects of GnIH on GnRHs and gonadotropins in fish are contradictory (inhibitory, stimulatory or ineffective) depending on the species, the administered peptide, the dose and route of administration, the nature of the assay (*in vivo* or *in vitro*), the sex or the physiological stage of animals (Muñoz-Cueto et al. 2017), reinforcing the need of additional studies in this group of vertebrates.

Taken together, our results have revealed the existence of a functional GnIH system in the sole brain, profusely innervating different brain areas and the pituitary gland, and that could represent an important inhibitory factor in the neuroendocrine control of flatfish reproduction. Whether this inhibitory neurohormone is involved in the reproductive dysfunctions that occur in the F1 generation of aquacultured sole will require further investigation in the near future.

5. Kisspeptins

In mammals, kisspeptin is a neuropeptide that plays an important role in reproduction through the stimulation of GnRH neurons, by activating its cognate receptor GPR54 (Oakley et al. 2009, Kotani et al. 2001). In a number of teleost fish species, two kisspeptin genes, namely kiss1 and kiss2, have been identified (Kitahashi et al. 2009, Felip et al. 2009, Li et al. 2009, Lee et al. 2009), whereas placental mammals possess only the kiss1 gene. Similarly, two kisspeptin receptor genes, named kiss1r and kiss2r, were also identified in several fish species (Li et al. 2009, Biran et al. 2008), suggesting that two Kiss/Kissr systems coexist in teleosts. However, this situation is not common among all fish species as only one kisspeptin gene, kiss2, and one receptor, kiss2r, are present in some other fish species, including the Senegalese sole (see below; Shi et al. 2010, Shahjahan et al. 2010, Mechaly et al. 2010), indicating that the kiss1 and kiss1r genes have been lost during evolution in these species (Mechaly et al. 2010, 2013). Tissue distribution analyses in different fish species indicated that both kiss1 and kiss2 and kisspeptin receptors mRNAs are expressed in the brain, the pituitary and the gonads (Kitahashi et al. 2009, Li et al. 2009, Lee et al. 2009, Biran et al. 2008, Shahjahan et al. 2010, Yang et al. 2010), suggesting autocrine/paracrine actions of kisspeptin in the pituitary and the gonads. In addition, several morphological and functional studies is fish species with two kisspeptin genes have indicated that the kiss2 system most probably plays an important role in the regulation of reproduction through the stimulation of GnRH (mainly GnRH-1) and gonadotropin secretion, while the kiss1 system seems to be implicated in the modulation of serotonergic system rather than in reproductive functions (Ogawa et al. 2012, Kitahashi et al. 2009, Shahjahan et al. 2010, Felip et al. 2009, Shi et al. 2010).

Recent studies carried out by Mechaly and collaborators (2009, 2011, 2012) in Senegalese sole have leaded to the identification of kiss2 and kiss2r genes in this species, as well as to the description of their brain mRNA levels during a short fasting period and during the annual reproductive cycle of sole breeders. These studies constitute the only information available up to date about kisspeptins in Senegalese sole and should be the basis for further research focused on the elucidation of the specific roles of the kisspeptin system in the control of reproduction of this species. As indicated above, research works carried out in Senegalese sole have only found one kisppeptin (kiss2) and one kisspeptin receptor (kiss2r) gene. This suggests that Kiss1 system was most probably lost in this species, likely as a consequence of the genome reduction that occurred during evolution (Smith and Gregory 2009) in the absence of stable adaptive advantages due to subfunctionalization of the two kiss genes.

The predicted Senegalese sole Kiss2 prepropeptide exhibits 129 amino acids, including a N-terminal signal peptide of 19 residues, the 78 amino acids of the Kiss2 peptide (Kp2-78) ending with the 10 residues corresponding to the Kp10 region (Kp2-10, which is of the FF form), a putative processing site (G^{98}-K^{99}-R^{100}) for proteolytic cleavage, and 29 amino acids corresponding to the C-terminal flanking peptide. Despite the Senegalese sole Kp2-78 peptide lacks internal cleavage sites, as observed in mammals and other fish (Kotani et al. 2001, Ohtaki et al. 2001, Yang et al. 2010), it is assumed that, similarly to these species, it would yield, probably by degradation (Popa et al. 2008), different shorter peptides in addition to the Kp2-10 peptide, for instance, Kp56, Kp27, Kp16, Kp13 and Kp10 which are detected in goldfish. The predicted Senegalese sole Kiss2r peptide consists of 378 amino acids including seven putative trans-membrane domains (TMDs) and four potential N-glycosylation sites.

RT-PCR analyses of Senegalese sole kiss2 and kiss2r indicated their ubiquitous expression along the brain-pituitary-gonad axis, a feature that opens the possibility of direct and local actions of kisspeptin at the level of the pituitary and the gonad. Although direct evidences are not yet available, the mRNA expression analyses performed by Mechaly et al. (2012) along the annual reproductive cycle of Senegalese sole male and female breeders suggest the involvement of the Kiss2 system in the regulation of gametogenesis and reproduction in this species. In males, forebrain and midbrain expression of both kiss2 and kiss2r reached the maximum levels towards the end of the winter, coinciding with the highest spermatogenetic activity and before the beginning of the spawning season (spring), and correlated well with the peaks in pituitary fshβ and lhβ transcript levels, plasma T and 11-KT concentration, gonadosomatic index (GSI) and testis fshr and lhcgr expression. In females, although these associations were not so evident, highest mRNA levels of both kiss2 and kiss2r were detected in the forebrain and midbrain during late winter and/or spring, in coincidence with the maximum GSI, pituitary fshβ mRNA levels, plasma E_2 concentration, and ovarian lhcgr mRNA expression. Taken together, the information provided by Mechaly et al. (2012) indicates that the expression of Kiss2 system in Senegalese sole males agrees with what one would expect according to its proposed role as a major regulator of the onset of reproduction. However, in females the situation was not so clear, as Kiss2 system expression was highest either before or during the spawning season. Finally, additional indirect evidence for a possible link between Kiss2 system and gonadotropin signalling in Senegalese sole has been obtained by submitting mixed-sex breeders to a 15-day fasting period (Mechaly et al. 2011), which tried to simulate the suppression of feeding that precedes the spawning season observed in this and other fish species (Costas et al. 2011, Milla et al. 2009). After fasting, hypothalamic kiss2 and kiss2r transcript levels increase significantly compared to control fish, concomitantly with the significant rise in the mRNA contents of pituitary fshβ and lhβ.

6. Neuropeptide Y (NPY)

The neuropeptide tyrosine or NPY is a 36-amino acid peptide of the pancreatic peptide family that is implicated both in the regulation of feeding and growth, but also in the control of reproductive processes in fish, by acting at the level of the brain and the pituitary (Zohar et al. 2010, Shahjahan et al. 2014, Volkoff 2016). Regarding feeding and growth regulation, NPY has potent and consistent orexigenic activity across fish species and thus, its brain mRNA levels increase during fasting but decreases after re-feeding. Moreover, NPY administration enhances food intake and pituitary mRNA content and release of growth hormone which, ultimately, leads to the stimulation of growth rates (Peng and Peter 1997, Carpio et al. 2006, Shahjahan et al. 2014, Conde-Sieira and Soengas 2017, Li et al. 2017). Concerning the neuroendrocrine control of reproduction, NPY treatment has been shown to increase both GnRH and LH release both *in vivo* and *in vitro* in a number of fish species (Zohar et al. 2010, Shahjahan et al. 2014). Due to its multiple roles in modulating feeding and GH, GnRH and gonadotropin release, it has been postulated that NPY could be one of the factors linking the growth, feeding and reproductive axes in fish (Zohar et al. 2010).

The complete cDNA sequence for the open reading frame of a putative Senegalese sole NPYa can be retrieved from the SoleaDB bioinformatic platform (http://www.juntadeandalucia.es/agriculturaypesca/ifapa/soleadb_ifapa/; Benzekri et al. 2014) under the Unigene name solea_v4.1_unigene466117 or solea_v4.1_unigene20107 (Global assembly: *Solea senegalensis* v4.1). The predicted putative Senegalese sole NPYa peptide consisted of a 99 amino acids precursor (prepro-NPYa), including an initial signal peptide of 28 amino acids, the biologically active NPY mature peptide between Y^{29} and Y^{64} (36 residues), a group of three amino acids (G^{65}-K^{66}-R^{67}) required for the processing of the mature peptide by several enzymes (such as prohormone convertase, cysteine protease, carboxypeptidase and peptidylglycine alpha-amidating monooxygenase), and a carboxyl-terminal flanking peptide (CPON) of 32 amino acids (Fig. 2). The comparative analysis of the Senegalese sole NPYa precursor protein shows very high levels of identity and similarity with the respective peptide in other Pleuronectiformes (identity: 92.9%–94.9%, similarity: 97.0%–99.0%, respectively) and fish species (similarity: 62.7%–93.9%, similarity: 75.5%–97.0%, respectively), as well as a relatively lower homology with prepro-NPY

A

Senegalese sole prepro-NPYa

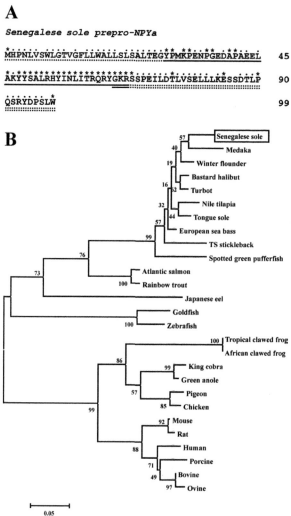

```
   *  *   *      *   *        *  ** **.**.** .**.**
MHPNLVSWLGTVGFLLWALLSLSALTEGYPMKPENPGEDAPAEEL    45

  *.********.*****.*******.****.*
AKYYSALRHYINLITRQRYGKRSSPEILDTLVSELLLKESSDTLP    90

QSRYDPSLW                                        99
```

Fig. 2. Senegalese sole Neuropeptide Ya (NPYa) prepro-protein sequence and phylogeny. (A). Senegalese sole prepro-NPYa sequence (Unigene name solea_v4.1_unigene466117; SoleaDB global assembly: *Solea senegalensis* v4.1) depicting the signal peptide (underlined with dotted line), the mature NPY peptide (underlined with solid line), the site for enzymatic processing of the mature peptide (double underlined with solid line), and the C-terminal flanking peptide (double underlined with dotted line). The identical amino acids among all the prepro-NPY sequences used for constructing the phylogenetic tree are marked by asterisks, while conserved residues among the Pleuronectiformes species analysed are labelled with a dot on top. (B). Unrooted phylogenetic tree of prepro-NPYa peptides in Senegalese sole (solea_v4.1_unigene466117; boxed) and other vertebrate species (African clawed frog, NP_001081300.1; Atlantic salmon, BAH24101.1; Bastard halibut, BAB62409.1; Bovine, NP_001014845.1; Chicken, NP_990804.1; European sea bass, CAB64932.1; Goldfish, AAA49186.1; Green anole, XP_003222076.1; Human, NP_000896.1; Japanese eel, AFN84517.1; King cobra, ETE61024.1; Medaka, ABU42130.1; Mouse, NP_075945.1; Nile tilapia, XP_003448902.1; Ovine, NP_001009452.1; Pigeon, NP_001269740.1; Porcine, NP_001243296.1; Rainbow trout, AAQ13835.1; Rat, NP_036746.1; Spotted green pufferfish, ALD51278.1; Three-spined [TS] stickleback, ALD51279.1; Tongue sole, XP_008329496.1; Tropical clawed frog, NP_001072530.1; Turbot, AIT12656.1; Winter flounder, ACH42755.1; Zebrafish, NP_571149.1). The tree was constructed using the Neighbor-Joining method. The percentage of replicate trees in which the associated taxa clustered together in the bootstrap test (1000 replicates) are shown next to the branches. The tree is drawn to scale, with branch lengths in the same units as those of the evolutionary distances used to infer the phylogenetic tree. The evolutionary distances were computed using the Poisson correction method and are in the units of the number of amino acid substitutions per site. Evolutionary analyses were conducted in MEGA7.

from reptiles, amphibians, birds and mammals (identity: 55.0%–61.0%; similarity: 70.1%–78.0%). These results are consistent with the phylogenetic tree constructed with a selected set of NPY prepro-proteins from multiple animal species from fish to mammals (Fig. 2). Despite the decreasing percentage of identity of NPY prepro-peptide sequences from fish to mammals, it is interesting to note that the last seventeen amino acids of the NPY mature peptide (Y^{48}YSALRHYINLITRQRY^{64}), which correspond to the receptor binding region of the peptide, were identical in all these vertebrate species, suggesting that the physiological functions for NPY are mostly conserved among vertebrates. In addition, Senegalese sole NPYa mature peptide shows as well the seven conserved residues involved in the dimerization of the peptide (L^{45}-Y^{49}-L^{52}-I^{56}-L^{58}-I^{59}-T^{60}).

The first information about the NPY system in Senegalese was obtained by Rodríguez-Gómez et al. (2001), providing an exhaustive description of NPY-immunoreactive cells and fibres within the brain and pituitary of this species (Fig. 1). Using an antiserum raised against porcine NPY, these authors identified NPY-immunoreactive cells in the ventral and dorsal telencephalon (Fig. 1B, C), caudal preoptic area (Fig. 1E, F), ventrocaudal hypothalamus (Fig. 1H), optic tectum (Fig. 1I, J), torus longitudinalis (Fig. 1I), synencephalon (Fig. 1I) and isthmic region (Fig. 1K) of Senegalese sole brain. Some of these brain areas, such as the ventral telencephalon, the preoptic area, and the hypothalamus, also presented GnRH immunoreactive neurons (Rodríguez-Gómez et al. 1999) and thus, possible interactions between NPY and GnRH cells are feasible from a morphofunctional point of view. Furthermore, NPY-immunoreactive fibres were also detected in Senegalese sole neurohypophysis reaching subsequently the proximal *pars distalis* of the adenohypophysis, where gonadotropin- and growth hormone (GH)-producing cells were found in this species (Rendón et al. 1997). This feature reinforces the possible role of NPY in the regulation of gonadotropin and GH secretion in Senegalese sole, as described in other fish species (see above), although no direct evidence for this has been obtained to date. In addition, the occurrence of NPY-immunoreactive fibres in the *pars intermedia* of Senegalese sole suggests that NPY may be also involved in the control of melanotropin secretion from the pituitary (Rendón et al. 1997, Rodríguez-Gómez et al. 1999).

Different neuroanatomical features regarding the localization of NPY-immunoreactive cells and fibres along the Senegalese sole brain could also support the role of this neuropeptide in the control of metabolic processes and food intake in this species. First, NPY-immunoreactive fibres were also abundant in the rhombencephalon, especially in viscerosensory areas such as the glossopharyngeal and vagal lobes, where primary gustatory centres have been reported in fish (Wullimann 1998). Second, NPY-immunoreactive cells in contact with the cerebrospinal fluid and in the proximity of blood vessels, were commonly observed in the nucleus of the posterior recess (Fig. 1H) within the ventral hypothalamus (that also contained abundant NPY-immunoreactive fibres). These cells seem to be involved in monitoring the homeostatic state of the organism, taking chemical information from the cerebrospinal fluid and blood (for instance, glucose, fatty acids or amino acids levels) or secreting NPY into these fluids (Rodríguez-Gómez et al. 1999). In line with these findings, the participation of hypothalamic NPY-dependent signalling in appetite control, as an orexigenic factor, has been recently reported in Senegalese sole post-larvae, because hypothalamic NPY mRNA levels decreased after the oral administration of diets enriched with different fatty acids (Velasco et al. 2017).

Finally, NPY-immunoreactive neurons were found in Senegalese sole brain areas where abundant catecholaminergic (tyrosine hydroxylase-immunoreactive) cells were also present, such as the posterior periventricular nucleus from the preoptic area (Fig. 1E, F) and the rhombencephalic locus coeruleus (Fig. 1K) (Rodríguez-Gómez et al. 1999, 2000d). This finding provides neuroanatomical support to the coexistence and potential interactions between NPY and catecholaminergic systems, both in the central and peripheral nervous system of Senegalese sole, as reported in other vertebrates (Everitt et al. 1984, Smeets and Reiner 1994).

7. Serotonin

The monoamine serotonin or 5-hydroxytryptamine is one of the major neurotransmitters in the central nervous system and thus, it has been involved in multiple neuroendocrine regulatory networks. Moreover, serotonin was found in the pineal gland as a precursor of melatonin, as well as in enterochromaffin and myenteric cells of the digestive system, in pancreatic beta cells, in parafollicular cells of the thyroid, in ovarian cumulus cells, in the dorsal root ganglia and in taste buds (Trowbridge et al. 2011). Like classical neurotransmitters, serotonin is released into the synaptic cleft during neuronal activity and then binds to its receptors on the post-synaptic membrane. In addition, serotonin is often released outside the synapse, i.e., acting by "volume transmission", and thereby exerting more systemic effects on surrounding cells (Fuxe et al. 2010). In mammals, numerous behaviours and physiological functions, including mood, sleep, aggressiveness, fear, stress coping, appetite, vascular function, pain and reproduction, have been demonstrated to be influenced by serotonin (Jacobs and Azmitia 1992, Lucki 1998, Puglisi-Allegra and Andolina 2015). This wide functional spectrum of serotonin in mammals seems to be conserved in teleost as well, and several studies have shown the involvement of serotonin signalling in the regulation of locomotion, behaviour, aggression, fear, anxiety, stress, reproduction, growth, and food intake in fish (Lillesaar 2011, Prasad et al. 2015, Winberg and Thörnqvist 2016, Kuz'mina 2015).

Rodríguez-Gómez et al. (2000b) have reported the distribution of serotonin-immunoreactive structures in the brain of the adult Senegalese sole (Fig. 1). In this study, they found that whereas serotonin-immunoreactive cell bodies appeared concentrated around the diencephalic recesses (Fig. 1H) and in the rhombencephalic reticular formation, raphe and inferior olivary area (Fig. 1J-L), serotonin-immunoreactive fibres were widely distributed throughout the entire brain. This wide distribution of serotonin fibres across the brain may explain why serotonin has been involved in the neuroendocrine regulatory networks of many behaviours in both fish and mammals (Lucki 1998, Lillesaar 2011). Interestingly, there is a profuse serotonin-immunoreactive innervation throughout ventral forebrain areas such as the ventral, central and lateral parts of the ventral telencephalon or the preoptic area. These areas represent identified neuroendocrine regions in Senegalese sole (Rodríguez-Gómez et al. 1999) and, thus, a possible role of serotonin in modulating the secretions of some neurohormones and/or releasing factors in neurosecretory cells might also be expected. This is strongly supported by different studies associating serotonin signalling with both CRH/CRF and GnRH neurosecretory systems in fish species (Norris and Hobbs 2006, Prasad et al. 2015). In addition, although serotonin fibres were not found in the neurohypophysis, serotonin-immunoreactive cells have been identified within the adenohypophysis (Fig. 1D–F) of Senegalese sole (Rodríguez-Gómez et al. 2000b). In several fish species, stimulatory and inhibitory effects of serotonin have been reported in gonadotropin release and growth hormone secretion, respectively (Prasad et al. 2015). The absence of serotonin fibres in Senegalese sole neurohypophysis indicates that, if these regulatory functions are conserved in this species, they are not controlled by direct pituitary projections from central serotonin neurons. Conversely, the presence of serotonin-immunoreactive cells in the adenohypophysis of sole may sustain paracrine actions of locally produced serotonin in this species, although experimental data supporting this assumption are still lacking. Other possibilities for serotonin signalling in Senegalese sole pituitary include its uptake from the blood, as it has been suggested in some species (Kah et al. 1993), or indirect effects through the mediation of other neuroendocrine/ neuromodulator factors (e.g., GnRH or catecholamines). The presence of serotonin has also been reported in melatonin-synthesizing cells from the pineal organ of Senegalese sole (Confente et al. 2008), which agrees with the reported role of serotonin as a precursor in the synthesis of melatonin in the pineal photoreceptor cells (Falcón et al. 2010).

Experimental data on the functionality of serotonin in Senegalese sole have only been obtained for the control of behavioural and physiological stress responses. Both in fish and mammals, brain serotonin seems to play a pivotal role in a complex neuroendocrine loop serving to defend homeostasis and promote acclimation during physiological or environmental challenges (Winberg and Thörnqvist 2016). Accordingly, different chronic (e.g., high stocking density, low water replacement or repeated handling) and acute (e.g., exposure to exogenous ammonia, air exposure, net handling, or chasing) stress situations in Senegalese sole juveniles have been shown to increase significantly the cerebral serotonergic activities (Costas et al. 2012, Weber et al. 2012, López-Patiño et al. 2013, Gesto et al. 2016), an increase that can be partially or totally counteracted by the presence of melatonin in the water (López-Patiño et al. 2013, Gesto et al. 2016).

Finally, partial cDNA sequences predicting for Senegalese sole serotonin receptors and tryptophan hydroxylase (an enzyme involved in serotonin synthesis) are available in the SoleaDB bioinformatic platform (e.g., Unigene name solea_v4.1_unigene1809 and Unigene name solea_v4.1_unigene207629, respectively; http://www.juntadeandalucia. es/agriculturaypesca/ifapa/soleadb_ifapa/; Benzekri et al. 2014). These sequences can be useful in the future as tools to elucidate the precise actions of the serotonergic system in the neural/neuroendocrine regulation of different physiological processes (reproduction, biological rhythms, food intake, stress response, behaviour, etc.) in this species.

8. Galanin

Galanin is a peptide present in the central nervous system and intestine of vertebrates, from mammals to fish. Galanin is initially synthesized as a precursor protein, known as prepro-galanin, containing a N-terminal signal peptide, the mature galanin peptide flanked by enzymatic cleavage sites and a C-terminal peptide named GMAP (Galanin Message-Associated Peptide). In mammals, there is only one variant of prepro-galanin, consisting of a 122–124 amino acids protein yielding a mature galanin peptide of 29–30 residues. The exception to this rule is the mouse, where a second variant of 141 amino acids with a truncated and extended GMAP has been identified, the mature peptide being identical to the first variant. On the other hand, birds, reptiles, amphibians and fish normally have two variants (and in some cases, up to five) of the galanin prepro-peptide with 115–120 and 140–144 amino acids, that give rise either a short galanin mature peptide of 27–31 residues (highly conserved with the mature galanin form reported in mammals) or a long galanin mature peptide of 49–55 amino acids. In all the cases, the long galanin mature peptide is formed by an insertion of 22–24 residues in the central part of the short galanin mature peptide (Fig. 3), and thus, they are believed to represent splice variants (Mensah et al. 2010).

The involvement of galanin in the control of reproductive axis was evidenced in mammals, in which galanin stimulated the *in vitro* GnRH release (Merchenthaler et al. 1990), and GnRH neurons expressed a galanin receptor (Gal-R1; Mitchell et al. 1999). In addition, galanin neuronal fibres can be found in close apposition with GnRH-1 neurons

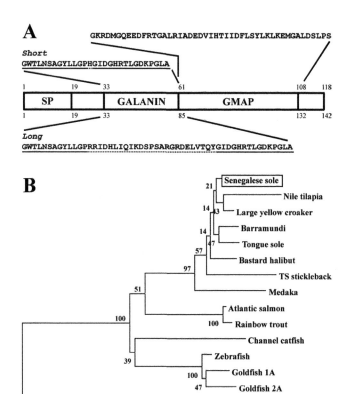

Fig. 3. Senegalese sole galanin prepro-protein sequence and phylogeny. (A). Representation of Senegalese sole short and long prepro-galanin protein sequences (Unigene names solea_v4.1_unigene25523 and solea_v4.1_unigene523609, respectively; SoleaDB global assembly: *Solea senegalensis* v4.1) depicting the signal peptide (SP), the mature galanin peptide (GALANIN) and the galanin message associated peptide (GMAP). The identical amino acids between the Senegalese sole short and the long galanin mature peptides are underlined with solid line, while the different residues are underlined with dotted line. (B). Unrooted phylogenetic tree of prepro-galanin peptides (short variants) in Senegalese sole (solea_v4.1_unigene25523; boxed) and other vertebrate species (African clawed frog, OCT84127.1; Atlantic salmon, XP_014004298.1; Barramundi, XP_018534089.1; Bastard halibut, XP_019962480.1; Bovine, NP_776339.1; Channel catfish, XP_017329225.1; Chicken, NP_001138861.1; Goldfish 1A, AAO65775.1; Goldfish 2A, AAO65778.1; Green anole, XP_008106933.1; Human, NP_057057.2; Japanese quail, AAD37348.1; Large yellow croaker, XP_019113407.1; Medaka, XP_004086257.1; Mouse, NP_034383.1; Nile tilapia, XP_003453629.1; Ovine, XP_011957774.1; Porcine, NP_999399.1; Rainbow trout, CDQ59534.1; Rat, NP_150240.1; Three-spined [TS] stickleback, ALD51505.1; Tongue sole, XP_008309765.1; Tropical clawed frog, XP_017948725.1; Zebra finch, XP_012429554.1; Zebrafish, XP_005163049.1). For further information about the construction of the tree, see Fig. 2.

in mice, rats and humans (Rajendren and Li 2001, Merchenthaler et al. 1991, Dudas and Merchenthaler 2004). In fish, it has been reported that galanin fibres establish contacts with gonadotropes in the adenohypophyseal proximal *pars distalis* (PPD) of several fish species (Batten et al. 1990, Anglade et al. 1994, Power et al. 1996). An *in vitro* study performed in goldfish pituitary cells has shown that galanin stimulates LH secretion (Prasada Rao et al. 1996), supporting a possible link between galanin and gonadotropin systems also in fish. Moreover, the *in vivo* treatment with estradiol and methyltestosterone, an aromatizable androgen that stimulates spermatogenesis in eels, increased the galanin immunoreactivity in the brain of European and American eels, suggesting the existence of a sex steroid positive feedback in galanin neurons (Olivereau and Olivereau 1991). Finally, a recent study of Martins et al. (2014) has shown that 11-ketotestorene (a non-aromatizable androgen directly involved in regulating fish reproductive processes) is able to up-regulate the mRNA levels of galanin type 1 receptor in the testis (but not in the brain) of male pre-pubertal European sea bass, supporting an additional link between galanin and androgen-signalling at gonadal level during fish puberty and reproduction. In addition to its roles in reproduction, galanin acts as an orexigenic factor and stimulates food intake, regulates gut motility decreasing transit time to increase food intake and modulates somatotropic and lactotropic cell activity in the pituitary of fish and mammals, being also involved in fish osmoregulation (Mensah et al. 2010, Shahjahan et al. 2014, Volkoff 2016). All these evidences indicate that galanin might be a major player in the neuroendocrine integration of various physiological systems in fishes (Mensah et al. 2010).

Concerning the Senegalese sole, Rodríguez-Gómez et al. (2000a) have identified galanin-immunoreactive cells in the anterior part of the parvocellular preoptic nucleus (Fig. 1 C), and in the ventral part of the nucleus of the lateral recess of the hypothalamus (Fig. 1F). In this study, abundant galanin-immunoreactive fibres and terminals were observed both in the neurohypophysis and the PPD of the adenohypophysis, where gonadotropin-, growth hormone- and thyrotropin-producing cells are found in this species (Rendón et al. 1997), suggesting that galanin may modulate the activity and/or secretion of these endocrine cells. Interestingly, the galanin-immunoreactive fibres entering the hypophysis of the Senegalese sole seem to originate in galanin-labelled cell bodies of the parvocellular preoptic nucleus, a region also displaying abundant GnRH-immunoreactive neurons (Rodríguez-Gómez et al. 1999). Thus, possible interactions among galanin neurons, GnRH cells and gonadotrophs, somatotrophs and thyrotrophs are feasible from a structural point of view in Senegalese sole. In addition, widespread galanin-immunoreactive fibres were found throughout the entire brain of Senegalese sole, with very abundant galanin projections reaching the caudal olfactory bulbs, ventral telencephalon, preoptic area, diencephalic ventricular recesses, ventricular mesencephalic area, median reticular formation, and viscerosensory rhombencephalon (Rodríguez-Gómez et al. 2000a). The extensive distribution of these immunoreactive fibres in the brain of Senegalese sole suggests an important role for galanin in the neuroendocrine regulation of brain functions.

Recently, Bonacic et al. (2016) have evaluated food intake and appetite regulation mechanisms in Senegalese sole larvae and post-larvae fed with live preys enriched with different oils, analysing the mRNA levels of a wide range of putative anorexigenic and orexigenic factors, including galanin. However, this study shows inconsistent variations according to the time before/after feeding and oil enrichment in the transcript levels of galanin (as well as in other analysed genes), indicating that the appetite regulatory system is still underdeveloped at early larval stages (Bonacic et al. 2016).

While no specific studies have been published reporting the molecular characterization of galanin in Senegalese sole, we have identified by homology searches in the SoleaDB bioinformatic platform (http://www.juntadeandalucia.es/agriculturaypesca/ifapa/soleadb_ifapa/; Benzekri et al. 2014) a complete cDNA sequence encoding for the open reading frame of a putative short galanin mature peptide, as well as a sequence encoding the insertion to form the putative long galanin mature peptide (Unigene names solea_v4.1_unigene25523 and solea_v4.1_unigene523609, respectively; Global assembly: *Solea senegalensis* v4.1). The predicted putative Senegalese sole short galanin peptide consisted of a 118 amino acids precursor (prepro-galanin), including an initial signal peptide of 19 amino acids, the biologically active short galanin peptide between G^{33} and A^{61} (29 amino acids), and the galanin message associated peptide (GMAP) between K^{62} and S^{108} (Fig. 3). In the long galanin mature peptide of Senegalese sole, the H^{46} amino acid present in the short galanin variant is substituted by 25 new residues, giving rise a final peptide of 53 amino acids (Fig. 3). The comparative analysis of the Senegalese sole short galanin precursor protein showed very high levels of identity and similarity with the respective proteins of other Pleuronectiformes (identity: 94.1%–94.9%; similarity: 97.5%–98.3%, respectively) and fish species (identity: 73.3%–94.9%; similarity: 85.8%–100%, respectively), as well as decreasing levels of homology with reptiles (identity: 55.0%; similarity: 72.5%), amphibians (identity: 54.2–55.8%; similarity: 71.7–73.3%), birds (identity: 54.0–55.8%; similarity 71.0–73.3%) and mammals, which exhibited the lowest percentages of identity (45.5%–50.0%) and similarity (61.1%–66.9%). These results are consistent with the phylogenetic tree constructed with a selected set of galanin prepro-proteins from diverse animal species from fish to mammals (Fig. 3). Despite the decreasing percentage of identity in the galanin prepro-peptide sequence from fish to mammals, it is interesting to note that the first thirteen amino acids of the galanin mature peptide (G^{33}WTLNSAGYLLGP^{45}), which correspond to the receptor binding region of the peptide (Mensah et al. 2010), are almost identical in all these vertebrate species,[5] suggesting that the physiological functions for galanin are mostly conserved among vertebrates.

In addition to the complete peptide sequences for both short and long galanin variants, partial cDNA sequences predicting for Senegalese sole galanin type 1 and type 2 receptors peptides were also found in the SoleaDB platform (e.g., Unigene name solea_v4.1_unigene8193 for a type 1 receptor, and Unigene name solea_v4.1_unigene 115642 for a type 2 receptor; http://www.juntadeandalucia.es/agriculturaypesca/ifapa/soleadb_ifapa/; Benzekri et al. 2014).

9. Cocaine- and amphetamine-regulated transcripts (CARTs)

In mammals, converging lines of evidence strongly demonstrate a role of the hypothalamic neuropeptide CART in the regulation of feeding, acting as an anorexigenic factor, and energy balance, promoting energy expenditure (Subhedar et al. 2014). Moreover, CART is co-expressed with peptides that are involved in the regulation of feeding, pituitary hormone secretion and energy metabolism, such as pro-opiomelanocortin (POMC), melanin concentrating hormone (MCH), thyrotropin-releasing hormone (TRH), corticotrophin-releasing factor (CRF) and neuropeptide Y (Vicentic and Jones 2007, Lau and Herzog 2014). Other emerging roles of CART are its implication in sensory processing, endocrine regulation, stress and anxiety, cardiovascular function and bone remodelling (Rogge et al. 2008). In this respect, CART mRNA and peptides are abundant along the hypothalamic-pituitary-adrenal (HPA) axis and sympathetic-adrenal system, where peptide expression can be altered by stress (Balkan et al. 2012). In addition, CARTs have been involved in the control of GnRH neuronal activity, by stimulating GnRH pulsatile release in rodents (Lebrethon et al. 2000, Parent et al. 2000), a role that is supported by the close appositions of CART fibres on hypothalamic GnRH neurons (Leslie et al. 2001).

In fish, different studies have pointed out that CART may produce anorexia (reviewed by Shahjahan et al. 2014) and, for instance, intra-cerebroventricular injections of CART peptides have been found to inhibit food intake in goldfish (Volkoff and Peter 2000). In addition, prolonged fasting and feeding was associated to a significant reduction or up-regulation, respectively, of the expression of some CART variants in multiple fish species (reviewed by Bonacic et al. 2015). Also, in Philippine catfish *Clarias batrachus* L., the projections of CART fibres were observed in the PPD of the pituitary (Singru et al. 2007) and CART was also expressed in pituitary LH cells during sexual maturation period (Barsagade et al. 2010), suggesting its local function in the sexual maturation process of this species. Finally, the localization of CART cells in the retina of different fish species suggests that CART could also represent an important signalling factor in several components of the visual circuitry of this group of vertebrates (reviewed by Bonacic et al. 2015).

Recently, Bonacic et al. (2015) have reported the molecular characterization of seven CARTs in Senegalese sole, that were named as cart1a, cart1b, cart2a, cart2b, cart3a, cart3b, and cart4. The high number of cart genes seems to be a common feature among highly evolved perciform and pleuronectiform teleosts, as 6 different CART peptides have been described in multiple fish species. These authors proposed that their retention along evolution could be the consequence of either a "subfunctionalization", i.e., the partition of old functions that in mammals are performed by a single CART form, or their "neofunctionalization", i.e., the adoption of new functions (Bonacic et al. 2015).

The CART proteins are initially translated from mRNA as pre-propeptides of variable length, which are then processed into multiple smaller active forms, in a tissue-specific manner. At least six CART active peptides, which result from this post-translational processing of the CART proprotein, have been identified so far in mammals, which seem to have different anorectic potencies (Kuhar and Yoho 1999, Dylag et al. 2006). Regarding Senegalese sole CARTs, they are translated into 96-107 amino acids pre-propeptides and are all predicted to contain a N-terminal signal peptide of 19 to 30 amino acids. While the signal peptides of Senegalese sole CARTs are quite divergent when compared to mammalian counterparts, a high degree of conservation was found in the C-terminal end of the CART proproteins (last 37 amino acids), where the 6 cysteine residues forming the disulfide bridges responsible for the protein tertiary structure were strictly conserved. Finally, the two sites for prohormone convertase processing typically present in mammalian CARTs are only conserved in Senegalese sole CART2b, while the remaining ones only exhibited the second site.

The expression analysis of the CART mRNAs in different brain regions of Senegalese sole indicated that both cart2a and cart4 showed a ubiquitous expression in the brain, cart1a, cart1b and cart3a had higher transcript levels in the mesencephalon and the diencephalon, cart2b was almost exclusively found in the olfactory bulbs, and cart3b was predominantly expressed in the spinal cord. The general pattern of distribution of the different Senegalese sole cart transcripts in the brain suggests that, although subfunctionalization probably occurs, overall, CART peptides are likely to have similar roles in fish as in mammals, acting in the regulation of feeding and energy homeostasis, but also in neuroendocrine and hypophysiotropic functions, and in the processing of peripheral sensory information.

In addition to the molecular and brain distribution characterization of the of the seven CART genes present in Senegalese sole, Bonacic et al. (2015) also analysed their brain mRNA levels just before and at different times after feeding. The results of this experiment showed that only cart1a, cart2a and cart4 mRNA levels are significantly up-regulated in the brain after feeding suggesting that, at least, these three CART proteins behave as anorexigenic factors in Senegalese sole. However, additional experimental data obtained by Conde-Sieira et al. 2015, showing an up-regulation of hypothalamic mRNA levels of cart2b in Senegalese sole 3 h after intraperitoneal administration of several unsaturated fatty acids such as oleate, α-linolenate and eicosapentanoate, suggest that this CART peptide may also participate in the general anorexigenic balance that is usually associated with the decrease in food intake.

10. Thyrotropin-releasing hormone (TRH)

Thyrotropin-releasing hormone (TRH) is a tripeptide hormone (pyro-G-H-P-NH$_2$) synthesized as a larger precursor protein (prepro-TRH) containing multiple copies of the TRH progenitor sequence flanked by paired basic residues (-K/R-R-Q-H-P-G-K/R-R-) (reviewed in Galas et al. 2009, Joseph-Bravo et al. 2017). The active tripeptides are released after different enzymatic cleavages and post-translational modifications, including cleavage at the flanking paired basic residues by prohormone convertases, removal of the C-terminal basic amino acids by carboxypeptidase enzymes, N-terminal acetylation, pyroglutamate formation, and C-terminal amidation (Galas et al. 2009, Joseph-Bravo et al. 2017). TRH sequence is well conserved across vertebrates with the number of TRH progenitors per prepro-protein ranging from 4 to 6 in mammals and from 6 to 8 in fish (Galas et al. 2009, Joseph-Bravo et al. 2017).

TRH mRNA is expressed in various brain regions and a high number of peripheral organs (Galas et al. 2009) and in mammals, TRH has been demonstrated to be critical for normal regulation of the hypothalamic-pituitary-thyroid (HPT) axis by stimulating thyroid stimulating hormone (TSH) synthesis and release from the anterior pituitary which, in turn, promotes thyroid hormone (TH) synthesis and release from the thyroid gland. In addition to thyrotrophs, TRH also regulates the functionality and activities of the lactotrophs and a fraction of somatotrophs, all of them expressing the TRH receptor-1, being therefore involved in the control of prolactin and growth hormone synthesis and release from the pituitary (Joseph-Bravo et al. 2017). In fish, the role of TRH in the regulation of the HPT axis is controversial since some studies have shown the ability of TRH to stimulate TSH and T4 release or TSH β transcription while other reports failed to find such stimulatory effects (reviewed in Galas et al. 2009, Iziga et al. 2010). On the other hand, multiple evidences in different fish species indicate that TRH regulates the synthesis and release from the pituitary of other important hormones such as growth hormone (GH), prolactin (PRL), adrenocorticotropic hormone (ACTH) and α-melanocyte-stimulating hormone (α-MSH) (reviewed in Galas et al. 2009, Iziga et al. 2010).

In Senegalese sole, Iziga et al. (2010) have recently reported the molecular characterization of the cDNA encoding the full-length TRH pre-propeptide. This precursor contains 276 amino acids, including a 22 amino acid signal peptide followed by the TRH propetide containing eight copies of the TRH progenitor sequence QHPG flanked by paired basic amino acids. While the first five copies of Senegalese sole TRH progenitor sequences are consistently conserved among vertebrates (although copies 2 or 3 are absent in cyprinid fishes), the remaining ones shows higher variability. Gene expression analysis in different tissues indicates that the brain is the main source of TRH in Senegalese sole, although its expression was also found at very low levels in multiple peripheral tissues. Within the brain, no information is available about the distribution of TRH expressing neurons and fibres in Senegalese sole, although in sea trout *Salmo trutta*, TRH immunoreactivity is mainly located in the olfactory bulbs, telencephalon, preoptic nucleus, vascular organ of the lamina terminalis, posterior tubercle/posterior hypothalamic lobe, and vagal motor nucleus (Díaz et al. 2001).

Regarding the functionality of TRH within Senegalese sole neuroendocrine systems, attempts to find a relationship between TRH-dependent signalling and metamorphosis or light adaptation were made in the species (Iziga et al. 2010, Campinho et al. 2015). However, these studies did not find any significant modulation of TRH mRNA levels during metamorphosis or after the adaptation of Senegalese sole larvae to different photoperiods (including constant darkness and illumination). In addition, different treatments with the goitrogen thiourea, T4, and/or the glucocorticoid dexamethasone in Senegalese sole pre- and post-metamorphic larvae were not able to significantly modify the mRNA levels of TRH (Iziga et al. 2010, Salas-Leiton et al. 2012).

11. Corticotropin-releasing hormone (CRH/CRF)

One of the two major neuroendocrine pathways involved in the primary response to stress in fish is represented by the hypothalamus-pituitary-interrenal (HPI) axis (Wendelaar Bonga 1997, Pankhurst 2011). In this cascade, corticotropin-releasing hormone (CRH) synthesized in the hypothalamus activates the production and release of adrenocorticotropic hormone (ACTH) from the pituitary corticotrophs, which in turn will stimulate the production and release of cortisol in the interrenal cells (Mommsen et al. 1999). At the hypothalamic level, a CRH-binding protein (CRH-BP) with antagonistic roles to CRH has also been described (Huising et al. 2004).

Fish, either in the nature or in aquaculture conditions, meet a variety of challenges, such as food competition, social aggression, handling and disturbance, or poor water quality (hypoxia). As a response to these conditions, the specimens react making neuroendocrine and behavioural adjustments, which are usually mediated by cortisol or are accompanied by changes in plasma cortisol levels (Barton 2002, Kulczykowska and Sánchez-Vázquez 2010). Thus, high circulating levels of cortisol are commonly used as indicators of fish acute stress, though there is no consensus on the endocrine profile for chronically stressed animals or how to assess it without invoking further stress (Pankhurst 2011, Dickens and Romero 2013). Besides its involvement in stress response, cortisol is also relevant in the modulation of multiple processes in fish such as the regulation of metabolism, food intake, behaviour, growth, reproduction, and osmoregulation (Mommsen et al. 1999, Bernier and Peter 2001).

The full cDNA sequences encoding for Senegalese sole CRH and CRH-BP have been recently obtained by Wunderink et al. (2011). The CRH cDNA contains an open reading frame encoding a 181 amino acid protein, including a 24 residues signal peptide, a cryptic binding motif between R^{61} and E^{73}, a typical N-terminal dibasic cleavage site at R^{137}-R^{138}, the CRH mature peptide comprising from S^{139} to Y^{179}, and a C-terminal amidation site at the two last amino acids of the prepro-protein. On the other hand, Senegalese sole CRH-BP is a 321 amino acids protein containing a signal peptide of 23 amino acids, two conserved amino acids (R^{53} and D^{59}) probably implicated in ligand binding with CRH, and ten conserved cysteine residues involved in the formation of five C-C disulfide bridges that are important for ligand binding activity. Regarding spatial distribution of CRH transcripts in Senegalese sole (no information is available for CRH-BP), a study using whole-mount *in situ* hybridization with pre- and post-metamorphic larvae has shown that CRH expression is restricted to the brain, with the CRH-positive cells being distributed in the hindbrain, pretectum, preoptic area, posterior tuberculum and hypothalamic regions (Campinho et al. 2015). This widespread distribution of CRH signal within the Senegalese brain agrees with the expression patterns of both CRH and CRH-BP reported for other fish species (Lu et al. 2004, Alderman and Bernier 2007, Grone and Maruska 2015, Amano et al. 2016). However, the restriction of CRH expression to the brain in Senegalese sole larvae is in disagreement with the information available in other fish species, where transcripts for both CRH and CRH-BP have been detected in other central and peripheral tissues, including the retina, gills, kidney, liver, intestine, and pituitary (only for CRH-BP) (Lu et al. 2004, Chen and Fernald 2008, Endsin et al. 2017).

The involvement of CRH signalling system in the neuroendocrine regulatory network controlling Senegalese sole stress responses has been demonstrated in several studies. For instance, Wunderink et al. (2011) showed that both plasma cortisol concentrations and brain CRH mRNA levels were elevated in Senegalese sole juveniles maintained at high culture densities (i.e., a chronic stressor), whereas up-regulated brain CRH and CRH-BP transcript levels were found one day after the transfer of animals to hypersaline water (i.e., an acute stressor). Similar results have also been obtained by Gesto et al. (2016), showing that chasing the fish with a net for 5 minutes (i.e., an acute stressor) induced an up-regulation of hypothalamic CRH mRNA levels just 15 to 60 minutes post-stress, an up-regulation that could be partially counteracted by the presence of melatonin in the water. In Senegalese sole post-larvae (50 days post-hatching), other chronic stress situations, such as food deprivation, have also been reported to enhance plasma cortisol and CRH mRNA levels and down-regulate CRH-BP transcription (Wunderink et al. 2012), indicating that HPI axis is already fully active in this species at this developmental stage. This finding was also supported by the results of Salas-Leiton et al. (2012) in 20–21 days post-hatching Senegalese sole larvae, showing that short- and/or long-term treatments (3 days and 2 weeks, respectively) with the glucocorticoid dexamethasone were able to up-regulate the transcription of both CRH and CRH-BP.

A recent study using juveniles of Senegalese sole, which is a species with nocturnal habits, has also shown that the hypothalamic expression of CRH and CRH-BP mRNA, as well as plasma cortisol levels, displayed significant daily rhythms, with respective acrophases being located at the beginning of the dark phase for CRH and cortisol and at the end of the dark phase for CRH-BP (López-Olmeda et al. 2013). Interestingly, these authors also found that depending on the time of day, the response of Senegalese sole HPI axis to a given acute stressor, for instance 30 seconds of air exposure, was different. Thus, when fish were stressed at the beginning of light period plasma cortisol levels were significantly elevated as compared to controls, although brain CRH mRNA amounts and locomotor activity remained unchanged. On the other hand, the opposite situation was observed when the air-exposure stress treatment was given at the beginning of the dark phase, i.e., CRH mRNA levels and locomotor activity were enhanced while plasma cortisol levels were unaffected.

Other possible neuroendocrine functions of CRH in Senegalese sole have been explored, for instance in the hypothalamic control of TH production during metamorphosis or in the regulation of feeding in post-larvae as a putative anorexigenic factor, but results obtained were inconclusive and no significant relations between CRH signalling and these mechanisms have been found (Bonacic et al. 2015, Campinho et al. 2015).

Table 1. Abbreviations used for cell masses in the Senegalese sole brain.

A	anterior thalamic nucleus
ACo	anterior commissure
CC	cerebellar crest
CCe	corpus of the cerebellum
CM	mammillary bodies
CN	cuneate nucleus
CP	central posterior thalamic nucleus
CZ	central zone of the optic tectum
Dc	central part of the dorsal telencephalon
Dd	dorsal part of the dorsal telencephalon
Dld	dorsolateral part of the dorsal telencephalon
Dlp	posterolateral part of the dorsal telencephalon
Dlv	ventrolateral part of the dorsal telencephalon
Dm1	subdivision 1 of the medial part of the dorsal telencephalon
Dm2	subdivision 2 of the medial part of the dorsal telencephalon
Dm3	subdivision 3 of the medial part of the dorsal telencephalon
Dm4	subdivision 4 of the medial part of the dorsal telencephalon
DON	descending octaval nucleus
DOT	dorsal optic tract
Dp	posterior part of the dorsal telencephalon
DT	dorsal tegmental nucleus
DTr	descending trigeminal tract
DWZ	deep white zone of the optic tectum
E	entopeduncular nucleus
ECL	external cellular layer of olfactory bulbs
FLL	lateral longitudinal fascicle
FR	fasciculus retroflexus
G	granular layer of the cerebellum
HaCo	habenular commissure
HCo	horizontal commissure
I	intermediate thalamic nucleus
ICL	internal cellular layer
IO	inferior olivary nucleus
IP	interpeduncular nucleus
IR	inferior nucleus of the raphe
LC	nucleus of the locus coeruleus
LFB	lateral forebrain bundle
LSO	lateral septal organ
M	molecular layer of the cerebellum
MCo	minor commissure
MF	medial funicular nucleus
MFB	medial forebrain bundle
MLF	medial longitudinal fascicle
MON	medial octavolateral nucleus
NAPv	anterior periventricular nucleus
NAT	anterior tuberal nucleus
NC	cortical nucleus
NCLI	central nucleus of the inferior lobe
NDLI	diffuse nucleus of the inferior lobe
NGp	posterior part of the glomerular nucleus
NGS	secondary gustatory nucleus
NGT	tertiary gustatory nucleus
NHd	dorsal habenular nucleus
NHv	ventral habenular nucleus
NI	isthmic nucleus
NLTd	dorsal part of the lateral tuberal nucleus

NLTi	inferior part of the lateral tuberal nucleus
NLTlc	lateral-caudal part of the lateral tuberal nucleus
NLTlr	lateral-rostral part of the lateral tuberal nucleus
NLTm	medial part of the lateral tuberal nucleus
NLTv	ventral part of the lateral tuberal nucleus
NLV	lateral nucleus of the valvula
nMLF	nucleus of the medial longitudinal fascicle
NMLI	medial nucleus of the inferior lobe
NPC	central pretectal nucleus
NPGa	anterior preglomerular nucleus
NPGc	commissural preglomerular nucleus
NPGl	lateral preglomerular nucleus
NPGm	medial preglomerular nucleus
NPOpc	parvocellular part of the parvocellular preoptic nucleus
NPPv	posterior periventricular nucleus
NPT	posterior tuberal nucleus
nPVO	nucleus of the paraventricular organ
NRLd	dorsal part of the nucleus of the lateral recess
NRLl	lateral part of the nucleus of the lateral recess
NRLv	ventral part of the nucleus of the lateral recess
NRP	nucleus of the posterior recess
NSC	suprachiasmatic nucleus
NSV	nucleus of the vascular sac
NT	nucleus taenia
nTE	nucleus of the thalamic eminence
OC	optic chiasm
OT	optic tectum
P	pituitary
PCo	posterior commissure
PE	preeminential nucleus
PGZ	periventricular gray zone of the optic tectum
PLl	lateral part of the perilemniscular nucleus
PLm	medial part of the perilemniscular nucleus
PMgc	gigantocellular part of the magnocellular preoptic nucleus
POA	preoptic area
PPv	ventral periventricular pretectal nucleus
PSi	intermediate superficial pretectal nucleus
PSm	magnocellular supercial pretectal nucleus
PSp	parvocellular superficial pretectal nucleus
PT	posterior thalamic nucleus
PVO	paraventricular organ
RI	inferior reticular nucleus
RL	lateral reticular nucleus
RS	superior reticular nucleus
SCO	subcommissural organ
SOF	secondary olfactory fibers
SR	superior nucleus of the raphe
SV	vascular sac
SWGZ	superficial white and grey zone of the optic tectum
TCo	tectal commissure
TEL	telencephalon
TGS	secondary gustatory tract
TLa	nucleus of the lateral torus
TNgc	terminal nerve ganglion cells
TS	semicircular torus
TSc	central part of the semicircular torus
TSld	lateral-dorsal part of the semicircular torus
TSv	ventral part of the semicircular torus

Vc	central nucleus of the ventral telencephalon
VCe	valvula of the cerebellum
Vd	dorsal nucleus of the ventral telencephalon
Vi	intermediate nucleus of the ventral telencephalon
VL	ventrolateral thalamic nucleus
Vl	lateral nucleus of the ventral telencephalon
VLo	facial-vagal lobe
VM	ventromedial thalamic nucleus
VOT	ventral optic tract
Vp	postcommissural nucleus of the ventral telencephalon
Vs	supracommissural nucleus of the ventral telencephalon
Vv	ventral nucleus of the ventral telencephalon
Xm	facial-vagal visceromotor column

Acknowledgements

These works were supported by grants from the Spanish Ministry of Education and Science (MEC, DGICYT PB93-1209) and the Spanish Ministry of Science and Innovation (MICINN, AGL2010-22139-C03/03).

References

Adams, B.A., E.D. Vickers, C. Warby, M. Park, W.H. Fischer, A. Grey Craig et al. 2002. Three forms of gonadotropin-releasing hormone, including a novel form, in a basal salmonid, *Coregonus clupeaformis*. Biol. Reprod. 67: 232 239.

Agulleiro, M.J., V. Anguis, J.P. Cañavate, G. Martínez-Rodríguez, C.C. Mylonas and J. Cerdá. 2006. Induction of spawning of captive reared Senegalese sole (*Solea Senegalensis*) using different delivery systems for gonadotropin-releasing hormone agonist. Aquaculture 257: 511–524.

Agulleiro, M.J., A.P. Scott, N. Duncan, C.C. Mylonas and J. Cerdá. 2007. Treatment of GnRHa-implanted Senegalese sole (*Solea senegalensis*) with 11-ketoandrostenedione stimulates spermatogenesis and increases sperm motility. Comp. Biochem. Physiol. A 147: 885–892.

Alderman, S.L. and N.J. Bernier. 2007. Localization of corticotropin-releasing factor, urotensin I, and CRF-binding protein gene expression in the brain of the zebrafish, *Danio rerio*. J. Comp. Neurol. 502: 783–793.

Aliaga-Guerrero, M., J.A. Paullada-Salmerón, V. Piquer, E.L. Mañanós and J.A. Muñoz-Cueto. 2018. Gonadotropin-inhibitory hormone in the flatfish, *Solea senegalensis*: Molecular cloning, brain localization and physiological effects. J. Comp. Neurol. 526: 349–370.

Amano, M., K. Okubo, T. Yamanome, H. Yamada, K. Aida and K. Yamamori. 2004. Changes in brain GnRH mRNA and pituitary GnRH peptide during testicular maturation in barfin flounder. Comp. Biochem. Physiol. B 138: 435–443.

Amano, M., N. Amiya, T. Yokoyama, K. Onikubo, N. Yamamoto and A. Takahashi. 2016. Immunohistochemical detection of corticotropin-releasing hormone (CRH) in the brain and pituitary of the hagfish, *Eptatretus burgeri*. Gen. Comp. Endocrinol. 236: 174–180.

Andersson, E., P.G. Fjelldal, U. Klenke, E. Vikingstad, G.L. Taranger, Y. Zohar et al. 2001. Three forms of GnRH in the brain and pituitary of the turbot, *Scophthalmus maximus*: immunological characterization and seasonal variation. Comp. Biochem. Physiol. B 129: 551–558.

Anglade, I., Y. Wang, J. Jensen, G. Tramu, O. Kah and J.M. Conlon. 1994. Characterization of trout galanin and its distribution in trout brain and pituitary. J. Comp. Neurol. 350: 63–74.

Balkan, B., A. Keser, O. Gozen, E.O. Koylu, T. Dagci, M.J. Kuhar et al. 2012. Forced swim stress elicits region-specific changes in CART expression in the stress axis and stress regulatory brain areas. Brain Res. 1432: 56–65.

Barsagade, V.G., M. Mazumdar, P.S. Singru, L. Thim, J.T. Clausen and N. Subhedar. 2010. Reproductive phase-related variations in cocaine- and amphetamine-regulated transcript (CART) in the olfactory system, forebrain, and pituitary of the female catfish, *Clarias batrachus* (Linn.). J. Comp. Neurol. 518: 2503–2254.

Barton, B.A. 2002. Stress in fishes: a diversity of responses with particular reference to changes in circulating costicosteroids. Integr. Comp. Biol. 42: 517–525.

Batten, T.F.C. and P.M. Ingleton. 1987. The hypothalamus and the pituitary gland. pp. 285–409. *In*: Chester-Jones, I., P.M. Ingleton and J.G. Phillips [eds.]. Fundamentals of Comparative Vertebrate Endocrinology. Plenum Press, New York, NY, USA.

Batten, T.F., L. Moons, M. Cambre and F. Vandesande. 1990. Anatomical distribution of galanin-like immunoreactivity in the brain and pituitary of teleost fishes. Neurosci. Lett. 111: 12–17.

Benzekri, H., P. Armesto, X. Cousin, M. Rovira, D. Crespo, M.A. Merlo et al. 2014. *De novo* assembly, characterization and functional annotation of Senegalese sole (*Solea senegalensis*) and common sole (*Solea solea*) transcriptomes: integration in a database and design of a microarray. BMC Genomics 15: 952.

Bernier, N.J. and R.E. Peter. 2001. The hypothalamicpituitary interrenal axis and the control of food intake in teleost fish. Comp. Biochem. Physiol. B 129: 639–644.

Biran, J., S. Ben-Dor and B. Levavi-Sivan. 2008. Molecular identification and functional characterization of the kisspeptin/kisspeptin receptor system in lower vertebrates. Biol. Reprod. 79: 776–786.

Bogerd, J., K.W. Li, J. Janssen-Dommerholt and H.J.Th. Goos. 1992. Two gonadotropin-releasing hormones from African catfish (*Clarias gariepinus*). Biochem. Biophys. Res. Commun. 187: 127–134.

Bonacic, K., A. Martínez, Á.J. Martín-Robles, J.A. Muñoz-Cueto and S. Morais. 2015. Characterization of seven cocaine- and amphetamine-regulated transcripts (CART) differentially expressed in the brain and peripheral tissues of *Solea senegalensis* (Kaup). Gen. Comp. Endocrinol. 224: 260–272.

Bonacic, K., C. Campoverde, J. Gómez-Arbones, E. Gisbert, A. Estevez and S. Morais. 2016. Dietary fatty acid composition affects food intake and gut-brain satiety signaling in Senegalese sole (*Solea senegalensis*, Kaup 1858) larvae and post-larvae. Gen. Comp. Endocrinol. 228: 79–94.

Cabrita, E., F. Soares, J. Beirao, Á. García-López, G. Martínez-Rodríguez and M.T. Dinis. 2011. Endocrine and milt response of Senegalese sole, *Solea senegalensis*, males maintained in captivity. Theriogenology 75: 1–9.

Campinho, M.A., N. Silva, J. Román-Padilla, M. Ponce, M. Manchado and D.M. Power. 2015. Flatfish metamorphosis: A hypothalamic independent process? Mol. Cell. Endocrinol. 404: 16–25.

Canosa, L.F., J.P. Chang and R.E. Peter. 2007. Neuroendocrine control of growth hormone in fish. Gen. Comp. Endocrinol. 151: 1–26.

Carolsfeld, J., J.F. Powell, M. Park, W.H. Fischer, A.G. Craig, J.P. Chang et al. 2000. Primary structure and function of three gonadotropin-releasing hormones, including a novel form, from an ancient teleost, herring. Endocrinology 141: 505–512.

Carpio, Y., J. Acosta, A. Morales, F. Herrera, L.J. González and M.P. Estrada. 2006. Cloning, expression and growth promoting action of Red tilapia (*Oreochromis* sp.) neuropeptide Y. Peptides 27: 710–718.

Chang, J.P. and R.M. Jobin. 1994. Regulation of gonadotropin release in vertebrates: a comparison of GnRH mechanisms of action. pp. 41–51. *In*: Davey, K.B., R.E. Peter and S.S. Tobe [eds.]. Perspectives in Comparative Endocrinology. National Research Council of Canada, Otawa, Canada.

Chang, J.P. and J.G. Pemberton. 2018. Comparative aspects of GnRH-Stimulated signal transduction in the vertebrate pituitary - Contributions from teleost model systems. Mol. Cell. Endocrinol. 463: 142–167.

Chen, C.C. and R.D. Fernald. 2008. Sequences, expression patterns and regulation of the corticotropin-releasing factor system in a teleost. Gen. Comp. Endocrinol. 157: 148–155.

Conde-Sieira, M., K. Bonacic, C. Velasco, L.M.P. Valente, S. Morais and J.L. Soengas. 2015. Hypothalamic fatty acid sensing in Senegalese sole (*Solea senegalensis*): response to long-chain saturated, monounsaturated, and polyunsaturated (n-3) fatty acids. Am. J. Physiol. Regul. Integr. Comp. Physiol. 309: R1521–R1531.

Conde-Sieira, M. and J.L. Soengas. 2017. Nutrient sensing systems in fish: impact on food intake regulation and energy homeostasis. Front. Neurosci. 10: 603.

Confente, F., A. El M'Rabet, A. Ouarour, P. Voisin, W.J. De Grip, M.C. Rendón et al. 2008. The pineal complex of Senegalese sole (*Solea senegalensis*): Anatomical, histological and immunohistochemical study. Aquaculture 285: 207–215.

Costas, B., C. Aragão, I. Ruiz-Jarabo, L. Vargas-Chacoff, F.J. Arjona, M.T. Dinis et al. 2011. Feed deprivation in Senegalese sole (*Solea senegalensis* Kaup, 1858) juveniles: effects on blood plasma metabolites and free amino acid levels. Fish Physiol. Biochem. 37: 495–504.

Costas, B., C. Aragão, J.L. Soengas, J.M. Míguez, P. Rema, J. Dias et al. 2012. Effects of dietary amino acids and repeated handling on stress response and brain monoaminergic neurotransmitters in Senegalese sole (*Solea senegalensis*) juveniles. Comp. Biochem. Physiol. A 161: 18–26.

Delgado, M.J., J.M. Cerdá-Reverter and J.L. Soengas. 2017. Hypothalamic integration of metabolic, endocrine, and circadian signals in fish: involvement in the control of food intake. Front. Neurosci. 11: 354.

Díaz, M.L., M. Becerra, M.J. Manso and R. Anadon. 2001. Development of thyrotropin-releasing hormone immunoreactivity in the brain of the brown trout *Salmo trutta fario*. J. Comp. Neurol. 429: 299–320.

Dickens, M.J. and L.M. Romero. 2013. A consensus endocrine profile for a chronically stressed wild animal does not exist. Gen. Comp. Endocrinol. 191: 177–189.

Dudas B. and I. Merchenthaler. 2004. Bi-directional associations between galanin and luteinizing hormone-releasing hormone neuronal systems in the human diencephalon. Neuroscience 127: 695–707.

Dufour, S., F.A. Weltzien, M.E. Sebert, N. Le Belle, B. Vidal, P. Vernier et al. 2005. Dopaminergic inhibition of reproduction in teleost fishes: ecophysiological and evolutionary implications. Ann. NY Acad. Sci. 1040: 9–21.

Dylag, T., J. Kotlinska, P. Rafalski, A. Pachuta and J. Silberring. 2006. The activity of CART peptide fragments. Peptides 27: 1926–1933.

Endsin, M.J., O. Michalec, L.A. Manzon, D.A. Lovejoy and R.G. Manzon. 2017. CRH peptide evolution occurred in three phases: Evidence from characterizing sea lamprey CRH system members. Gen. Comp. Endocrinol. 240: 162–173.

Everitt, B.J., T. Hökfelt, L. Terenius, K. Tatemoto, V. Mutt and M. Goldstein. 1984. Differential co-existence of neuropeptide Y (NPY-like) immunoreactivity with catecholamines in the central nervous system of the rat. Neuroscience 11: 443–462.

Falcón, J., H. Migaud, J.A. Muñoz-Cueto and M. Carrillo. 2010. Current knowledge on the melatonin system in teleost fish. Gen. Comp. Endocrinol. 165: 469–482.

Felip, A., S. Zanuy, R. Pineda, I. Pinilla, M. Carrillo, M. Tena-Sempere et al. 2009. Evidence for two distinct KiSS genes in non-placental vertebrates that encode kisspeptins with different gonadotropin-releasing activities in fish and mammals. Mol. Cell. Endocrinol. 312: 61–71.

Fernald, R.D. and R.B. White. 1999. Gonadotropin-releasing hormone genes: phylogeny, structure, and functions. Front. Neuroendocrinol. 20: 224–240.

Flik, G., P.H. Klaren, E.H. Van den Burg, J.R. Metz and M.O. Huising. 2006. CRF and stress in fish. Gen. Comp. Endocrinol. 46: 36–44.

Fuxe, K., A.B. Dahlstrom, G. Jonsson, D. Marcellino, M. Guescini, M. Dam et al. 2010. The discovery of central monoamine neurons gave volume transmission to the wired brain. Prog. Neurobiol. 90: 82–100.

Galas, L., E. Raoult, M.C. Tonon, R. Okada, B.G. Jenks, J.P. Castaño et al. 2009. TRH acts as a multifunctional hypophysiotropic factor in vertebrates. Gen. Comp. Endocrinol. 164: 40–50.

García-López, Á., E. Couto, A.V.M. Canario, C. Sarasquete and G. Martínez-Rodríguez. 2007. Ovarian development and plasma sex steroid levels in cultured female Senegalese sole (*Solea senegalensis*). Comp. Biochem. Physiol. A 146: 342–354.

Gesto, M., R. Alvarez-Otero, M. Conde-Sieira, C. Otero-Rodino, S. Usandizaga, J.L. Soengas et al. 2016. A simple melatonin treatment protocol attenuates the response to acute stress in the sole *Solea senegalensis*. Aquaculture 452: 272–282.

Gonzalez-Martinez, D., N. Zmora, E. Mañanós, D. Saligaut, S. Zanuy, Y. Zohar et al. 2002. Immunohistochemical localization of three different prepro-GnRHs in the brain and pituitary of the European sea bass (*Dicentrarchus labrax*) using antibodies to the corresponding GnRH associated peptides. J. Comp. Neurol. 446: 95–113.

Gothilf, Y., J.A. Muñoz-Cueto, C.A. Sagrillo, M. Selmanoff, T.T. Chen, O. Kah et al. 1996. Three forms of gonadotropin-releasing hormone in a perciform fish (*Sparus aurata*): complementary deoxyribonucleic acid characterization and brain localization. Biol. Reprod. 55: 636–645.

Grone, B.P. and K.P. Maruska. 2015. Divergent evolution of two corticotropin-releasing hormone (CRH) genes in teleost fishes. Front. Neurosci. 9: 365.

Guzmán, J.M., M. Rubio, J.B. Ortiz-Delgado, U. Klenke, K. Kight, I. Cross et al. 2009a. Comparative gene expression of gonadotropins (FSH and LH) and peptide levels of gonadotropin-releasing hormones (GnRHs) in the pituitary of wild and cultured Senegalese sole (*Solea senegalensis*) broodstocks. Comp. Biochem. Physiol. A 153: 266–277.

Guzmán, J.M., J. Ramos, C.C. Mylonas and E.L. Mañanós. 2009b. Spawning performance and plasma levels of GnRHa and sex steroids in cultured female Senegalese sole (*Solea senegalensis*) treated with different GnRHa-delivery systems. Aquaculture 291: 200–209.

Guzmán, J.M., R. Cal, Á. García-López, O. Chereguini, K. Kight, M. Olmedo et al. 2011a. Effects of *in vivo* treatment with the dopamine antagonist pimozide and gonadotropin-releasing hormone agonist (GnRHa) on the reproductive axis of Senegalese sole (*Solea senegalensis*). Comp. Biochem. Physiol. A 158: 235–245.

Guzmán, J.M., J. Ramos, C.C. Mylonas and E.L. Mañanós. 2011b. Comparative effects of human chorionic gonadotropin (hCG) and gonadotropin-releasing hormone agonist (GnRHa) treatments on the stimulation of male Senegalese sole (*Solea senegalensis*) reproduction. Aquaculture 316: 121–128.

Holland, M.C.H., Y. Gothilf, I. Meiri, J.A. King, K. Okuzawa, A. Elizur et al. 1998a. Levels of the native forms of GnRH in the pituitary of the gilthead seabream, *Sparus aurata*, at several stages of the gonadal cycle. Gen. Comp. Endocrinol. 112: 394–405.

Holland, M.C., S. Hassin and Y. Zohar. 1998. Effects of long-term testosterone, gonadotropin-releasing hormone agonist, and pimozide treatments on gonadotropin II levels and ovarian development in juvenile female striped bass (*Morone saxatilis*). Biol. Reprod. 59: 1153–1162.

Holland, M.C., S. Hassin and Y. Zohar. 2001. Seasonal fluctuations in pituitary levels of the three forms of gonadotropin-releasing hormone in striped bass, *Morone saxatilis* (Teleostei), during juvenile and pubertal development. J. Endocrinol. 169: 527–538.

Huising, M.O., J.R. Metz, C. van Schooten, A.J. Taverne-Thiele, T. Hermsen B.M. Verburg-van Kemenade et al. 2004. Structural characterisation of a cyprinid (*Cyprinus carpio* L.) CRH, CRH-BP and CRH-R1, and the role of these proteins in the acute stress response. J. Mol. Endocrinol. 32: 627–648.

Ikemoto, T. and M.K. Park. 2005. Chicken RFamide-related peptide (GnIH) and two distinct receptor subtypes: identification, molecular characterization, and evolutionary considerations. J. Reprod. Dev. 51: 359–377.

Iziga, R., M. Ponce, C. Infante, L. Rebordinos, J.P. Cañavate and M. Manchado. 2010. Molecular characterization and gene expression of thyrotropin-releasing hormone in Senegalese sole (*Solea senegalensis*). Comp. Biochem. Physiol. B 157: 167–174.

Jacobs, B.L. and E.C. Azmitia. 1992. Structure and function of the brain serotonin system. Physiol. Rev. 72: 165–229.

Joseph-Bravo, P., L. Jaimes-Hoy and J.L. Charli. 2017. Advances in TRH signaling. Rev. Endocr. Metab. Disord. (in press).

Kah, O., I. Anglade, E. Leprétre, P. Dubourg and D. Monbrison. 1993. The reproductive brain in fish. Fish Physiol. Biochem. 11: 85–98.

Kim, M.H., Y. Oka, M. Amano, M. Kobayashi, K. Okuzawa, Y. Hasegawa et al. 1995. Immunocytochemical localization of sGnRH and cGnRH-II in the brain of goldfish, *Carassius auratus*. J. Comp. Neurol. 356: 72–82.

King, J.A. and R.P. Millar. 1997. Coordinated evolution of GnRHs and their receptors. pp. 51–77. *In:* Parhar, I.S. and Y. Sakuma [eds.]. GnRH Neurons: Gene to Behavior. Brain Shuppan, Tokyo, Japan.

Kitahashi, T., S. Ogawa and I.S. Parhar. 2009. Cloning and expression of kiss2 in the zebrafish and medaka. Endocrinology 150: 821–31.

Kotani, M., M. Detheux, A. Vandenbbogaerde, D. Communi, J.M. Vanderwinden, E. Le Poul et al. 2001. The metastasis suppressor gene KiSS-1 encodes kisspeptins, the natural ligands of the orphan G protein coupled receptor GPR54. J. Biol. Chem. 276: 34631–34636.

Kuhar, M.J. and L.L. Yoho. 1999. CART peptide analysis by Western blotting. Synapse 33: 163–171.

Kulczykowska, E. and F.J. Sánchez Vázquez. 2010. Neurohormonal regulation of feed intake and response to nutrients in fish: Aspects of feeding rhythm and stress Aquacult. Res. 41: 654–667.

Kuz'mina, V.V. 2015. Effect of serotonin on exotrophy processes in fish. pp. 89–121. *In:* M.D. Li [ed.]. New developments in serotonin research. Nova Science Publishers, Hauppauge, New York, USA.

Lau, J. and H. Herzog. 2014. CART in the regulation of appetite and energy homeostasis. Front. Neurosci. 8: 313.

Lebrethon, M.C., E. Vandersmissen, A. Gerard, A.S. Parent and J.P. Bourguignon. 2000. Cocaine and amphetamine-regulated-transcript peptide mediation of leptin stimulatory effect on the rat gonadotropin-releasing hormone pulse generator *in vitro*. J. Neuroendocrinol. 12: 383–385.

Lee, Y.R., K. Tsunekawa, M.J. Moon, H.N. Um, J.I. Hwang, T. Osugi et al. 2009. Molecular evolution of multiple forms of kisspeptins and GPR54 receptors in vertebrates. Endocrinology 150: 2837–2846.

Leslie, R.A., S.J. Sanders, S.I. Anderson, S. Schuhler, T.L. Horan and F.J. Ebling. 2001. Appositions between cocaine and amphetamine-related transcript- and gonadotropin releasing hormone-immunoreactive neurons in the hypothalamus of the Siberian hamster. Neurosci. Lett. 314: 111–114.

Levavi-Sivan, B., A. Avitan and T. Kanias. 2003. Characterization of the inhibitory dopamine receptor from the pituitary of tilapia. Fish Physiol. Biochem. 28: 73–75.

Li, M., X. Tan, Y. Sui, S. Jiao, Z. Wu, L. Wang et al. 2017. The stimulatory effect of neuropeptide Y on growth hormone expression, food intake, and growth in olive flounder (*Paralichthys olivaceus*). Fish Physiol. Biochem. (in press).

Li, S., Y. Zhang, Y. Liu, X. Huang, W. Huang, D. Lu et al. 2009. Structural and functional multiplicity of the kisspeptin/GPR54 system in goldfish (*Carassius auratus*). J. Endocrinol. 201: 407–418.

Lillesaar, C. 2011. The serotonergic system in fish. J. Chem. Neuroanat. 41: 294–308.

López-Olmeda, J.F., B. Blanco-Vives, I.M. Pujante, Y.S. Wunderink, J.M. Mancera and F.J. Sanchez-Vazquez. 2013. Daily rhythms in the hypothalamus-pituitary-interrenal axis and acute stress responses in a teleost flatfish, *Solea senegalensis*. Chronobiol. Int. 30: 530–539.

López-Patiño, M.A., M. Conde-Sieira, M. Gesto, M. Librán-Pérez, J.L. Soengas and J.M. Míguez. 2013. Melatonin partially minimizes the adverse stress effects in Senegalese sole (*Solea senegalensis*). Aquaculture 388–391: 165–172.

Lovejoy, D.A., W.H. Fischer, S. Ngamvongchon, A.G. Craig, C.S. Nahorniak, R.E. Peter et al. 1992. Distinct sequence of gonadotropin-releasing hormone (GnRH) in dogfish brain provides insight into GnRH evolution. Proc. Natl. Acad. Sci. USA 89: 6373–6377.

Lu, W., L. Dow, S. Gumusgoz, M.J. Brierley, J.M. Warne, C.R. McCrohan et al. 2004. Coexpression of corticotropin-releasing hormone and urotensin i precursor genes in the caudal neurosecretory system of the euryhaline flounder (*Platichthys flesus*): a possible shared role in peripheral regulation. Endocrinology 145: 5786–5797.

Lucki, I. 1998. The spectrum of behaviors influenced by serotonin. Biol. Psychiatry 44: 151–162.

Martins, R.S.T., P.I.S. Pinto, P.M. Guerreiro, S. Zanuy, M. Carrillo and A.V.M. Canário. 2014. Novel galanin receptors in teleost fish: Identification, expression and regulation by sex steroids. Gen. Comp. Endocrinol. 205: 109–120.

Mechaly, A.S., J. Viñas and F. Piferrer. 2009. Identification of two isoforms of the Kisspeptin-1 receptor (kiss1r) generated by alternative splicing in a modern teleost, the Senegalese sole (*Solea senegalensis*). Biol. Reprod. 80: 60–69.

Mechaly, A.S., J. Viñas, C. Murphy, M. Reith and F. Piferrer. 2010. Gene structure of the Kiss1 receptor-2 (Kiss1r-2) in the Atlantic halibut: insights into the evolution and regulation of Kiss1r genes. Mol. Cell. Endocrinol 317: 78–89.

Mechaly, A.S., J. Viñas and F. Piferrer. 2011. Gene structure analysis of kisspeptin-2 (Kiss2) in the Senegalese sole (*Solea senegalensis*): characterization of two splice variants of Kiss2, and novel evidence for metabolic regulation of kisspeptin signaling in non-mammalian species. Mol. Cell. Endocrinol. 339: 14–24.

Mechaly, A.S., J. Viñas and F. Piferrer. 2012. Sex-specific changes in the expression of kisspeptin, kisspeptin receptor, gonadotropins and gonadotropin receptors in the Senegalese sole (*Solea senegalensis*) during a full reproductive cycle. Comp. Biochem. Physiol. A 162: 364–371.

Mechaly, A.S., J. Viñas and F. Piferrer. 2013. The kisspeptin system genes in teleost fish, their structure and regulation, with particular attention to the situation in Pleuronectiformes. Gen. Comp. Endocrinol. 188: 258–268.

Meek, J. 1994. Catecholamines in the brains of Osteichthyes bony fishes. pp. 49–76. *In*: Smeets, W.J.A.J. and A. Reiner [eds.]. Phylogeny and Development of Catecholamine Systems in the CNS of Vertebrates, Part I. Cambridge University Press, Cambridge, United Kingdom.

Mensah, E.T., H. Volkoff and S. Unniappan. 2010. Galanin systems in non-mammalian vertebrates with special focus on fishes. Experientia 102: 243–262.

Merchenthaler, I., F.J. Lopez and A. Negro-Vilar. 1990. Colocalization of galanin and luteinizing hormone-releasing hormone in a subset of preoptic hypothalamic neurons: anatomical and functional correlates. Proc. Natl. Acad. Sci. USA 87: 6326–6330.

Merchenthaler, I., F.J. Lopez, D.E. Lennard and A. Negro-Vilar. 1991. Sexual differences in the distribution of neurons coexpressing galanin and luteinizing hormone-releasing hormone in the rat brain. Endocrinology 129: 1977–1986.

Milla, S., N. Wang, S.N.M. Mandiki and P. Kestemont. 2009. Corticosteroids: friends or foes of teleost fish reproduction? Comp. Biochem. Physiol. A 153: 242–251.

Mitchell, V., S. Bouret, V. Prevot, L. Jennes and J.C. Beauvillain. 1999. Evidence for expression of galanin receptor Gal-R1 mRNA in certain gonadotropin releasing hormone neurons of the rostral preoptic area. J. Neuroendocrinol. 11: 805–812.

Mommsen, T.P., M.M. Vijayan and T.W. Moon. 1999. Cortisol in teleosts: dynamics, mechanisms of action, and metabolic regulation. Rev. Fish Biol. Fish. 9: 211–268.

Montero, M., B. Vidal, F. Vandesande, J.A. King, G. Tramu, S. Dufour et al. 1994. Comparative distribution of mammalian GnRH (gonadotrophin-releasing hormone) and chicken GnRH-II in the brain of the European eel. J. Chem. Neuroanat. 7: 227–241.

Muñoz-Cueto, J.A., J.A. Paullada-Salmerón, M. Aliaga-Guerrero, M.E. Cowan, I.S. Parhar and T. Ubuka. 2017. A Journey through the Gonadotropin-Inhibitory Hormone System of Fish. Front. Endocrinol. 8: 285.

Muske, L.E. and F.L. Moore. 1994. Antibodies against different forms of GnRH distinguish different populations of cells and axonal pathways in a urodele amphibian, *Taricha granulosa*. J. Comp. Neurol. 345: 139–147.

Nardocci, G., C. Navarro, P.P. Cortés, M. Imarai, M. Montoya, B. Valenzuela, P. Jara, C. Acuña-Castillo and R. Fernández. 2014. Neuroendocrine mechanisms for immune system regulation during stress in fish. Fish Shellfish Immunol. 40: 531–538.

Nikolics, K., A.J. Mason, E. Szonyi, J. Ramachandran and P.H. Seeburg. 1985. A prolactin-inhibiting factor within the precursor for human gonadotropin-releasing hormone. Nature 316: 511–517.

Norris, D.O. and S.L. Hobbs. 2006. The HPA axis and functions of corticosteroids in fishes. pp. 721–766. *In*: M. Reinecke, G. Zaccone and B.G. Kapoor [eds.]. Fish Endocrinology. Science Publishers, Enfield, NH, USA.

Oakley, A.E., D.K. Clifton and R.A. Steiner. 2009. Kisspeptin signaling in the brain. Endocr. Rev. 30: 713–743.

Ogawa, S., K.W. Ng, P.N. Ramadasan, F.M. Nathan and I.S. Parhar. 2012. Habenular Kiss1 neurons modulate the serotonergic system in the brain of zebrafish. Endocrinology 153: 2398–2407.

Ogawa, S. and I.S. Parhar. 2014. Structural and functional divergence of gonadotropin-inhibitory hormone from jawless fish to mammals. Front. Endocrinol. 5: 177.

Ogawa, S., M. Sivalingam, J. Biran, M. Golan, R.S. Anthonysamy, B. Levavi-Sivan and I.S. Parhar. 2016. Distribution of LPXRFa, a gonadotropin-inhibitory hormone ortholog peptide, and LPXRFs receptor in the brain and pituitary of the tilapia. J. Comp. Neurol. 524: 2753–2775.

Ohtaki, T., Y. Shintani, S. Honda, H. Matsumoto, A. Hori, K. Kanehashi et al. 2001. Metastasis suppressor gene KiSS-1 encodes peptide ligand of a G-protein-coupled receptor. Nature 411: 613–617.

Okubo, K., M. Amano, Y. Yoshiura, H. Suetake and K. Aida. 2000. A novel form of gonadotropin-releasing hormone in the medaka, *Oryzias latipes*. Biochem. Biophys. Res. Commun. 276: 298–303.

Okubo, K., H. Suetake and K. Aida. 2002. Three mRNA species for mammalian-type gonadotropin-releasing hormone in the brain of the eel *Anguilla japonica*. Mol. Cell. Endocrinol. 192: 17–25.

Olivereau, M. and J. Olivereau. 1991. Galanin-like immunoreactivity is increased in the brain of estradiol- and methyltestosterone-treated eels. Histochemistry 96: 487–497.

Osugi, T., D. Daukss, K. Gazda, T. Ubuka, T. Kosugi, M. Nozaki, S.A. Sower and K. Tsutsui. 2012. Evolutionary origin of the structure and function of gonadotropin-inhibitory hormone: insights from lampreys. Endocrinology 153: 2362–2374.

Pankhurst, N.W. 2011. The endocrinology of stress in fish: An environmental perspective. Gen. Comp. Endocrinol. 170: 265–275.

Parent, A.S., M.C. Lebrethon, A. Gerard, E. Vandersmissen and J.P. Bourguignon. 2000. Leptine ffects on pulsatile gonadotropin releasing hormone secretion from the adult rat hypothalamus and interaction with cocaine and amphetamine regulated transcript peptide and neuropeptide Y. Regul. Pept. 92: 17–24.

Parhar, I.S. and M. Iwata. 1994. Gonadotropin releasing hormone (GnRH) neurons project to growth hormone and somatolactin cells in the steelhead trout. Histochemistry 102: 195–203.

Parhar, I.S. 1997. GnRH in tilapia: three genes, three origins and their roles. pp. 99–122. *In*: I.S. Parhar and Y. Sakuma [eds.]. GnRH Neurons: Gene to Behavior. Brain Shuppan, Tokyo, Japan.

Paullada-Salmerón, J.A., M. Cowan, M. Aliaga-Guerrero, A. Gómez, S. Zanuy, E. Mañanós and J.A. Muñoz-Cueto. 2016. LPXRFa peptide system in the European sea bass: a molecular and immunohistochemical approach. J. Comp. Neurol. 524: 176–198.

Peng, C. and R.E. Peter. 1997. Neuroendocrine regulation of growth hormone secretion and growth in fish. Zool. Studies 36: 79–89.

Perry, S.F. and N.J. Bernier. 1999. The acute humoral adrenergic stress response in fish: facts and fiction. Aquaculture 177: 285–295.

Pham, K.X., M. Amano, N. Amiya, Y. Kurita and K. Yamamori. 2006. Distribution of three GnRHs in the brain and pituitary of the wild Japanese flounder *Paralichthys olivaceus*. Fish. Sci. 72: 89–94.

Planas, J., H.A. Bern and R.P. Millar. 1990. Effects of GnRH associated peptide and its component peptides on prolactin secretion from the tilapia pituitary *in vitro*. Gen. Comp. Endocrinol. 77: 386–396.

Popa, S.M., D.K. Clifton and R.A. Steiner. 2008. The role of kisspeptins and GPR54 in the neuroendocrine regulation of reproduction. Annu. Rev. Physiol. 70: 213–238.

Popesku, J.T., C.J. Martyniuk, J. Mennigen, H. Xiong, D. Zhang, X. Xia et al. 2008. The goldfish (*Carassius auratus*) as a model for neuroendocrine signaling. Mol. Cell. Endocrinol. 293: 43–56.

Powell, J.F., Y. Zohar, A. Elizur, M. Park, W.H. Fischer, A.G. Craig et al. 1994. Three forms of gonadotropin-releasing hormone from brains of one species. Proc. Natl. Acad. Sci. USA 91: 12081–12085.

Power, D.M., A.V. Canario and P.M. Ingleton. 1996. Somatotropin release-inhibiting factor and galanin innervations in the hypothalamus and pituitary of seabream (*Sparus aurata*). Gen. Comp. Endocrinol. 101: 264–274.

Prasad, P., S. Ogawa and I.S. Parhar. 2015. Role of serotonin in fish reproduction. Front. Neurosci. 9: 195.

Prasada Rao, P.D., C.K. Murthy, H. Cook and R.E. Peter. 1996. Sexual dimorphism of galanin-like immunoreactivity in the brain and pituitary of goldfish, *Carassius auratus*. J. Chem. Neuroanat. 10: 119–135.

Prat, F., S. Zanuy and M. Carrillo. 2001. Effect of gonadotropin-releasing hormone analogue GnRHa and pimozide on plasma levels of sex steroids and ovarian development in sea bass *Dicentrarchus labrax* L. Aquaculture, 198: 325–338.

Puglisi-Allegra, S. and D. Andolina. 2015. Serotonin and stress coping. Behav. Brain Res. 277: 58–67.

Rajendren, G. and X. Li. 2001. Galanin synaptic input to gonadotropin-releasing hormone perikarya in juvenile and adult female mice: implications for sexual maturity. Dev. Brain Res. 131: 161–165.

Rendón, C, F.J. Rodríguez-Gómez, J.A. Muñoz-Cueto, C. Piñuela and C. Sarasquete. 1997. An immunocytochemical study of pituitary cells of the Senegalese sole, *Solea senegalensis* (Kaup, 1858). Histochem. J. 29: 813–822.

Rodríguez, L., M. Carrillo, L.A. Sorbera, M.A. Soubrier, E. Mañanós, M.C. Holland et al. 2000. Pituitary levels of three forms of GnRH in the male European sea bass (*Dicentrarchus labrax*, L.) during sex differentiation and first spawning season. Gen. Comp. Endocrinol. 120: 67–74.

Rodríguez-Gómez, F.J. 1999. Study of the pineal organ and melatonin receptors along the brain-pituitary-gonadal axis of sole, *Solea Senegalensis*. Ph.D. Thesis, University of Cádiz, Cádiz, Spain.

Rodríguez-Gómez, F.J., C. Rendón, M.C. Sarasquete and J.A. Muñoz-Cueto. 1999. Distribution of gonadotropin-releasing hormone (GnRH) immunoreactive systems in the brain of the Senegalese sole, *Solea senegalensis*. Histochem. J. 31: 695–703.

Rodríguez-Gómez, F.J., M.C. Rendón-Unceta, C. Sarasquete and J.A. Muñoz-Cueto. 2000a. Localization of galanin-like immunoreactive structures in the brain of the Senegalese sole, *Solea senegalensis*. Histochem. J. 32: 123–131.

Rodríguez-Gómez, F.J., M.C. Rendón-Unceta, C. Sarasquete and J.A. Muñoz-Cueto. 2000b. Distribution of serotonin in the brain of the Senegalese sole, *Solea senegalensis*: an immunohistochemical study. J. Chem. Neuroanat. 18: 103–115.

Rodríguez-Gómez, F.J., M.C. Sarasquete and J.A. Muñoz-Cueto. 2000c. A morphological study of the brain of *Solea senegalensis*. I. The telencephalon. Histol. Histopathol. 15: 355–364.

Rodríguez-Gómez, F.J., C. Rendón, M.C. Sarasquete and J.A. Muñoz-Cueto. 2000d. Localization of tyrosine hydroxylase immunoreactivity in the brain of the Senegalese sole, *Solea senegalensis*. J. Chem. Neuroanat. 19: 17–32.

Rodríguez-Gómez, F.J., C. Rendón-Unceta, M.C. Sarasquete and J.A. Muñoz-Cueto. 2001. Distribution of neuropeptide Y-like immunoreactivity in the brain of the Senegalese sole (*Solea senegalensis*). Anat. Rec. 262: 227–237.

Rogge, G., D. Jones, G.W. Hubert, Y. Lin and M.J. Kuhar. 2008. CART peptides: regulators of body weight, reward and other functions. Nat. Rev. Neurosci. 9: 747–758.

Salas-Leiton, E., O. Coste, E. Asensio, C. Infante, J.P. Cañavate and M. Manchado. 2012. Dexamethasone modulates expression of genes involved in the innate immune system, growth and stress and increases susceptibility to bacterial disease in Senegalese sole (*Solea senegalensis* Kaup, 1858). Fish Shellfish Immunol. 32: 769–778.

Sawada, K., K. Ukena, H. Satake, E. Iwakoshi, H. Minakata and K. Tsutsui. 2002. Novel fish hypothalamic neuropeptide. Cloning of a cDNA encoding the precursor polypeptide and identification and localization of the mature peptide. Eur. J. Biochem. 269: 6000–6008.

Shahjahan, M., E. Motohashi, H. Doi and H. Ando. 2010. Elevation of Kiss2 and its receptor gene expression in the brain and pituitary of grass puffer during the spawning season. Gen. Comp. Endocrinol. 169: 48–57.

Shahjahan, M., T. Ikegami, T. Osugi, K. Ukena, H. Doi, A. Hattori, K. Tsutsui and H. Ando. 2011. Synchronised expressions of LPXRFamide peptide and its receptor genes: seasonal, diurnal and circadian changes during spawning period in grass puffer. J. Neuroendocrinol. 23: 39–51.

Shahjahan, M., T. Kitahashi and I.S. Parhar. 2014. Central pathways integrating metabolism and reproduction in teleosts. Front. Endocrinol. 5: 36.

Sherwood, N., L. Eiden, M. Brownstein, J. Spiess, J. Rivier and W. Vale. 1983. Characterization of a teleost gonadotropin-releasing hormone. Proc. Natl. Acad. Sci. USA 80: 2794–2798.

Shi, Y., Y. Zhang, S. Li, Q. Liu, D. Lu, M. Liu et al. 2010. Molecular identification of the Kiss2/Kiss1ra system and its potential function during 17alpha-methyltestosterone-induced sex reversal in the orange-spotted grouper, *Epinephelus coioides*. Biol. Reprod. 83: 63–74.

Singru, P.S., M. Mazumdar, A.J. Sakharkar, R.M. Lechan, L. Thim, J.T. Clausen et al. 2007. Immunohistochemical localization of cocaine- and amphetamine-regulated transcript peptide in the brain of the catfish, *Clarias batrachus* (Linn.). J. Comp. Neurol. 502: 215–235.

Smeets, W.J.A.J. and A. Reiner. 1994. Catecholamines in the CNS of vertebrates: Current concepts of evolution and functional significance. pp. 463-481. *In*: Smeets, W.J.A.J. and A. Reiner [eds.]. Phylogeny and development of catecholamine systems in the CNS of vertebrates. Cambridge University Press, Cambridge, UK.

Smith, E.M. and T.R. Gregory. 2009. Patterns of genome size diversity in the ray-finned fishes. Hydrobiologia 625: 1–25.

Sower, S.A., Y.C. Chiang, S. Lovas and J.M. Conlon. 1993. Primary structure and biological activity of a third gonadotropin-releasing hormone from lamprey brain. Endocrinology 132: 1125–1131.

Sower, S.A. 2018. Landmark discoveries in elucidating the origins of the hypothalamic-pituitary system from the perspective of a basal vertebrate, sea lamprey. Gen. Comp. Endocrinol. 264: 3–15.

Subhedar, N.K., K.T. Nakhate, M.A. Upadhya and D.M. Kokare. 2014. CART in the brain of vertebrates: circuits, functions and evolution. Peptides 54: 108–130.

Trowbridge, S., N. Narboux-Neme, P. Gaspar. 2010. Genetic models of serotonin (5-HT) depletion: what do they tell us about the developmental role of 5-HT? Anat. Rec. 294: 1615–1623.

Trudeau, V. L. 1997. Neuroendocrine regulation of gonadotrophin II release and gonadal growth in the goldfish, *Carassius auratus*. Rev. Reprod. 2: 55–68.

Tsutsui, K., E. Saigoh, K. Ukena, H. Teranishi, Y. Fujisawa, M. Kikuchi S. Ishii and P. J. Sharp. 2000. A novel avian hypothalamic peptide inhibiting gonadotropin release. Biochem. Biophys. Res. Commun. 275: 661–667.

Tsutsui, K. 2009. A new key neurohormone controlling reproduction, gonadotropin-inhibitory hormone (GnIH): Biosynthesis, mode of action and functional significance. Prog. Neurobiol. 88: 76–88.

Tsutsui, K., T. Ubuka, Y.L. Son, G.E. Bentley and L.J. Kriegsfeld. 2015. Contribution of GnIH Research to the Progress of Reproductive Neuroendocrinology. Front. Endocrinol. 6: 179.

Ubuka, T., Y.L. Son, Y. Tobari and K. Tsutsui. 2012. Gonadotropin-inhibitory hormone action in the brain and pituitary. Front. Endocrinol. 3: 148.

Ukena, K., H. Satake, E. Iwakoshi, H. Minakata and K. Tsutsui. 2002. A novel rat hypothalamic RFamide-related peptide identified by immunoaffinity chromatography and mass spectrometry. FEBS Lett. 512: 255–258.

Velasco, C., K. Bonacic, J.L. Soengas and S. Morais. 2017. Orally administered fatty acids enhanced anorectic potential but did not activate central fatty acid sensing in Senegalese sole post-larvae. J. Exp. Biol. (in press).

Vicentic, A. and D.C Jones. 2007. The CART (cocaine- and amphetamine-regulated transcript) system in appetite and drug addiction. J. Pharmacol. Exp. Ther. 320: 499–506.

Volkoff, H. 2016. The neuroendocrine regulation of food intake in fish: a review of current knowledge. Front Neurosci. 10: 540.

Volkoff, H. and R.E. Peter. 2000. Effects of CART peptides on food consumption, feeding and associated behaviors in the goldfish, *Carassius auratus*: actions on neuropeptide Y- and orexin A-induced feeding. Brain Res. 887: 125–133.

Wang, B., Q. Liu, X. Liu, Y. Xu and B. Shi. 2018a. Molecular characterization and expression profiles of LPXRFa at the brain-pituitary-gonad axis of half-smooth tongue sole (*Cynoglossus semilaevis*) during ovarian maturation. Comp. Biochem. Physiol. B Biochem. Mol. Biol. 216: 59–68.

Wang, B., G. Yang, Q. Liu, J. Qin, Y. Xu, W. Li, X. Liu and B. Shi. 2018b. Characterization of LPXRFa receptor in the half-smooth tongue sole (*Cynoglossus semilaevis*): Molecular cloning, expression profiles, and differential activation of signaling pathways by LPXRFa peptides. Comp. Biochem. Physiol. A Mol. Integr. Physiol. 223: 23–32.

Weber, R.A., J.J.P. Maceira, M.J. Mancebo, J.B. Peleteiro, L.O.G. Martín and M. Aldegunde. 2012. Effects of acute exposure to exogenous ammonia on cerebral monoaminergic neurotransmitters in juvenile *Solea senegalensis*. Ecotoxicology 21: 362–369.

Weber, R.A., J.J.P. Pérez Maceira, M.J. Aldegunde, J.B. Peleteiro, L.O. García Martín and M. Aldegunde. 2015. Effects of acute handling stress on cerebral monoaminergic neurotransmitters in juvenile Senegalese sole *Solea senegalensis*. J. Fish Biol. 87: 1165–1175.

Weltzien, F.A., E. Andersson, Ø. Andersen, K. Shalchian-Tabrizi and B. Norberg. 2004. The brain–pituitary–gonad axis in male teleost, with special emphasis on flatfish (Pleuronectiformes). Comp. Biochem. Physiol. A 137: 447–477.

Wendelaar Bonga, S.E. 1997. The stress response in fish. Physiol. Rev. 77: 591–625.

White, S.A., T.L. Kasten, C.T. Bond, J.P. Adelman and R.D. Fernald. 1995. Three gonadotropin-releasing hormone genes in one organism suggest novel roles for an ancient peptide. Proc. Natl. Acad. Sci. USA 92: 8363–8367.

White, S.A. and R.D. Fernald. 1998. Genomic structure and expression sites of three gonadotropin-releasing hormone genes in one species. Gen. Comp. Endocrinol. 112: 17–25.

Winberg, S. and P.-O. Thörnqvist. 2016. Role of brain serotonin in modulating fish behavior. Curr. Zool. 62: 317–323.

Wullimann, M.F. 1998. The central nervous system. pp. 245–282. *In*: Evans, D.H. [ed.]. The Physiology of Fishes. CRC Marine Science Series. CRC Press, Boca Raton, Florida, USA.

Wunderink, Y.S., S. Engels, S. Halm, M. Yúfera, G. Martínez-Rodríguez, G. Flik et al. 2011. Chronic and acute stress responses in Senegalese sole (*Solea senegalensis*): the involvement of cortisol, CRH and CRH-BP. Gen. Comp. Endocrinol. 171: 203–210.

Wunderink, Y.S., G. Martinez-Rodriguez, M. Yúfera, I.M. Montero, G. Flik, J.M. Mancera et al. 2012. Food deprivation induces chronic stress and affects thyroid hormone metabolism in Senegalese sole (*Solea senegalensis*) post-larvae. Comp. Biochem. Physiol. A 162: 317–322.

Yamamoto, K. and P. Vernier. 2011. The evolution of dopamine systems in chordates. Front. Neuroanat. 29: 21.

Yamamoto, N., Y. Oka, M. Amano, K. Aida, Y. Hasegawa and S. Kawashima. 1995. Multiple gonadotropin-releasing hormone (GnRH) immunoreactivity system in the brain of the dwarf gourami, *Colisa lalia*: Immunohistochemistry and radioimmunoassay. J. Comp. Neurol. 355: 354–368.

Yamamoto, N., Y. Oka and S. Kawashima. 1997. Lesions of gonadotropin releasing hormone-immunoreactive terminal nerve cells: effects on the reproductive behavior of male dwarf gouramis. Neuroendocrinology 65: 403–412.

Yang, B., Q. Jiang, T. Chan, W.K.W. Ko and A.O.L. Wong. 2010. Goldfish kisspeptin: molecular cloning, tissue distribution of transcript expression, and stimulatory effects on prolactin, growth hormone and luteinizing hormone secretion and gene expression via direct actions at the pituitary level. Gen. Comp. Endocrinol. 165: 60–71.

Yaron, Z., G. Gur, P. Melamed, H. Rosenfeld, A. Elizur and B. Levavi-Sivan. 2003. Regulation of fish gonadotropins. Int. Rev. Cytol. 225: 131–185.

Zohar, Y., A. Elizur, N.M. Sherwood, J.F. Powell, J.E. Rivier and N. Zmora. 1995. Gonadotropin-releasing activities of the three native forms of gonadotropin releasing hormone present in the brain of gilthead seabream, *Sparus aurata*. Gen. Comp. Endocrinol. 97: 289–299.

Zohar, Y., J.A. Munoz-Cueto, A. Elizur and O. Kah. 2010. Neuroendocrinology of reproduction in teleost fish. Gen. Comp. Endocrinol. 165: 438–455.

B-1.3

Reproductive Physiology and Broodstock Management of Soles

Evaristo L. Mañanós Sánchez,[1,*] *Neil Duncan,*[2]
Carmen Sarasquete Reiriz,[3] *Angel García López,*[4]
José M. Guzmán[1,5] and *Jose Antonio Muñoz Cueto*[6]

1. Introduction

The understanding of a species reproductive biology is one of the corner stones that supports a sustainable aquaculture industry. Sustainable aquaculture can only be achieved when reproduction is controlled to provide good quality eggs that are used in hatcheries to provide the industry with a supply of juveniles for grow out to market-size. The complete control of reproduction provides good quality eggs all-year-round from all generations (Mañanós et al. 2009, Duncan et al. 2013). Initially eggs are obtained from wild caught breeders and subsequently eggs are from generations hatched and reared in captivity to close the life cycle. Once the life cycle is closed, genetic breeding programs and domestication

[1] Institute of Aquaculture of Torre la Sal (IATS-CSIC), Spanish Council for Scientific Research (CSIC), 12595-Ribera de Cabanes, Castellón, Spain.
[2] IRTA Sant Carles de la Rapita, AP200, 43540-Sant Carles de la Rápita, Spain.
[3] Institute of Marine Sciences of Andalucia (ICMAN-CSIC), Spanish Council for Scientific Research (CSIC), 11510-Puerto Real, Cádiz, Spain.
[4] Department of Biomedicine, Biotechnology and Public Health, Faculty of Sciences, University of Cádiz, E11510-Puerto Real, Cádiz, Spain.
[5] School of Aquatic and Fishery Sciences, University of Washington, Seattle, WA 98195-5020 USA.
[6] Department of Biology, Faculty of Marine and Environmental Sciences, CEIMAR, University of Cádiz, E11510-Puerto Real, Cádiz, Spain.
* Corresponding author: evaristo@iats.csic.es

can be initiated. With properly managed breeding programs, the industry operates independently from the natural resources of a species and the capture of wild fish is not required. In addition to achieving a closed life cycle and breeding programs, if necessary, reproduction can be controlled to provide a supply of eggs all-year-round or when market demand dictates (Mañanós et al. 2009, Duncan et al. 2013). However, many species exhibit reproductive dysfunctions in captivity that hinder achieving the reproductive control required for sustainable aquaculture (Zohar and Mylonas 2001, Mañanós et al. 2009). The cause of these reproductive dysfunctions is the culture environment, which may have important differences from the natural environment and these differences can arrest the progress of maturation. Therefore, the understanding of a species reproductive biology is essential to achieve reproductive control in captivity. In this respect, the aquaculture industry has been a constant driving force that stimulates high quality research into the reproduction of commercial species with good aquaculture potential.

In Europe, the soles are important commercial species that are highly appreciated for their organoleptic qualities and that consequently command high market prices despite of a good fisheries production that peaked in the early 1990s with annual catches of 54,000 to 67,000 t (FAO 2018). These attributes of a large market with high demand and price made soles amongst the marine species that attracted early interest as a potential aquaculture species. During the 1970s and 1980s research began to focus on reproduction of common sole (*Solea solea*), together with a low level of aquaculture production (FAO 2018). After initial success with common sole, interest in Senegalese sole (*Solea senegalensis*) initiated similar research into reproduction during the 1980s and 1990s. These early research efforts essentially established that wild caught breeders spawned spontaneously in captivity, with no apparent reproductive dysfunction, to provide the quantity and quality of eggs required for aquaculture of common sole (Devauchelle et al. 1987, Baynes et al. 1993, Imsland et al. 2003) and Senegalese sole (Dinis et al. 1999, Imsland et al. 2003, Anguis and Cañavate 2005). However, as work continued to examine reproduction in the first generations produced in captivity it was found that common and Senegalese sole reared in captivity failed to produce viable fertilized eggs. Although, this situation is not unique to these fish species, the sharp contrast from large qualities of fertilized eggs from wild caught breeders that are relatively easy to establish as a spawning broodstock to the first generation of cultured broodstocks that spawn considerable quantities of eggs that are not fertilized is remarkable. This contrasting problem and the need for a solution to make sole aquaculture sustainable has stimulated an enormous research effort backed by a participating aquaculture industry. Essentially the research effort has focused on two goals, (1) repairing the reproductive dysfunction to provide spontaneous tank spawning of fertilized eggs from cultured broodstock and (2) applying *in vitro* fertilization techniques on a commercial scale. Working towards these two goals has addressed almost every aspect of reproductive development of sole and has achieved probably the most complete description of this type of reproductive dysfunction and undoubtedly the most complete description of the uncommon semi-cystic testis development that is a characteristic of these sole species. This chapter provides a state-of-the-art review of sole reproduction including the anatomy, morphology and maturation of the gonads, the annual reproductive cycles, the endocrinology of reproduction and the hormonal stimulation of reproduction. The chapter covers work on both sole species, but especially on the Senegalese sole that has received a far greater research effort, as it has become the favoured species for aquaculture development due to better growth rates compared to common sole and perhaps a greater industrial interest in sole culture in the Iberian Peninsula.

2. Anatomy, morphology and maturation of the gonads

The anatomical and morphological characteristics of the gonads and the processes of development and maturation has been studied in detail in Senegalese sole. The characteristics of the ovary are similar to those of other fish but, on the contrary, testis development shows significant peculiarities in structure and mostly, on the type of development, which is semi-cystic in Senegalese sole as opposed to the typical cystic type of development in most teleosts. Special emphasis has been put on investigating the male reproductive system and the process of spermatogenesis in Senegalese sole, to obtain valuable comparative information with other fish models.

2.1 Gonad development and sex differentiation

The undifferentiated gonads of fish, like other vertebrates, differentiate into ovaries or testes during ontogenesis, a process that is driven by genetic and environmental factors (Piferrer 2001, Devlin and Nagahama 2002, Penman and Piferrer 2008, Piferrer and Guiguen 2008, Nakamura 2013, Martinez et al. 2009, 2014). Sexual characteristics and gonad development vary greatly between fish species. In addition to gonochorism, several types of hermaphroditism (i.e., protandry, protogyny, synchronous) are found in fish (Devlin and Nagahama 2002). Interestingly, the formation of gonad primordia during ontogenesis has important consequences for sex differentiation and careful histological observations of the process of morphogenesis of the gonad area is of primary importance for a precise understanding of the mechanisms of gonad sex differentiation (Nakamura et al. 1998, Devlin and Nagahama 2002, Nakamura 2013).

It has long been stated in fish, that gem cells in putative ovaries outnumber those in putative testes just after hatching and mitosis of germ cells takes place earlier in ovaries (before hatching) than in testes. Meiotic nuclear changes also occur far earlier in the ovary than in testes, during initial phases of gonad differentiation. Therefore, the early appearance of pre-meiotic germ cells is one of the criteria for morphological judgement of initial ovarian differentiation in fish (Nakamura 2013). In addition, the formation of the ovarian cavity, derived from the development of stromal tissue and located at the cranial region of the ovary during early development, is another sign of ovarian development and this cavity can appear coinciding with the presence of pre-meiotic germ-cells or synaptic oocytes, taking place after ovarian differentiation (Yamamoto 1969, Satoh and Egami 1972, Nakamura et al. 1998, Devlin and Nagahama 2002). For most teleost fish, a noticeable increase in the number of germ cells, prior to their meiotic changes, is one of the essential signs of ovarian differentiation. On another hand, it is difficult to detect positively morphological testicular differentiation by the developmental characteristics of germ cells, because they usually remain quiescent for a long period of time before starting the process of spermatogenesis (i.e., first transition of spermatogonia (spg) to spermatocytes (spc)). Furthermore, the formation of efferent ducts and increases of both vascular and connective tissue, are important criteria by which testicular differentiation can be recognized (Devlin and Nagahama 2002, Nakamura 2013).

Soles are differentiated gonochoristic species; sexually undifferentiated gonads develop directly into either testes or ovaries and thus, no transitional ovarian-like to testis gonads are observed. The process and timing of sex differentiation has not been yet studied in detail in any *Solea* spp. There are only two studies providing histological information on gonad differentiation in Senegalese sole, reporting that in this species gonad differentiation occurs sometime around the third month of life (Viñas et al. 2013, Pacchiarini et al. 2013).

Viñas et al. (2013) described the histology of individuals from 98 d post-fertilization (dpf) onwards (rearing temperature was 20 ± 1°C) and showed that, at first sampling (TL 33.31 ± 0.89 mm), 40 percent of the examined fish were sexually undifferentiated and 60 percent were differentiating females, whereas differentiating males (i.e., gonads with clearly discernible spg) only appeared in the next sampling point at 127 dpf (TL 44.50 ± 0.83 mm). These authors observed the presence of undifferentiated gonads until 169 dpf (TL 48.32 ± 1.03 mm), but not at 199 dpf. Considering these data, the authors concluded that gonad differentiation in this species begin earlier in females than in males and that sex differentiation started sometime before 98 dpf and finished sometime between 169 and 199 dpf. They also observed that an ovarian cavity is clearly visible in juveniles at 246 dpf, but barely identified at 98–127 dpf, when oocytes were already present in the developing gonads. Anatomical differentiation of female gonads, consisting in the appearance of an ovarian cavity, before oogonia enter meiosis, is a common sign used to identify ovarian differentiation in fish, including flatfishes (Hendry et al. 2002, Radonic and Macchi 2009, Nakamura 2013). The study by Pacchiarini et al. (2013), identified genital ridges and/or differentiating males (e.g., proliferative germinal cells (PGCs), undifferentiated germinal cells (GCs) and spg) at 120–130 d post-hatching (dph) (TL 40–50 mm; rearing temperature 19 ± 1°C) and a clear differentiation of females (e.g., presence of PGCs, oogonia and chromatin nuclear oocytes) at 150–160 dph (TL 55–65 mm). Pachiarini (2012) showed, by *in situ* hybridization and immunohistochemistry, that all the proliferating and differentiating sex cells, from both males and females, were positive to germ and sex markers (i.e., Vasa gene, ATP-Helicases, DEAD family), similarly as previously described in other fish (Nagahama and Devlin 2002).

Some gonochoristic fish start the process of sex differentiation, in both sexes, at a similar age or developmental time, while in others it starts earlier in one sex than in the other. The differentiation and development of the gonads in Senegalese sole continues during juvenile growth and, at around 250 dph, ovaries with late perinucleolar oocytes and testes containing spc can be readily found. According to this information, the process of sex differentiation in Senegalese sole follows a similar pattern to that observed in other flatfish species (Devlin and Nagahama 2002, Nakamura 2013) and it can be considered roughly that it starts at around 90 dph and it is practically completed, in both males and females, at around the fifth month of life. It should be pointed that the available information in Senegalese sole does not permit to ensure that gonad differentiation occurs earlier in one sex than in the other.

In addition to a genetic determination of sex, different zootechnical and environmental factors have been shown to influence and/or to be related to sex determination in fish, such as, temperature, light, nutrition, origin of broodstook, synthesis of steroids and exposure to endocrine disruptors (Piferrer 2001, Devlin and Nagahama 2002, Penman and Piferrer 2008, Piferrer and Guiguen 2008, Martínez et al. 2009, 2014, Nakamura 2013, Portela-Bens et al. 2017). Studies in Senegalese sole have shown an influence of rearing water temperature on sex determination, with high temperatures having a masculinizing effect (around 60 percent of males at 20°C) compared to the 1:1 sex ratio normally obtained at 16°C (Blanco-Vives et al. 2011, Viñas et al. 2013), which is in agreement with previous observations in other Pleuronectiform species (Ospina-Álvarez and Pifferrer 2008). The mechanism of sex determination has not been established in any *Solea* species (Howell et al. 1995), but it appears to be regulated by both genetic and environmental factors, as is the case in many fish. Also, the process of sex differentiation in fish has been related to the synthesis of different hormones and the size of the individuals (Piferrer and Guiguen 2008) and particularly in Senegalese sole, a minimum size of larvae of around 33–40 mm has been correlated with the initiation of the process of sex differentiation (Blanco-Vives et al. 2011, Pacchiarini 2012, Viñas et al. 2013, Martinez et al. 2014). It is well known that Senegalese sole, as happens in other fish, exhibits sex related differences in growth rates, with females growing faster than males. Thus, development of environmental protocols to produce monosex or female-predominant populations could have important technological implications for the aquaculture industry, reducing significantly the time to reach market size (Howell et al. 1995).

2.2 Morphoanatomy of the testis

The testes of Senegalese sole are among the smallest compared to other flatfishes of similar body size. The Senegalese sole is a dextral flatfish, with the gonads located at the upper-posterior region of the body cavity. Characteristic features of the male reproductive system are the rounded shape and the small size of the testicular lobes. The testis is composed of two whitish and asymmetric lobes attached to the visceral cavity by mesenteric tissue. The testicular lobe from the ocular side is round- and flat-shaped and located over the anterior region of the central skeletal portion, while the lobe at the blind side is round and conic-shaped and situated in a ventral position from the beginning of the central skeletal portion.

The round shape of the Senegalese sole testis lobes is quite different to the elongate shaped lobes observed in other Pleuronectiformes (Sol et al. 1998, Weltzien et al. 2002) and seems to be characteristic of the genus *Solea*, as observed in common sole (Baynes et al. 1994, Bromley 2003). The ocular side testicular lobe is normally larger than the blind side lobe and overall, both testis lobes account for less than 0.2 percent of the total weight of the animal; the gonadosomatic index (GSI) in male Senegalese sole adults ranges from 0.05 to 0.17 during the year (Rodríguez 1984, García-López et al. 2005, 2006a, Chauvigné et al. 2015), similar to values described also in common sole and Egyptian sole *Solea aegyptiaca* (Baynes et al. 1994, Bromley 2003, Ahmed et al. 2010). This low GSIs are an exception, because other flatfishes have GSI values of 2–5 during the spawning season (Barr 1963, Weltzien et al. 2002, Koya et al. 2003). These differences are probably related to the different type of spermatogenesis and testicular maturation dynamics observed in Senegalese sole compared to other flatfishes.

The deferent ducts emerging from each testicular lobe fuse in the anterior limit of the central skeletal portion constituting the spermatic duct that opens through the urogenital pore, located in the ocular side very close to the pelvic fins. From a structural point of view, the testis of Senegalese sole is organized in two main regions, cortex and medulla (García-López et al. 2005). In the cortical region, the tunica albuginea, a thin connective capsule that covers the entire testis lobe, can be observed, as well as the seminiferous lobules, the main location of the spermatocysts, formed by germ cells surrounded by the cytoplasmic extensions of the somatic Sertoli cells. The spermatocysts are distributed all along the seminiferous lobules, as corresponds to the unrestricted spermatogonial type (Grier 1993, Schulz et al. 2010). However, the majority of spermatocysts containing spg type A (spg-A) are found at the distal part of the lobules of Senegalese sole testis, which may be reported as an intermediate type between the restricted (i.e., all the spg cysts are located in the periphery of the testis) and the unrestricted types (Selman and Wallace 1986, Grier 1993). In the medullar region, developing spermatocysts are scarce or inexistent and it is possible to find the efferent duct system that collects and stores the spermatozoa (spz). The inner surface of these efferent ducts is fully covered by epithelial cells, which as suggested in other fish species (Lahnsteiner et al. 1990, Grier 1993, Manni and Rasotto 1997), are thought to be derived from Sertoli cells. Interestingly, it has been shown that these Sertoli-like epithelial cells possess a complex glycosylation pattern, which may support their role in

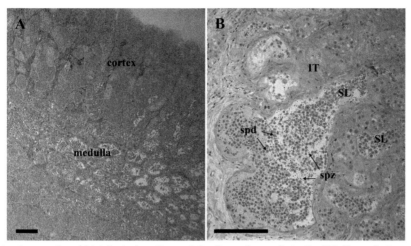

Fig. 1. Photomicrographs of histological sections of the Senegalese sole testis. (A) General view of the testis showing the regionalization in the outer cortex and the inner medulla. (B) The semi-cystic type of testis development in *Solea* species is characterized by the presence in the lumen of the seminiferous lobules (SL) of both spermatids (spd) and spermatozoa (spz). IT, interstitial tissue. Scale bars: 100 μm (Photos by Evaristo L. Mañanós).

the synthesis and secretion of glycoproteins into the lumen, required for final germ cell maturation and storage (Desantis et al. 2010).

Despite the differences in external morphology between the Senegalese sole testis and the vast majority of teleosts, some similarities are evident when observing histological sections. The testis of Senegalese sole presents a radial or reniform structure with two main regions, the cortical one where germ cells are produced and the medullar one, with the efferent duct system that collects and stores the sperm (Fig. 1); a structure that is similar to that reported in other fish (Yoneda et al. 1998, Brown-Peterson et al. 2002, Abascal et al. 2004). Therefore, the function changes from germ cell production in the outer region to sperm collection and storage in the central medulla.

The seminiferous lobules radiate from the medullar efferent ducts (without crossing each other) terminating blindly beneath the tunica albuginea at the cortex. The seminiferous lobules and the efferent ducts are surrounded and structurally supported by the testicular interstitial tissue (or stroma), based on branches of connective tissue that run from the tunica albuginea into the cortex and medulla. Within the interstitial tissue, myoid cells, collagen fibres, Leydig cells, amyelinic nerves and capillaries can be observed. Within the germinal compartment of Senegalese sole testis, the cytoplasmic extensions of Sertoli cells only enclose spg and spc, since at a certain time during the end of meiosis, the spermatocysts open and the spermatids (spd) are released into the seminiferous lobule lumen, where they differentiate into spz (Fig. 1). This feature makes that Senegalese sole spermatogenesis should be classified as of the semi-cystic type (no information is available for other species of the genus *Solea*) as termed by Mattei et al. (1993) for *Ophidion* spp. The semi-cystic type of testis has only been described for other two fish species, *Scorpaena notata* (Muñoz et al. 2002) and *Lophiomus setigerus* (Yoneda et al. 1998), while all other studied teleosts exhibit the cystic type of spermatogenesis, in which the developing cysts open at the transition between spd and spz transformation. This specific type of spermatogenesis found in Senegalese sole determines the dynamics of testicular development and maturation along the annual reproductive cycle. A description of the different cell types found in the testis of Senegalese sole is given in Table 1, as modified from García-López et al. 2005.

Table 1. Morphological features of Senegalese sole male germ cells, Sertoli cells and Leydig cells. Data of diameter are given as mean ± SEM.

Cell type	Nucleus diameter (μm)	Cell characteristics
Type A spermatogonia (spg-A)	8.44 ± 0.48	Cells are oval-shaped with a small cytoplasm and are individually surrounded by Sertoli cell processes. The nucleus is slightly basophilic and contains a single prominent and strongly basophilic central nucleolus.
Type B spermatogonia (spg-B)	5.86 ± 0.21	Similar aspect to spg-A, but the nucleus is smaller and more basophilic and contains several nucleoli. Found in small/intermediate Sertoli cell-enclosed groups resulting from successive mitosis of spg-A.
Primary spermatocytes (spc-I)	5.65 ± 0.11	Similar shape and size than spg-B and found inside spermatocysts. Clear cytoplasm with faint cellular limits. At the prophase of the first meiotic division, the nucleus has an irregular/granular texture, due to the distribution of the genetic material.
Secondary spermatocytes (spc-II)	3.46 ± 0.09	Rarely observed, as they enter rapidly in the second meiotic division. Found inside spermatocysts. The nucleus contains condensed and basophilic chromatin.
Spermatids (spd)	2.31 ± 0.05	Round-shaped, with scarce cytoplasm and a highly basophilic nucleus. They are released into the lobule lumen, where they differentiate into spz. It is the most abundant germ cell type in the testis.
Spermatozoa (spz)	1.04 ± 0.04	Small, rounded and strongly basophilic head. Tail formed by one large acidophilic flagellum.
Sertoli cells	not measured	Triangular and slightly basophilic nucleus, containing a single nucleolus. Clear cytoplasm difficult to distinguish. Usually in contact with the basement membrane and surrounding germ cells (spg, spc).
Leydig cells	3.18 ± 0.14	Spherical and basophilic nucleus with one or several nucleoli. Clear cytoplasm with a fine acidophilic dotted pattern. Found in small groups near blood capillaries within the interstitial compartment.

2.3 Testicular development

After sex differentiation, when a testicular tissue containing fully differentiated spg can be distinguished histologically (around 3-month old fish), the testis remains in a quiescent stage until puberty which, in Senegalese sole, occurs at 2 or 3-years old fish. Adult males are spermiating all year round, with highest maturation and fluency attained in spring and autumn, correlating with the female spawning periods (Rodríguez 1984, Anguis and Cañavate 2005, García-López et al. 2005).

Based on the histological changes and the relative abundance of the different germ cell types in the gonad (García-López et al. 2006a), the testicular development of Senegalese sole can be classified in five stages (I–V):

Stage I (early spermatogenesis): the cortex is occupied by numerous germinal cysts containing spg surrounded by Sertoli cells. Few spd are present in the lumen of the seminiferous lobules. Few spd and spz in the medullar efferent ducts, which diameter is quite short. Abundant empty spaces observed in both regions. This stage is found from summer to autumn and characterized by the lowest GSI and sperm release capacity.

Stage II (mid spermatogenesis): germ cells at all developmental stages are present. Meiosis is initiated in the cortex, as evidenced by the declined number of spg and the appearance of a big population of spc. Germinal cysts containing spc distributed in the periphery of the seminiferous lobules, leaving a small central lumen filled by spd (there are no empty spaces). Some spd and spz observed in the medullar efferent ducts. This stage is prevalent from autumn to late winter and characterized by intermediate GSI and sperm release capacity.

Stage III (late spermatogenesis): spd are the most abundant cell type inside the testis; the spd fill the lumen of the cortical seminiferous lobules. There are few cysts containing spc. The spz become more abundant in the medullar efferent ducts. This stage is found from winter to early spring and characterized by intermediate GSI and sperm release capacity.

Stage IV (full maturation): spd are still very abundant in the cortical seminiferous lobules at initial phases of maturation, but their number decreases progressively as successive batches transform into spz. The diameter of the medullar efferent ducts increases considerably, as ripe spz accumulate in large quantities. The spg and associated Sertoli cells start to proliferate again at the distal part of the cortical lobules, beneath the tunica albuginea. This stage is found in spring and is characterized by the highest GSI and sperm release capacity.

Stage V (recovery): there is a strong reduction of spd in the cortex and a clear decrease in the number of spz in the medullar efferent ducts, compared to stage IV. Numerous groups of spg and Sertoli cells appears at the distal part of the cortical seminiferous lobules, indicating active proliferation. No spc are observed, indicating the absence of meiosis. Stage found in spring and summer and characterized by low GSI and sperm release capacity.

Comparing the testicular development of Senegalese sole with the testis recrudescence dynamics of other flatfish species, it is possible to identify important differences, especially at late spermatogenesis and full maturation stages, which are probably related to the semi-cystic type of spermatogenesis reported in Senegalese sole. In flatfishes exhibiting the cystic type of spermatogenesis, e.g., winter flounder *Pseudopleuronectes americanus* (Harmin et al. 1995), English sole *Parophrys vetulus* (Sol et al. 1998), Atlantic halibut *Hippoglossus hippoglossus* (Weltzien et al. 2002), spotted halibut *Verasper variegatus* (Koya et al. 2003), spd are only abundant during a short period of time, since they transform into spz (i.e., spermiogenesis) being released into the lumen of seminiferous and spermatic ducts massively (i.e., spermiation). Thus, at the final phases of testicular development and during the spermiation period, the testis accumulates a large quantity of sperm, that can be easily stripped; the size and weight of the organ increases until values accounting for up to 5–6 percent of total body weight of the fish (Barr 1963, Harmin et al. 1995, Sol et al. 1998, Weltzien et al. 2002, Koya et al. 2003). On the contrary, in the testis of Senegalese sole, meiotic activity leads to the appearance of a large population of spd that accumulate in the cortical seminiferous lobules at mid and late spermatogenesis. At full maturation, spermiogenesis appears to happen gradually in successive batches. This mechanism explains the large number of spd that are present through all the stages of spermatogenesis and the low quantity of sperm (few tens of microlitres) that could be stripped by abdominal pressure at a particular point in time in mature Senegalese sole (García-López et al. 2005, 2006b). In addition, this low accumulation of sperm in the testicular efferent ducts makes the increase in size/weight during full maturation to be quite modest and explain the reduced testis size during the whole reproductive cycle of Senegalese sole. These observations are also in agreement with those reported for the common sole, in which the testis appear to store only a few hundred microlitres of semen (Baynes et al. 1994).

2.4 Morphoanatomy of the ovary

The macroscopic morphology of the Senegalese sole ovary corresponds with the general description of teleost cyst-ovarian type (Dodd 1977) and has been described in detail by different authors (Dinis 1986, 1992, Rodríguez 1984, García-López 2005). The ovary consists of two independent ovarian lobes, an upper lobe in the ocular side and a lower lobe in the blind side, which run longitudinally all along the visceral cavity at both sides of the central skeletal portion. The color of the ovary ranges from pale or whitish tones at immature stages, to orange at advanced maturation stages. The two lobes are asymmetrical, with the ocular side lobe being normally around 1.5-fold bigger than the blind side lobe. As a whole, the ovary may account for up to 15–20 percent of the total body weight of the fish during late maturation and spawning periods (Rodríguez 1984, Anguis and Cañavate 2005, García-López et al. 2007, Tingaud-Sequeira et al. 2009). Both lobes remain fully enclosed by the peritoneum, which forms a deferent duct in the ventro-cephalic region of each lobe. Both ducts fuse in the anterior limit of the central skeletal portion constituting the oviduct, which goes along this limit in ventro-cephalic sense until opening outwards through the genital pore, located very close to the pelvic fins in the blind side. The ovarian artery, which ramifications irrigate the ovary, runs longitudinally along the dorsal surface of each ovarian lobe, between the peritoneum and the ovarian wall. Numerous nerves can be observed in the ovarian artery vicinity. Each ovarian lobe has a hollow space inside it, the ovarian cavity, which is located in the ventral region of the lobe. Numerous ovigerous folds or lamellae project into the ovarian cavity. These lamellae consist of connective tissue lined by germinal epithelium in which oocytes develop. During their development, the oocytes are enclosed individually by the ovarian somatic follicular cells, forming as a whole the so-called follicular complex. The follicle itself is composed of two cellular monolayers named granulosa, the inner layer directly in contact with the oocyte and secreting the basal membrane and theca, the outer layer organized around the basal membrane (Grier 2000).

2.5 Ovarian development

Puberty in Senegalese sole females occurs at 3 to 4-year old fish, depending on size and individuals. The adult breeders then enter in the annual reproductive cycles, with a main spawning period in the spring and a minor spawning period in autumn, although in some culture conditions spawning can be a continuous event from early spring to mid-autumn (Rodríguez 1984, Dinis et al. 1999, Anguis and Cañavate 2005, García-López 2005, Guzmán 2010). During this long spawning period, female Senagelese sole spawns repetitively and the ovary contains oocytes at all developmental stages. The oocytes are in clearly defined groups of development, including a group of advanced mature oocytes ready for ovulation and a group of post-vitellogenic oocytes ready to enter into maturation and replace the group of present mature oocytes (Fig. 2). These characteristics correspond to a group-synchronous multiple spawning fish (Rodríguez 1984, Mañanós et al. 2009, Guzmán 2010).

The morphology and characteristics of the different oocyte cell types has been studied in detail in Senegalese sole (García-López 2005, García-López et al. 2007, Guzmán 2010) and showed to follow the general features of oocyte development described in the fish ovary (Tyler and Sumpter 1996, Grau et al. 1996, Núñez and Duponchelle 2009, Grier 2012). Accordingly, eleven oocyte developmental stages have been identified in the ovary of Senegalese sole, as described in Table 2 (modified from García-López et al. 2007).

The development and maturation of the ovary has been described in Senegalese sole (García-López 2006, García-López et al. 2007, Guzmán 2010). Based on macroscopic

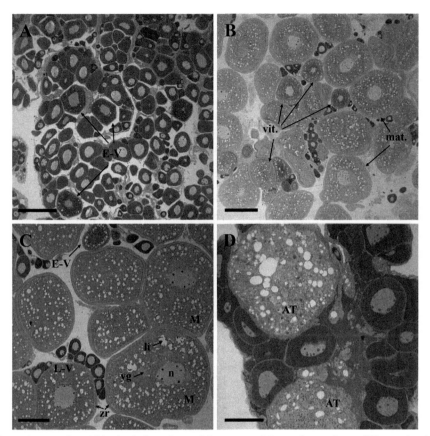

Fig. 2. Photomicrographs of histological sections of the ovary of Senegalese sole, at different developmental stages. (A) Immature ovary, containing pre-vitellogenic oocytes (oogonia, chromatin nucleolar, perinucleolar) and some scattered oocytes entering the process of vitellogenesis (E-L; early-vitellogenic). (B) general view of a fully developed ovary illustrating the group-synchronous type of development, where pre-vitellogenic and vitellogenic oocytes (vit.) at different developmental stages coexist with an advanced batch of post-vitellogenic oocytes initiating the process of maturation (mat.). (C) Mature ovary, characterized by the presence of a batch of mature oocytes (M; see the eccentric position of the nucleus (n) and the coalescence of the yolk granules (yg) and lipid inclusions (li)) together with late-vitellogenic oocytes (L-V; centrally located nucleus, smaller cell size and thinner zona radiate (zr)). (D) Regressed ovary, typical at the post-spawning period, containing immature oocytes and in some cases, scattered atretic oocytes. Scale bars: 100 μm (D), 200 μm (A, C), 400 μm (B) (Photos by Evaristo L. Mañanós).

characteristics of the ovaries and the microscopic observation (histology) of the most advanced oocyte stage present within the ovary and following similar criteria than previous studies in fish (Núñez and Duponchelle 2009, Mañanós et al. 2009), the ovarian development in Senegalese sole can be classified in five stages (1–5):

Immature/recovery (stage 1): undeveloped small ovaries containing nests of oogonia (Oog) and some chromatin nucleolar (CN) and perinucleolar (PN) oocytes.

Vitellogenic (stage 2): still small ovaries but clearly developing, with the presence of cortical alveoli (CA) and early vitellogenic (E-V) oocytes.

Late vitellogenic (stage 3): fully developed and voluminous ovaries containing oocytes at different vitellogenic stages but predominantly late vitellogenic (L-V) oocytes.

Table 2. Oocyte developmental stages in Senegalese sole and morphological features of the different cell types. Data are given in microns (mean ± SEM).

Cell type	Cell diameter range	Cell characteristics
Oogonia (Oog)	8 – 12 (10 ± 0.4)	Large and slightly basophilic nucleus with a prominent central nucleolus. Clear and scarce cytoplasm. Found isolated or in small nests in the ovarian epithelium.
Chromatin nucleolar (CN)	12 – 23 (16 ± 0.8)	Moderately basophilic cytoplasm. Nucleus containing a prominent nucleolus surrounded by chromatin threads.
Early perinucleolar (E-PN)	17 – 51 (32 ± 0.6)	Highly basophilic cytoplasm, with zones less basophilic (pallial substance of Balbiani's body) close to the nucleus. One big highly basophilic nucleolus.
Late perinucleolar (L-PN)	32 – 184 (101 ± 2.3)	The cytoplasmic basophilia starts to decrease. Nucleus with numerous, relatively large, basophilic nucleoli, peripherally distributed. Developing follicular layer.
Cortical alveoli (CA)	124 – 287 (193 ± 3.4)	Numerous vacuoles (cortical alveoli) appear in the periphery of the cytoplasm, also occupied by lipid inclusions. Zona radiata is visible for the first time. Clearly distinguished follicular layer surrounding the oocyte.
Early vitellogenic (E-V)	240 – 413 (309 ± 3.8)	Appearance of small and acidophilic yolk granules in the outer cortex of the cytoplasm. Numerous larger lipid inclusions or globules scattered within the cytoplasm. The zona radiata becomes thicker.
Late vitellogenic (L-V)	325 – 569 (458 ± 4.5)	Yolk granules spread inwards as they increase in number and size, occupying most of the cytoplasm. Lipid inclusions and the zona radiata reach their maximum development.
Early mature (E-M)	443 – 597 (528 ± 6.0)	Migration of the nucleus towards the animal pole and coalescence of the yolk granules and lipid inclusions.
Late mature (L-M)	580 – 1000 (880 ± 45)	Breakdown of the nucleus (GVBD), which is no longer observed. Partial to total coalescence of yolk granules and lipid. Oocyte hydration.
Atresia (AT)	not measured	Fragmentation of the nuclear membrane and zona radiata. Disorganized and highly vacuolated cytoplasm. Hypertrophied follicular layer invading the cytoplasm.
Post-ovulatory follicle (POF)	not measured	Follicular layers that remain after ovulation. Irregular shaped containing a large, convoluted and thick cell layer.

Mature (stage 4): corresponding to highest GSI and characterized by the presence in the ovary of batches of oocytes entering the process of maturation (migrating or peripheral position of the germinal vesicle, coalescence of the cytoplasmic yolk and lipid inclusions and hydration), together with L-V oocytes (as well as other previous vitellogenic stages).

Spawning (stage 5): similar to the previous stage, but the ovarian lobes become less compact due to hydration and increase in ovarian fluid; it is characteristic the presence of post-ovulatory follicles (POF) within the ovary, indicating previous spawning events.

The analysis of the spatial distribution of the different oocyte developmental stages along each ovarian lobe and between ovarian lobes shows that follicles are distributed homogeneously within the ovary. In Senegalese sole, primary oocytes, including CN and PN stages, are present in the ovary during the whole year, although with higher relative proportions from early summer to early winter. This indicates the continuous availability of germ cells for recruitment, which is of primary importance in a multiple-spawning species and also, that oogenesis (i.e., transformation of oog into primary oocytes) occurs cyclically and is mostly concentrated just at the end of the spawning season. Follicles at CA stage and E-V are observed in the ovary (i.e., vitellogenic females) of Senegalese sole from autumn to spring, indicating the main period of follicular growth; during this period, the number of late vitellogenic follicles increase gradually as an anticipation of the main spawning season in spring. In this regard, the transition from CA to L-V follicles in the ovary of Senegalese sole seems to proceed quickly, as denoted by the much lower relative proportions of oocytes at CA and E-V compared to L-V follicles, a feature probably related to the continuous and rapid follicular recruitment and growth required to produce multiple spawns over the reproductive season. During the spawning period, the presence and prevalence of mature oocytes varies greatly between females and mostly between populations, depending on the origin (wild *versus* cultured) and culture conditions (Rodriguez 1984, Anguis and Cañavate 2005, García-López 2005). Under appropriate culture conditions, spawning will occur and the ovary will contain two main batches of oocytes, one consisting of vitellogenic and post-vitellogenic oocytes and another one corresponding to mature oocytes. Under inappropriate rearing conditions, reduced or absence of egg release is observed and the ovary will be characterized by a great proportion of vitellogenic and post-vitellogenic oocytes, but low or none number of mature oocytes and also, a number of atretic oocytes, which increase concomitantly with the progression of the spawning period.

The histochemical composition of different macromolecules (proteins, lipids, glucids, glycoproteins, vitellogenin, etc.), that are major constituents of the granular cytoplasm of previtellogenic, vitellogenic and maturating oocytes, as well as of the oocyte envelopes (follicular and zona radiata) and the composition of yolk and histochemical changes during oogenesis of Senegalese sole have been well described (Rodriguez 1984, Gutiérrez et al. 1985, García-López 2005, Arellano and Sarasquete 2005, Accogli et al. 2012). These studies provide information on the specific lipid composition of the globules (or vacuoles) and the nature of the yolk-granules (e.g., rich in glycolipophosphoproteins), as well as the neutral-acid glycoproteins (i.e., lectin-binding affinity) present within the CA, which content must be released into the perivitelline space at the fertilization process.

3. Annual reproductive cycles

3.1 Reproduction in captivity: Wild versus F1

The culture of soles is presently based on wild broodstocks, which are adult fish captured from the wild and held in captivity. The principal difference between wild and cultured broodstocks (adult fish reared from eggs in captivity) is fertilization success. Wild broodstocks have sufficient fertilization success to provide eggs that formed the basis of the industrial production of 1510 t of Senegalese sole and 11 t of common sole in 2016 (FAO 2018). Differences in fertilization success were less pronounced in common sole compared to Senegalese sole. Wild common sole broodstocks have a percentage mean fertilization success of 60 with a range of mean percentages for an entire season from 24 to 78 (Baynes et al. 1993) compared to cultured broodstocks that had fertilization success (actual percentage success was not reported) from four different broodstocks (Palstra et al. 2015). Little work has been completed to examine differences between wild and cultured common sole broodstocks as differences appear to be minor and there is less interest to culture this slower growing species. The situation is very different for Senegalese sole, with large differences in fertilization success and a rapidly increasing interest in industrial production.

Wild Senegalese sole broodstocks have fertilization success of percentages ranging from 44.9 ± 18 to 78.46 ± 2.93 (Anguis and Cañavate 2005, Martin et al. 2014) compared to cultured broodstocks that have fertilization success of zero, both without hormones (Guzmán et al. 2008) and when stimulated with hormones (Agulleiro et al. 2006, Guzmán et al. 2011a,b). Just one fertilized spawn has been reported from a cultured broodstock (Guzmán et al. 2011a). The spawn was from a broodstock in which the males were treated with weekly injections of 1,000 IU/kg of human chorionic gonadotropin (hCG) and females with 40 µg/kg implants of gonadotropin releasing hormone analogue (GnRHa) and out of 20 spawns that were obtained, a single spawn had a fertilization success percentage of 100 (Guzmán et al. 2011a), whilst 19 spawns had zero fertilisation success. Therefore, the difference in fertilization success between wild and cultured Senegalese sole broodstocks is extreme and, despite of the lack of a properly controlled comparison with broodstocks under identical conditions, has been accepted as the state of the art. The reason for the low fertilization success from cultured broodstocks appears to be a reproductive dysfunction in cultured males that do not participate in courtship and spawning to fertilise eggs (Mañanós et al. 2007, Carazo 2013, Martin 2016; see also Behaviour chapter).

A second difference to which reference has been made is that wild broodstocks have a higher fecundity (both size and number of spawns) compared to cultured broodstocks. However, this does not appear to exist in common sole that had fecundities of 11,000 to 14,100 eggs/kg in wild (Baynes et al. 1993) and 77,700 eggs/kg in cultured broodstocks (Palstra et al. 2015). In addition, it is questionable in Senegalese sole, where wild broodstocks have given annual relative fecundities of $29,600 \pm 21,600$ egg/kg (Anguis and Cañavate 2005) compared to $22,370 \pm 12,500$ egg/kg from cultured broodstocks (Guzmán et al. 2008). Similarly, little difference was observed in fecundities of wild and cultured broodstocks held in similar conditions. In two studies performed in the same installation, 23–29 wild breeders were held in 14 m^3 tanks (Martin et al. 2014) compared to 16 cultured breeders held in a 7 m^3 tank (Martin 2016). The wild broodstocks gave mean annual relative fecundities of $6,510 \pm 200$ egg/kg compared to $6,550 \pm 540$ egg/kg from cultured broodstocks. As suggested, these comparisons are compromised as these fecundities almost certainly included varying numbers of females that did not spawn and the different studies on wild or cultured broodstocks had different characteristics and different holding conditions. Characteristics that were different between broodstocks include, older or unknown age of wild breeders, environment (temperature, density, water quality/exchange), nutrition and husbandry routines and these could explain or partly explain the differences in fecundity. Martin (2016), also described the spawning of a mixed origin broodstock that held wild males with cultured females and observed that the fecundities of this mixed broodstock ($6,510 \pm 460$ egg/kg) were similar to fecundities of wild broodstocks ($6,510 \pm 200$ eggs/kg) (Martin et al. 2014) and superior to a broodstock of cultured males with wild females ($2,860 \pm 390$ eggs/kg) (Martin 2016). This would suggest that as with fertilization success, any differences in fecundity between wild and cultured broodstocks was caused by the reproductive dysfunction of cultured males that did not stimulate spawning from the females that have the capacity to produce fecundities similar to wild females.

Consistent with the suggestion that differences between Senegalese sole wild and cultured broodstock are related to cultured male behavioural dysfunctions, is the observation that cultured broodstocks produce viable gametes. The gametes have been stripped from cultured broodstocks and fertilised *in vitro* (Rasines et al. 2012, 2013). Females were induced to ovulate with 25 µg/kg of GnRHa and had fecundities of 134,600 eggs/kg from a single ovulation, which is superior to the mean fecundities reported per female over an entire reproductive season. The sperm from cultured males was cryopreserved and used in pools to fertilize the eggs and obtain fertilization success percentages of 31 ± 3.7 when injected at 12:00 hr (Rasines et al. 2013). The quantity of sperm stripped from sole males was consistently low and wild males have been reported to produce more sperm compared to cultured males (Cabrita et al. 2006). In wild broodstocks, sperm collected was 10 to 80 µl/male with cell density of $1-2 \times 10^9$ spz/ml and sperm production (total spz per stripping per animal) of $40-60 \times 10^6$ spz/male compared to 5 to 20 µl/male, a density of $0.7-1.2 \times 10^9$ spz/ml and production of $20-60 \times 10^6$ spz/male in cultured broodstocks. However, although this difference was attributed to the origin of males it is probable that differences were also related to another aspect other than origin *per se*. For example, Cabrita et al. (2006) used cultured males that were 4-years old that had always been kept in captivity compared to wild males of unknown age that had 6 months in captivity prior to capture and the diet would have been different compared to cultured fish. In comparison a recent study on small young (354 ± 7 g and 2+ years) cultured males obtained 100×10^6 spz/male (281×10^6 spz/kg) (Chauvigne et al. 2017), which would suggest that cultured males have a potential sperm production that is similar to wild males. Sperm production or quality has not been studied in common sole.

3.2 Spermiation of males

Spermiation in Senegalese sole has been observed all-year-round in every month of the year (Anguis and Cañavate 2005, García-López et al. 2006, Beirão et al. 2011). The proportion of males with sperm was observed to vary through the year, with the highest proportion (percentage range 70–100) during late winter and early spring (January to April) and the lowest proportion (percentage range 15–50) during the summer (June to August). There was a clear association of the highest proportion producing sperm during the spawning periods of spring (February–May) and autumn (October–November). However, as mentioned above the quantities of sperm produced were always low and wild fish sperm volume varied without significant differences from 18.9 ± 3.4 µl/male in July to 32.6 ± 8.6 µl/male in December (Beirão et al. 2011). The sperm quality did vary significantly during the year with peaks in quality associated to the spawning seasons (Beirão et al. 2011). Sperm motility was highest during the cooler months (October to March) with peaks in the spawning seasons (March and October) and lowest during the warmer months May to July. Linear velocity was highest during the spring and decreased from March to July before increasing slightly from August to October and the second spawning season. Fragmentation of DNA and the percentage of cells resistant to seawater exposure also presented peaks related to the coolest months and spawning seasons.

3.3 Spawning of females

Under natural photoperiod and temperature, spawning and egg collection from broodstocks is highly seasonal in both sole species. Common sole exhibit a single spawning period in the spring, from early March through to early June (Baynes et al. 1993). The females have spawned successfully with males in a wide range of tanks with water depths ranging from 0.7 to 1 m, volume from 9 to 25 m^3, light intensity from 20 to 1,500 lx, stocking density of 1 to 6 per m^2 and a ratio of 0.5 to 3 males to females. Seasonal egg production ranged from 11,000 to 141,000 eggs/kg as indicated above, with daily production from 0 to over 120 g of eggs. Egg quality was variable and seasonal and mean fertilization percentage ranged from 24 to 78. Egg diameter ranged from 1 to 1.6 mm and declined during the spawning season. Senegalese sole have two periods of egg production with one period in the spring and a second in the autumn (Dinis et al. 1999, Anguis and Cañavate 2005). The spring spawning period is more important providing 94.6 percent of the eggs spawned during the year and just a few isolated spawns were collected in the autumn usually during October (Anguis and Cañavate 2005). The spring period is similar to common sole and begins in February or March and extends through to June to make a period of 4–5 months (Dinis et al. 1999, Anguis and Cañavate 2005). The females have spawned successfully with males in a wide range of tanks with water depths ranging from 0.7 to 1.4 m and volume from 3 to 28 m^3, light intensity from 50 to 200 lx, stocking density of 0.6 to 4.6 kg/m^2 and ratio of 0.7 to 2.3 males to females (Howell et al. 2011, Morais et al. 2016). Daily egg production was variable including many days without spawns, but generally, production is highest in the middle of the season with a roughly normal distribution in fecundities. Egg quality also presents highest values in the middle of the season, but variation is low with mean monthly fertilisation success varying from 40 to 65 percent (Anguis and Cañavate 2005). Egg diameter had means from 0.96 to 0.97 mm (Dinis et al. 1999, Anguis and Cañavate 2005) and diameter decreased from the start to the end of the spawning period (Dinis et al. 1999). In comparison to these studies on the Southern coast of Portugal (Dinis et al. 1999) and Spain (Anguis and Cañavate 2005) under natural conditions, spawning in a centre on the north coast of Spain extended from March/April through to October/November without stopping during the summer months (Martin et al. 2014). This extended spawning period was related to the lower temperatures through the summer months and possibly a constant photoperiod of 16 hr of light and 8 hr of darkness.

3.4 Spawning behaviour

Spawning behaviour is very similar in the two species, common sole and Senegalese sole. In both species after a period of interactions between males and females, a female swims from the bottom of the tank and a male swim under the female (Baynes et al. 1994, Carazo et al. 2016; Behaviour chapter). The female and male swim to the surface in a coupled swim with the male's dorsal side pushed close to the female's ventral (blind) side. The coupled swim arrives to the water surface and the pair release gametes with the genital pores held close together to fertilize the eggs. Egg release in both species is during the first part of the night 17:00 to 01:00 (Baynes et al. 1993, 1994, Olivera et al. 2010, Carazo et al. 2016). Spawning has only been observed in pairs (Baynes et al. 1994, Carazo et al. 2016) and microsatellite analysis of paternity supports that spawning is in pairs with 61.7 per cent of Senegalese sole spawns coming from a single pair (Martin et al. 2014). A few pairs dominate the spawning from a broodstock and show fidelity both during a spawning season and between years (Martin et al. 2014). The percentage participation of individuals in a broodstock ranged from 8 to 57 and the contributions were variable from each individual with a few individuals or pairs dominating the families obtained from a broodstock (Porta et al. 2006a,b, 2007, Martin et al. 2014). This dominance of the spawning and hence production of families by a few individuals has resulted in a high loss of genetic variability in the first generation produced in captivity (Porta et al. 2006a,b, 2007). The described spawning of these dominant pairs demonstrated more clearly the spawning behaviour of individual breeders over time as compared to egg production from a broodstock. Martin et al. (2014) described the spawning of 4 dominant Senegalese sole females from 3 different broodstocks. The spawning period for a female lasted 96 to 252 days and females could spawn daily (as many as 6 days consecutively) with a mean of 7.3 ± 0.6 days between spawns. Individual fecundity was also high compared to fecundities calculated for the entire broodstock with females often spawning in excess of

100,000 eggs/kg/day. Therefore, sole and particularly Senegalese sole, spawn in loosely monogamous pairs, with a high frequency (every 7 days) and high fecundity. However, a large proportion of the broodstock do not contribute or have a low contribution to spawning, which results in a low effective population size, few families and underestimates relative fecundities.

3.5 Influence of temperature on reproduction

Temperature has a strong influence on the maturation and spawning of sole. Although the evidence is largely observational it would appear that sole need low winter temperatures to stimulate gametogenesis, an increasing temperature profile to stimulate the progress of gametogenesis towards the final stages of maturation and small variations in temperature to stimulate ovulation and spawning. Under conditions of natural temperature variation, the completion of gametogenesis and spawning has been observed and reported for both common (Baynes et al. 1993) and Senegalese sole (Anguis Cañavate 2005). Common sole spawning was reported under natural temperature cycles that ranged from winter lows of 3°C to summer highs of 20°C, after a natural increase from temperatures below 10°C spawning began at temperatures ranging from 8–10°C and finished as temperatures raised above 12°C (Baynes et al. 1993). Senegalese sole spawning was reported under similar, but slightly warmer temperature cycles ranging from lows of 11°C to highs of 23°C for this warmer water species (Anguis and Cañavate 2005). Spawning began between 13–15°C and finished as temperatures raised above 23°C. Maximum egg production was between 17–20°C. The collection of spawns was associated to temperature variations with significantly more eggs being collected when the temperature rose by 0.2–2.5°C in the 1–3 days prior to egg collection (Anguis and Cañavate 2005). Deviations from these temperature cycles has been observed to have negative impacts on spawning and egg quality. Winter temperatures of above 10°C resulted in low egg production in common sole (Devauchelle et al. 1987) and 10°C was suggested to be a temperature that aided the stimulation of spermatogenesis (Agulleiro et al. 2007).

Senegalese sole maintained under constant temperature (18–20°C) and natural photoperiod had disrupted gonadal development with less males and females reaching advanced stages of maturation (García-López et al. 2006c). Martin et al. (2014), applied and demonstrated how a simulated natural temperature cycle can be used to control and stimulate maturation in Senegalese sole. After a period of low winter temperatures (≤ 13°C) an increase in temperature of 0.5°C per week was initiated in late January. When the temperature reached 16°C, weekly oscillations of two degrees were applied (16–18°C) to stimulate ovulation and spawning. Optimal temperatures (16–20°C) for spawning were maintained from February through to November and spawning was stimulated during the entire period. In November, temperatures were reduced to below 13°C to provide a resting period to finish and subsequently initiate gametogenesis.

3.6 Influence of light (photoperiod) on reproduction

Although few studies have addressed the effect of photoperiod and temperature on sole maturation, it seems that photoperiod is the controlling factor, as observed in most temperate fish species (Bromage et al. 2001, Wang et al. 2010). The spawning period was advanced in common sole (Devauchelle et al. 1987) and Senegalese sole (García-López et al. 2006c) using an advance in the photoperiod. In common sole, the photoperiod with associated temperature cycle was advanced gradually over 3 years to obtain a cycle that was advanced by 6 months compared to the natural photoperiod (Devauchelle et al. 1987). Under the 6-month advanced photo-thermal cycle, spawning was obtained from August through to November, which represented an advance in spawning of 6–7 months. In Senegalese sole an increase in photoperiod in March from light:dark (LD) 11:13 to LD24:0 advanced maturation in both males and females (García-López et al. 2006c) and the disruption of maturation was greater under constant photoperiod and temperature compared to natural photoperiod and constant temperature. However, under a continuous LD16:8 photoperiod with a simulated natural temperature cycle, maturation progressed normally with good quality egg, indicating that maturation free run in the absence of photoperiod cues or even entrained to temperature cycles (Martin et al. 2014). In conclusion, photoperiod appears to be an over-riding factor that can phase shift reproductive timing in sole, but it is essential to combine manipulations with the temperature cycle and although no data has been published it would appear that the Senegalese sole culture industry uses combined photo-thermal cycles to secure egg production throughout the year.

4. Endocrinology of reproduction

The reproductive processes in fish, including sex determination, sex differentiation, puberty and adult reproductive cycles (gametogenesis and spawning), are controlled by multiple neuronal and endocrine systems, within the so-called brain-pituitary-gonad (BPG) axis. The functioning of the BPG axis starts with the integration in the brain of the external environmental signals and their transduction and activation of different neuroendocrine systems, from which the gonadotropin releasing hormone (GnRH) is the primary regulatory system. Neuronal inputs controls in the pituitary the synthesis and release of the major reproductive hormones, the gonadotropins (i.e., follicle stimulating hormone (FSH) and luteinizing hormone (LH)), that are secreted into the bloodstream and act on the gonads to regulate the synthesis of sex steroids (androgens, estrogens and progestins) and growth factors, ultimately driving the development and maturation of the gonads (Carnevali 2007, Mañanos et al. 2009, Zohar et al. 2010).

The general scheme of the BPG axis is conserved among vertebrates, but some differences in the nature and bioactivity of the hormones can be found in different vertebrate classes, orders or even species, which are normally associated to different reproductive strategies. Thus, for a given species, it is essential to have knowledge on the components and functioning of the BPG axis, because this will provide not only information to understand potential causes of reproductive problems but also to give the possibility of developing tools and methods (e.g., environmental, hormonal, etc.) for the solution of these disorders. In Senegalese sole, there is a good deal of knowledge accumulated in the last fifteen years on reproductive endocrinology, prompted by the interest to develop the aquaculture of soles and the necessity to understand and solve the observed reproductive dysfunctions of cultured (F1) sole breeders in captivity (Morais et al. 2016).

4.1 The brain, pituitary and gonadal hormones of reproduction

The hormones of the BPG axis act coordinately to regulate the reproductive processes, in a cascade of hormonal events where each component stimulate the synthesis of the following one but also exert feedback regulation on the synthesis of previous components. There are key hormones of the axis with pivotal roles in reproduction, that are found in all vertebrate species with a quite conserved biological activity, such as the neuronal GnRH system, the pituitary gonadotropins and some gonadal sex steroids, while the nature and bioactivity of other "minor", but not less important, components can vary between species.

4.1.1 Brain neurohormones

The brain integrates all the signals coming from the outside, both abiotic (e.g., water temperature, photoperiod) and biotic (e.g., presence of males/females, predators), as well as the internal signals coming from the different organs of the individual (e.g., pituitary, gonads, liver, muscle, adipose tissue). The integration and processing of this information involves multiple neuronal and endocrine systems which altogether conform complex and highly interconnected regulatory networks. The primary neuroendocrine system regulating reproduction in all vertebrates is the GnRH, which exerts a stimulatory action on pituitary gonadotropin production. In addition, a number of other neuroendocrine systems have been identified in fish to regulate reproduction, exerting either positive (e.g., kisspeptin (kiss), neuropeptide Y (NPY), serotonin, galanin, gamma-aminobutyric acid (GABA), neurokinin B, secretoneurin (SN)) or negative (gonadotropin inhibitory hormone (GnIH) and dopamine) actions (Zohar et al. 2010, Trudeau et al. 2012, Zmora et al. 2017). The major neuroendocrine reproductive systems have been identified and characterized to some extent in Senegalese sole, including GnRH, kiss, serotonin, GnIH and dopamine.

The GnRH is a decapeptide identified in the brain of all vertebrate classes as several isoforms, named GnRH-1, GnRH-2 and GnRH-3. The GnRH-1 system innervates profusely the neurohypophysis and is believed to be the most relevant form in the control of pituitary gonadotropin secretion (i.e., the hypophysiotrophic form), whereas the GnRH-2 and GnRH-3 play relevant roles in the control of other reproductive-related aspects, such as courtship, locomotion and sensory transduction (Lethimonier et al. 2004, Zohar et al. 2010). In Senegalese sole, the existence of GnRH neurons and their brain distribution has been studied by immunohistochemistry and showed the presence of GnRH immunoreactive (ir) cells and fibers in specific reproductive relevant brain areas and a profuse innervation of the pituitary (Rendon et al. 1997). This study did not identify the specific nature of GnRH peptides, but in a subsequent study, using specific antibodies for GnRH-1, GnRH-2 and GnRH-3 and specific immunoassays (ELISA), Guzmán et al. (2009b) detected the presence of all three GnRH isoforms in the pituitary of reproductively active male and female breeders, demonstrating the existence of all GnRH forms in Senegalese sole. Nevertheless, studies on the specific biological actions of each GnRH form are still lacking in Senegalese sole.

The kisspeptin has recently emerged as a key player in the neuroendocrine control of reproduction in vertebrates, exerting stimulatory actions over the BPG axis (de Roux et al. 2003, Seminara et al. 2003, Funes et al. 2003, Pinilla et al. 2012). The existence of kisspeptin has been demonstrated in several teleost fish, in which one or two kiss variants (kiss1 and kiss2) have been identified (Servili et al. 2013, Pasquier et al. 2014). The kiss system has been characterized in Senegalese sole (Mechaly et al. 2009, 2011) and a predominant role of kiss2 over kiss1, in the regulation of Senegalese sole reproduction, has been inferred by the analysis of mRNA expression levels through the reproductive cycle (Mechaly et al. 2009, 2011, 2012, Marín-Juez et al. 2013).

From other reported stimulatory neuroendocrine systems, only serotonin (5-hydroxytryptamine (5-HT)) (Rodríguez-Gomez et al. 2000a) and galanin (Rodríguez-Gomez et al. 2000b) have been identified in Senegalese sole. Serotonin is an indoleamine that, together with cathecolamines (dopamine, norepinephrine and epinephrine), constitute the group of monoamine neurotransmitters (Yu et al. 1991). Galanin is a neuropeptide that mediates multiple physiological processes in vertebrates, including learning, food intake and reproduction (Lang et al. 2007). A role of both serotonin and galanin in the neuroendocrine regulation of fish reproduction has been demonstrated (Khan and Thomas 1992, Trudeau 1997, Martins et al. 2014). The works performed in Senegalese sole, studied the brain distribution of serotonin and galanin structures by immunohistochemistry and found, in both cases, a widespread distribution of serotonin and galanin ir-cells and fibers through the entire brain, mainly in reproduction-related areas and a profuse innervation of the pituitary, suggesting that both systems might play relevant roles in the regulation of Senegalese sole reproduction (Rodriguez et al. 2000a,b), which remains to be elucidated.

Conversely to the cited stimulatory systems, dopamine has long been identified as a major inhibitory system in fish reproduction (Peter and Yu 1997, Trudeau 1997, Yaron et al. 2003, Dufour et al. 2005, Popesku et al. 2008). Inhibitory effects of dopamine have been demonstrated in freshwater fish, including cyprinids (Yaron 1995, Trudeau 1997), silurids (Silverstein et al. 1999) and salmonids (Saligaut et al. 1999) but not in marine fish, in which dopamine inhibition is supposed to be absent or weakly expressed (Dufour et al. 2005, 2010). Interestingly, two studies performed in Senegalese sole have demonstrated a dopaminergic inhibition in this species, suggesting that in fact, dopamine inhibition might exist in marine fish. First, an immunohistochemical study using antibodies against tyrosine hydroxylase (TH), the rate-limiting enzyme in the synthesis of dopamine, showed the presence of TH-ir cells in reproductive relevant brain areas and TH-ir fibers innervating the adenohypophysis, providing an anatomical basis for the potential role of dopamine in the regulation of pituitary hormone secretion (Rodríguez-Gomez et al. 2000c). Second, *in vivo* treatment of Senegalese sole breeders with a dopamine antagonist (pimozide), influenced positively the endocrine BPG axis and the maturation of the gonad, indicating that blockage of an endogenous inhibitory dopaminergic system would cause positive actions over the reproductive axis of Senegalese sole (Guzmán et al. 2011b).

The GnIH is a dodecapeptide that has been recently discovered in the brain of birds (Tsutsui et al. 2000) and demonstrated to be a relevant inhibitory signal in the reproduction of many vertebrates (Ubuka et al. 2016), including fish (Paullada-Salmeron et al. 2016, Muñoz et al. 2017). This system has been recently characterized in Senegalese sole, through molecular, immunological and pharmacological techniques (Aliaga-Guerrero et al. 2018, see Chapter B-1.2). In this study, a Gnih precursor encoding for three putative Gnih peptides (ssGnih-1, ssGnih-2, ssGnih-3) was cloned from the brain and further analysis by qPCR detected the gene expression of Gnih's in several parts of the brain and in the pituitary. The presence and distribution of Gnih-ir cells and fibers was demonstrated by immunohistochemistry, in relevant reproductive related areas of the brain, with profuse projections to the pituitary, providing anatomical support for a role of Gnih in the regulation of reproduction in sole. The potential bioactivity of Gnih on reproduction was supported by a preliminary experiment showing that treatment of adult Senegalese sole with ssGnih-3 decreased gnrh-3 and lh gene expression (Aliaga-Guerrero et al. 2018).

These few cited studies indicate the existence of several active neuroendocrine systems, both stimulatory and inhibitory, in Senegalese sole and constitute a highly relevant area of future research to understand the neuronal and endocrine regulation of sole reproduction and subsequently, to provide the bases for the development of new hormonal therapies and approaches for the control of sole reproduction in captivity.

4.1.2 Pituitary hormones (gonadotropins)

The second level of the BPG axis is the pituitary, which produce and secrete a wide range of hormones that regulates many physiological processes. Reproduction is under the control of the gonadotropin hormones, FSH and LH, in all vertebrates, including fish (Swanson et al. 2003, Levavi-Sivan et al. 2010). Both FSH and LH are heterodimers composed of a hormone-specific beta subunit (FSHβ or LHβ) and a common glycoprotein alpha subunit (GPα), shared by other members of the glycoprotein hormone family (i.e., choriogonadotropin and thyrotropin). The gonadotropins have been identified and characterized in Senegalese sole (Cerdá et al. 2008, Guzmán et al. 2009b) and their synthesis and biological functions studied in Senegalese sole (Guzmán et al. 2009b,c, 2011b, Chauvigné et al. 2010, 2012, 2014a,b, 2015, 2016, Mechally et al. 2011, 2012) and common sole (Palstra et al. 2015). Molecular studies have shown that the predicted Senegalese sole mature FSHβ, LHβ and GPα peptides consist of 99, 148 and 94 amino acids, respectively, all of them including a signal peptide of variable length. The comparative analysis of their primary structure indicates the high conservation in Senegalese sole of all characteristic features found in other teleost gonadotropins, including potential glycosylation sites and cysteine residues forming cross-linked disulphide bonds needed for proper folding of the protein and for ligand-receptor interactions (Cerdá et al. 2008, Guzmán et al. 2009b). The presence and localization of gonadotropin-producing cells has been studied in the pituitary of Senegalese sole, by immunohistochemistry (Rendón et al. 1997) and *in situ* hybridization (Cerdá et al. 2008, Guzmán et al. 2009b).

The gonadotropins exert their actions on the gonads via interaction with their cognate receptors, the FSH receptor (FSHR) and the LH/choriogonadotropin receptor (LHCGR), both members of the family of glycoprotein hormone receptors (GpHRs) and were recently cloned in Senegalese sole (Chauvigné et al. 2010). The FSHR was localized in Leydig and Sertoli cells, whereas the LHR was localized in Leydig cells and in the spd (Chauvigné et al. 2010, 2012, 2014a,b). In the ovary, there is no information on the cellular localization of gonadotropin receptors in Senegalese sole, but it is assumed that both receptors are expressed in follicular cells as reported in other fish (Miwa et al. 1994, Oba et al. 2001, Wang and Ge 2003). The biological activity of FSH and LH and their synthesis and secretion patterns has been studied in Senegalese sole breeders during the reproductive cycle, providing important information on their physiological functions regulating gametogenesis and spawning in both males and females (see following sections).

4.1.3 Gonadal hormones (sex steroids and growth factors)

The third level of the BPG axis is the gonad, which under the influence of the pituitary gonadotropins, produce sex steroids (androgens, estrogens and progestins) and growth factors (e.g., activins, inhibins, amh, gdf9, bmp15, gsdf, igfs, insl3) to ultimately driving germ cell proliferation and differentiation as well as exerting feedback effects over the brain and pituitary (Nagahama and Yamashita 2008, Lubzens et al. 2010, Schulz et al. 2010, 2015, Zohar et al. 2010, Schulz and Nóbrega 2011). All steroids derive from cholesterol, which is converted into pregnenolone by the action of the enzyme cytpchrome p450scc and subsequently, through the action of multitude enzymes along the steroidogenic pathway, give rise to the rest of C21 (progestins), C19 (androgens) and C18 (estrogens) steroids (Tokarz et al. 2015). The main androgen in male fish is 11ketotestosterone (11KT) and the main estrogen in female fish is estradiol (E2), while the androgen testosterone (T) serve as a precursor molecule for both steroids, in addition to its own biological functions in both sexes (Borg 1994, Nelson and Habibi 2013). Several progestins have been identified in fish to be potentially relevant for the control of gonad maturation, including the 17α,20β,21-trihydroxy-4-pregnen-3-one (20β-S) and mostly, the 17α,20β-dihydroxy-4-pregnen-3-one (DHP) (Scott et al. 2010).

The steroidogenic pathway has been studied in some detail in Senegalese sole males by immunological and molecular techniques, providing a broad information on the identity and activity of several enzymes and receptors involved in the biosynthesis of sex steroids (García-López et al. 2005, Chauvigné et al. 2014b). On another hand, the recent application of ultra-high performance liquid chromatography tandem mass spectrometry (UHPLC-MS/MS) for the analysis of steroids in Senegalese sole opens up the possibility of obtaining new and valuable information on the identity and bioactivity of the relevant steroids in this species (Beltrán et al. 2015, Mañanós et al. unpublished). This technique allows the simultaneous analysis of multiple steroids in a single sample, which will permit to enlarge the information previously obtained by immunological analysis (ELISA), but also to identify and quantify potentially relevant new steroids. Recent UHPLC-MS/MS analysis of samples from mature Senegalese sole breeders have shown the presence in the gonads of, (1) progestins, including DHP and 20β-S, (2) androgens, including T, 11KT, androstenedione (A4), 11β-hydroxy-androstenedione (11OHA) and 11ketoandrostenedione (11KA) and, (3) estrogens, including E2 and estrone (E1). In addition to the presence in the gonads, all these hormones were found in plasma and urine, indicating that a variety of sex steroids are synthesized and released in Senegalese sole during the process of gonad maturation (Mañanós et al. unpublished). Finally, a good deal of information is available in Senegalese sole for the most relevant steroids (E2, T, 11KT and DHP) by ELISA analysis, mainly regarding the fluctuation of tissue/blood levels during the reproductive cycle and under different culture conditions (García-López et al. 2006a,b,c, 2007, 2009, Agulleiro et al. 2006, 2007, Cerdà et al. 2008, Guzmán et al. 2008, 2009b, 2011a,b, Oliveira et al. 2009, 2010, 2011, Cabrita et al. 2011, Marín-Juez et al. 2011, Mechaly et al. 2012, Chauvigné et al. 2012, 2014b, 2015, 2016, Norambuena et al. 2013); some of this information is described in the following sections.

4.2 Endocrine regulation of testicular development and spermiation

In the last fifteen years, a good deal of knowledge has been accumulated on the neuronal and endocrine regulation of the reproductive cycle in males of Senegalese sole, in an effort to understand the physiological causes underlying the reproductive dysfunctions observed in F1 male breeders. In addition, the Senegalese sole has become an interesting research model to study the mechanisms of testicular maturation in fish, due to the uncommon semi-cystic type of spermatogenesis exhibited by this species, in contrast to the common cystic type found in most fish species.

4.2.1 Neuronal systems regulating male reproduction

From the identified neuronal systems in the brain of Senegalese sole, biological evidences for a regulatory role on male reproduction have only been obtained, directly or indirectly, for GnRH, kiss and dopamine.

The GnRH peptides have been detected in the pituitary of mature male Senegalese sole by ELISA, with a relative abundance of GnRH-2>GnRH-1>GnRH-3, suggesting that all GnRH variants reach the pituitary of male soles and participate to some extent in the regulation of pituitary secretion (Guzmán et al. 2009b), although their specific functions remain to be elucidated. To date, the bioactivity of the GnRH system in males is inferred by the stimulatory action of GnRHa treatments on androgen (T and 11KT) secretion, spermatogenesis and sperm production (Agulleiro et al. 2006, 2007, Guzmán et al. 2011a,b, Cabrita et al. 2011) (described in following sections).

The involvement of the kiss system has been suggested by the analyses of mRNA expression levels during the reproductive cycle of Senegalese sole male breeders (Mechaly et al. 2012, Marín-Juez et al. 2013). These studies showed that brain mRNA levels of *kiss2* and *kiss2r* increase from summer to peak at the end of winter, in coincidence with highest GSI and spermatogenetic activity and highest levels of pituitary *lhβ* mRNA, gonad *fshr* and *lhcgr* mRNA and plasma levels of T and 11KT, indicating a good correlation between the activity of the kiss2 system and other reproductive endocrine systems in male Senegalese sole.

The bioactivity of the dopamine system was determined in males of Senegalese sole by *in vivo* treatment with a dopamine antagonist, which showed a clear stimulatory effect on spermatogenesis and sperm production, as well as in the transcription of pituitary *gpα* and *lhβ* mRNAs (Guzmán et al. 2011b). No more studies have been performed on dopamine in sole, but the available information on both the dopaminergic and the GnIH systems, makes them good candidates to understand potential inhibitory mechanisms underlying the reproductive dysfunctions observed in male sole breeders.

4.2.2 Annual hormone profiles during the reproductive cycle of males

The first developmental stages (spermatogenesis) of the male reproductive cycle are regulated predominantly by FSH and androgens (mainly T and 11KT), whereas the final stages of maturation and spermiation are regulated mainly by LH and progestins (Borg 1994, Mañanós et al. 2009, Scott et al. 2010). An initial basic approach to understand the reproductive cycle of a given species, is the analysis of plasma levels of reproductive hormones along the reproductive cycle and its correlation with the progress of gonad maturation. Specific immunoassays (ELISA) have been developed and validated for the analysis of the major reproductive hormones in Senegalese sole and used to describe the annual plasma levels of FSH and LH (Chauvigne et al. 2016), the androgens T and 11KT (García-López et al. 2006a,b, 2009, Bayarri et al. 2011) and the progestin DHP (García-López et al. 2006a,b, Bayarri et al. 2011) in male breeders. Based on this information, a representation of the fluctuating plasma profiles of these hormones along the year is provided in Fig. 3. The availability of hormone values along the year and its correlation with the reproductive cycle is useful information for fish managers and researchers working on Senegalese sole, because plasma hormone analysis can be used as a quick accurate indication of the maturational stage of the individuals, similar to other more invasive (biopsy) or lethal (gonad histology) methods.

Adult Senegalese sole males are spermiating all year round but with two periods of highest spermiation activity, coincident with the female spawning periods; the major one in the spring and a secondary minor one in autumn. Plasma FSH levels increase progressively during winter to reach maximum levels (20 ng/ml) at the beginning of the spring spawning season, decreasing thereafter to minimum levels in the summer (July), at the post-spawning/recovery period. Then, levels of FSH rise again during the 2 months (August-September) preceding the second reproductive cycle in autumn, to reach peak levels at the beginning of this period. On the other hand, plasma LH levels are low during winter and rise drastically in the spring, to peak levels (38 ng/ml) in the middle of the spawning period, decreasing thereafter to minimum levels in July, which are maintained low through the rest of the year. These profiles reflect, (1) a delayed and sharper peak of LH with respect to FSH during the spring spawning period and, (2) a lack of an LH rise in the autumn spawning period. The first aspect is in accordance with the known role of FSH regulating early developmental stages of testicular growth in fish, which is predominant

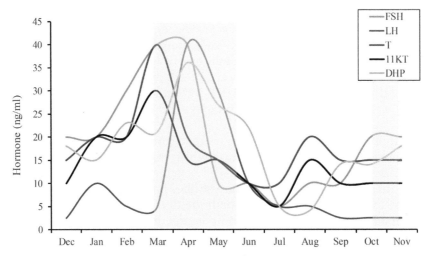

Fig. 3. Representative annual fluctuation of plasma levels of reproductive hormones in male Senegalese sole. Hormone levels are corrected to optimize fitting to the Y axis: FSH (x2), LH (x1), T (x20), 11KT (x1), DHP (x10). Shaded grey areas indicate the natural spawning periods of Senegalese sole.

during the 2–3 months preceding spawning and the known role of LH regulating final stages of maturation and spermiation, taken place later in the cycle and being predominant within the spawning period of the females. The second aspect, related to the absence of an observed rise of LH associated to the secondary spawning period in autumn is unexpected; it cannot be discarded that in fact it exists an LH rise in autumn, which would be normally associated to an activation of testicular maturation, but a sharp short-lived LH peak in autumn could be easily missed in studies performing monthly blood samplings.

Annual plasma levels of T and 11KT fluctuate in parallel, as correspond to their biological and biosynthetic relatedness and also, they correlate with the profile of FSH, as expected by their coordinated regulatory action on spermatogenesis. Both androgens increase during winter to peak at the beginning of the spring spawning period, decreasing progressively during the spawning period to reach basal low levels in the summer (July) and then showing a second minor increase preceding the autumn reproductive period.

The good correlation of plasma T and 11KT levels with testicular development and spermiation, as described in males of Senegalese sole, has been observed in many other fish and indicates the appropriateness of using the concentration of androgens in the blood as an adequate biomarker to estimate the maturity stage of an individual. In the case of Senegalese sole, 11KT analysis is the most accurate value, because the increase of 11KT in the blood of maturing males is several orders of magnitude higher than those of T, maybe as a result of the predominant role of this androgen in regulating testicular development. Plasma levels of T increase from 0.5 (basal levels at resting/recrudescence) to around 2 ng/ml (peak levels at full maturation), whereas those of 11KT increase from 4 to around 30 ng/ml.

The annual profile of the progestin DHP is similar to that of androgens but delayed on time, probably as a reflection of its role regulating later stages of development; both the peak levels at the spring spawning time (around 2 ng/ml) and the further increase for the autumn reproductive period, are delayed by approximately one month compared to those of androgens. This profile would suggest that in Senegalese sole, as described in other fish, DHP would be the relevant maturation inducing steroid. This role as maturation inducing steroid has also been indicated by studies on testicular steroidogenic pathways (see following section) and UHPLC-MS/MS analysis. Analysis by UHPLC-MS/MS of samples obtained from spermiating male Senegalese soles showed that DHP, but not 20β-S, is present in testicular extracts and also in plasma and urine, indicating that this steroid is synthesized and released into the bloodstream and excreted through the urine, maybe also acting as pheromone in combination with other steroid compounds (Mañanós et al., unpublished).

In addition to the circulating plasma levels, some information is available on the annual levels of *fshβ* and *lhβ* expression in the pituitary of male Senegalese sole and showed to diverge to a great extent with the described annual profiles of plasma FSH and LH levels (Cerdá et al. 2008, Mechaly et al. 2012). These divergences might be related to differences in protein translation and secretion patterns from the pituitary and/or clearance rates from the circulatory system and indicated that mRNA levels are probably not good indicators of the reproductive stage of the fish as they are the plasma levels of FSH and LH, as previously suggested in other fish (Swanson et al. 2003). This limitation is also extensible to the mRNA and protein levels of the gonadotropin receptors in the testis of Senegalese sole, which patterns according to season or testicular development stage do not parallel the plasma levels of their ligands (Chauvigné et al. 2010, 2014b, Mechaly et al. 2012, Marín-Juez et al. 2013).

4.2.3 Regulation of testicular steroidogenesis

A good deal of knowledge has been accumulated in the last years in the biosynthesis of steroids in the testis of male Senegalese sole. Direct evidences supporting the steroidogenic role of Leydig cells have been obtained by the detection of 3β-hydroxysteroid dehydrogenase enzymatic activity by enzyme-histochemistry (García-López et al. 2005) and the localization of mRNAs encoding for several steroidogenic enzymes and StAR protein by *in situ* hybridization (Chauvigné et al. 2014b). The latter approach has also identified the mRNAs of these enzymes (i.e., cbr1, cyp11a1, cytochrome P450 19a1) in spd, indicating that steroids may potentially be produced locally by germ cells. The mRNA expression of other factors involved in steroid-signaling within the testis, such as nuclear receptors for androgens (ara and arb), estrogens (erb) and progestins (pgr), have also been localized in Senegalese sole germ cells (ara, erb, and pgr), Leydig cells (erb) and Sertoli cells (arb and pgr) (Chauvigné et al. 2014b).

Different studies in Senegalese sole males have demonstrated the ability of both FSH and LH to stimulate testicular *in vitro* production of androgens and to increase androgen plasma levels *in vivo* (Chauvigné et al. 2012, 2014b), an action that is mimicked by treatment with human chorionic gonadotropin (hCG), a gonadotropin that also binds and activates the LHCGR (Guzmán et al. 2011a, Marín-Juez et al. 2011). Both FSH and LH exhibit the same potency for stimulating the *in vitro* androgen release at early to mid stages of testicular development, whereas at more advanced stages of recrudescence the relative potency of FSH is lower than that of LH (Chauvigné et al. 2012, 2014b). Considering the differences in androgenic potency and the steady increase of plasma FSH, but not of LH, from autumn to late winter, it seems reasonable to assume that the increase in plasma androgen levels preceding the spring spawning season are predominantly regulated by FSH. On the other hand, the role of LH during this period is intriguing since, despite its relatively high androgenic potency in the testis, the LH plasma surge during spring is associated with a strong reduction of the circulating levels of androgens (García-López et al. 2006a,b, Cerdá et al. 2008, Mechaly et al. 2012, Chauvigné et al. 2012, 2014b). It is possible that during autumn and winter, the basal plasma LH levels are sufficient to activate the LHCGR in the Leydig cells, contributing to FSH-signaling and to the stimulation of androgen production, not only during early to mid stages of testicular development but also during late spermatogenesis. The cooperation between FSH and LH seems to be also extensible to the production of progestins, since both FSH and LH (as well as hCG) are able to up-regulate the testicular mRNA expression of *cbr1*, an enzyme responsible for the conversion of 17a-hydroxyprogesterone into DHP, both at early/mid and late spermatogenesis. The analysis of circulating levels of DHP along the annual reproductive cycle of wild-caught (García-López et al. 2006b) and F1 and F2 (Bayarri et al. 2011) Senegalese sole male breeders shows the presence of this steroid in the plasma all the year round but with higher levels at spring and autumn and thus, it seems plausible to think that the testis produces DHP under different conditions of gonadotropin stimulation.

The expression of nuclear androgen, estrogen and progestin receptors is also controlled by gonadotropin-dependent steroid-mediated signaling in the testis of Senegalese sole (Marín-Juez et al. 2011, Chauvigné et al. 2014b). Both FSH and LH up-regulate the expression of *ara*, *arb*, *erb* and *pgr* at early/mid stages of spermatogenesis, while at late testicular development only FSH is able to increase the mRNA levels of *arb* and *pgr*. These results, together with the expression of the different sex steroid receptors in germ cells and Sertoli and Leydig cells, highlight the importance of sex steroids in the regulatory network mediating gonadotropin stimulation during spermatogenesis in Senegalese sole, acting either directly on the germ cells and/or by mediation of Sertoli cells (Schulz et al. 2010, 2015, Schulz and Nóbrega 2011).

The specific actions of sex steroids in the regulation of cystic spermatogenesis in diverse fish models include the stimulation of the slow spermatogonial proliferative self-renewal (estrogens), the rapid spermatogonial clonal proliferation and differentiation towards meiosis (androgens), the entry into and progression of meiosis, spermiogenesis and spermiation (androgens and progestins) (Schulz et al. 2010, 2015, Schulz and Nóbrega 2011). Unfortunately, these actions have not been demonstrated to date in Senegalese sole (nor in any other fish species with semi-cystic spermatogenesis). Nevertheless, based on the levels of circulating androgens and considering that, until the completion of meiosis, the spermatogenic process in Senegalese sole is apparently similar to that of other fish with cystic spermatogenesis, it is reasonable to assume that androgens are also involved in the regulation of the rapid spermatogonial proliferation and differentiation towards meiosis and the entry into and progression of meiosis in Senegalese sole. In addition, the elevated plasma levels of androgens in coincidence with the massive formation and accumulation of spd in the cortical region of Senegalese sole testis suggest that androgens may also be involved in the spermatocysts opening and release of spd into the lumen of seminiferous lobules of Senegalese sole (García-López et al. 2006a, Chauvigné et al. 2014b). Finally, different studies have shown a correlation between GnRHa- and hCG-induced androgen release and the concomitant stimulation of spermatogonial proliferation and differentiation, meiotic activity and accumulation of spd in the testis of Senegalese sole (Agulleiro et al. 2007, Guzmán et al. 2011a,b). In contrast, the involvement of progestins (e.g., DHP) in regulating the entry into meiosis and spd formation and release in Senegalese sole is unclear. The role of the progestin DHP at this stage has been inferred by the analysis of plasma levels by ELISA, although with slightly divergent results. Analysis of plasma DHP in F1 males showed a significant increase during the winter and autumn periods of testicular growth (Fig. 3; Bayarri et al. 2011) but data on wild-caught males, did not reveal significant changes between early development stages and highest meiotic activity and spermatocysts opening (García-López et al. 2006a). Also, the mRNA expression of *pgr* in the testis significantly increase from early to mid testicular development stages (Marín-Juez et al. 2013) and *in vivo* hormonal treatments eliciting a stimulation of meiotic activity and spermatid formation and release seems to stimulate DHP synthesis, conjugation and/or metabolization in the testis (Agulleiro et al. 2007), supporting the involvement of sex steroids in the regulation of meiotic activity in Senegalese sole male germ cells.

The contribution of gonadotropins and sex steroids in controlling the differentiation of spd into spz (spermiogenesis) in Senegalese sole has been well characterized (Chauvigné et al. 2014a,b). The formation of spz from spd in the lumen of seminiferous lobules is triggered by LH acting through LHCGR expressed in the spd, in a cAMP/PKA-mediated but steroid-independent pathway and by promoting germ cell *de novo* protein synthesis. Additionally, the significant surge of LH plasma levels during the spring spawning period of Senegalese sole supports the direct involvement of LH in the progressive formation of spz from the great reservoir of spd accumulated in the testis cortex during the autumn and winter spermatogenic periods (García-López et al. 2006a, Chauvigné et al. 2016).

One of the most characteristic features of Senegalese sole testicular development is the prevalence of spd and to lesser extent spz within the testis, as well as the ability of males to release sperm during the whole year, including the summer (García-López et al. 2006a,b). The histological analysis of Senegalese sole testis shows that, particularly during summer and early autumn, the spd and spz are accompanied in the testis by spg-A, but more differentiated spg-B and spc are absent, indicating no meiotic activity during mid and late spermatogenic stages (García-López et al. 2006a). This indicates that the supply of spz during these periods would come from the progressive differentiation of spd that greatly accumulate in the cortical region of the testis, mainly during late fall and winter, prior to the spring spawning season. Considering that plasma LH levels during summer and early autumn are low, the signaling mechanisms triggering the differentiation process of spd into spz during these periods are unclear. It is possible that basal plasma LH levels are sufficient to activate the LHCGR in certain populations of spd, considering that the constitutive activity of the LHCGR in the spd is able to drive their differentiation into spz, or assuming the existence of other LH-independent endocrine, paracrine or autocrine factor(s) regulating this transformation.

Together with sex steroids, multiple growth factors produced in Sertoli and Leydig cells have been reported to participate in the paracrine regulation of proliferation and differentiation of germ cells in male teleost fish (Loir and Le Gac 1994, Miura et al. 1995, 2002, Nader et al. 1999, Sawatari et al. 2007, Schulz et al. 2010, 2015, Schulz and Nóbrega 2011, Assis et al. 2016), either positively (e.g., activins, gsdf, Igfs, Insl3) or negatively (e.g., amh and inhibins). In Senegalese sole, there are no studies on this respect, although FSH and to a lesser extent LH, were shown to up-regulate testis mRNA expression of *amh*, *gsdf* and *igf1a* (Chauvigné et al. 2014b), an action that, in the case of amh, was mediated by the production of sex steroids, which is contradictory with studies in other fishes describing a strong down-regulation of *amh* expression by both FSH and androgens (Miura et al. 2002, Rolland et al. 2013, Sambroni et al. 2013, Schulz et al. 2015).

Using an oligonucleotide microarray technique combined with bidimensional difference gel electrophoresis (2D-DIGE), up to 400 genes and 49 proteins in the testis of Senegalese sole have been shown to be differentially expressed during the progression of spermatogenesis, providing a limited but comprehensive picture on the major molecular changes taking place during testicular development in this species (Forné et al. 2011). At the transcriptomic level, the main changes included, (1) the down-regulation of peptidase-related genes from early to late spermatogenesis, which suggests an active inhibition of proteolysis, (2) significant changes in genes involved in the organization of the cytoskeleton, (3) increased expression of genes involved in sperm maturation and egg binding from late spermatogenesis to full maturation, (4) up-regulation of transcriptional and translational regulators from late spermatogenesis to functional maturation in relation to the required turnover of proteins associated with chromatin arrangement and, (5) up- or down-regulation of genes involved in protein modification and remodeling. At the proteomic level, changes included, (1) increase of a set of proteins, from mid to late spermatogenesis and then decreased at full maturation stage, related to the cytoskeleton and the ubiquitin-proteasome system and, (2) decrease of a set of proteins, from mid to late spermatogenesis stage and then increase again at full maturation, related to proteasome subunits and tissue remodeling processes.

4.2.4 Endocrine parameters in wild versus F1 males

Several neuronal and endocrine parameters have been analyzed comparatively in wild *versus* hatchery-produced (F1) Senegalese sole broodstock, in an attempt to identify variations in the endocrine axis that could underlie the different reproductive performance of both populations. The comparative analysis performed by Guzmán et al. (2009b) over full spermiating males, using wild and F1 individuals reared under similar culture/ environmental and feeding conditions for several years, showed no differences in the pituitary levels of GnRH-1 and GnRH-2 and pituitary mRNA levels of *fsh*β and *lh*β, but differences in the pituitary levels of GnRH-3 (wild > F1), pituitary *gpα* mRNA levels (F1 > wild) and plasma levels of T and 11KT (F1 > wild). Another study using similar type of fish, showed no differences in plasma levels of FSH, T and 11KT between males from both origins, but differences in plasma LH levels (wild > F1) (Chauvigné et al. 2016). Another study in Senegalese sole used a transcriptomic approach to analyze the testis of mature males captured from the wild compared to testis of mature F1's reared in captivity, and showed lower abundance of proteins related to protease inhibition, iron and glucose metabolism, cystoskeleton, sperm motility and protection against oxidative stress in F1 than in wild males (Forné et al. 2009, 2011). These studies have detected some endocrine differences between Senegalese soles from different origin, that could correlate the observed reproductive dysfunctions of F1 male breeders (i.e., reduced sperm production and inhibited reproductive behavior) to alterations in the operation of the reproductive axis, particularly at the level of GnRH-3 and LH synthesis, although results are not conclusive and much further research will be necessary in the future to get a comprehensive comparative picture between breeders from both origins.

Other comparative studies tried to determine if potential endocrine differences could be related to differences in nutrition. Analyzing samples obtained from wild and F1 male Senegalese sole breeders, these last fed with commercial dry pellets, differences on cyclooxygenase (cox-2; enzyme involved in the synthesis of prostaglandins from fatty acids) mRNA expression levels in the gills and spermduct, prostaglandin concentration in the gills, testis and kidney and fatty acid composition in muscle, liver and testis, were observed and were associated with detrimental effects of the commercial dry food compared to the natural feeding of wild fish, mainly regarding unbalanced and inadequate amounts of ARA and the ratio EPA/ARA on the artificial diets (Norambuena et al. 2012a,b). This would be particularly important on the altered production and release of prostaglandins, which are relevant regulators of testicular steroid synthesis, milt production and courtship behavior in teleost fish (Norambuena et al. 2012a,b, 2013).

4.3 Endocrine regulation of ovarian development and spawning

Soles are annual spawners, producing multiple spawning events through a quite prolonged spawning season. Accordingly, gonadal development can be classified as the group-synchronous multiple batch type, in which batches of developing oocytes can be found in the ovary (Mañanós et al. 2009). During the reproductive season, the ovary contains vitellogenic oocytes at different developmental stages and a predominant group of more advanced oocytes, corresponding to the batch of oocytes that will go through maturation and ovulation in the next spawning event. Female Senegalese sole breeders may produce up to 60 spawning events during the spawning period that can expand for up to 8 months (Martin et al. 2014). The intra-ovarian dynamics of oocyte batch development and maturation is repeated through the whole spawning period, until the end of the period when the last batch of mature oocytes is ovulated. The spawning period is also characterized by a progressive presence of intra-ovarian post-ovulatory follicles and atretic oocytes and the disappearance of further vitellogenic oocytes. This is the most extended type of ovarian development in teleost fish and the corresponding regulatory endocrine mechanisms have been widely studied in several fish species (Lubzens et al. 2010).

The reproductive cycle of female fish is characterized by the specific process of vitellogenesis, referred to the synthesis of vitellogenin (VTG). The VTG is a complex lipophosphoglycoprotein, synthesized in the liver under the stimulation of E2 and released into the bloodstream to be incorporated into the growing oocytes via specific receptors (Mañanós et al. 1994a,b, 1997, Babin et al. 2007). Inside the oocyte, the VTG is progressively cleaved into smaller components (phosvitin, lipovitellin and β-component) giving rise to the vitellin reserves of the egg, the yolk or vitellus (Carnevali et al. 1993, 2006). The optimal accumulation and processing of VTG is of vital importance for egg quality and further survival of the hatched larvae, as this constitute the nutritional reserves of the larva, until mouth opening and external feeding, several days after fertilization (Carnevali 2007). The VTG is a female specific protein and its circulating blood levels correlates well with the initiation and progression of the period of gametogenesis (so-called vitellogenesis in females). Because of these characteristics, VTG is used as a biomarker in reproductive endocrinology and ecotoxicology. Analysis of VTG in blood samples is used for gender determination, in those species with no external signs of sexual dimorphism, because a positive reaction in VTG immunoassays identify samples collected from females. The VTG immunoassays are also used to estimate the stage of gonadal development, as VTG blood levels correlates well with oocyte growth and development (Mañanos et al. 1994a,b). In sole, the VTG molecule has been characterized in three species, the Senegalese sole (Guzmán et al. 2008), the common sole (Núñez-Rodríguez et al. 1989) and the English sole *Parophrys vetulus* (Roubal et al. 1997) and the usefulness of VTG immunoassays as a method for gender determination and/or the estimation of ovarian development in female breeders, determined (Núñez-Rodríguez et al. 1989, Lomax et al. 1998, Guzmán et al. 2008).

4.3.1 Neuronal systems regulating female reproduction

From the identified neuronal systems in the brain of Senegalese sole, evidences for a direct biological role in female reproduction have only been obtained for the GnRH and kiss systems. All the three GnRH variants have been detected in the pituitary of mature females by ELISA, with a relative abundance of GnRH-2>GnRH-1>GnRH-3, demonstrating the active synthesis of all GnRHs in this species (Guzmán et al. 2009b). Nevertheless, the precise involvement of each GnRH form in the regulation of gonadotropin secretion and ovarian development in sole females remains to be studied. To date, the bioactivity of the

GnRH system in females is inferred from the ability of GnRHa treatments to induce T and E2 secretion and to stimulate oocyte maturation, ovulation and spawning (Agulleiro et al. 2006, Guzmán et al. 2009a, 2011a,b).

The involvement of the Kiss system in female Senegalese sole reproduction has been suggested by the good correlation between increased mRNA expression levels of brain *kiss2* and *kiss2r* with maximum levels of GSI, pituitary *fshβ* mRNA expression, ovarian *lhcgr* mRNA expression and plasma E2 in mature females (Mechaly et al. 2012). Also, in a study where Senegalese sole breeders were fasted for 15 days, in an attempt to simulate the suppression of feeding that precedes the spawning season (Milla et al. 2009, Costas et al. 2011), hypothalamic *kiss2* and *kiss2r* mRNA levels increased compared to controls, concomitantly with significant rises in pituitary mRNA expression of *fshβ* and lhβ, suggesting a link between kiss2 and gonadotropin signaling (Mechaly et al. 2011). On another hand, a role of kiss2 in the regulation of puberty in female Senegalese sole was suggested by a study showing high expression levels of *kiss2r* in the brain and *kiss2* in the ovary of females during puberty (Mechaly et al. 2009, 2011).

To date, there are no studies demonstrating the activity of any inhibitory neuronal system in female sole reproduction. There is no research on the GnIH system and the only study that investigated the dopamine system, failed to demonstrate a direct action of dopamine on the female reproductive axis (Guzmán et al. 2011b). In this study, treatment of Senegalese sole females with a dopamine antagonist did not affect the endocrine axis (GnRH content, gonadotropin gene expression, steroid plasma levels) neither ovarian maturation and spawning. Nevertheless, it is possible that the lack of observed effects in this study could be a consequence of, (1) the sampling point for analysis was not the adequate and effects could be observed at other times post-treatment or, (2) the stage of maturation of the treated females was not the adequate and effects might be observed at other developmental stages. These two hypotheses should be checked and future research will be focused to verify the lack or not of an active dopaminergic system in female soles, as well as the potential bioactivity of other inhibitory systems (e.g., GnIH) in the regulation of female reproduction.

4.3.2 Annual hormone profiles during the reproductive cycle of females

The reproductive cycle of females is regulated by the coordinated action of gonadotropins and sex steroids, with a major role of estrogens (C18 steroids) and progestins (C21 steroids) regulating early (ovarian growth) and late (ovarian maturation and ovulation) developmental stages, respectively. The major estrogen in female fish is E2 and the major progestin DHP, while androgens (e.g., T) play secondary additional roles. The synthesis and release of each reproductive hormone is linked to its biological function and thus, tissue/plasma levels for a given hormone correlates well with a given stage of ovarian development. Thus, data on hormone levels in plasma, as well as VTG levels, are good indicators of the progression of gonadal development and maturation and the associated periods of spawning and recovery.

In Senegalese sole, specific immunoassays have been developed and validated for the analysis of the major reproductive hormones in females and used to describe the annual plasma levels of FSH and LH (Chauvigné et al. 2016), T, E2 and DHP (García-López et al. 2006b, 2007, Guzmán et al. 2008, Bayarri et al. 2011) and VTG (Guzmán et al. 2008). Based on this information, a representation of the fluctuating plasma profiles of these hormones along the year is provided in Fig. 4.

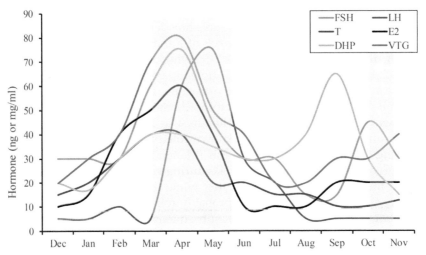

Fig. 4. Representative annual fluctuation of plasma levels of reproductive hormones in female Senegalese sole. Levels are in ng/ml for gonadotropins and steroids and in mg/ml for VTG. Hormone levels are corrected to optimize fitting to the Y axis: FSH (x3), LH (x1), T (x100), E2 (x10), DHP (x100), VTG (x20). Shaded grey areas indicate the natural spawning periods of Senegalese sole.

In female Senegalese sole, plasma levels of FSH increase at late winter and peak (27 ng/ml) at the beginning of the spring spawning season and decrease thereafter to lowest levels in the summer (July); then, FSH levels rise again at early autumn for a second minor peak at the beginning of the autumn spawning season, decreasing again until the end of the year. Plasma LH levels remain low during winter and rise drastically at early spring to peak in the middle of the spawning period (80 ng/ml) and then, decrease gradually to reach low basal levels in summer (July), with no further increase. Two interesting points characterize theses profiles. First, the sharpness and displacement of the LH rise/peak with respect to that of FSH during the spring spawning period. This is probably related to the differential biological functions of both gonadotropins, with FSH mainly regulating the initiation and progression of ovarian growth (vitellogenesis) and LH mainly regulating final stages of oocyte maturation and ovulation, as widely described in other fish species (Levavi-Sivan et al. 2010, Lubzens et al. 2010, Zhang et al. 2015a,b, Takahashi et al. 2016). The second aspect is that, while FSH follows similar profiles in the two spawning periods, plasma LH levels do not show any increase during the second autumn spawning period. This is unexpected, but similar to that observed in the males (Fig. 3) and we can again hypothesize that an elevation of plasma LH in maturing females was missed by monthly blood samplings.

Other studies in Senegalese sole females have analyzed the mRNA expression levels of pituitary gonadotropins and ovarian gonadotropin receptors, information that can be correlated to the described plasma levels of the gonadotropin proteins (Chauvigné et al. 2010, 2015, 2016, Mechaly et al. 2012). The obtained information showed that, while the temporal evolutions of plasma FSH and LH levels are somehow paralleled by the ovarian mRNA expression level patterns of their respective receptors (*fshr* and *lhdgr*), they are not paralleled by the mRNA expression levels of *fshβ* and *lhβ* subunits in the pituitary, as these levels only increase significantly at late spring (*fshβ*) or remain unchanged during the year (*lh*β), suggesting that pituitary gonadotropin mRNA levels are not good indicators of the stage of ovarian development in Senegalese sole breeders.

The temporal evolution of circulating levels of T and E2 mirrors to a great extent that of FSH, increasing during winter to peak at the beginning of the spring spawning period. This would be in agreement with the known relationship between these hormones, with FSH stimulating the synthesis of androgens and estrogens, T being the precursor molecule for E2 synthesis and the three of them regulating coordinately the first stages of oocyte growth and development (Mañanós et al. 2009). There is also a positive correlation between the E2 and VTG profiles, as expected by the known role of E2 to stimulate the hepatic VTG synthesis during the process of vitellogenesis (Mañanós et al. 1994a,b). Accordingly, the increasing levels of plasma FSH, E2, T and VTG during winter would reflect the recruitment and growth of the first main batch of oocytes, while the elevated plasma levels of FSH, E2, T and VTG during the spring would probably reflect the development of successive oocyte waves entering the growth phase for subsequent maturation and spawning events. On the other hand, the peak of LH during the spring spawning season is concurrent with intermediate or already declining plasma levels of FSH, E2, T and VTG, which could reflect altogether the steroidogenic shift from estrogen (E2) to progestin (DHP) production that occurs in the follicles when entering into the maturation process preceding ovulation and spawning (Lubzens et al. 2010, Asturiano et al. 2002).

The profile of plasma levels of the progestin DHP correlates quite well with that of the other steroids (T, E2), with rises of DHP preceding both spawning periods. A significant difference between profiles would be the maintenance of high DHP levels during the entire spawning period and even at post-spawning, in contrast to the reduction of T and E2 and their low levels at post-spawning. These data would be in accordance with a role of DHP in regulating the sustained maturation and ovulation of batches of oocytes along the spawning period and also with a role of DHP in the resumption of ovarian activity during the recovery post-spawning period, as described in other fish (Scott et al. 2010). Nevertheless, it should be pointed that some divergences on plasma DHP levels in Senegalese sole have been observed in different studies, in contrast to the consistent and similar data obtained by different authors on T and E2 analysis. The profile represented in the figure was based on the data provided by Bayarri et al. (2001) but, for example, another study by García-López et al. (2006b) showed that despite maximum levels of DHP were found in mature females, there was not a clear annual variation on plasma DHP levels. Other studies in Senegalese sole females did not show a clear relationship between gonad maturation and increased DHP levels (García-López et al. 2007). These discrepancies might be associated to particularities in the releasing and metabolic rates of the progestins compared to those of androgens/estrogens. It has been shown in other fishes that the elevation of circulating DHP levels can be of very short duration (Rinchard et al. 1997), as the free steroid may be quickly cleared from the circulation by metabolization or conjugation in the gonad (Scott and Canario 1992, Scott et al. 1997, Agulleiro et al. 2007) and thus, peak levels in plasma can easily be missed in studies performing monthly samplings. The rapid clearance of DHP from the circulation has been related to the excretion of this progestin and its metabolites in the female urine to serve as pre-ovulatory pheromone (Scott et al. 2010). Recently, the putative role of DHP and/or its metabolites as pheromone in Senegalese sole has been suggested by Cabrita et al. (2011), who showed that treatment of females with DHP induces significant increases in sperm viability, velocity and motility in surrounding un-treated males.

4.3.3 Regulation of ovarian steroidogenesis

In contrast to the abundant research performed in males, no information is available on the steroidogenic pathways in female Senegalese sole and the involvement of granulosa and theca follicular cells in the synthesis of steroids, but it is assumed that these somatic cells are the source of steroids, as reported in many other fish (Nagahama and Yamashita 2008, Lubzens et al. 2010). There is also no information on the localization of steroid receptors in the Senegalese sole follicle complex, but information in other fish indicates that they are expressed in the follicular cells and/or in the oocytes (Thomas et al. 2007, Thomas 2012, Zapater et al. 2013, Nagasawa et al. 2014). Also, there are no experimental data about the production and functionality of any growth factor in the ovary of Senegalese sole, despite these factors play well-known roles in the regulation of follicular development, maturation and spawning in female teleost fish (Ge 2005, Lubzens et al. 2010).

A study performed in Senegalese sole, using an oligonucleotide microarray (Tingaud-Sequeira et al. 2009) identified up to 118 differentially expressed transcripts in the ovary during oocyte growth (vitellogenesis), maturation and atresia. The ovarian transcripts up-regulated from pre-vitellogenesis (primary growth) to vitellogenesis (secondary growth) include several ones with putative mitochondrial function/location, suggesting high energy production (NADH dehydrogenase subunits, cytochromes) and increased antioxidant protection (selenoprotein W2a), whereas other regulated transcripts are related to cytoskeleton and zona radiata organization (zona glycoprotein 3, alpha and beta actin, keratin 8), intracellular signaling pathways (heat shock protein 90, Ras homolog member G), cell-to-cell and cell-to-matrix interactions (beta 1 integrin, thrombospondin 4b) and the maternal RNA pool (transducer of ERBB2 1a, neurexin 1a). In Senegalese sole females undergoing extensive follicular atresia, these authors found that despite few transcripts were up-regulated, most of them were remarkably localized in follicular cells of atretic follicles; these transcripts had inferred roles in lipid transport (apolipoprotein C-I, fatty acid-binding protein 11), chemotaxis (leukocyte cell-derived chemotaxin 2), angiogenesis (thrombospondin) and prevention of apoptosis (S100a10 calcium binding protein).

4.3.4 Endocrine parameters in wild versus F1 females

A comparative analysis performed by Guzmán et al. (2009b) on wild-caught and F1 Senegalese sole female breeders, both maintained under similar culture and feeding conditions for several years, showed no differences in pituitary levels of GnRH-1, GnRH-2 and GnRH-3, pituitary *lhβ* mRNA and plasma levels of T and E2, but differences in pituitary mRNA levels of *gpα* and *fshβ* and plasma VTG (wild > F1). Another study showed similar plasma levels of all reproductive hormones (FSH, LH, T and E2) between mature females from both origins (Chauvigné et al. 2016), concluding that they do not seem to display any apparent dysfunction in the operation of the reproductive axis. Nevertheless, other comparative studies using wild (captured and directly sampled, i.e., not stocked in captivity) and F1 Senegalese sole female breeders (maintained under natural photo- and thermoperiods and fed with broodstock dry pellets) have shown that F1 breeders have differential cyclooxygenase (cox-2; an enzyme involved in the synthesis of prostaglandins from fatty acids) mRNA expression levels in the gills and oviduct, prostaglandin concentrations in the gills and blood and fatty acid compositions in muscle, liver and ovary as compared to wild females (Norambuena et al. 2012a,b). According to these changes, the authors concluded that the lipid composition of dry pellets used to feed Senegalese sole captive broodstocks (particularly the amounts of ARA and the ratio EPA/ARA) must be adequately balanced to fit as much as possible the diet consumed by the fish in the nature, as unbalanced diets may have negative effects on the female reproductive axis (Cabrita et al. 2011, Carazo et al. 2011, Norambuena et al. 2012a,b). On this regard, it deserves special attention the altered production and release of prostaglandins, which have been reported to regulate intra-ovarian steroid production as well as to serve as spawning pheromone in teleost fish (Norambuena et al. 2012a,b).

5. Hormonal stimulation of reproduction

The use of hormonal therapies is a common practice in aquaculture, not only for the treatment of reproductive disorders, but also to optimize the reproductive performance of broodstock. Hormonal treatments are effective to stimulate spermatogenesis and sperm production in males and oocyte maturation, ovulation and spawning in females in a variety of fish species (Zohar et al. 1990, Donaldson 1996, Mañanós et al. 2009, Mylonas et al. 2010, 2017). There are two major types of hormonal treatments, the so-called "first generation", based on the use of drugs derived from pituitary hormones (specifically gonadotropins) and the "second generation", based on drugs derived from brain neurohormones (Mañanós et al. 2009). The main difference between them is the target organ; the gonadotropin preparations acting directly over the gonad and the neurohormone preparations acting mainly over the pituitary, to stimulate the release of the fish´s endogenous gonadotropins that will then act over the gonad.

First generation preparations include, pituitary homogenates, commercially available pituitary extracts, purified fish gonadotropins and mammalian gonadotropins, mostly hCG. Also, recombinant gonadotropins, produced *in vitro* through DNA recombinant techniques, would be included in this type; those from mammalian origin are widely available for veterinary and clinical use, while fish recombinant gonadotropins are still under research and remains at an experimental level. The hCG, although from human origin, is the most widely used "gonadotropin" treatment in aquaculture, because of its effectiveness, wide availability and standardized activity. Nevertheless, there is an important disadvantage on the use of gonadotropins, including hCG, which relates to the complex structure of the molecule; gonadotropins are large and species-specific molecules and thus, they can cause immune response when administered to a non-related species, which may reduce the responsiveness of fish to repetitive treatments (Zohar and Mylonas 2001).

Second generation treatments include all drugs derived from neurohormones, being the most important ones the agonists/analogues of GnRH (GnRHa). The GnRHa's are highly effective to induce spermiation and ovulation in many fish species (Mylonas and Zohar 2001) and has become the most widely used hormonal treatment in aquaculture. The GnRHa-based therapies have important advantages over gonadotropin preparations, mainly related to the simple structure of the molecule; they are easy to synthesize and due its small size, do not induce immune response in the treated animals and thus, repeated treatments can be applied with no desensitization problems. Also, GnRH's are generic and thus, GnRHa treatments can be used in all fish species and produce similar physiological responses.

A limitation of the GnRHa's would be the short half-life of the molecule; the effect of a GnRHa administered via injection may last for around 12–72 hr. This problem has been overlapped with the development of slow release delivery systems (Zohar et al. 1990, Mylonas and Zohar 2001), which cause a sustained release of the GnRHa into the bloodstream and thus, a prolonged stimulation of endogenous gonadotropin release from the pituitary, which can last for several weeks, depending on the type of delivery system used (Mylonas et al. 1995). The use of implants is specially recommended for multiple-batch spawning fish and have shown to be highly effective to induce ovulation and spawning and also sperm production in a wide variety of fish species (Mylonas et al. 2004, 2007, Marino et al. 2003, Guzmán et al. 2009a).

Apart from the GnRHa's, other commercially available drugs within the "second generation" type of treatments, are the dopamine antagonists; drugs that blocks the brain dopaminergic inhibitory system (e.g., pimozide, domperidone, metoclopramide). These drugs are normally administered with GnRHa as a combined treatment and are used in those fish exhibiting a strong endogenous inhibitory dopaminergic tone, i.e., cyprinids and other freshwater fish species (Dufour et al. 2005, 2010). Other neurohormones are investigated for their potential use as drugs for hormonal induction in fish, including GnIHs, kisspeptins and neurokinin B (Muñoz-Cueto et al. 2017, Zmora et al. 2017), but this is still at an experimental level and needs further research for development and commercial application.

Many cultured fish exhibit some kind of reproductive dysfunction, that normally requires the application of hormonal treatments. In the case of flatfishes, common reproductive disorders are, the inhibition of spawning in females, after completion of oocyte maturation and ovulation and, diminished sperm production and production of abnormally viscous milt in males, which has negative consequences for the fertilization capacity, both in tank-spawning and artificial fertilization (Vermeirssen et al. 2000, Mañanós et al. 2009). Reproductive problems affect predominantly cultured broodstocks (F1 and successive generations) in comparison to a better reproductive performance of wild-caught broodstocks (Farquharson et al. 2018). A good deal of knowledge on reproductive biology and responses to different hormonal therapies has been accumulated for several decades for the most relevant aquaculture flatfish species (e.g., Japanese flounder *Paralichthys olivaceus*, Atlantic halibut, turbot *Scophthalmus maximus*). For these species, hormonal treatments have demonstrated to be effective to induce and synchronize ovulation in females and to enhance spermiation and milt fluidity in males, allowing the development of efficient broodstock management and reproductive control technologies and further establishment of successful aquaculture industries (Mañanós et al. 2009).

In the case of *Solea* species, the recent interest in developing sole aquaculture, has encouraged intense research on the reproductive biology of soles, especially in Senegalese sole and the development of hormonal therapies to stimulate and control reproduction in captivity (Morais et al. 2016). Similar to other flatfish, major reproductive problems of cultured Senegalese soles are diminished sperm production in males and reduced spawning in females and most importantly, the absence of fertilized spawning from F1 broodstock.

5.1 *Hormonal induction of spermiation*

As mentioned before, a common reproductive dysfunction of cultured males of flatfish species, is the production of reduced amounts of highly viscous milt, especially in hatchery produced cultured individuals (F1 generation). The amount of expressible milt that can be obtained from a wild adult Senegalese sole male, although highly variable between individuals, is around 0.1 ml for a 1.5 kg fish, while for F1 males, this is reduced to around 0.04 ml. Only few studies have tested the efficiency of hormonal treatments to induce spermiation in males of *Solea* species and these were restricted to Senegalese sole. Depending on the study, the effectiveness of hormonal induction has been determined by analyzing, (1) sperm quantity (volume of expressible milt) and quality (percentage of motile cells, cell density and viability, spz linearity and velocity), (2) testicular development and maturation (i.e., spermiogenesis, spermatogenesis) by histology and, (3) activation of the endocrine system, by measuring gene expression and/or synthesis of reproductive hormones, mainly plasma levels of androgens (T and 11KT) and progestins (DHP and 20βS). These studies have been focused on F1 broodstock, as these individuals are more affected by reproductive dysfunctions than their wild counterparts. There is only one published work on hormonal induction of wild-caught Senegalese sole males acclimated to captivity (Cabrita et al. 2011). In this study, males treated with a single injection of GnRHa (20 µg/kg) showed increased 11KT and T plasma levels (65 and 62 percentage increase, respectively) at 2 days post-treatment (dpt), but with no significant effect on expressible milt volume (around 35 µl of milt per kg BW fish), neither on some sperm quality parameters (e.g., cell percentage of motile cells, spz linearity), although a slight increase was observed at 2 dpt in cell viability and sperm velocity.

In F1 Senegalese sole males, several drugs and administration protocols have been tested and most of them showed to be effective, to some extent, to induce spermatogenesis and spermiation and affect positively the endocrine axis. Treatment with GnRHa has been studied under, (1) different doses (range 5–50 µg/kg BW), (2) different administration protocols (single injection, multiple injection, slow release delivery systems) and, (3) different combined treatments with additional drugs. Regarding the administration method, it should be remarked that the administration of GnRHa via injection, usually applied into the dorsal musculature and dissolved in saline, produce an acute effect on the treated fish that lasts for a short period of time (few days). This is normally not a recommended treatment for male fish, as the asynchronous development of the testis and the prolonged period of spermiation normally requires a prolonged and sustained stimulation of the process and thus, multiple injections or slow delivery implants are normally more effective therapies, as widely demonstrated in several fishes, including the Senegalese sole (Agulleiro et al. 2006, Mañanós et al. 2009, Guzmán et al. 2011a).

Treatment of F1 Senegalese sole males with GnRHa EVAc implants (dose 40 µg/kg; 2 implants at 0–21 days) have demonstrated to effectively stimulate spermatogenesis, effect that was not observed in males treated with GnRHa injection (dose 25 µg/kg; 2 injections at 0–21 days) (Guzmán et al. 2011a). In this study, the histological analysis of the testis (at 42 dpt) showed that, compared to control and GnRHa injected males, the testis of EVAc implanted males showed higher proliferation of germinal cysts containing spc, higher abundance and size of clutches of spd transforming into spz, increased lumen size of the medullar efferent ducts and higher amount of spz within the medullar duct system, indicating stimulated spermatogenesis and spermiogenesis in implanted males. Correlated to this information, EVAc implants stimulated androgen secretion, with increased plasma levels of T and 11KT compared to controls or injected males (Agulleiro et al. 2006, Guzmán et al. 2011a). Nevertheless, effects on sperm production (volume of expressible milt) and sperm quality are highly variable between fish and difficult to demonstrate. Treatment with EVAc implants have shown to have either no effect on gonad size (GSI) and sperm production (EVAc implant 25 ug/kg dose; Agulleiro et al. 2006) or to have a slight stimulatory effect on both GSI (increment from 0.06 to 0.1) and sperm production (1.5-fold increase) (EVAc implant 40 ug/kg dose; Guzmán et al. 2011a).

Treatment of F1 Senegalese sole males with hCG have demonstrated to have higher stimulatory effects than GnRHa EVAc implants. A multiple hCG injection protocol (dose 1,000 IU/kg; 6 weekly injections) induced androgen secretion (plasma levels of T and 11KT) for longer time and at higher levels than males treated with GnRHa (EVAc implants and injection) and produced the highest stimulation of testicular development and maturation, which was correlated with stimulated sperm production (2-fold compared to controls) and increased GSI (2-fold comparted to controls and in this case, similar effect to EVAc implants) (Guzmán et al. 2011a). The efficiency of hCG to stimulate steroidogenesis and indirectly, spermatogenesis, has also been demonstrated in juvenile Senegalese sole (60 g BW), in which a single injection of hCG (5,000 IU/kg BW) increased (24 hr pt) plasma levels of T and 11KT and testicular gene expression of steroidogenic enzymes and multiple genes potentially involved in steroidogenesis, spermatogenesis and germ cell maturation (qPCR and microarray analysis) (Marin-Juez et al. 2011).

Other than hCG, another "gonadotropin" treatment tested in Senegalese sole are recombinant gonadotropins, which although at an experimental research level offer a method that can increase sperm production (Chauvigne et al. 2017, 2018). In these studies, several administration schedules were tested, on pubescent (300 g BW) and adult (1000 g BW) sole. Pubescent sole that were administered weekly injections (10–11 injections) of combined FSH (dose range 3–16 µg/kg) and LH (dose range 2–17 µg/kg) showed stimulation of gonad weight (2-fold), plasma 11KT levels, spermatogenesis (germ cell proliferation and maturation, spz differentiation), expression of spermatogenesis-related genes, sperm production (7-fold) and slightly sperm quality (Chavigne et al. 2017). Whilst, adult sole that were administered 9 µg/kg FSH during six successive weeks followed by a single injection of 9 µg/kg LH during the autumn showed stimulation of 11KT (4-fold) and sperm production (4-fold) (Chauvigne et al. 2018). Nevertheless, further research is necessary for optimization of this protocol as the effects on sperm quantity and quality were highly variable between fish and season dependent and also, a harmful effect of higher dose treatments over the testis (Leydig cell survival and steroidogenic capacity) was observed (Chauvigne et al. 2017, 2018).

Two studies tested the use of combined treatments to improve the efficiency of GnRHa treatments, by the co-administration of GnRHa with androgens (i.e., 11-ketoandrostenedione (OA)) (Agulleiro et al. 2007) or dopamine antagonists (i.e., pimozide) (Guzmán et al. 2011a). The combined treatment with OA (GnRHa EVAc implants (50 µg/kg) + OA injection (2 or 7 mg/kg)) did not show to potentiate GnRHa stimulated sperm production (2-fold increase in milt volume) but increased the effects over the testis and the endocrine system (Agulleiro et al. 2007). Histological analysis (28 dpt) showed that GnRHa alone slightly stimulated spermatogenesis and this was enhanced by the combined GnRHa+OA treatment, that showed a higher number of spd than fish treated with GnRHa alone. Also, GnRHa alone transiently increased (7 dpt) plasma 11KT and 5β-reduced metabolite(s) of DHP (28 dpt), whereas co-treatment GnRHa+OA caused a higher elevation of both plasma 11KT and free+sulphated 5β-reduced metabolites of DHP (Agulleiro et al. 2007). The study testing the potential effects of the dopamine antagonist (pimozide; D2-receptor antagonist), either alone (3 injections of 5 mg/kg; 0-10-24 dpt) or combined with GnRHa (EVAc implant 40 µg/kg at 0 dpt and injection of 25 ug/kg at 24 dpt), showed positive stimulatory effects at different levels (Guzmán et al. 2011b). Milt volume was increased, with respect to controls (25 dpt), by all treatments, 2-fold by GnRHa alone, 3.5-fold by pimozide alone and 5.2-fold by the combined treatment, indicating an stimulatory effect of pimozide over sperm production and this was correlated with increased gonad size (GSI from 0.07 in controls to 0.09 in all treated fish) and stimulated spermatogenesis (histology); both GnRHa alone and pimozide alone increased similarly the presence of cysts containing early developing germ cells (spg and spc), reduced the number of spd and increased the abundance of spz in the medullar efferent ducts and all these parameters were increased significantly in males receiving the combined treatment. At the endocrine level, none of the treatments modify (25 dpt) the pituitary content of GnRHs (GnRH-1, GnRH-2, GnRH-3) or gene expression of gonadotropins (*fshβ*, *lhβ*), but both GnRHa alone and pimozide alone increased T and 11KT plasma levels (25 dpt), while the combined treatment showed lower androgen levels, probably as a consequence of shifted steroidogenesis through the synthesis of maturation inducing steroids. Taken together, the results of the study demonstrated an inhibitory dopaminergic tone over the reproductive axis of male Senegalese soles and opens the possibility of optimizing hormonal therapies through the use of anti-dopaminergic drugs (Guzmán et al. 2011b).

In conclusion, therapies based on GnRHa implants, combined GnRHa treatments (GnRHa with 11KT precursors or dopamine antagonists), recombinant gonadotropins and mostly hCG multiple injections, are effective to stimulate spermatogenesis and sperm production in males of Senegalese sole, with no detrimental effects on sperm quality, but the effects are shown to be variable between fish and reduced (around 2-fold increase in milt volume) and thus, their impact and application on aquaculture protocols is still limited.

5.2 Hormonal induction of spawning

A common reproductive dysfunction in flatfishes is the inhibition of egg release, after completion of final oocyte maturation and ovulation. This problem can be overlapped by stripping of the ovulated eggs (natural or hormonally induced) and artificial fertilization, a protocol that is commonly used in the aquaculture of many flatfish species, such as turbot, Japanese flounder or halibut (Mañanós et al. 2009). For a given species, reproductive problems use to affect F1 females to a greater extent than wild females. For example, spontaneous tank spawning of captive adapted wild broodstock has been described for Japanese flounder (Tsujigado et al. 1989), turbot (Devauchelle et al. 1988), Atlantic halibut (Holmefjord et al. 1993) and common sole (Devauchelle et al. 1987), but further establishment of captive reared broodstock generations (F1 and successive) have shown that, in most cases, hand stripping of gametes and artificial fertilization is required.

The two main cultured species of the family Soleidae (*Solea senegalensis* and *Solea solea*), reproduce spontaneously in captivity and in contrast to other flatfishes, egg spawning is obtained in the tank, without egg stripping (Baynes et al. 1993, Guzmán et al. 2008, Mañanós et al. 2009). However, for the Senegalese sole this is only true for wild breeders, because F1 broodstock show an almost total absence of fertilized tank spawning. This is a critical problem because hatcheries continue to depend on captures and acclimation of fish from the wild, which limits the application of breeding selection programs and the consolidation of a Senegalese sole aquaculture industry (Howell et al. 2009, 2011, Morais et al. 2016).

Hormonal treatments to induce and synchronize ovulation and spawning in females of flatfish have been extensively tested for the most consolidated species, including the Japanese flounder (Tsujigado et al. 1989, Tanaka et al. 1997), Atlantic halibut (Norberg et al. 1991, Vermeirssen et al. 2000, 2004) and turbot (Devauchelle et al. 1988, Howell and Scott 1989, Forés et al. 1990, Mugnier et al. 2000). When looking for a quick response, such as the induction of mature oocytes for ovulation, an acute treatment of GnRHa through single injection is a very effective therapy, with a latency period or around 1–2 days. When looking for tank spawning, the most effective therapy for flatfishes is the single administration of GnRHa implants, which are especially effective in female fish exhibiting group-synchronous multiple batch ovarian development, to induce a long period of multiple spawns, as is the case of most flatfishes, including soles (Mañanós et al. 2009).

For soles, scientific information on reproductive biology and reproductive performance in captivity, including some studies on the effects of hormonal treatments on cultured broodstock, has been performed in the last fifteen years. Depending on the study, the effectiveness of hormonal induction has been determined by analyzing effects on, (1) the number of spawns and egg quantity and quality (floatability, fertilization rates), (2) oocyte development and maturation (histology, biopsies) and, (3) the activation of the endocrine reproductive system.

Most of the studies have been focused in Senegalese sole, with only few published works in common sole. In this last species, the pioneer studies of Ramos tested the effects of hCG (Ramos 1986a) and GnRHa (Ramos 1986b) treatments on wild-caught breeders. In a first study, this author showed that injection of hCG was highly effective to induce tank spawning in mature females. Control females spawned spontaneously, with a total egg production of 256,000 eggs (6 spawns) and females injected with hCG (single injection of 250 or 500 IU/kg) produced higher number of spawning events (16 and 12 spawns, respectively) and higher total egg production (355,000 and 358,000 eggs, respectively), with high hCG doses (750–1,000 IU/kg) being less effective than lower ones (spawns ranged on 2–8 and egg production on 30,000–340,000 eggs). The author observed a high variability in egg quality between spawns (fertilization rate percentage ranged 15–80) and in the responsiveness between females, with lapsed times to ovulation varying between 24 and 96 hr, which was associated to different stages of oocyte development of the females at the time of treatment (Ramos 1986a). In a second study, Ramos tested the effects of GnRHa injection (10 µg/kg) and observed induced spawning in all females at 48 hr post-injection, with further spawning events in the following weeks, with a total of 4-6 spawning events per female and an egg production in the range of 8,000–21,000 eggs per spawning (average percentage fertilization rates 20–70). Comparatively, control females did spawn spontaneously during the 5-week period of the study, producing 3–5 spawning events per female and 9,000–21,000 eggs per spawn, with similar fertilization rates. The author concluded that GnRHa induction did not modify total egg production or the number of spawning events in common sole broodstock but was highly effective to accelerate oocyte maturation and synchronize ovulation, as all GnRHa treated females spawned at 48 hr after injection compared to the first spawning in control females observed 3 weeks later (Ramos 1986b).

Another study in wild-caught common sole, also tested the effects of GnRHa injection (Berlotto et al. 2006). In this study, control broodstock spawned spontaneously in captivity (February/March to May), under natural temperature and photoperiod, but spawning performance was highly variable between years and females. Experiments performed during seasons of failed spontaneous spawning, showed that a single GnRHa injection was effective to induce spawning for several weeks, with low doses (20 µg/kg) being more effective (egg production and egg quality) than high doses (40 µg/kg); but again, a high variability was observed between females and spawns, with fertilization rate percentages that ranged between 0 and 80.

Studies on Senegalese sole used F1 broodstock and focused on the GnRHa therapy, determining optimal doses and administration protocols. The acute treatment via injection was efficient to induce ovulation and spawning in Senegalese sole, but only in females at advanced stages of oocyte maturation. Administration of GnRHa via either multiple injections or sustained release delivery systems does not require the females to be as advanced as for the injection method, as the prolonged (weeks) effect of these systems is adequate to stimulate not only ovulation but also the progression through oocyte maturation. In any case, for successful stimulatory effects, an accurate estimation of the maturity stage of the female should performed before treatment (Fig. 5). This evaluation can be done by examination of ovarian biopsies under the binocular, to determine oocyte diameter and morphological features of the oocytes, which are useful parameters to determine the developmental stage of the different batches of intra-ovarian oocytes (Fig. 5). Nevertheless, an additional useful approach is the assessment of the female maturity stage by external visual examination (Anguis and Cañavate 2005, García-López et al. 2006b, 2007, Guzmán et al. 2008). This is possible because, (1) the size of the ovaries can be easily estimated by abdominal palpation, (2) a very patent abdominal swelling can be seen in mature females and, (3) the orange coloration of the mature ovary can be seen on the white skin of the blind side of the fish. This procedure has led to the elaboration of different

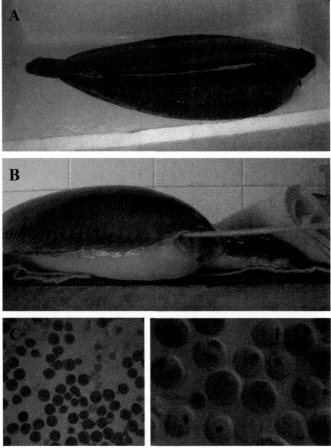

Fig. 5. Photographs illustrating the estimation of the maturity stage in female Senegalese sole, by external observation of abdominal swelling (A) or by cannulation (B) and observation of ovarian biopsies (C, D). (A) Mature female Senegalese sole showing high abdominal swelling along the ventral region, due to the growth of the ovarian lobes. (B) Cannulation of a mature female through the oviduct and collection of intraovarian oocytes. (C) Observation under the binocular of a fresh ovarian biopsy (maintained in Ringer solution), obtained from a female during the spawning period; see the presence of dark oocytes of different sizes (different developmental stages), some translucent ovulated eggs (diameter around 1050 μm) and some post-ovulatory follicular envelopes (sign of previous spawning events). (D) Similar ovarian biopsy observed under the binocular after immersion in clarification solution (ethanol:formol:acetic acid, 6:3:1) for the observation of intra-oocyte structures; see a dominant batch of largest mature oocytes characterized by the peripheral position of the germinal vesicle (animal pole), together with smaller vitellogenic oocytes with a centrally located nucleus (Photos by Evaristo L. Mañanós).

subjective scales of apparent maturation stages, either based on discrete observations or scorings of females at a particular time (Anguis and Cañavate 2005, Guzmán et al. 2008) or, if fish are individually identified by pit-tags, on multiple observations of the same female during consecutive time periods (e.g., month to month; García-López et al. 2006b,c, 2007, 2009). In both cases, it has been confirmed that, if the observations are performed by experienced personnel and/or combining the scoring of different observers, the estimation of the apparent ovarian maturation match to a great extent with the pattern of follicular development determined by histological analysis or the evolution of the GSI.

Hormonal treatment of F1 Senegalese sole with a single injection of GnRHa (dose 25 ug/kg), has been shown to be effective to induce ovulation in 70–90 per cent of the females, with a lapsed time to ovulation of 35–48 hr pt and fecundities around 125,000 eggs/kg (Rasines et al. 2013). In a study testing different GnRHa doses and determining tank spawning, injection of 1 µg/kg induced 2 spawns (25,688 eggs/kg), 5 µg/kg induced 5 spawns (65,476 eggs/kg) and 30 µg/kg was ineffective, with a similar lapsed time to first spawning of 3–4 dpt (Agulleiro et al. 2006). Induced spawning showed an egg viability (percentage of buoyant eggs) around 40, but all eggs were unfertilized (Agulleiro et al. 2006).

Testing a multiple GnRHa injection protocol, this was effective to induce multiple ovulations (Rasines et al. 2012) and spawning (Agulleiro et al. 2006) in F1 Senegalese soles. Four weekly injections of GnRHa (5 µg/kg) or eight injections given every 3 days, induced 21 (151,684 eggs/kg) and 34 spawns (256,228 eggs/kg), respectively, with similar egg quality (buoyancy percentage around 50) and a total absence of fertilization (Agulleiro et al. 2006). Four weekly injections of GnRHa (25 ug/kg) induced ovulation (checked by abdominal pressure and hand stripping of eggs) at 41–44 hr pt, with some females responding to all 4 injections, others to 3 and others to only the first 2 injections, and providing an average of 574,000 eggs/kg in 6.3 ovulations per female (Rasines et al. 2012).

Finally, slow release delivery systems have been tested in F1 Senegalese sole broodstock and showed to be the most effective system to induce multiple tank spawning and increase egg production (Agulleiro et al. 2006, Guzmán et al. 2009a). In a trial testing EVAc implants loaded with different GnRHa doses (5-25-50 µg/kg), single administration of 5 µg/kg implants produced no effect (no spawning), whereas 25 and 50 µg/kg implants induced 8 consecutive spawns starting at 7 dpt (egg production 6,000 eggs/kg) or 14 spawns starting at 3 dpt (41,000 eggs/kg), respectively, with again no fertilization in all spawnings (Agulleiro et al. 2006). The results, in terms of egg production, were similar to those obtained through a multiple GnRHa injection protocol (3 weekly injections of 5 µg GnRHa/kg), which induced 3 spawns and a total egg production of 29,082 eggs/kg (Agulleiro et al. 2006), but with the inherent disadvantage of repetitive manipulation of the broodstock. Guzmán et al. (2009a) tested the effects of two different GnRHa loaded slow-release delivery systems, EVAc implants (Zohar et al. 1990) and biodegradable microspheres (Mylonas et al. 1995), on the stimulation of oocyte maturation, spawning and steroid secretion in F1 Senegalese soles. This is the only study where the kinetics of *in vivo* release of GnRHa has been evaluated in Senegalese sole, by analyzing blood levels (ELISA) of the administered GnRHa. The study showed that GnRHa lasted in the bloodstream (detectable levels) for 3, 7 and 14 days, after administration of GnRHa via injection, microspheres and EVAc implant, respectively (Guzmán et al. 2009a). All GnRHa treatments induced a transient elevation of T and E2 plasma levels, that lasted for 1–3 days. Non-treated control broodstock of this study spawned sporadically during the 2-month experimental period (5 spawns, 9,720 eggs/kg), with a mean percentage egg viability of 58 and no fertilization. The analysis of ovarian biopsies (weekly) showed maturation of only few females in the GnRHa injection, whereas treatment with both GnRHa implants and microspheres caused a clear differentiation of the oocyte size distribution, beginning at 7 dpt and developing for several weeks, with a high population of oocytes in the groups of 600, 700 and 800 microns, indicating stimulated progression of batches of oocytes through the process of maturation. Correlating this information, the GnRHa injection caused no apparent immediate effect on spawning, although spawning events (13 spawns) and egg production (41,960 eggs/kg) was slightly higher than controls. The absence of a rapid response to the GnRHa injection (5 or 25 µg/kg) in this study was associated to the low maturation developmental stage of the females at the time of treatment, a critical parameter on the effectiveness of acute treatments. In contrast, the two sustained-release GnRHa-delivery systems were very effective inducing tank spawning, with spawns starting at 4–5 dpt and continuing daily for 3 weeks, with an important increase in egg production (total fecundity 208,800 and 141,880 eggs/kg for implants and microspheres, respectively, in 26 spawns). This is a clear reflection not only on the effectiveness of slow release delivery systems in multiple-batch spawning fish, such as Senegalese sole, but also an indication of their effectiveness on females that are not necessarily at very advanced stages of oocyte maturation, when the injection treatment might fail. Egg quality, assumed by floatability and observation under the binocular, was variable between spawns and similar between groups and ranged from 0 to 95 per cent (mean around 40–60) but again, no fertilization was obtained in any group. Thus, the GnRHa slow release systems were highly efficient to induce oocyte maturation and spawning, presumably through the long-term elevation of pituitary LH release, but the final effectiveness of the treatment was negated by possible dysfunctions in reproductive behavior as, again, a total absence of fertilized spawning was observed. An example of induced spawning caused by treatment of female Senegalese sole with GnRHa injection and implants is given in Fig. 6.

The potential existence of a dopamine inhibition over female reproduction was investigated in Senegalese sole in a study by Guzmán et al. (2011b). In this study, F1 females were treated with the dopamine antagonist pimozide (3 injections of 5 mg/kg; 0-10-24 dpt), GnRHa (EVAc implant 40 µg/kg at 0 dpt and injection of 25 ug/kg at 24 dpt) or the combined GnRHa+pimozide treatment and the effects evaluated on spawning and the endocrine system. Over the endocrine axis, GnRHa implants increased gonadotropin (*lhβ* and *gpα*) gene transcription in the pituitary (25 dpt) and this effect was not modified by the combined treatment, with pimozide alone having no effect. In this study, the control F1 broodstock did not spawn during the 1-month experimental period, while pimozide-treated fish showed some sporadic spawning (6 spawns, 12,000 eggs/kg), indicating a slight stimulatory effect of the treatment. The GnRHa treatment induced, as observed in other studies, daily egg release from day 3 to 15 and this kinetic was not modified by the combined treatment. The combined treatment did not either modified GnRHa effects on the number of spawns (13 spawns) and total fecundity (260,000 and 189,000 eggs/kg, respectively). Egg quality (floatability percentage and morphology) was similar between groups and to other studies and again, all spawnings were unfertilized. With these results, the authors could not demonstrate the activity of a dopamine inhibition in female Senegalese sole, but some data would suggest its existence and maybe its potential higher activity at other sampling points or maturation stages, which deserves to be further investigated for a definitive characterization in this and other related species.

In conclusion, GnRHa treatments are highly effective to induce ovulation and spawning in F1 Senegalese sole, with no detrimental effects on egg quality. For the purpose of artificial fertilization, the method of choice would be the single injection of GnRHa at doses of 15–25 ug/kg. This method requires an accurate determination of the maturity stage of

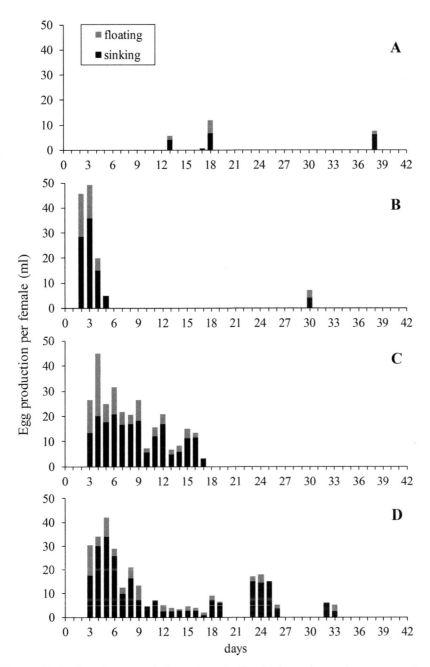

Fig. 6. Hormonal induction of spawning in Senegalese sole. Graphs shows the spontaneous spawning of a control broodstock (A) and the hormone-induced spawning of broodstock treated with, (B) single GnRHa injection (day 0; 25 µg/kg), (C) single GnRHa implant (day 0; 40 µg/kg) and (D) double GnRHa implant (days 0 and 21; 40 µg/kg). Broodstock tanks contained 16–20 fish of around 1500 g BW (1:1 sex ratio).

the females (ovarian biopsy) and, if performed over females at final stages of oocyte maturation, the GnRHa injection effectively induce ovulation at 35–48 hr pt (mean response at 41 hr pt) and eggs can be easily hand-stripped, with a fecundity of around 250,000 eggs/kg and an egg viability around 50 percent, although highly variable between fish (percentage range 0–90). For the purpose of tank spawning, the method of choice are slow release delivery implants, which induce daily spawning for 2–3 weeks and a high amount of good quality eggs. The response could be similar to that obtained through multiple GnRHa injections, but avoiding detrimental effects caused by repetitive handling of the broodstock. Also, implants have the advantage of stimulating oocyte maturation through sustained stimulation of the endogenous pituitary LH release and thus, they are effective on females at several stages of oocyte development (early, mid, late maturation). Finally, it should be remarked that the major problem regarding reproduction of F1 Senegalese sole, which is the total absence of fertilized tank spawning, remains unsolved. This is caused by the absence of reproductive behavior and courtship in both sexes. It is true that hormonal treatments are useful to induce sperm production in males and to induce ovulation and spawning of "good quality" eggs in females, but they are ineffective to stimulate reproductive behaviour. Thus, further research would have to focus on environmental and hormonal protocols that can stimulate reproductive behaviour in soles, probably through the potentiation of the synthesis and secretion of pheromonal compounds.

References

Abascal, F.J., C. Megina and A. Medina. 2004. Testicular development in migrant and spawning bluefin tuna (*Thunnus thynnus* L.) from the eastern Atlantic and Mediterranean. Fish. Bull. 102: 407–417.

Accogli, G., S. Zizza, A. García-López, C. Sarasquete and S. Desantis. 2012. Lectin-binding pattern of the Senegalese sole, *Solea senegalensis* oogenesis. Micr. Res. Tech. 75: 1124–1135.

Agulleiro, M.J., V. Anguis, J.P. Cañavate, G. Martínez-Rodríguez, C.C. Mylonas and J. Cerdá. 2006. Induction of spawning of captive reared Senegalese sole (*Solea Senegalensis*) using different delivery systems for gonadotropin-releasing hormone agonist. Aquaculture 257: 511–524.

Agulleiro, M.J., A.P. Scott, N. Duncan, C.C. Mylonas and J. Cerdá. 2007. Treatment of GnRHa-implanted Senegalese sole (*Solea senegalensis*) with 11-ketoandrostenedione stimulates spermatogenesis and increases sperm motility. Comp. Biochem. Physiol. Part A. 147: 885–892.

Ahmed, A.I., M.M. Sharaf and H.A. Laban. 2010. Reproduction of the Egyptian sole, *Solea aegyptiaca* (Actinopterygii: Pleuronectiformes: Soleidae), from Port Said, Egypt, Mediterranean Sea. Acta Ichthyol. Piscat. 40: 161–166.

Aliaga-Guerrero, M., J.A. Paullada-Salmerón, V. Piquer, P. Terry-Castro, V. Gallego-Recio, E.L. Mañanós and J.A. Muñoz-Cueto. 2018. Gonadotropin-inhibitory hormone in the flatfish, *Solea Senegalensis*: molecular cloning, brain localization and physiological effects. J. Comp. Neurol. 526: 349–370.

Anguis, V. and J.P. Cañavate. 2005. Spawning of captive Senegal sole (*Solea senegalensis*) under a naturally fluctuating temperature regime. Aquaculture 243: 133–145.

Arellano, J.M. and C. Sarasquete. 2005. Atlas Histomorfológico del lenguado senegalés, *Solea senegalensis*. Publicaciones CSIC, Madrid, Spain.

Assis, L.H.C., D. Crespo, R.D.V.S. Morais, L.R. França, J. Bogerd and R.W. Schulz. 2016. INSL3 stimulates spermatogonial differentiation in testis of adult zebrafish (*Danio rerio*). Cell Tissue Res. 363: 579–588.

Asturiano, J.F., L.A. Sorbera, J. Ramos, D.E. Kime, M. Carrillo and S. Zanuy. 2002. Group-synchronous ovarian development, ovulation and spermiation in the European sea bass (*Dicentrarchus labrax* L.) could be regulated by shifts in gonadal steroidogenesis. Sci. Mar. 66: 273–82.

Babin, P.J., O. Carnevali, E. Lubzens and W.J. Schneider. 2007. Molecular aspects of oocyte vitellogenesis in fish. pp. 39–76. *In*: Babin, P.J., J. Cerdà and E. Lubzens [eds.]. The Fish Oocyte: From Basic Studies to Biotechnological Applications. Springer, Dordrecht, Holland.

Barr, W.A. 1963. The endocrine control of the sexual cycle in the plaice, *Pleuronectes platessa* (L.). III. The endocrine control of spermatogenesis. Gen. Comp. Endocrinol. 3: 216–225.

Bayarri, M.J., J.M. Guzmán, J. Ramos, V. Piquer and E. Mañanós. 2011. Annual variations of maturation inducing steroid in two cultured generations of Senegalese sole, *Solea senegalensis*. S8. pp. 120–121. *In*: Joy, K.P., R.M. Inbaraj, R. Kirubagaran and R. Chaube [eds.]. Reproductive Physiology of Fish. Indian Society for Education and Environment, India.

Baynes, S.M., B.R. Howell and T.W. Beard. 1993. A review of egg production by captive sole, *Solea solea* (L.). Aquac. Fish Manag. 24: 171.

Baynes, S.M., B.R. Howell, T.W. Beard and J.D. Hallam. 1994. A description of spawning behaviour of captive dover sole, *Solea solea* (L.). Neth. J. Sea Res. 32: 271–275.

Beirao, J., F. Soares, M.P. Herraez, M.T. Dinis and E. Cabrita. 2011. Changes in *Solea senegalensis* sperm quality throughout the year. Anim. Reprod. Sci. 126: 122–129.

Beltrán, E., V. Piquer, J.M. Guzmán, P. Swanson, E. Mañanós and R. Serrano. 2015. Quantification of plasma steroids in sole (*Solea senegalensis*) by ultra high performance liquid chromatography coupled to tandem mass spectrometry. pp. 170–172. *In*: Calduch, J., J.M. Cerdá-Reverter and J. Pérez [eds.]. Advances in Comparative Endocrinology, vol. VIII. Publications of the University Jaume I, Castellón, Spain.

Bertotto, D., J. Barbaro, A. Francescon, J. Richard, A. Libertini and A. Barbaro. 2006. Induced spawning in common sole (*Solea solea* L.). Aq. Res. 37(4): 423–427.

Blanco-Vives, B., L.M. Vera, J. Ramos, M.J. Bayarri, E. Mañanós and F.J. Sánchez-Vázquez. 2011. Exposure of larvae to daily thermocycles affects gonad development, sex ratio, and sexual steroids in *Solea senegalensis*, Kaup. J. Exp. Zool. 315: 162-169.

Borg, B. 1994. Androgens in teleost fishes. Comp. Biochem. Physiol. Part C. 109(3): 219–245.

Bromage, N., M. Porter and C. Randall. 2001. The environmental regulation of maturation in farmed finfish with special reference to the role of photoperiod and melatonin. Aquaculture 197: 63–98.

Bromley, P.J. 2003. The use of market sampling to generate maturity ogives and to investigate growth, sexual dimorphism and reproductive strategy in central and south-western North Sea sole (*Solea solea* L.). ICES J. Mar. Sci. 60: 52–65.

Brown-Peterson, N.J., H.L. Grier and R.M. Overstreet. 2002. Annual changes in germinal epithelium determine male reproductive classes of the cobia. J. Fish. Biol. 60: 178–202.

Cabrita, E., F. Soares and M.T. Dinis. 2006. Characterization of Senegalese sole, *Solea senegalensis*, male broodstock in terms of sperm production and quality. Aquaculture 261: 967–975.

Cabrita, E., F. Soares, J. Beirao, A. García-López, G. Martínez-Rodríguez and M.T. Dinis. 2011. Endocrine and milt response of Senegalese sole, *Solea senegalensis*, males maintained in captivity. Theriogenology 75: 1–9.

Carazo, I., I. Martin, P. Hubbard, O. Chereguini, E. Mañanós, A. Canario et al. 2011. Reproductive behavior, the absence of reproductive behavior in cultured (G1 generation) and chemical communication in the Senegalese sole (*Solea senegalensis*). Indian J. Sci. Technol. 4: 96–97.

Carazo, I. 2013. Reproductive behaviour and physiology of Senegalese sole (*Solea senegalensis*) broodstock in captivity. Ph.D. Thesis, University of Barcelona, Barcelona, Spain.

Carazo, I., O. Chereguini, I. Martín, F. Huntingford and N. Duncan. 2016. Reproductive ethogram and mate selection in captive wild Senegalese sole (*Solea senegalensis*). Spanish J. Agric. Res. 14(4): e0401.

Carnevali, O., G. Mosconi, A. Roncarati, P. Belveder, E. Limatola and A.M. Polzoneni-Magni. 1993. Yolk protein changes dying oocyte growth in European sea bass *Dicentrarchus labrax* L. J. Appl. Ichthyol. 9(3-4): 175–184.

Carnevali, O., C. Cionna, L. Tosti, E. Lubzens and F. Maradonna. 2006. Role of cathepsins in ovarian follicle growth and maturation. Gen. Comp. Endocrinol. 146(3): 195–203.

Carnevali, O. 2007. Reproductive endocrinology and gamete quality. Gen. Comp. Endocrinol. 153(1-3): 273–274.

Cerdá, J., F. Chauvigné, M.J. Agulleiro, E. Marin, S. Halm, G. Martínez-Rodríguez et al. 2008. Molecular cloning of Senegalese sole (*Solea senegalensis*) follicle-stimulating hormone and luteinizing hormone subunits and expression pattern during spermatogenesis. Gen. Comp. Endocrinol. 156: 470–481.

Chauvigné, F., A. Tingaud-Sequeira, M.J. Agulleiro, M. Calusinska, A. Gómez, A., R.N. Finn et al. 2010. Functional and evolutionary analysis of flatfish gonadotropin receptors reveals cladal- and lineage-level divergence of the teleost glycoprotein receptor family. Biol. Reprod. 82: 1088–1102.

Chauvigné, F., S. Verdura, M.J. Mazón, N. Duncan, S. Zanuy, A. Gómez et al. 2012. Follicle-stimulating hormone and luteinizing hormone mediate the androgenic pathway in Leydig cells of an evolutionary advanced teleost. Biol. Reprod. 87: 35.

Chauvigné, F., C. Zapater, J.M. Gasol and J. Cerdà. 2014a. Germ line activation of the luteinizing hormone receptor directly drives spermiogenesis in a non-mammalian vertebrate. Proc. Natl. Acad. Sci. USA 111: 1427–1432.

Chauvigné, F., C. Zapater, D. Crespo, J.V. Planas and J. Cerdá. 2014b. Fsh and Lh direct conserved and specific pathways during flatfish semicystic spermatogenesis. J. Mol. Endocrinol. 53: 175–190.

Chauvigné, F., S. Verdura, M.J. Mazón, M. Boj, S. Zanuy, A. Gómez et al. 2015. Development of a flatfish-specific enzyme-linked immunosorbent assay for Fsh using a recombinant chimeric gonadotropin. Gen. Comp. Endocrinol. 221: 75–85.

Chauvigné, F., E. Fatsini, N. Duncan, J. Ollé, S. Zanuy, A. Gómez et al. 2016. Plasma levels of follicle-stimulating and luteinizing hormones during the reproductive cycle of wild and cultured Senegalese sole (*Solea senegalensis*). Comp. Biochem. Physiol. A 191: 35–43.

Chauvigne, F., J. Olle, W. Gonzalez, N. Duncan, I. Gimenez and J. Cerda. 2017. Toward developing recombinant gonadotropin-based hormone therapies for increasing fertility in the flatfish Senegalese sole. PLoS ONE 12(3): e0174387.

Chauvigné, F., W. González, S. Ramos, C. Ducat, N. Duncan, I. Giménez et al. 2018. Seasonal-and dose-dependent effects of recombinant gonadotropins on sperm production and quality in the flatfish *Solea senegalensis*. Comp. Biochem. Physiol. Part A. 225: 59–64.

Costas, B., C. Aragão, I. Ruiz-Jarabo, L. Vargas-Chacoff, F.J. Arjona, M.T. Dinis et al. 2011. Feed deprivation in Senegalese sole (*Solea senegalensis* Kaup, 1858) juveniles: effects on blood plasma metabolites and free amino acid levels. Fish Physiol. Biochem. 37: 495–504.

Desantis, S., S. Zizza, Á. García-López, V. Sciscioli, E. Mañanós, V.G. De Metrio et al. 2010. Lectin-binding pattern of Senegalese sole *Solea senegalensis* (Kaup) testis. Histol. Histopathol. 25: 205–216.

Devauchelle, N., J.C. Alexandre, N. Le Corre and Y. Letty. 1987. Spawning of sole (*Solea solea*) in captivity. Aquaculture 66: 125.

Devauchelle, N., J.C. Alexandre, N. Le Corre and Y. Letty. 1988. Spawning of turbot (*Scophthalmus maximus*) in captivity. Aquaculture 69: 159.

Devlin, R.H. and Y. Nagahama. 2002. Sex determination and sex differentiation in fish: an overview of genetic, physiological, and environmental influences. Aquaculture 208: 191–364.

Dinis, M.T. 1986. Quatre Soleidae de lÉstuaire du Tage. Reproduction et Croissance Essai dʹElevage de *Solea senegalensis* Kaup 1858. Ph.D. Thesis, Université de Bretagne Occidentale, France.

Dinis, M.T. 1992. Aspects of the potential of *Solea senegalensis* Kaup for Aquaculture: larval rearing and weaning to artificial diets. Aquac. Fish. Manag. 23: 515–520.

Dinis, M.T., L. Ribeiro, F. Soares and M.C. Sarasquete. 1999. A review on the cultivation potential of *Solea senegalensis* in Spain and Portugal. Aquaculture 176: 27–38.

Dodd, J.M. 1977. The structure of the ovary of nonmammalianvertebrates. pp. 219–263. *In*: Zuckerman, S. and B. Weir [eds.]. The ovary. Academic Press, New York, USA.

Donaldson, E.M. 1996. Manipulation of reproduction in farmed fish. Anim. Reprod. Sci. 42: 381.

Dufour, S., F.A. Weltzien, M.E. Sebert, N. Le Belle, B. Vidal, P. Vernier et al. 2005. Dopamine inhibition of reproduction in teleost fishes. Ann. NY Acad. Sci. 1040: 9–21.

Dufour, S., M.E. Sebert, F.A. Weltzien, K. Rousseau and C. Pasqualini. 2010. Neuroendocrine control by dopamine of teleost reproduction. J. Fish Biol. 76(1): 129–160.

Duncan, N.J., A.K. Sonesson and H. Chavanne. 2013. Principles of finfish broodstock management in aquaculture: control of reproduction and genetic improvement. *In*: Allan, G. and G. Burnell [eds.]. Advances in Aquaculture Hatchery Technology. Woodhead Publishing Limited, Cambridge, UK.

FAO. 2018. Fishery and Aquaculture Statistics. Global aquaculture production 1950–2016 (FishstatJ). In: FAO Fisheries and Aquaculture Department [online]. Rome. Updated 2018. www.fao.org/fishery/statistics/software/fishstatj/en.

Farquharson, K.A., C.J. Hogg and C.E. Grueber. 2018. A meta-analysis of birth-origin effects on reproduction in diverse captive environments. Nature Comm. 9: 1055.

Forés, R., J. Iglesias, M. Olmedo, F.J. Sanchez and J.B. Peleteiro. 1990. Induction of spawning in turbot (*Scophthalmus maximus* L.) by a sudden change in the photoperiod. Aquac. Eng. 9: 357.

Forné, I., M.J. Agulleiro, E. Asensio, J. Abián and J. Cerdá. 2009. 2-D DIGE analysis of Senegalese sole (*Solea senegalensis*) testis proteome in wild-caught and hormone-treated F1 fish. Proteomics 9: 2171–2181.

Forné, I., B. Castellana, R. Marín-Juez, J. Cerdá, J. Abián and J.V. Planas. 2011. Transcriptional and proteomic profiling of flatfish (*Solea senegalensis*) spermatogenesis. Proteomics 11: 2195–2211.

Funes, S., J.A. Hedrick, G. Vassileva, L. Markowitz, S. Abbondanzo, A. Golovko et al. 2003. The KiSS-1 receptor GPR54 is essential for the development of the murine reproductive system. Biochem. Biophys. Res. Comm. 312: 1357–1363.

García-López, A. 2005. Efectos del fotoperiodo y del termoperiodo sobre el ciclo reproductor en cautividad del lenguado senegalés, *Solea senegalensis* (Kaup, 1858). Ph.D. Thesis. University of Cádiz, Cádiz, Spain.

García-López, A., G. Martínez-Rodríguez and C. Sarasquete. 2005. Male reproductive system in Senegalese sole *Solea senegalensis* (Kaup): anatomy, histology and histochemistry. Histol. Histopathol. 20: 1179–1189.

García-López, A., V. Fernández-Pasquier, E. Couto, A.V.M. Canario, C. Sarasquete and G. Martínez-Rodríguez. 2006a. Testicular development and plasma sex steroid levels in cultured male Senegalese sole *Solea senegalensis* Kaup. Gen. Comp. Endocrinol. 147: 343–351.

García-López, A., V. Anguis, E. Couto, A.V.M. Canario, J.P. Cañavate, C. Sarasquete et al. 2006b. Non-invasive assessment of reproductive status and cycle of sex steroid levels in a captive wild broodstock of Senegalese sole *Solea senegalensis* (Kaup). Aquaculture 254: 583–593.

García-López, A., E. Pascual, C. Sarasquete and G. Martínez-Rodríguez. 2006c. Disruption of gonadal maturation in cultured Senegalese sole *Solea senegalensis* Kaup by continuous light and/or constant temperature regimes. Aquaculture 261: 789–798.

García-López, A., E. Couto, A.V.M. Canario, C. Sarasquete and G. Martínez-Rodríguez. 2007. Ovarian development and plasma sex steroid levels in cultured female Senegalese sole *Solea senegalensis*. Comp. Biochem. Physiol. A 146: 342–354.

García-López, A., C. Sarasquete and G. Martínez-Rodríguez. 2009. Temperature manipulation stimulates gonadal maturation and sex steroid production in Senegalese sole *Solea senegalensis* Kaup kept under two different light regimes. Aquac. Res. 40: 103–111.

Ge, W. 2005. Intrafollicular paracrine communication in the zebrafish ovary: The state of the art of an emerging model for the study of vertebrate folliculogenesis. Mol. Cell. Endocrinol. 237: 1–10.

Grau, A., S. Crespo, F. Riera, S. Pou and M.C. Sarasquete. 1996. Oogenesis in the amberjack *Seriola dumerili* Risso, 1810. An histological, histochemical and ultrastructural study of oocyte development. Sci. Mar. 60: 391–406.

Grier, H.J. 1993. Comparative organization of Sertoli cells including the Sertoli cell barrier. pp 704–739. *In*: Russel, L.D. and M.D. Griswold [eds.]. The Sertoli cell. Cache River Press, Clearwater, FL, USA.

Grier, H. 2000. Ovarian germinal epithelium and folliculogenesis in the common snook, *Centropomus undecimalis* (Teleostei: Centropomidae). J. Morphol. 243: 265–281.

Grier, H. 2012. Development of the Follicle Complex and Oocyte Staging in Red Drum, *Sciaenops ocellatus* Linnaeus, 1776 (Perciformes, Sciaenidae). J. Morphol. 273: 801–829.

Guzmán, J.M., B. Norberg, J. Ramos, C.C. Mylonas and E. Mañanós. 2008. Vitellogenin, steroid plasma levels and spawning performance of cultured female Senegalese sole (*Solea senegalensis*). Gen. Comp. Endocrinol. 156: 285–297.

Guzmán, J.M., J. Ramos, C.C. Mylonas and E.L. Mañanós. 2009a. Spawning performance and plasma levels of GnRHa and sex steroids in cultured female Senegalese sole (*Solea senegalensis*) treated with different GnRHa-delivery systems. Aquaculture 291: 200–209.

Guzmán, J.M., M. Rubio, J.B. Ortiz-Delgado, U. Klenke, K. Kight, I. Cross et al. 2009b. Comparative gene expression of gonadotropins (FSH and LH) and peptide levels of gonadotropin-releasing hormones (GnRHs) in the pituitary of wild and cultured Senegalese sole (*Solea senegalensis*) broodstocks. Comp. Biochem. Physiol. A 153: 266–277.

Guzmán, J.M., M.J. Bayarri, J. Ramos, Y. Zohar, C. Sarasquete, E.L. Mañanós. 2009c. Follicle stimulating hormone (FSH) and luteinizing hormone (LH) gene expression during larval development in Senegalese sole (*Solea senegalensis*). Comp. Biochem. Physiol. A 154: 37–43.

Guzmán, J.M. 2010. Estudio fisiológico de la reproducción del lenguado (*Solea senegalensis*) y desarollo de terapias hormonales para el control de la espermación y puesta en cautividad. Ph.D. thesis, University of Cádiz, Cádiz, Spain.

Guzmán, J.M., J. Ramos, C.C. Mylonas and E.L. Mañanós. 2011a. Comparative effects of human chorionic gonadotropin (hCG) and gonadotropin-releasing hormone agonist (GnRHa) treatments on the stimulation of male Senegalese sole (*Solea senegalensis*) reproduction. Aquaculture 316: 121–128.

Guzmán, J.M., R. Cal, A. García-López, O. Chereguini, K. Kight, M. Olmedo et al. 2011b. Effects of *in vivo* treatment with the dopamine antagonist pimozide and gonadotropin-releasing hormone agonist (GnRHa) on the reproductive axis of Senegalese sole (*Solea senegalensis*). Comp. Biochem. Physiol. Part A. 158: 235–245.

Harmin, S.A., L.W. Crim and M.D. Wiegand. 1995. Plasma sex steroid profiles and the seasonal reproductive cycle in male and females winter flounder, *Pleuronectes americanus*. Mar. Biol. 121: 601–610.

Hendry, C.I., D.J. Martin-Robichaud and T.J. Benfey. 2002. Gonadal sex differentiation in Atlantic halibut. J. Fish Biol. 60: 1431–1442.

Holmefjord, I., J. Gulbrandsen, I. Lein, T. Refstie, P. Leger, T. Harboe et al. 1993. An intensive approach to Atlantic halibut fry production. J. World Aquacult Soc. 24: 275.

Howell, B.R. and A.P. Scott. 1989. Ovulation cycles and post-ovulatory deterioration of eggs of the turbot, *Scophthalmus maximus* L. Rap. CIEM. 191: 21.

Howell, B., S.M. Baynes and D. Thompson. 1995. Progress towards the identification of the sex-determining mechanism of the sole, *Solea solea* (L.), by the induction of diploid gynogenesis. Aquac. Res. 26(2): 135–140.

Howell, B., L. Conceição, R. Prickett, P. Cañavate and E. Mañanos. 2009. Sole farming: nearly there but not quite? A report of the 4th workshop on the cultivation of soles. Aquaculture Europe. 34(1): 24–27.

Howell, B., R. Pricket, J.P. Cañavate, E. Mañanos, M.T. Dinis, L. Conceiçao et al. 2011. Sole farming: there or thereabouts! A report of the 5th workshop on the cultivation of soles. Aquaculture Europe. 36: 42–45.

Imsland, A.K., A. Foss, L.E.C. Conceiçao, M.T. Dinis, D. Delbare, E. Schram et al. 2003. A review of the culture potential of *Solea solea* and *Solea senegalensis*. Rev. Fish Biol. Fish. 13: 379–407.

Khan, I.A. and P. Thomas. 1992. Stimulatory effects of serotonin on maturational gonadotropin release in the Atlantic croaker, *Micropogonias undulates*. Gen. Comp. Endocrinol. 88(3): 388–396.

Koya, Y., H. Watanabe, K. Soyano, K. Ohta, M. Aritaki and T. Matsubara. 2003. Testicular development and serum steroid hormone levels in captive male spotted halibut *Verasper variegatus*. Fish. Sci. 69: 792–798.

Lahnsteiner, F., U. Ritcharski and R.A. Patzner. 1990. Functions of the testicular gland in two blenniid fishes, *Salaria* (*Blennius*) *pavo* and *Lypophrys* (*Blennius*) *dalmantinus* (Blenniidae, Teleostei) as revealed by electron microscopy and enzyme histochemistry. J. Fish Biol. 37: 85–97.

Lang, R., A.L. Gundlach and B. Kofler. 2007. The galanin peptide family: receptor pharmacology, pleiotropic biological actions, and implications in health and disease. Pharmacol. Ther. 115 (2): 177–207.

Lethimonier, C., T. Madigou, J.A. Muñoz-Cueto, J.J. Lareyre and O. Kah. 2004. Evolutionary aspects of GnRHs, GnRH neuronal systems and GnRH receptors in teleost fish. Gen. Comp. Endocrinol. 135(1): 1–16.

Levavi-Sivan, B., J. Bogerd, E.L. Mañanós, A. Gómez and J.J. Lareyre. 2010. Perspectives on fish gonadotropins and their receptors. Gen. Comp. Endocrinol. 165: 412–437.

Loir, M. and F. Le Gac. 1994. Insulin-like growth factor-I and -II binding and action on DNA synthesis in rainbow trout spermatogonia and spermatocytes. Biol. Reprod. 51: 1154–1163.

Lomax, D.P., W.T. Roubal, J.D. Moore and L.L. Johnson. 1998. An enzyme linked immunosorbent assay (ELISA) for measuring vitellogenin in English sole (*Pleuronectes vetulus*): development, validation and crossreactivity with other pleuronectids. Comp. Biochem. Physiol. B. 121: 425–436.

Lubzens, E., G. Young, J. and J. Cerdà. 2010. Oogenesis in teleosts: How fish eggs are formed. Gen. Comp. Endocrinol. 165: 367–389.

Mañanós, E., S. Zanuy, F. Le Menn, M. Carrillo and J. Núñez. 1994a. Sea bass (*Dicentrarchus labrax* L.) vitellogenin. I: Induction, purification and partial characterization. Comp. Biochem. Physiol. B. 107: 205.

Mañanós, E., J. Núñez, S. Zanuy, M. Carrillo and F. Le Menn. 1994b. Sea bass (*Dicentrarchus labrax* L.) vitellogenin. II: Validation of an enzyme-linked immunosorbent assay (ELISA). Comp. Biochem. Physiol. B. 107(2): 217–223.

Mañanós, E.L., J. Núñez Rodríguez, F. Le Menn, S. Zanuy and M. Carrillo. 1997. Identification of vitellogenin receptors in the ovary of a teleost fish, the Mediterranean sea bass (*Dicentrarchus labrax*). Reprod. Nutr. Develop. 37: 51–61.

Mañanós, E., I. Ferreiro, D. Bolón, J.M. Guzmán, C.C. Mylonas and A. Riaza. 2007. Different responses of Senegalese sole *Solea senegalensis* broodstock to a hormonal spawning induction therapy, depending on their wild or captive-reared origin. Proc. Annual Meeting Aquaculture Europe. 25–27 October 2007, Istanbul, Turkey.

Mañanós, E., N. Duncan and C.C. Mylonas. 2009. Reproduction and control of ovulation, spermiation and spawning in cultured fish. pp. 3–80. *In*: Cabrita, E., V. Robles and M.P. Herráez [eds.]. Methods in Reproductive Aquaculture: Marine and Freshwater Species. CRC Press, Taylor and Francis Group, Boca Raton, FL, USA.

Manni, L. and M.B. Rasotto. 1997. Ultrastructure and histochemistry of the testicular efferent duct system and spermiogenesis in *Opistognathus whitehurstii* (Teleostei, Trachinoidei). Zoomorphology 117: 93–102.

Marín-Juez, R., B. Castellana, M. Manchado and J.V. Planas. 2011. Molecular identification of genes involved in testicular steroid synthesis and characterization of the response to gonadotropic stimulation in the Senegalese sole (*Solea senegalensis*) testis. Gen. Comp. Endocrinol. 172: 130–139.

Marín-Juez, R., J. Viñas, A.S. Mechaly, J.V. Planas and F. Piferrer. 2013. Stage-specific gene expression during spermatogenesis in the Senegalese sole (*Solea senegalensis*), a fish with semi-cystic type of spermatogenesis, as assessed by laser capture microdissection and absolute quantitative PCR. Gen. Comp. Endocrinol. 188: 242–250.

Marino, G., E. Panini, A. Longobardi, A. Mandich, M.G. Finoia, Y. Zohar et al. 2003. Induction of ovulation in captive-reared dusky grouper, *Epinephelus marginatus* (Lowe, 1834) with a sustained-release GnRHa implant. Aquaculture 219: 841.

Martín, I.E. 2016. Advances in the reproductive biology and zootechnics of the Senegalese sole (*Solea senegalensis* Kaup, 1858). Ph.D. Thesis, University of Cantabria, Santander, Spain.

Martin, I., I. Rasines, M. Gomez, C. Rodriguez, P. Martinez and O. Chereguini. 2014. Evolution of egg production and parental contribution in Senegalese sole, *Solea senegalensis*, during four consecutive spawning seasons. Aquaculture 424-425: 45–52.

Martins, R.S.T., P.I.S. Pinto, P.M. Guerreiro, S. Zanuy, M. Carrillo and A.V.M. Canario. 2014. Novel galanin receptors in teleost fish: Identification, expression and regulation by sex steroids. Gen. Comp. Endocrinol. 205: 109–120.

Martínez, P., C. Bouza, M. Hermida, J. Fernández, A. Toro, M. Vera et al. 2009. Identification of the major sex-determining region of turbot (*Scophthalmus maximus*). Genetics 183: 1443–1452.

Martínez, P., A.M. Viñas, L. Sanchez, N. Diaz, L. Ribas and F. Piferrer. 2014. Genetic architecture of sex determination in fish: applications to sex ratio control in aquaculture. Front. Genetics. 5: 340.

Mattei, X., Y. Siau, O.T. Thiaw and D. Thiam. 1993. Peculiarities in the organization of testis of *Ophidion* sp. (Pisces, Teleostei). Evidence of two types of spermatogenesis in teleost fish. J. Fish Biol. 43: 931–937.

Mechaly, A.S., J. Viñas and F. Piferrer. 2009. Identification of two isoforms of the Kisspeptin-1 receptor (kiss1r) generated by alternative splicing in a modern teleost, the Senegalese sole (*Solea senegalensis*). Biol. Reprod. 80: 60–69.

Mechaly, A.S., J. Viñas and F. Piferrer. 2011. Gene structure analysis of kisspeptin-2 (Kiss2) in the Senegalese sole (*Solea senegalensis*): characterization of two splice variants of Kiss2, and novel evidence for metabolic regulation of kisspeptin signaling in non-mammalian species. Mol. Cell. Endocrinol. 339: 14–24.

Mechaly, A.S., J. Viñas and F. Piferrer. 2012. Sex-specific changes in the expression of kisspeptin, kisspeptin receptor, gonadotropins and gonadotropin receptors in the Senegalese sole (*Solea senegalensis*) during a full reproductive cycle. Comp. Biochem. Physiol. A 162: 364–371.

Mechaly, A.S., J. Viñas and F. Piferrer. 2013. The kisspeptin system genes in teleost fish, their structure and regulation, with particular attention to the situation in Pleuronectiformes. Gen. Comp. Endocrinol. 188: 258–268.

Milla, S., N. Wang, S.N.M. Mandiki and P. Kestemont. 2009. Corticosteroids: friends or foes of teleost fish reproduction?. Comp. Biochem. Physiol. A. 153: 242–251.

Miura, T., C. Miura, K. Yamauchi and Y. Nagahama. 1995. Human recombinant activin induces proliferation of spermatogonia *in vitro* in the Japanese eel (*Anguilla japonica*). Fish. Sci. 61: 434–437.

Miura, T., C. Miura, Y. Konda and K. Yamauchi. 2002. Spermatogenesis-preventing substance in Japanese eel. Development 129: 2689–2697.

Miwa, S., L. Yan and P. Swanson. 1994. Localization of two gonadotropin receptors in the salmon gonad by *in vitro* ligand autoradiography. Biol. Reprod. 50: 629–642.

Morais, S., C. Aragão, E. Cabrita, L.E.C. Conceição, M. Constenla, B. Costas et al. 2016. New developments and biological insights into the farming of *Solea senegalensis* reinforcing its aquaculture potential. Rev. Aquacult. 8: 227–263.

Mugnier, C., M. Guennoc, E. Lebegue, A. Fostier and B. Breton. 2000. Induction and synchronisation of spawning in cultivated turbot (*Scophthalmus maximus* L.) broodstock by implantation of a sustained-release GnRH-a pellet. Aquaculture 181: 241.

Muñoz, M., M. Casadevall and S. Bonet. 2002. Testicular structure and semicystic spermatogenesis in a specialized ovuliparous species: *Scorpaena notata* (Pisces, Scorpaenidae). Acta Zool. 83: 213–219.

Muñoz-Cueto, J.A., J.A. Paullada-Salmerón, M. Aliaga-Guerrero, M.E. Cowan, I.S. Parhar and T. Ubuka. 2017. A journey through the gonadotropin-inhibitory hormone system of fish. Front. Endocrinol. 8: article n° 285.

Mylonas, C.C., Y. Tabata, R. Langer and Y. Zohar. 1995. Preparation and evaluation of polyanhydride microspheres containing gonadotropin-releasing hormone GnRH, for inducing ovulation and spermiation in fish. J. Control. Release. 35: 23.

Mylonas, C.C. and Y. Zohar. 2001. Use of GnRHa-delivery systems for the control of reproduction in fish. Rev. Fish Biol. Fisher. 10(4): 463–491.

Mylonas, C.C., N. Papandroulakis, A. Smboukis, M. Papadaki and P. Divanach. 2004. Induction of spawning of cultured greater amberjack (*Seriola dumerili*) using GnRHa implants, Aquaculture 237: 141.

Mylonas, C.C., C.R. Bridges, H. Gordin, A. Belmonte-Ríos, A. García, F. De la Gándara et al. 2007. Preparation and administration of gonadotropin-releasing hormone agonist (GnRHa) implants for the artificial control of reproductive maturation in captive-reared Atlantic bluefin tuna (*Thunnus thynnus thynnus*). Rev. Fish. Sci. 15: 183.

Mylonas, C.C., A. Fostier and S. Zanuy. 2010. Broodstock management and hormonal manipulations of fish reproduction. Gen. Comp. Endocrinol. 165(3): 516–534.

Mylonas, C.C., N.J. Duncan and J.F. Asturiano. 2017. Hormonal manipulations for the enhancement of sperm production in cultured fish and evaluation of sperm quality. Aquaculture 472: 21–44.

Nader, M.R., T. Miura, N. Ando, C. Miura and K. Yamauchi. 1999. Recombinant human insulin-like growth factor I stimulates all stages of 11-ketotestosterone-induced spermatogenesis in the Japanese eel, *Anguilla japonica*, *in vitro*. Biol. Reprod. 61: 944–947.

Nagahama, Y. and M. Yamashita. 2008. Regulation of oocyte maturation in fish. Develop. Growth Differ. 50: S195–S219.

Nagasawa, K., C. Presslauer, L. Kirtiklis, I. Babiak and J.M.O. Fernandes. 2014. Sexually dimorphic transcription of estrogen receptors in cod gonads throughout a reproductive cycle. J. Mol. Endocrinol. 52: 357–371.

Nakamura, M. 2013. Morphological and physiological studies on gonadal sex differentiation in teleosts fish. Aqua-BioScience Monographs. 6(1): 1–47.

Nakamura, M., T. Kobayasi, X.T. Chang and Y. Nagahama. 1998. Gonadal sex differentiation in teleost fish. J. Exp. Zool. 281: 362–372.

Nelson, E.R. and H.R. Habibi. 2013. Estrogen receptor function and regulation in fish and other vertebrates. Gen. Comp. Endocrinol. 192: 15–24.

Norambuena, F., A. Estevez, G. Bell, I. Carazo and N. Duncan. 2012a. Proximate and fatty acid composition in muscle, liver and gonads of wild versus cultured broodstock of Senegalese sole (*Solea senegalensis*). Aquaculture 356: 176–185.

Norambuena, F., S. Mackenzie, J.G. Bell, A. Callol, A. Estevez and N. Duncan. 2012b. Prostaglandin (F and E, 2-and 3-series) production and cyclooxygenase (COX-2) gene expression of wild and cultured broodstock of Senegalese sole (*Solea senegalensis*). Gen. Comp. Endocrinol. 177: 256–262.

Norambuena, F., A. Estevez, E. Mañanós, J. Gordon Bell, I. Carazo and N. Duncan. 2013. Effects of graded levels of arachidonic acid on the reproductive physiology of Senegalese sole (*Solea senegalensis*): Fatty acid composition, prostaglandins and steroid levels in the blood of broodstock bred in captivity. Gen. Comp. Endocrinol. 191: 92–101.

Norberg, B., V. Valkner, J. Huse, I. Karlsen and G.L. Grung. 1991. Ovulatory rhythms and egg viability in the Atlantic halibut (*Hippoglossus hippoglossus*). Aquaculture 97: 365.

Nuñez-Rodriguez, J., O. Kah, M. Geffard and F. Le Menn. 1989. Enzymelinked immunosorbent assay (ELISA) for sole (*Solea vulgaris*) vitellogenin. Comp. Biochem. Physiol. 92B: 741–746.

Núñez, J. and F. Duponchelle. 2009. Towards a universal scale to assess sexual maturation and related life history traits in oviparous teleost fishes. Fish Physiol. Biochem. 35: 167–180.

Oba, Y., T. Hirai, Y. Yoshiura, T. Kobayashi and Y. Nagahama. 2001. Fish gonadotropin and thyrotropin receptors: the evolution of glycoprotein hormone receptors in vertebrates. Comp. Biochem. Physiol. B. 129: 441–448.

Oliveira, C., L.M. Vera, J.F. López-Olmeda, J.M. Guzmán, E. Mañanós, J. Ramos et al. 2009. Monthly day/night changes and seasonal daily rhythms of sexual steroids in Senegal sole (*Solea senegalensis*) under natural fluctuating or controlled environmental conditions. Comp. Biochem. Physiol. A. 152: 168–175.

Oliveira, C., N.J. Duncan, P. Pousão-Ferreira, E. Mañanós and F.J. Sánchez-Vázquez. 2010. Influence of the lunar cycle on plasma melatonin, vitellogenin and sex steroids rhythms in Senegal sole, *Solea senegalensis*. Aquaculture 306: 343–347.

Oliveira, C., E. Mañanós, J. Ramos and F.J. Sánchez-Vázquez. 2011. Impact of photoperiod manipulation on day/ night changes in melatonin, sex steroids and vitellogenin plasma levels and spawning rhythms in Senegal sole, *Solea senegalensis*. Comp. Biochem. Physiol. A. 159: 291–295.

Ospina-Álvaez, N. and F. Piferrer. 2008. Temperature dependent sex determination in fish revisited: prevalence, a single ratio response pattern and possible effect of climate change, PloS One. 3(7): e2837.

Pacchiarini, T. 2012. El factor materno vasa como marcador molecular para la caracterizión de las células germinales/CG en dos especies de peces planos: el lenguado senegalés (*Solea senegalensis*) y el rodaballo (*Scophthalmus maximus*). Aplicaciones biotecnológicas en acuicultura. Ph.D. Thesis, University of Cádiz, Cádiz, Spain.

Pacchiarini, T., I. Cross, R.B. Leite, P. Gavaia, J.B. Ortiz-Delgado, P. Pousão-Ferreira et al. 2013. *Solea senegalensis* vasa transcripts: molecular characterisation, tissue distribution and developmental expression profiles. Reprod. Fertil. Dev. 25(4): 646–660.

Palstra, A.P., M.C. Blok, J. Kals, E. Blom, N. Tuinhof-Koelma, R.P. Dirks et al. 2015. In- and outdoor reproduction of first generation common sole *Solea solea* under a natural photothermal regime: Temporal progression of sexual maturation assessed by monitoring plasma steroids and gonadotropin mRNA expression. Gen. Comp. Endocrinol. 221: 183–192.

Pasquier, J., N. Kamech, A.-G Lafont, H. Vaudry, K. Rousseau and S. Dufour. 2014. Molecular evolution of GPCRs: Kisspeptin/kisspeptin receptors. J. Mol. Endocrinol. 52(3): T101–T117.

Paullada-Salmerón, J.A., M. Cowan, M. Aliaga-Guerrero, A. Gomez, S. Zanuy, E. Mañanos et al. 2016. LPXRFa peptide system in the European sea bass: A molecular and immunohistochemical approach. J. Comp. Neurol. 524(1): 176–198.

Penman, D.J. and F. Piferrer. 2008. Fish gonadogenesis. Part 1. Genetic and environmental mechanisms of sex determination. Rev. Fisher. Sci. 16 (S1): 16–34.

Piferrer, F. 2001. Endocrine sex control strategies for the feminisation of teleosts fish. Aquaculture 197: 229–281.

Piferrer, F. and Y. Guiguen. 2008. Fish gonadogenesis. Part 2. Molecular biology and genomics of fish differentiation. Rev. Fisher. Sci. 16(S1): 35–55.

Pinilla, L., E. Aguilar, C. Dieguez, R.P. Millar and M. Tena-Sempere. 2012. Kisspeptins and reproduction: physiological roles and regulatory mechanisms. Physiol. Rev. 92: 1235–1316.

Popesku, J.T., C.J. Martyniuk, J. Mennigen, H. Xiong, D. Zhang, X. Xia et al. 2008. The goldfish (*Carassius auratus*) as a model for neuroendocrine signaling. Mol. Cell. Endocrinol. 293: 43–56.

Porta, J., J.M. Porta, G. Martinez-Rodriguez and M.C. Alvarez. 2006a. Development of a microsatellite multiplex PCR for Senegalese sole (*Solea senegalensis*) and its application to broodstock management. Aquaculture 256: 159–166.

Porta, J., J.M. Porta, G. Martinez-Rodriguez and M.C. Alvarez. 2006b. Genetic structure and genetic relatedness of a hatchery stock of Senegal sole (*Solea senegalensis*) inferred by microsatellites. Aquaculture 251: 46–55.

Porta, J., J.M. Porta, P. Cañavate, G. Martinez-Rodriguez and M.C. Alvarez. 2007. Substantial loss of genetic variation in a single generation of Senegalese sole (*Solea senegalensis*) culture: implications in the domestication process. J. Fish Biol. 71: 223–234.

Portela-Benz, S., A. Merlo, M.E. Rodríguez, I. Cross, M. Manchado, N. Kosyakova et al. 2017. Integrated gene mapping and synteny studies give insights into the evolution of a sex proto-chromosome in *Solea senegalensis*. Chromosoma 126: 261–277.

Radonic, M. and G.J. Macchi. 2009. Gonad sex differentiation in cultured juvenile flounder, *Paralichthys orbignyanus* (Valenciennes, 1839). J. World Aq. 40: 129–133

Ramos, J. 1979. Fisiología de la reproducción del lenguado (*Solea solea*, L). Ph.D. Thesis, Universidad Complutense de Madrid, Madrid, Spain.

Ramos, J. 1983. Contribución al estudio de la oogenesis del lenguado, *Solea solea* (Linneo, 1758) (Pisces, Soleidae). Inv. Pesq. 47: 241–251.

Ramos, J. 1986a. Induction of spawning in common sole (*Solea solea* L.) with human chorionic gonadotropin (HCG). Aquaculture 56(3-4): 239–242.

Ramos, J. 1986b. Luteinizing hormone-releasing hormone analogue (LH-RHa) induces precocious ovulation in common sole (*Solea solea* L.). Aquaculture 54: 185–190.

Rasines, I., M. Gómez, I. Martín, C. Rodríguez, E. Mañanós and O. Chereguini. 2012. Artificial fertilization of Senegalese sole (*Solea senegalensis*): Hormone therapy administration methods, timing of ovulation and viability of eggs retained in the ovarian cavity. Aquaculture 326-329: 129–135.

Rasines, I., M. Gómez, I. Martín, C. Rodríguez, E. Mañanós and O. Chereguini. 2013. Artificial fertilisation of cultured Senegalese sole (*Solea senegalensis*): Effects of the time of day of hormonal treatment on inducing ovulation. Aquaculture 392–395: 94–97.

Rendón, C, F.J. Rodríguez-Gómez, J.A. Muñoz-Cueto, C. Piñuela and C. Sarasquete. 1997. An immunocytochemical study of pituitary cells of the Senegalese sole, *Solea senegalensis* (Kaup, 1858). Histochem. J. 29: 813–822.

Rinchard, J., P. Kestemont and R. Heine. 1997. Comparative study of reproductive biology in single and multiple-spawner cyprinid fish: II. Sex steroid and plasma protein phosphorus concentrations. J. Fish Biol. 50: 169–180.

Rodríguez, R.B. 1984. Biología y cultivo de *Solea senegalensis* Kaup 1858 en el Golfo de Cádiz. Ph.D. Thesis, University of Seville, Seville, Spain.

Rodríguez-Gómez, F.J., M.C. Rendón-Unceta, C. Sarasquete and J.A. Muñoz-Cueto. 2000a. Distribution of serotonin in the brain of the Senegalese sole, *Solea senegalensis*: An immunohistochemical study. J. Chem. Neuroanat. 18(3): 103–115.

Rodríguez-Gómez, F.J., M.C. Rendón-Unceta, C. Sarasquete and J.A. Muñoz-Cueto. 2000b. Localization of galanin-like immunoreactive structures in the brain of the Senegalese sole, *Solea senegalensis*. Histochem. J. 32(2): 123–131.

Rodríguez-Gómez, F.J., M.C. Rendón-Unceta, C. Sarasquete and J.A. Muñoz-Cueto. 2000c. Localization of tyrosine hydroxylase-immunoreactivity in the brain of the Senegalese sole, *Solea senegalensis*. J. Chem. Neuroanat. 19(1): 17–32.

Rolland, A.D., A. Lardenois, A.S. Goupil, J.J. Lareyre, R. Houlgatte, F. Chalmel et al. 2013. Profiling of androgen response in rainbow trout pubertal testis: relevance to male gonad development and spermatogenesis. PLoS ONE 8: e53302.

Roubal, W.T., D.P. Lomax, M.L. Willis and L.L. Johnson. 1997. Purification and partial characterization of English sole (*Pleuronectes vetulus*) vitellogenin. Comp. Biochem. Physiol. B. 118(3): 613–622.

de Roux, N., E. Genin, J.C. Carel, F. Matsuda, J.L. Chaussain and E. Milgrom. 2003. Hypogonadotropic hypogonadism due to loss of function of the KiSS1-derived peptide receptor GPR54. PNAS. 100: 10972–10976.

Saligaut, C., B. Linard, B. Breton, I. Anglade, T. Bailhache, O. Kah et al. 1999. Brain aminergic system in salmonids and other teleosts in relation to steroid feedback and gonadotropin release. Aquaculture 177: 13–20.

Sambroni, E., A.D. Rolland, J.J. Lareyre and F. Le Gac. 2013. FSH and LH have common and distinct effects on gene expression in rainbow trout testis. J. Mol. Endocrinol. 50: 1–18.

Satoh, N. and N. Egami. 1972. Sex differentiation of germ cells in the teleost, *Oryzias latipes*, during normal embryonic development. J. Embryol. Exp. Morphol. 28: 385–395.

Sawatari, E., S. Shikina, T. Takeuchi and G. Yoshizaki. 2007. A novel transforming growth factor-b superfamily member expressed in gonadal somatic cells enhances primordial germ cell and spermatogonial proliferation in rainbow trout (*Oncorhynchus mykiss*). Dev. Biol. 301: 266–275.

Schulz, R.W., L.R. França, J.J. Lareyre, F. LeGac, H. Chiarini-Garcia, R.H. Nóbrega et al. 2010. Spermatogenesis in fish. Gen. Comp. Endocrinol. 165: 390–411.

Schulz R.W. and R.H. Nóbrega. 2011. Regulation of Spermatogenesis. pp. 627–634. *In*: Farrell, A.P. [ed.]. Encyclopedia of Fish Physiology: From Genome to Environment. Academic Press, San Diego, CA, USA.

Schulz R.W., R.H. Nóbrega, R.D.V.S. Morais, P.P. de Waal, L.R. França and J. Bogerd. 2015. Endocrine and paracrine regulation of zebrafish spermatogenesis: the Sertoli cell perspective. Anim. Reprod. 12: 81–87.

Scott, A.P. and A.V. Canario. 1992. 17α,20β-dihydroxi-4-pregnen-3-one 20-sulphate: a major new metabolite of the teleost oocyte maturation-inducing steroid. Gen. Comp. Endocrinol. 85: 91–100.

Scott, A.P., R.M. Inbaraj and E.L.M. Vermeirssen. 1997. Use of a radioimmunoassay which detects C21 steroids with a 17,20β-dihydroxyl configuration to identify and measure steroids involved in final oocyte maturation in female plaice *Pleuronectes platessa*. Gen. Comp. Endocrinol. 105: 62–70.

Scott, A.P., J.P. Sumpter and N. Stacey. 2010. The role of the maturation inducing steroid, 17,20β-dihydroxypregn-4-en-3-one, in male fishes: a review. J. Fish Biol. 76: 183–224.

Selman, K. and R.A. Wallace. 1986. Gametogenesis in *Fundulus heteroclitus*. Am. Zool. 26: 173–192.

Seminara, S.B., S. Messager, E.E. Chatzidaki, R.R. Thresher, J.S. Acierno, J.K. Shagoury et al. 2003. The GPR54 gene as a regulator of puberty. New England J. Med. 349: 1614–1627.

Servili, A., Y. Le Page, J. Leprince, A. Caraty, S. Escobar, I.S. Parhar et al. 2011 Organization of two independent kisspeptin systems delivered from evolutionary-ancient kiss genes in the brain of zebrafish. Endocrinol. 152: 1527–1540.

Shahjahan, M., T. Kitahashi and I.S. Parhar. 2014. Central pathways integrating metabolism and reproduction in teleosts. Front. Endocrinol. 5: 36.

Silverstein, J.T., B.G. Bosworth and W.R. Wolters. 1999. Evaluation of dual injection of LHRHa and the dopamine receptor antagonist pimozide in cage spawning of channel catfish (*Ictalurus punctatus*). J. World Aquat. Soc. 30: 263–268.

Sol, S.Y., O.P. Olson, D.P. Lomax and L.L. Johnson. 1998. Gonadal development and associated changes in plasma in plasma reproductive steroids in English sole, *Pleuronectes vetulus*, from Puget Sound, Washington. Fish. Bull. 96: 859–870.

Swanson, P., J.T. Dickey and B. Campbell. 2003. Biochemistry and physiology of fish gonadotropins. Fish Physiol. Biochem. 28: 53–59.

Tanaka, M., T. Ohkawa, T. Maeda, I. Kinoshita, T. Seikai and M. Nishida. 1997. Ecological diversities and stock structure of the flounder in the Sea of Japan in relation to stock enhancement. Bull. Natl. Res. Inst. Aquacult. 3: 77.

Takahashi, A., S. Kanda, T. Abe and Y. Oka. 2016. Evolution of the hypothalamic-pituitary-gonadal axis regulation in vertebrates revealed by knockout medaka. Endocrinol. 157: 3994–4002.

Thomas, P., C. Tubbs, H. Berg and G. Dressing. 2007. Sex steroid hormone receptors in fish ovaries. pp. 203–233. *In*: Babin, P.J., J. Cerdà and E. Lubzens [eds.]. The fish oocyte: From Basic Studies to Biotechnological Applications. Springer, Dordrecht, Holland.

Thomas, P. 2012. Rapid steroid hormone actions initiated at the cell surface and the receptors that mediate them with an emphasis on recent progress in fish models. Gen. Comp. Endocrinol. 175: 367–383.

Tingaud-Sequeira, A., F. Chauvigné, J. Lozano, M.J. Agulleiro, E. Asensio and J. Cerdà. 2009. New insights into molecular pathways associated with flatfish ovarian development and atresia revealed by transcriptional analysis. BMC Genomics 10: 434.

Tokarz, J., G. Möller, M. Hrabĕ De Angelis and J. Adamski. 2015. Steroids in teleost fishes: A functional point of view. Steroids 103: 123–144.

Trudeau, V.L. 1997. Neuroendocrine regulation of gonadotrophin II release and gonadal growth in the goldfish, *Carassius auratus*. Rev. Reprod. 2(1): 55–68.

Trudeau, V.L., C.J. Martyniuk, E. Zhao, H. Hu, H. Volkoff, W.A. Decatur et al. 2012. Is secretoneurin a new hormone?. Gen. Comp. Endocrinol. 175(1): 10–18.

Tsujigado, A., T. Yamakawa, H. Matsuda and N. Kamiya. 1989. Advanced spawning of the flounder, *Paralichthys olivaceus*, in an indoor tank with combined manipulation of water temperature and photoperiod. Int. J. Aquac. Fish. Technol. 1: 351.

Tsutsui, K, E. Saigoh, K. Ukena, H. Teranishi, Y. Fujisawa, M. Kikuchi et al. 2000. A novel avian hypothalamic peptide inhibiting gonadotropin release. Biochem. Biophys. Res. Comm. 275: 661–667.

Tyler, C.R. and J.P. Sumpter. 1996. Oocyte growth and development in teleosts. Rev. Fish Biol. Fish. 6: 287–318.

Ubuka, T., Y.L. Son and K. Tsutsui. 2016. Molecular, cellular, morphological, physiological and behavioral aspects of gonadotropin-inhibitory hormone. Gen. Comp. Endocrinol. 227: 27–50.

Vermeirssen, E.L.M., R.J. Shields, C. Mazorra de Quero and A.P. Scott. 2000. Gonadotrophin-releasing hormone agonist raises plasma concentrations of progestogens and enhances milt fluidity in male Atlantic halibut (*Hippoglossus hippoglossus*). Fish Physiol. Biochem. 22: 77.

Vermeirssen, E.L.M., C. Mazorra de Quero, R.J. Shields, B. Norberg, D.E. Kime and A.P. Scott. 2004. Fertility and motility of sperm from Atlantic halibut (*Hippoglossus hippoglossus*) in relation to dose and timing of gonadotrophin-releasing hormone agonist implant. Aquaculture 230. 547.

Viñas, J., E. Asensio, J.P. Cañavate and F. Piferrer. 2013. Gonadal sex differentiation in the Senegalese sole (*Solea senegalensis*) and first data on the experimental manipulation of its sex ratios. Aquaculture 384–387: 74–81.

Wang, Y. and W. Ge. 2003. Spatial expression patterns of activin and its signaling system in the zebrafish ovarian follicle: evidence for paracrine action of activin on the oocytes. Biol. Reprod. 69: 1998–2006.

Wang, N., F. Teletchea, P. Kestemont, S. Milla and P. Fontaine. 2010. Photothermal control of the reproductive cycle in temperate fishes. Rev. Aquacult. 2: 209–222.

Weltzien, F.A., G.L. Taranger, Ø. Karlsen and B. Norberg. 2002. Spermatogenesis and related plasma androgen levels in Atlantic halibut (*Hippoglossus hippoglossus* L.). Comp. Biochem. Physiol. A. 132: 567–575.

Yamamoto, T. 1969. Sex differentiation. pp. 117–175. *In*: Hoar, W.S. and D.J. Randall [eds.]. Fish Physiology Vol III. Academic Press, New York, USA.

Yaron, Z. 1995. Endocrine control of gametogenesis and spawning induction in the carp. Aquaculture 129(1–4): 49–73.

Yaron, Z., G. Gur, P. Melamed, H. Rosenfeld, A. Elizur and B. Levavi-Sivan. 2003. Regulation of fish gonadotropins. Int. Rev. Cytol. 225: 131–185.

Yoneda, M., M. Tokimura, H. Fujita, N. Takeshita, K. Takeshita, M. Matsuyama et al. 1998. Reproductive cycle and sexual maturity of the anglerfish *Lophiomus setigerus* in the East China Sea with a note on specialized spermatogenesis. J. Fish Biol. 53: 164–178.

Yu, K.L., P.Ñ. Rosenblum and R.E. Peter. 1991. *In vitro* release of gonadotropin-releasing hormone from the brain preoptic-anterior hypothalamic region and pituitary of female goldfish. Gen. Comp. Endocrinol. 81(2): 256–267.

Zapater, C., F. Chauvigné, B. Fernández-Gómez, R.N. Finn and J. Cerdà. 2013. Alternative splicing of the nuclear progestin receptor in a perciform teleost generates novel mechanisms of dominant-negative transcriptional regulation. Gen. Comp. Endocrinol. 182: 24–40.

Zhang, Z., B. Zhu and W. Ge. 2015a. Genetic analysis of zebrafish gonadotropin (FSH and LH) functions by TALEN-mediated gene disruption. Mol. Endocrinol. 9: 76–98.

Zhang, Z., S.W. Lau, L. Zhang and W. Ge. 2015b. disruption of zebrafish follicle-stimulating hormone receptor (fshr) but not luteinizing hormone receptor (lhcgr) gene by TALEN leads to failed follicle activation in females followed by sexual reversal to males. Endocrinol. 156: 3747–3762.

Zmora N., T.T. Wong, J. Stubblefield, B. Levavi-Sivan and Y. Zohar. 2017. Neurokinin B regulates reproduction via inhibition of kisspeptin in a teleost, the striped bass. J. Endocrinol. 233(2): 159–174.

Zohar, Y., G. Pagelson, Y. Gothilf, W.W. Dickhoff, P. Swanson, S. Duguay et al. 1990. Controlled release of gonadotropin releasing hormones for the manipulation of spawning in farmed fish. Contr. Rel. Bioact. Mater. 17: 51.

Zohar, Y. and C.C. Mylonas. 2001. Endocrine manipulations of spawning in cultured fish: from hormones to genes. Aquaculture 197(1–4): 99–136.

Zohar, Y., J.A. Munoz-Cueto, A. Elizur and O. Kah. 2010. Neuroendocrinology of reproduction in teleost fish. Gen. Comp. Endocrinol. 165: 438–455.

B-1.4

Sperm Physiology and Artificial Fertilization

Elsa Cabrita,[1,*] *Marta F. Riesco*[1] and *Evaristo L. Mañanós*[2]

1. Basic aspects of sperm physiology

1.1 General considerations

Sperm is composed by spermatozoa and seminal fluid which, in teleost fish, maintains the cell in a quiescent stage within the testes, providing all the metabolic resources needed for maintenance. The specific characteristics of sperm constituents have been used in research for several reasons, such as the inclusion as quality tests for the analysis of certain procedures, selection of good breeders or even for the characterization of phylogenetic relationships between genera. The next subsections will focus on the importance of sperm characteristics in several contexts.

1.2 Spermatozoa structure

The general structure of fish spermatozoon comprises a round head and long flagella, both covered by a highly polyunsaturated plasma membrane (for details in this structure, see Subsection 1.4). Both structures participate directly and indirectly in the fertilization event, and their integrity is crucial to maintain during this process. Most fish species have external fertilization and for that they have an aquasperm cell type, characterized by the loss of acrosome membrane as a result of evolution. This lost appears simultaneously with a specific characteristic present in these fish oocytes, the micropyle, which enables

---

[1] Centre of Marine Sciences (CCMAR), University of Algarve, Campus de Gambelas, 8005-139 Faro, Portugal.
    Email: mfriesco@ualg.pt
[2] Instituto de Acuicultura de Torre la Sal (IATS-CSIC), Torre la Sal s/n, 12595-Ribera de Cabanes, Castellón, Spain.
    Email: evaristo@iats.csic.es
* Corresponding author: ecabrita@ualg.pt

the sperm to enter and fuse with the oolemma. Other specific characteristics in the spermatozoa structure also evolved, such as the presence of a round head and nucleus, few mitochondria arranged in a single ring and a simple flagellum with 9+2 axonema fibers (Jamieson and Leung 1991). When the spermatozoa structure is compared, the head is the cellular compartment showing more variability in morphology and size amongst aquatic organisms, even within the same group (Gwo et al. 2002). In fish teleosts, the midpiece is frequently poorly developed due to the presence of a small number of rounded mitochondria, common in external fertilizers. With regard to the tail, the most common form is uniflagellate; however, some teleosts of the genus *Ictalurus* have biflagellate spermatozoa (Jamieson and Leung 1991).

There are some spermatozoon structures that have been used to distinguish between genera and used to reclassify fish species and other aquatic organisms (Billard et al. 1995, Lanhsteiner and Patzner 1997, Santos et al. 2016). The Pleuronectiformes, where Senegalese sole is included, are not an exception. Species in this order share specific characteristics in their spermatozoa structure, such as the presence of two lateral fins along the flagella plasma membrane (Jones and Butler 1988, Medina et al. 2000), but have also specific characteristics, such as the number of mitochondria in the midpiece (8–10 in *S. senegalensis*) or the orientation of centrioles, apomorphic in Solenidae and Scophtalmidae and plesiomorphic in Pleuronectidae, Paralichthydae and Citharidae (Medina et al. 2000).

1.3 Seminal plasma components

Seminal plasma plays a crucial role supporting spermatozoa and has physiological and endocrinological functions during the release of sperm from the testis into the sperm duct, and subsequently to the aquatic environment (Alavi et al. 2011). Its composition reflects the conditions in which spermatogenesis occurs, such as possible ageing processes, metabolic alterations, contamination or other factors that affect sperm quality (Cabrita et al. 2008a). Different studies on seminal plasma constituents and its composition such as metabolites and enzymes can provide useful information about cell status and integrity (Ruranwga et al. 2004). To characterize seminal plasma constituents, cell fraction must be separated from seminal plasma, usually using conventional centrifugation, avoiding always cell fragmentation and contamination with urine and other fluids that may mask the results. However, in some species, seminal plasma is more difficult to obtain through centrifugation due to the small sperm volume collected (e.g., *Solea senegalensis*), presence of low seminal fluid (e.g., *Dicentrarchus labrax*) or due to the presence of certain compounds in seminal plasma that difficult a pellet formation (cells) clearly distinguished from the supernatant fraction (seminal fluid) (e.g., *Sparus aurata)*.

Inorganic compounds present in seminal plasma are usually ions involved in the process of sperm motility activation (Cosson et al. 2008, Pérez et al. 2016). Organic compounds such as lipids, proteins, metabolites and enzymes, normally detected in the seminal plasma could be involved in energy metabolism. Attending to these important functions in the spermatozoa, the presence of these compounds and in general the seminal plasma composition has been related with sperm functionality in terms of sperm motility (Butts et al. 2011) and fertilization rate (Lahnsteiner et al. 1998). The importance of seminal plasma proteins on sperm quality has been demonstrated in some fish species, as Eurasian perch (*Perca fluviatis*) and Japanese eel (*Anguilla japonica*). Some of these proteins (phosphatases and superoxide dismutase) are involved in the oxido-reductase activity providing a higher resistance to reactive oxygen species (ROS) and improving sperm motility (Celino et al. 2011, Shaliutina et al. 2012).

It has been demonstrated that certain types of lipids (arachidic acid, linoleic acid, and glycerol trimyristate) and metabolites act as substrates for the tricarboxylic acid cycle, have a positive effect on the sperm viability in *Sparus aurata* sperm (Lahnstainer et al. 2010) and can have a great importance for the improvement of cryopreservation extenders in several species. In addition, it is known that free amino acid composition has an effect on sperm metabolism. In some fish species as rainbow trout (*Oncorhynchus mykiss*), carp (*Cyprinus carpio*), gilthead sea bream (*Sparus aurata*), and perch (*Perca fluviatilis*) the incubation of sperm during 48 h showed an amino acid metabolism process (Lahnsteiner 2009, 2010). In Senegalese sole, the seminal plasma characterization remains unknown probably due to the low recovered volumes of plasma during semen collection, in some cases less than 5 μl (E. Cabrita, personal communication).

1.4 Plasma membrane structure and integrity

The cell plasma membrane is composed of a thin bilayer of amphipathic lipids that separates and protects the cell from the extracellular environment, and it is responsible, as in other cells, for the control of substance traffic through the cell. The integrity of the plasma membrane and its selective permeability are crucial for cell survival. Membrane organization, phospholipid asymmetry and lipid composition are the most important factors affecting cell membrane permeability. Phospholipids are responsible for membrane structure and, together with cholesterol, will determine membrane fluidity.

Sperm content in phospholipids, cholesterol and fatty acids affects sperm quality in fish and influences its ability to successfully fertilize the eggs. Compared with somatic cells, sperm plasma membrane phospholipids are richer in polyunsaturated fatty acids (Waterhouse et al. 2006). Additionally, it is usually accepted that the majority of marine species have a much higher level of polyunsaturated acids present in their membranes than freshwater ones (Kopeika and Kopeika 2007). Even so, the sperm quality and composition in fatty acids is highly species-specific as reported by Bell et al. (1997) and highly affected by diets (Pustowka et al. 2000, Beirão et al. 2015). Moreover, some authors have demonstrated that fatty acids could affect sperm quality, such as motility and fertilization capacity (Lahnsteiner et al. 2009). Specifically, a significant correlation between arachidonic acid consumption in the testis and sperm motility in European eel males has been demonstrated (Baeza et al. 2015). Finally, attending to the high amount of polyunsaturated fatty acids that constitute the fish spermatozoa membrane (Mansour et al. 2006, Shaliutina et al. 2013), lipid peroxidation has become a valuable marker used to evaluate oxidative stress in fish sperm.

In addition to phospholipids, sperm plasma membrane is also composed by several proteins, where aquaporins (AQPs), a group of hydrophobic and integral membrane proteins, have a crucial role during sperm motility activation in different seawater species (Abascal et al. 2007, Cosson et al. 2008, Boj et al. 2015). These proteins allow water exchange across cell membrane regulating extracellular osmolality (King et al. 1996, Verkman 2002). These osmolality-dependent cell-signaling systems determine the motility of fish spermatozoa (Takai and Morisawa 1995, Zilli et al. 2009). The presence of different types of these proteins with different and coordinated roles in some fish species as gilthead seabream was demonstrated (Boj et al. 2015). In this species, authors showed that Aqp1aa activates sperm motility and drives Aqp8b trafficking to the mitochondria, whereas Aqp1ab and Aqp7 control the pattern of sperm motility.

The study of spermatozoa plasma membrane proteins in fish species has been developed in recent years with more information coming on the proteomic profile of this cell. Sperm proteins characterization provides a crucial tool for the understanding of fundamental molecular processes in fish spermatozoa, for the ongoing development of novel markers of sperm quality and for the optimization of short- and long-term sperm preservation procedures (Nynca et al. 2014).

1.5 Chromatin structure and integrity

The final objective of spermatozoa would be the transmission of male genetic information to the embryo; for this reason, the DNA integrity and stability should be one of the priorities.

In a general cell, in the chromatin structure, DNA is associated with proteins, mainly histones, forming a nucleoprotein complex in eukaryotic organisms. The basic unit of chromatin structure is the nucleosome that is formed by a core particle and a linker DNA region. This nucleosome is a histone octamer formed by different types of histones (Luger et al. 1997) that have different functions as the compaction of the chromatin. This structure allows the genome compaction in a limited space in the nucleus and regulates the gene expression. Spermatozoon presents an extreme example of compacted chromatin, where histones are replaced by protamines during spermatogenesis. In this process, the continuous mitotic and meiotic divisions transform spermatogonias into spermatozoa. During the subsequent divisions, chromatin becomes highly condensed and in most cases, this involves the substitution of somatic histones for other sperm-specific proteins in fish sperm, called generically sperm nuclear basic proteins (SNBPs). These SNBPs in fish were classified into three main groups: Histones (H-type), protamines (P-type), and protamine like (PL-type) (Ausió et al. 2014). Different types of chromatin folding in fish sperm has been described according to the three major SNBP types. Protamines were described in sea bass (*Dicentrarchus labrax*) and salmon (*Oncorhynchus* sp.) (Kasinsky et al. 2011). The H-type is present in *Sparus aurata* and *Danio rerio* (Ausió et al. 2014), and the PL-type is present in red mullet, *Mullus surmuletus* (Saperas et al. 2006) and winter flounder (*Pseudopleuronectes americanus*) (Watson and Davies 1998). However, the specific protein replacement during sperm chromatin compaction in Senegalese sole remains unknown.

It is known that specific broodstock management techniques can affect the chromatin stability. Modifying the seasonality of reproduction is a common practice in fish farming and is often accomplished by changes in the environmental factors. These procedures imply that the spermatogenesis takes place during an inappropriate thermal period that could alter the DNA packaging mechanisms, provoke and increase the oxidative stress (Pérez-Cerezales et al. 2010, 2011). The increase in reactive oxygen species (ROS) has been shown to cause modifications at the DNA level (Thomson et al. 2009). ROS are capable of producing abasic sites, cross-linking and double lesions that finally result in a modified nitrogen bases, strand breakages and DNA fragmentation (Pérez-Cerezales et al. 2009, Aitken et al. 2013). Moreover, the telomere shortening has also been described after cryopreservation procedures as consequence of this oxidative stress in sperm of some fish species (Pérez-Cerezales et al. 2011). The result of these DNA alterations can compromise offspring development at different levels (Fatehi et al. 2006) and show altered transcription rates in many genes involved in progeny growth and development (Pérez-Cerezales et al. 2011). The increase in DNA fragmentation has been described in Senegalese sole spermatozoa collected out of the natural spawing season comparing to natural one (Beirao et al. 2011). In a recent study in Senegalese sole sperm, the captivity effect on ROS production was evaluated, demonstrating that spermatozoa suffer high levels of oxidative stress regardless of whether the males were wild or born in captivity (Valcarce and Robles

2016). This fact can explain some DNA fragmentation levels shown in previous studies (Beirão et al. 2011).

Diverse susceptibilities to damage have been described in different DNA regions that could be at least partially attributed to the different position that these specific regions have within the nuclei (nuclear territories). It is known that regions closer to the nuclear envelope may be more susceptible to damage (Bermejo et al. 2012), affecting the stability of certain genes.

1.6 Presence of RNA transcripts

During spermatogenesis, mature spermatids transform into spermatozoa, which involves the condensation of the nucleus, becoming transcriptionally inactive as histones are replaced by protamines (Weinbauer et al. 2010). As mentioned before, an extensive remodeling of sperm chromatin occurs, and this process represents one of the most important changes in chromatin organization (Herráez et al. 2017). For this reason, it is well known that spermatozoa are transcriptionally inactive cells (Lalancette et al. 2008), with a RNA remnant population that was considered for several years as non-functional RNAs since they are residuals from the spermatogenesis process.

Nowadays, it is totally demonstrated that the residual mRNAs from spermatogenesis have important roles in successful fertilization and early embryonic development in humans, other mammalian species and in fish (Ostermeier et al. 2005, García-Herrero et al. 2011, Johnson et al. 2011, Guerra et al. 2013). These sperm mRNAs are necessary from the moment of the first embryo cleavage until the activation of the embryonic genome (Ostermeier et al. 2004) and may influence the phenotypic traits of the embryo (Meseguer et al. 2006) and offspring (Miller and Ostermeier 2006).

In recent years, novel methods provide information about different mRNA pattern profiles and were used to detect different spermatozoa susceptibilities between species such as gilthead seabream and zebrafish (Guerra et al. 2013). These studies demonstrated that certain transcripts (*bdnf* and *kita*) were more abundant in good breeders than bad breeders, correlating, positively, this information with the fertility rate (Guerra et al. 2013). In Senegalese sole sperm, a recent study demonstrated that the detection of some crucial transcripts related to fertilization ability and embryo development were able to predict better post-thaw sperm quality in some protocols used for sperm cryopreservation in this species (Riesco et al. 2017).

1.7 Cell functionality

Fish spermatozoa, when released into water, is activated and maintain their motility in order to reach and fertilize the oocyte. Spermatozoa motility is necessary for the fertilization capacity and depends on different aspects of the cell, such as mitochondria status and ATP production. Mitochondria are a very important organelle attending to ATP production through an oxidative phosphorylation mechanism. This process triggers different metabolic pathways occurring inside the mitochondrion, such as the citric acid cycle, oxidative decarboxylation of aceto acids, β-oxidation of fatty acids, metabolism of amino acids and pyrimidine synthesis. For this reason, mitochondria functionality is often employed as an alternative way to evaluate sperm quality in different fish species (Guthrie et al. 2008, Liu et al. 2006). At the same time, the mitochondria are actively involved in other processes, such as the generation of free radicals, apoptosis and calcium signaling (Figueroa et al. 2014). The redox mechanisms that occur in the mitochondria are the most important source of reactive oxygen species (ROS) (Kowaltowski et al. 2009). Oxidative stress can alter the mitochondrial membrane and respiratory efficiency provoking the release of ROS (Ferramosca et al. 2013). In human sperm, it was demonstrated that ROS liberation decreases the sperm mitochondrial respiration by an uncoupling between electron transport and ATP synthesis, and as a result, the mitochondrial respiratory efficiency is reduced which provokes a decrease in spermatozoa motility (Ferramosca et al. 2013). Concerning ATP synthesis, this molecule has an important role in flagellar beating. ATP is catalyzed by dynein ATPase which drives the sliding of adjacent doublets of microtubules in the flagellum, leading to the generation of flagellar beating (Gibbons 1968). ATP is therefore consumed and its hydrolysis has been detected during the sperm movement of some fish species such as trout (*Oncorhynchus mykiss*) (Christen et al. 1987), carp (*Cyprinus carpio*), turbot (*Psetta maxima*), (Perchec et al. 1995, Dreanno et al. 1999a), Siberian sturgeon (*Acipenser baerii*) (Billard et al. 1999) or European sea bass (*Dicentrarchus labrax*) (Dreanno et al. 1999b). Any damage in the ATP production could be translated in a decrease in cell functionality pointing out the importance of the analysis of this molecule.

1.8 Motility mechanism activation

Sperm motility and the factors triggering this mechanism have been extensively studied by several authors, although some aspects of the mechanism of motility activation are not well understood. Two major factors have been identified as motility triggers, environmental factors involving sperm after its release into water and factors present in the egg, coelomic fluid or seminal plasma. These factors may contribute dependently or independently to the depolarization of the cell membrane, initiating a signal cascade that stimulates flagellar movement (Morisawa et al. 1983). The mechanism for sperm motility activation is different in freshwater and marine species, and since Senegalese sole is a marine species, we will mostly focus on the events occurring in this species.

After sperm release into seawater, a hyperosmotic solution (1,100 mOsm/Kg) compared with sperm osmolarity, sperm motility is activated due to changes in intracellular ion concentration caused by internal water release through aquaporin channels, promoting an increased gradient of ions such as Na^+, K^+ and Ca^{2+} inside of spermatozoa. The initiation of motility in marine teleosts also appears to involve an increase in the internal concentration of Ca^{2+}, probably derived from intracellular stores (Darszon et al. 1999), although this aspect has been controversial. The concentration of ions inside the cell activates a dynein ATPase, which will be responsible for flagellar movement. At this point, flagella beating is of maximal velocity, but decreases with time because of two reasons: the ionic concentration becomes too high to sustain correct dynein activity and ATP concentration declines and becomes limiting for flagellar beating (Dzyuba and Cosson 2014). Intracellular concentrations of cAMP and ATP are crucial for motility duration as well as other environmental factors where sperm was released, such as water temperature, salinity and pH (Alavi and Cosson 2006). All these factors seem to be interrelated and will contribute to defining the pattern and the duration of spermatozoa motility.

In Senegalese sole, motility activation follows the mechanism pattern previously described. Some environmental factors prove to be relevant in maintaining motility or in increasing the number of motile cells in a sperm sample. Lower salinities (25% and 30%) promoted spermatozoa with higher velocity and linearity in contrast to samples activated with salinities around 35% (Diogo et al. 2010a), which are usually the husbandry conditions used to maintain the breeders in aquaculture facilities. This effect can be related to two aspects: firstly, the higher the difference in osmolarity from the inside to the outside of the cell, the higher is the increase in ions inside the cell in a short period of time. This first hypothesis can be confirmed by the fact that a higher percentage of motile cells and duration of movement was achieved with non-ionic hyperosmotic solutions (sucrose) compared with the activation of normal seawater at the same osmolality. Secondly, it has been demonstrated that sperm exposure to a drastic environment, such as seawater, also leads to local defects of sperm flagella, an additional limitation in the motility duration (Dzyuba and Cosson 2014). Although not confirmed, it is possible that this damage could be higher according to the higher concentration of salts in the environment, since the osmotic shock suffered by spermatozoa would be more pronounced. Water temperature during motility activation is another parameter that influences motility in Senegalese sole, as also demonstrated in other species. Higher temperatures (20°C when compared with 16°C) trigger higher number of motile cells in this species (Diogo et al. 2008). Senegalese sole males can produce sperm all year around even when maintained at lower temperatures as 12°C (Beirão et al. 2008), but sperm released during the spawning season usually occurs at temperatures higher than 18°C.

Most of the information obtained regarding sperm motility activation in fish species is made by an *in vitro* analysis upon activation in a microscope. There are several techniques that can be used (described in Subsection 3.3), but a prerequisite is the activation with seawater, neglecting other factors that may also contribute to this mechanism. In Senegalese sole, and under natural conditions, sperm is released closely in contact with the female, when there is a release of the oocytes together with ovarian fluid. Therefore, it seems that ovarian fluid could have a specific rule in this mechanism. Studies performed by Diogo et al. (2010b) showed that Senegalese sole sperm motility was enhanced in the presence of 25–50% of ovarian fluid compared with normal activation only with seawater. The same finding was not confirmed when using ovarian fluid in the activation media coming from another species, which indicates some intra-specificity effect of ovarian fluid.

2. How important is sperm quality?

Spermatozoa contributes to the generation of a new individual, these germ cells being capable of reaching the oocytes, fertilizing the eggs, delivering the paternal genome and participating in the stabilization processes in early embryonic development. To do so, sperm needs to attain certain characteristics (previously mentioned during this chapter) and any change will influence spermatozoa functions. Several metabolic and cellular characteristics of spermatozoa contribute to all these phenomena, which can be altered by different factors already during the process of gametogenesis (Fig. 1). Sperm quality in Senegalese sole has been widely studied in the past years, where several factors were identified has contributing to the poor quality of sperm observed in this species. Some of them are reviewed in the follow subsections.

2.1 Wild versus F1 males

The quality of males in Senegalese sole has been a subject of research interest. Several differences between wild captured fish and those born and reared in captivity (F1), mostly related with testis development in terms of germ cell production and quality, hormone secretions, proteome composition, etc., were found (Cabrita et al. 2006, Forné et al. 2009, Guzman et al. 2009). However, in this subsection, we will focus on the subject in terms of male fluency and sperm quality, although previous subjects may be implicated in these aspects.

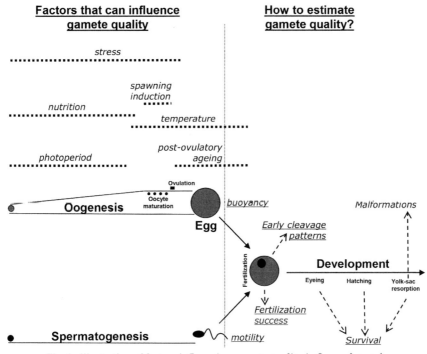

Fig. 1. Illustration of factors influencing gamete quality in Senegalese sole.

First evidences that showed sperm differences between males came up with the work performed by Cabrita et al. (2006), when the first report on Senegalese sole sperm quality was made. Authors suggested at the time a selection process based on male fluency, sperm production characteristics and provenance (wild-captured or F1). According to these authors who studied different sperm characteristics (motility, sperm production, density, volume throughout the spawning season), F1 males always registered a putative worst reproductive scenario. Sperm production and characteristics in wild males have been exhaustively studied (Beirão et al. 2008, 2009, 2011, 2015, Martínez-Pastor et al. 2008, Diogo et al. 2010a) compared with F1 stocks. Recently, Valcarce et al. (2016) found that the level of apoptotic cells was higher in F1 males than wild ones, also suggesting a slight but important decrease in sperm quality in this type of males. The factors behind this process are still unknown, but it has been associated with nutritional deficiencies or rearing conditions at previous stages (larval development or juvenile) or at gametogenesis process. All factors lead to a lower sperm production in these males which can now be stimulated by different hormonal therapies (see Chapter B-1.3. in this book), even when a normal gametogenesis process was not accomplished (Cabrita et al. 2015).

2.2 *Factors affecting sperm quality*

Several factors have been reported to affect sperm quality in different fish species (Fig. 2). There are several extensive reviews on sperm quality highlighting the potential factors that may contribute to an improvement of sperm quality. Most reviews deal with aspects related with broodstock improvement (Cabrita et al. 2011a, 2014) while others directly identify factors capable of affecting sperm quality traits (Rurangwa et al. 2004, Cabrita et al. 2008a, Bobe and Labbé 2010, Fauvel et al. 2010) or relate broodstock management

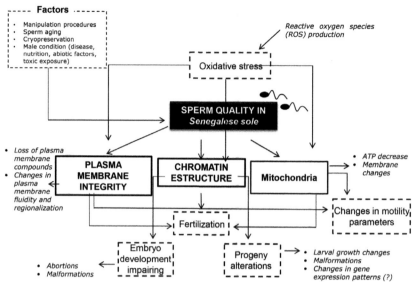

Fig. 2. Diagram illustrating the factors and variables involved in sperm quality as well as possible damage occurring in Senegalese sole sperm.

procedures with gamete features (Migaud et al. 2013). In this subsection, we will focus on the main topics specifically affecting sperm quality in Senegalese sole and on possible consequences for these cells and their use (Fig. 2).

2.2.1 *Husbandry conditions, reproductive period and period of sperm collection*

The period for sperm collection is decisive to obtain mature sperm of high quality. Sperm motility and the total volume and density of expressible milt undergo seasonal changes in many species. These changes do not only occur throughout the year, but also within the reproductive season (Cabrita et al. 2011b) and recent evidences indicate that in some species there are daily rhythms on sperm quality (Oliveira et al. 2013). Temperature and photoperiod are the two main factors changing between seasons and it has been known to affect spermiation period and sperm characteristics (Cabrita et al. 2011b).

In Senegalese sole, although the species present two defined spawning seasons (spring and autumn) (Dinis et al. 1999), sperm can be collected throughout the year (Cabrita et al. 2006, Beirão et al. 2011). However, quality changes have been registered specially in terms of sperm motility, DNA fragmentation and cell resistance to osmotic shock (Beirão et al. 2011). According to these authors, out of the two reproductive periods, both sperm velocity and progressive cells decreased and spermatozoa DNA fragmentation increased, especially in the summer months, associated with temperature rise above the normal values for reproduction in this species. Although sperm can be found in some males all-year-round exposed to temperatures near 12°C to 24°C, the percentage of fluent males is higher during the two normal reproductive seasons for this species (Cabrita et al. 2006). Another point to take in consideration is the high variability between males, both during the year and in the spawning season, a fact that influences sperm quality but that still does not have any explanatory reason. This subject needs further research to be conducted in order to help fish farms with their broodstock management.

Other factors related with husbandry conditions such as the presence of females, sex ratio, among others may be important for sperm quality. Cabrita et al. (2011b) found that wild-captured males maintained separately from females for several months do not spermiate and that this spermiation phenomenon can be stimulated by contact with females in the same tank.

2.2.2 *Hormonal treatments*

One of the principal problems in sperm collection in Senegalese sole is the difficulty in obtaining sperm due to their low volume and concentration, a characteristic specific in this species (wild males) but more pronounced in F1 males, as previously mentioned (Cabrita et al. 2006). Some hormonal treatments already proved to be efficient in slightly increasing the sperm volume (Aguillero et al. 2007, Guzmán et al. 2011a,b) but there are still some gaps on the evaluation of this contribution. Until recently, no reports on the quality of sperm have been delivered that guarantee a safety use of this material. Recently, an exhaustive evaluation of sperm quality in Senegalese sole F1 males exposed to different hormonal treatments showed that motility can be enhanced progressively with weekly hormonal injections (hCG or GnRHa). Sperm velocity was also slightly increased (Cabrita et al. 2015). Previous studies by the same group, but in wild-captured males, demonstrated that GnRHa injection improved, in a transitory way (just 2 days after treatment), cell viability and sperm velocity (Cabrita et al. 2011b). No detrimental effects have been mentioned in terms of lipid peroxidation and DNA fragmentation, although F1 males hormonally treated presented higher sperm DNA fragmentation compared with wild-captured males (Cabrita et al. 2006, Beirão et al. 2008, 2009), but no conclusions can be drawn until further studies.

2.2.3 *Nutrition*

Breeders' nutrition has proved to ameliorate the quality of sperm by improving several spermatozoa traits in several fish species, including Senegalese sole. In fact, seminal or plasma membrane composition may be positively modulated through the incorporation of certain compounds in the feeds, although most of the reports deal with an increase of DHA (docosahexaenoic acid) and ARA (arachidonic acid) in plasma membrane of fish feed on polyunsaturated fatty acids (PUFA) enriched diets (Asturiano et al. 2001, Nandi et al. 2007, Beirão et al. 2015). Several constituents (phospholipids, cholesterol) may affect the plasma membrane fluidity altering sperm regionalization, important for fertilization process, osmotic control, important during motility activation or resistance to certain process, such as cryopreservation (Muller et al. 2008, Wassall and Stillwell 2009). Therefore, their modulation has a direct impact on sperm physiology and functionality. Several authors have investigated the phospholipids, cholesterol and fatty acids' contents in sperm and their interaction to understand how its composition affects sperm quality and influences its ability to successfully fertilize the eggs (Muller et al. 2008, Lahnsteiner et al. 2009, Henrotte et al. 2010, Beirão et al. 2012a). These compounds are also inducers of sex steroid and eicosanoid production, particularly prostaglandins, and thereby their involvement in gonad development may also potentiate sperm quality (Izquierdo et al. 2001, Norambuena et al. 2013).

In Senegalese sole, the supplementation of ARA in the diets increased prostaglandins and sex steroid levels leading to a probable increase in sperm production (Norambuena et al. 2013). This fact was recently corroborated by Serradeiro et al. (2015), where a tendency of fish fed on higher ARA levels (0.45% ARA compared with 0.10% ARA in dry matter of feed) showed higher sperm volume and an increase in sperm velocity. No other parameters were affected (DNA fragmentation, cell viability or the percentage of motile cells). Supplementation of other PUFA on diet of this species increased the percentage of progressive spermatozoa and sperm velocity, probably as a consequence of DHA, other fatty acids and phospholipids' (phosphatidilcoline, phospohatidiletanolamine) increase in sperm membranes (Beirão et al. 2015). This effect was particularly noticed when this supplementation was complemented by an addition of antioxidants such as selenium (Se) and vitamin E (Beirão et al. 2015). Although in this work, the addition of antioxidants had the main objective of avoiding lipid peroxidation in feeds supplemented with DHA, antioxidants are commonly known to increase sperm traits and ameliorate sperm quality. There are several reports on the incorporation of antioxidant substances into feeds, proving to have a favourable effect on sperm quality, especially when it needs to be reinforced. However, in Senegalese sole, apart from this study, no other work was conducted for testing the effects of Se and vit E on sperm quality and in the light of these results, it should be an interesting step forward.

3. Sperm Management Techniques

3.1 Collection and sperm processing

Sperm handling is one of the most relevant factors that could affect the sperm quality. In hatcheries that perform artificial fertilization, it is a common procedure to maintain sperm in short or long term storage to optimize gametes management.

Survival of fish sperm is strongly predetermined by their sensitivity to osmotic changes in extracellular media. Specifically, in Senegalese sole, previous studies on sperm osmotic features revealed a low resistance of sole spermatozoa plasma membrane to seawater (Beirão et al. 2009). In fact, a high percentage of sole spermatozoa are extremely sensitive to seawater, decreasing viability due to osmotic shock. This fact might be related to some type of damage in the osmoregulatory mechanisms of the cells in this species or to some kind of damage in captive breeders affecting membrane osmoregulatory mechanisms.

Moreover, to improve sperm quality of collected sperm during an experimental procedure, male broodstock should be maintained under stress-free conditions. Stress disrupts growth and reproduction, possibly affecting sperm traits. In Senegalese sole, milt quality in low and high cortisol responsive males was analysed and no correlation was found between the cortisol plasma levels and the frequency of milting or sperm production. However, the volume of sperm varied within both groups, registering higher sperm production in the high stress responsive fish during several months (Cabrita et al. 2008b).

Sperm collection should be performed during the spawning season, where a pick in sperm quality is obtained, in healthy and mature males and under anesthetic conditions to minimize stress during handling. In Senegalese sole, Beirao et al. (2011) found significant differences in sperm motility and DNA fragmentation between the spawning seasons and the rest of the year. These authors described a tendency for the sole semen to reach a quality peak within the spawning season followed by a pronounced decrease probably caused by high temperatures, as mentioned before, indicating a possible effect of temperature on sperm quality during this period (Beirão et al. 2011). Therefore, special attention needs to be taken in order to guarantee that sperm is being collected, if possible, during the peak of quality production.

During sperm collection, other fluids such as urine or faeces should be avoided and for that, urogenital opening area needs to remain clean and dry, expelling urine and faecal contaminants before sperm collection (Cabrita et al. 2008a). In Senegalese sole, individuals can be anesthetized with 300 ppm 2-phenoxyethanol during 5 min before sperm collection.

The urogenital pore should be dried; sperm should be collected with a syringe or with a 20 μl micropipette by gently pressing the testes on the fish's blind side. After collection, sperm is introduced in tubes and can be stored at 4°C until use, discarding samples contaminated with urine. Contaminated samples will lose motility in a very short period of time. If there is any doubt about the contamination of sperm with urine, sperm can be diluted 10-fold in a non-activating medium: Ringer solution (20 mM HEPES, 5 mM KH_2PO_4, 1 mM $MgSO_4$, 1 mM $CaCl_2$, 136 mM NaCl, 4.7 mM KCl, 300 mOsm/Kg) and then centrifuged (300 g, 15 min at 10°C) to eliminate seminal plasma and contaminants. Cell concentration is restored by diluting the pellet cells again in Ringer solution. It is important to remark that Senegalese sole males have a very low sperm volume in testes compared to higher volumes produced by other species with the same gonadosomatic index. Since the sperm concentration is very low and the time of movement after activation with seawater is also short, these characteristics make the sperm handling in this species difficult.

3.2 Preservation versus Cryopreservation

Improving artificial fertilization success in captive Senegalese sole broodstocks has been a challenge in the last years due to the absence of natural reproduction in fish born in captivity (F1 individuals) (Rasines et al. 2012). As mentioned before, this species presents variable semen production and quality, thus reducing the chances of successful fertilization, which makes difficult the process of artificial fertilization. Some authors reported that the use of some extenders in Senegalese sole sperm could help in the preservation of samples at 4°C at least for 24 hr post-collection, with no detrimental effects on motility parameters (Pérez et al. 2015). These extenders provided to Senegalese sole samples the ionic composition required to maintain sperm in a quiescent stage prior to motility activation. However, this method of preservation is limited to a short period of time, depending always on the species, cell concentration, extenders' composition, among others. Therefore, the possibility of sperm cryopreservation would guarantee the continuous availability of sperm and quality control in this species. However, it is well known that the freezing/thawing process can damage spermatozoa at several levels such as plasma membrane, mitochondria and cell morphology (Cabrita et al. 2005, 2014). This damage can ultimately affect cell viability and fertilization capacity.

In Senegalese sole, cryopreserved sperm has traditionally been evaluated only by fertilization ability and the protocol used was directly applied from another flatfish species without any further assays (Rasines et al. 2012). In this protocol, the extender used in the sperm-freezing process was the Mounib solution (Mounib 1978), which included 10% bovine serum albumin (BSA) and 10% dimethyl sulphoxide (DMSO) as cryoprotectant. DMSO is the most widely used cryoprotectant for the cryopreservation of fish sperm and provides good protection (Yavaş et al. 2014). However, there are still a lot of steps to improve the cryopreservation of Senegalese sole sperm. Fertilization rates obtained with cryopreserved sperm are still very low (approximately 30%), which indicates the need to perform an exhaustive analysis of samples prior and after the procedure, to select samples and identify possible damage. In a recent study in Senegalese sole sperm cryopreservation, Riesco et al. (2017) demonstrated that specific analyses after sperm cryopreservation should be performed, since the production of ROS could be a potential inductor of DNA damage in this sperm. DNA is also prone to suffer cryodamage and its alteration is not always associated with a decrease in cell viability. This fact underlines the importance of evaluating DNA after cryopreservation, particularly for gamete banking purposes. The protocol developed by this group has slight differences from the previous one reported by

Rasines et al. (2012), since egg yolk incorporation as a membrane stabilizer showed better protection of cells from freezing-thawing.

3.3 How to analyze sperm quality?

The factors affecting sperm quality in fish have been widely described in the bibliography. In this subsection, we will focus on the analyzed parameters in sperm quality in Senegalese sole (Fig. 3). As mentioned before, Senegalese sole represents poor and variable semen quality (Cabrita et al. 2006) and a low sperm volume; these characteristics condition the type and quantity of quality tests performed in this species.

Definitely, motility has been described as the most used quality parameter in any species, particularly in fish sperm and widely employed in Senegalese sole sperm research. Motility is normally determined using computer assisted sperm analysis (CASA) and several softwares are available to determine motility parameters after sperm activation (Cabrita et al. 2008a). The settings for CASA software need to be adapted for each species. The most employed CASA parameters are the percentage of motile cells and velocity according to the actual path (VCL; µm/s). In Senegalese sole, since the duration of movement lasts around 1 min, motility parameters are usually recorded at 15 s intervals until 1 min post-activation. Analysis of data can be done by using mean values or due to the heterogeneity present in sperm population, and characterization of samples has been done using subpopulation analysis of data obtained from CASA analysis (Martinez-Pastor et al. 2008, Diogo et al. 2010a, Beirão et al. 2011). This last approach allows the identification of potentially interesting sperm subpopulations in order to correlate these subpopulations with fish sperm traits, characterizing males according to their reproductive potential.

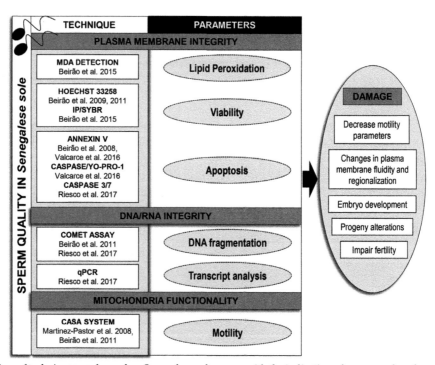

Fig. 3. Type of techniques performed on Senegalese sole sperm with the indication of sperm analyzed parameters.

In fish farms, the motility is also used to check sperm quality, especially if artificial fertilization is being performed. In Senegalese sole, visual inspection of sperm after activation with seawater is the method commonly used due to its simplicity and low cost (Morais et al. 2016). However, not always are the results accurate due to the subjectivity of the method and special attention should be taken when cryopreserved sperm is being analyzed.

The cell integrity and permeability represents one of the easiest methods to evaluate the sperm quality. The integrity of the plasma membrane and its selective permeability are crucial for cell survival, and therefore, several assays have been developed to analyze spermatozoa viability (which are directly related with plasma membrane integrity). Different protocols and dyes have been established for testing the viability of spermatozoa (Horváth et al. 2008, Beirão et al. 2012b).

Several dyes have been used to discriminate between viable and non-viable cells such as trypan blue (Lubzens et al. 1997), eosin (Lahnsteiner et al. 1996) and Hoechst 33258 (pentahydrate, bis-benzimide) (Cabrita et al. 1998), which only enter the cell through altered plasma membranes. This technique was successfully employed in Senegalese sole sperm to evaluate sperm membrane resistance to seawater throughout the year and during the reproductive season (Beirão et al. 2009, 2011) (Fig. 3).

Nowadays, the use of dual fluorescent staining SYBR-14 and propidium iodide together with flow cytometry, allowed a faster and more accurate evaluation of plasma membrane integrity (Cabrita et al. 2005, Hórvath et al. 2008). In sole, this technique has been widely applied to evaluate sperm quality after different diet supplementations (Beirão et al. 2015).

Despite the fact that the described parameters are useful in the determination on sperm quality, nowadays, novel techniques based on oxidative stress detection have been developed. As mentioned before, the production of ROS can cause sperm dysfunction due to plasma membrane lipid peroxidation and, consequently, a decrease in fertilization ability (Li et al. 2010). Lipid peroxidation has become a valuable marker used to evaluate oxidative stress in fish sperm. Lipid peroxidation is usually measured by the quantification of the final product of lipid oxidation, the malondialdehyde (MDA). In Senegalese sole, this determination was performed to evaluate the effect of certain diets on sperm quality (Beirão et al. 2015). In this study, malondialdehyde values were significantly higher in sperm of males fed with a specific experimental diet with a higher DHA content to improve sperm plasma membrane resistance. These oxidative effects were attenuated by the incorporation of antioxidant compounds.

Moreover, several studies demonstrated that ROS attacks the chromatin integrity causing nucleotide modifications, DNA strand breaks and cross-linking (Cartón-García et al. 2013). In addition, production of ROS, which can damage DNA, proteins, and lipids (Halliwell 1994), can ultimately lead to apoptosis or necrosis in living cells (Walter et al. 1998).

Concerning the sperm chromatin integrity, there are different approaches for its evaluation: the Comet Assay (single cell gel electrophoresis), the Sperm Chromatin Structure Assay (SCSA), the Terminal transferase dUTP Nick End Labeling (TUNEL), the Sperm Chromatin Dispersion test (SCD or halo test) and the analysis of specific DNA sequences. Comet assay is the widely employed technique in the analysis of DNA damage in Senegalese sole. This technique showed significant differences in sperm from captive males between the two spawning seasons (Beirão et al. 2011). Moreover, this technique allowed to detect an increase in sperm DNA fragmentation in Senegalese sole out of the natural spawning seasons (Beirao et al. 2011). No other techniques for sperm DNA analysis have been used in Senegalese sole sperm.

Apoptosis is a common phenomenon during spermatogenesis and it has already been detected in Senegalese sole spermatozoa during the reproductive season using Annexin-V FITC (Beirão et al. 2008). The Annexin-V assay allowed the identification of an early apoptotic population, since it detects the translocation of phosphatidylserine from the inner to the outer plasma membrane layer. Other methods based on nucleic acid dyes, as YO-PRO 1, were employed to detect membrane permeability changes usually correlated with apoptosis in gilthead seabream (*Sparus aurata*) cryopreserved sperm (Beirão et al. 2010). Other cellular apoptosis markers such as caspases activity (Ortega-Ferrusola et al. 2008) and activation of specific genes (Jeong et al. 2009) may contribute to provide additional information on spermatozoa apoptosis determination. In Senegalese sole, it has been demonstrated that caspase detection is more specific than YO-PRO-1 in the identification of apoptotic seminal samples (Valcarce et al. 2016). In a recent study, apoptosis activity detection by Caspase 3/7 was the key to select an optimized cryopreservation protocol for Senegalese sole sperm, allowing to select the best cryopreservation media for this species (Riesco et al. 2017).

Nowadays, molecular techniques have been implemented in sperm quality studies in different fish species. Using a qPCR approach, several transcripts were identified in zebrafish and gilthead sea bream sperm as molecular markers due to the fact that they have different expression profile in good and bad breeders (Guerra et al. 2013). Moreover, qPCR approach was also applied to detect specific lesions in different genome regions or genes, quantifying the delay of the DNA polymerase progression as a consequence of lesions in the DNA template, which results in a decrease in DNA polymerase fidelity and amplification efficiency (Rothfuss et al. 2010). This method was applied after cryopreservation protocols for the evaluation of seabream and rainbow trout sperm integrity (Cartón-García et al. 2013, González-Rojo et al. 2014) and to assess damage in zebrafish germinal cells (Riesco and Robles 2013).

4. Artificial fertilization

Artificial fertilization (AF) in sole has become feasible after the standardization of hormone-based protocols to induce ovulation in females (see Chapter B-1.3. in this book), which allowed obtaining eggs through manual stripping (Liu et al. 2008, Howell et al. 2009, 2011, Rasines et al. 2012, 2013). An AF protocol has been developed for Senegalese sole and may represent an alternative to obtain fertilized eggs from F1 broodstock, considering that F1's are unable to produce fertilized tank spawning. This could be an interesting approach for the aquaculture industry of Senegalese sole, which is actually based on seed obtained from wild broodstock adapted to captivity. In this regard, it is interesting to mention that the absence of fertilized tank spawning from F1's is mostly due to inhibited reproductive behaviour and the corresponding inability of F1 breeders to liberate gametes synchronously, but not to a bad gamete quality, which is roughly similar in breeders from both origins. This is, for example, indicated by AF trials performed in Senegalese sole that used different crosses of gametes, which showed similar fertilization and hatching rates using eggs from F1 females fertilized with sperm from wild or F1 males or using eggs from wild females fertilized with sperm from wild or F1 males (Mañanós, unpublished).

A major limitation for the development of AF in Senegalese sole, and probably other soles, is the programmed availability of large amounts of good quality gametes, mostly sperm. In contrast to other flatfishes, the type of testicular development in Senegalese sole (semi-cystic) implies a low gonadosomatic index and a low production of milt. In addition, both spermiation and ovulation are inhibited or reduced in F1 breeders compared to wild fish, all of which represent serious difficulties for obtaining gametes from F1 breeders. Advances on the development and optimization of hormonal treatments to stimulate spermiation and ovulation will parallel the development and optimization of AF protocols in soles.

4.1 Obtaining sperm for artificial fertilization

Availability of sperm is probably the most critical aspect for the development of a viable AF protocol for sole, considering the low amount of milt produced by male breeders. Senegalese sole males can produce motile sperm all year round, with specific peaks of high spermiation and high percentage of fluent males usually matching the female breeding season. There is a large variability in terms of sperm profiles in males maintained under the same conditions. Sperm volume collected usually ranges from 5 to 30 µl in F1 males and from 10 to 80 µl in wild-captured males. Cell density and sperm production (total spermatozoa per stripping) ranges from $0.7–1.2 \times 10^9$ spz/ml and 20×10^6 spz in F1 males to values of $1–2 \times 10^9$ spz/ml and $40–60 \times 10^6$ spz for wild males, respectively. These values indicate that sperm production in this species is very low and variable according to the type of males and, that wild males produce a larger amount of sperm, usually of higher quality (Cabrita et al. 2006).

The quality of sperm is also highly variable between males, which is an additional difficulty for AF, because a large amount of males should be manipulated to select good quality samples and get enough volume of pooled milt for AF trials. Also, sperm quality varies throughout the year, with a tendency for the semen to attain a quality peak between the beginning and the middle of the spawning season (Beirão et al. 2009, 2011). The collection of sperm for AF should take this into account and use the best periods of sperm quality. Before its use in AF trials, the quality of semen should be checked and this can be quickly estimated by analyzing the motility of spermatozoa, after activation with sea water, using a simple five stage classification (Chereguini et al. 2003), based on the percentage of motile spermatozoa within the population: (1) 0–20, (2) 20–40, (3) 40–60, (4) 60–80 and, (5) 80–100 percent of motility. Usually, sperm samples staged 1–3 are discarded for use in AF trials.

Fig. 4. Collection of sperm from male Senegalese sole. One operator applies gentle pressure on the area were the testes lobules are located, to push out the milt through the spermiduct (indicated with an arrow), located within the two pelvic fins that are maintained, separated with the left hand for easiness of manipulation. Another operator collects the emerging milt with a syringe.

Cryopreservation is a good alternative to overlap the difficulties associated with the low quantity of milt and variable milt quality in male Senegalese sole. Sperm cryopreservation can be used as a tool to guarantee the necessary quality control of sperm during the fertilization trials and to ensure that fish farms are able to store this material and have enough sperm to fertilize the egg batches when necessary. Several protocols have been used, mostly adapted from other species such as turbot. Sperm from Senegalese sole has been stored using the Mounib extender with 10% DMSO, packaged in 0.5 ml straws and cryopreserved using a slow cooling rate (straws set at 5 cm from the surface of liquid nitrogen) (Rasines et al. 2012) or packaged in 0.25 ml straws and cooled at a faster rate (2 cm from the N_2 surface) (Riesco et al. 2017). Both protocols yield good results in terms of fertility and sperm motility, but more research on this field is required to guarantee the optimization of the procedure, ensuring that no progeny quality interference is produced by this technique. Although there are no scientific evidences, field observations on different AF trials in Senegalese sole have suggested worse results when using cryopreserved sperm compared to fresh sperm, indicating the necessity to optimize cryopreservation and thawing protocols for sole.

A helpful approach to increase the quantity of sperm is the previous stimulation of spermiation in males by hormonal treatment. This would be helpful, not only for the purpose of cryobanking, but also to obtain the required volume of fresh sperm for immediate AF trials. Hormonal induction of spermiation has never been reported to have deleterious effects on sperm quality parameters, neither in Senegalese sole nor in other fish species (Mañanós et al. 2009). Several hormonal treatments have been tested in Senegalese sole and demonstrated to stimulate sperm production, although the final output of this approach is limited (see Chapter B-1.3. in this book). Hormonal methods are usually based on prolonged treatments with GnRHa or hCG (Mañanós et al. 2009) because acute short-lived treatments are usually not enough for a significant stimulation of sperm production, as observed in Senegalese sole (Cabrita et al. 2011, Agulleiro et al. 2007, Guzmán et al. 2007, 2011a). Treatment of F1 Senegalese sole with GnRHa EVAc implants (50 µg GnRHa/kg) or with multiple injections of hCG (6 weekly injections of 1,000 IU/kg) have shown to increase sperm production around 2-fold and most importantly, to increase the number of fluent males, from 50 to 85–100 percent (Guzmán et al. 2007, 2011a). Apart from these classical therapies, some experimental protocols have been tested in F1 Senegalese sole males that might be applied in the future after appropriate development and optimization. Treatment with recombinant gonadotropins (multiple weekly injections of combined FSH (dose range 3–16 µg/kg) and LH (dose range 2–17 µg/kg)) increases sperm production, but the response is highly variable between individuals (0- to 7-fold increase) and in some cases, harmful for testicular development (Chauvigne et al. 2017). Treatment with the dopamine antagonist pimozide (injection; 5 mg/kg BW) stimulates sperm production (3.5-fold over controls) and enhances GnRHa stimulation of sperm production in co-treatment (5.2-fold over controls), indicating that the blockage of the dopamine system could be an adequate approach to stimulate spermiation in Senegalese sole (Guzmán et al. 2011b).

4.2 Egg stripping and artificial fertilization

Collection of eggs by hand stripping and further development of an AF protocol was possible in Senegalese sole after the development of hormonal therapies to induce ovulation (see Chapter B-1.3 in this book). Collection of eggs by hand striping was not feasible from naturally ovulated Senegalese sole females. Female Senegalese sole breeders, both wild and F1, ovulate and spawn spontaneously in captivity, but the day and even the time of natural ovulation, which normally occurs sometime from sunset to sunrise, is unpredictable and highly variable between individuals.

The standardized therapy to induce ovulation in Senegalese sole is the single injection of GnRHa (dose 25 µg/kg; dissolved in saline and applied in the dorsal musculature), with 70 to 94 percent of the females responding to the treatment, depending on the study (Liu et al. 2008, Rasines et al. 2012, 2013). Treatment with hCG might also be effective, but it has never been tried for the purpose of egg stripping and AF. In this regard, the works by Ramos (1986) on common sole showed that a single injection of hCG (250 or 500 IU/kg) was effective to induce tank spawning at 24–96 hr post treatment.

The effectiveness of hormonal induction depends highly on the appropriate selection of females that should be at advanced stages of oocyte maturation. This evaluation should be performed previously, by examination of ovarian biopsies under the binocular, to determine oocyte diameter and/or morphological features of the oocytes, which are useful parameters to estimate the ovarian maturational stage (Mañanós et al. 2009), though it has not been extensively used in sole, as the ovary duct is located in a kind of cloaca with the anus and is not easily found. Another complementary option is the external examination of abdominal swelling and palpation of the ovary, as an indication of the ovarian stage of development (Anguis and Cañavate 2005, Guzmán et al. 2008).

The time elapsed until ovulation after GnRHa induction is 38–50 hr, depending on water temperature, individuals and the maturational stage of the fish (Liu et al. 2008, Mañanós et al. 2009, Rasines et al. 2012). The mean lapsed time for ovulation is 41-43 hr and this is not affected by the time of the day when the hormonal treatment is applied, as demonstrated in assays were GnRHa was applied at 06:00, 12:00 or 19:00 hr (Rasines et al. 2013). The time of the day for hormonal treatment does not greatly affect fecundity, fertilization and hatching rates after AF, although further investigation should be performed to definitively demonstrate the flexibility for timing hormonal induction in this species.

Relative fecundity of Senegalese sole females after hormonal induction and egg stripping varies between studies, ranging from 100,000–150,000 eggs/kg BW (Rasines et al. 2013) to around 250,000 eggs/kg BW (Liu et al. 2008). Female Senegalese sole might be induced to ovulate several times if needed, as demonstrated in a study using weekly GnRHa injections (25 µg/kg), which showed females responding to 4 injections, others to 3 and others to the first 2 injections, providing an average relative fecundity of 574,000 eggs/kg (Rasines et al. 2012).

Egg quality is determinant for successful AF and is primarily dependent on the time elapsed between ovulation and egg collection. Collecting eggs later than the appropriate

 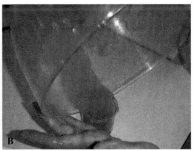

Fig. 5. Collection of eggs by hand stripping from female Senegalese sole. A: application of gentle abdominal pressure (see the important abdominal swelling of the hormone induced female) to force the ovulated eggs to come out easily through the oviduct, that appears expanded and reddish (indicated by an arrow). B: collection and manipulation of the stripped eggs.

time window after ovulation can lead to low or null fertilization rates, due to the loss of egg viability associated to over-ripening (Bromage et al. 1994, Flajshans et al. 2007). Preliminary studies in Senegalese sole indicate that over-ripping is a quick process in this species. A significant reduction in fertilization rates has been observed in eggs collected 6 hr after ovulation, compared to those collected at 0 or 3 hr after ovulation, indicating that over-ripping starts sometime at 3–6 hr and thus, AF should be performed within the first 3 hr after ovulation (Rasines et al. 2012). Even collecting eggs at the appropriate time window, egg quality in Senegalese sole is highly variable between females, the reason of which is unknown. Thus, a quick estimation of the quality of the stripped eggs should be performed for each batch before its use for AF. This can be easily done by taking subsamples of the egg batch and analyzing (1) morphology of the eggs under the binocular and, (2) buoyancy in sea water (Fig. 6). Viable eggs are floating and morphologically spherical, translucent, without a perivitelline space and with lipid droplets evenly spread throughout the egg, while dead eggs sink and become non-spherical, opaque and with agglomerations of lipid droplets. Egg batches containing a mid-high percentage of dead eggs should be discarded for AF.

The AF protocol developed for Senegalese sole is based on the dry method used for other flatfishes and has been shown to be successful for the production of fertilized eggs and viable larva in this species (Liu et al. 2008, Rasines et al. 2012, 2013). The general procedure is to place the stripped eggs into a dry plastic bowl, add sperm and seawater (34–37 ppt salinity) successively for sperm activation and fertilization, mix gently and let stand for several minutes (3–10 min), after which the fertilized eggs are transferred to incubation buckets. The success of the manipulation and an estimation of the percentage

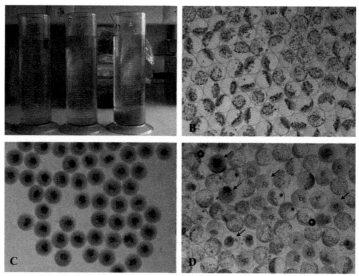

Fig. 6. Estimation of egg quality of stripped eggs, by determination of buoyancy in sea water (A) or morphological observation under the binocular (B, C, D). A: floatability of three batches of stripped eggs corresponding, from left to right, to two "good quality" batches (around 90 percent floating eggs) and one "bad quality" batch (100 percent sunken). B-C-D: morphological aspect of three batches of stripped eggs, corresponding to a good (B; 100% "good" eggs), bad (C; 100% dead eggs) or mid-quality (D; 40% good eggs) batch. In photo B, a green line indicates the measured diameter of an egg (1030 microns). In photo D, arrows indicate some dead ("bad") eggs, clearly distinguishable from the alive good eggs (identified by asterisks), because of the progressive loss of transparency and sphericity and coalescence of the lipid droplets (it can be seen at different degrees).

of fertilization should be performed and is the easiest to be done at around 90 min after fertilization, when the 2-cell or 4-cell stage is attained, depending on temperature (Fig. 7).

Before fertilization, a previous estimation of sperm density and the number of eggs should be done for each trial, to use an appropriate sperm to egg ratio. Sperm density (and also motility) is estimated previously under the microscope and, in Senegalese sole, is approximately 1×10^9 spz/ml. The number of stripped eggs can be estimated by counting the eggs in subsamples (e.g., triplicated 0.5 gr or 0.5 ml subsamples) and extrapolating to the weight/volume of the batch. Although it is best to perform the measurements for each trial, it can be roughly assumed that 1 ml of eggs weight approximately 1 gr and that 1 ml (or 1 gr) contains approximately 1,400 eggs, considering that the number of eggs per millilitre in Senegalese sole has been reported to be 1,278 (Liu et al. 2008) or 1,493 (Rasines et al. 2012).

Two AF trials have been published to date for Senegalese sole, one performed in China (Liu et al. 2008) and another one in Spain (Rasines et al. 2012, 2013). The Chinese trials used 23 and 10 females in two consecutive years and reported successful induction of ovulation of 91 percent of the females the first year and 100 percent the second year, which was associated to a better selection of females (correct maturation stage) in the second trial. In this study, the reported fecundity was 526,000 eggs per female (fish BW around 1.8 kg) with a mean egg quality (i.e. buoyancy) of 70 percent. Fertilization was performed with fresh sperm (no sperm to egg ratio is provided in the study) and provided mean fertilization and hatching rates of 64 and 80 percent, respectively. A production of 6 million of newly hatched larvae and 3,610,000 juveniles was reported from these trials (Liu et al. 2008).

The Spanish trials used similar F1 Senegalese sole breeders (fish of around 1.5 kg BW) and reported successful induction of ovulation in 67 to 100 percent of the females, depending on the trial and pointing out again to individual variations in the responsiveness and the importance for a careful selection of mature females before hormonal induction. Relative fecundity was 98,000 to 145,000 eggs/kg and the apparent viability rate was

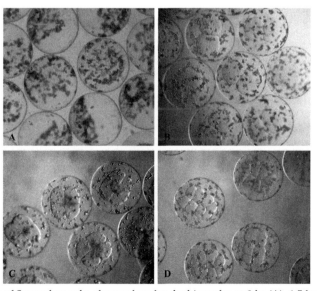

Fig. 7. Fertilized eggs of Senegalese sole, observed under the binocular at 0 hr (A), 1.5 hr (B; 2-cell stage), 2 hr (C; 4-cell stage) and 2.5 hr (D; 8-cell stage) after artificial fertilization (incubation at 17°C).

52–74 percent. The AF trials used cryopreserved sperm (diluted 1:2 in Mounib solution) and a ratio of 30 µl of the thawed cryopreserved sperm sample per 1 millilitre of eggs. The amount of sperm was probably used in excess, as suggested by further trials performed by these authors in which similar fertilization rates were obtained using lower amount of sperm and pointing out to the necessity of further optimization of the sperm to egg ratio for the AF protocol in Senegalese sole. The reported fertilization rates ranged from 11 to 37 percent and the hatching rates from 8 to 18 percent (Rasines et al. 2012, 2013). These values are low compared to those reported in the Chinese trials and could be associated to several factors, including the use of cryopreserved instead of fresh sperm.

The reported fertilization and hatching rates in the cited published works as well as in other unpublished studies are highly variable, between trials and between females and most importantly, unpredictable. Fertilization and hatching rates may vary from 15 to 95 percent, which confirms the potential of this technique and highlights that future experiments are necessary to determine and optimize the critical parameters influencing fertilization and hatching rates and to establish a predictable reliable AF protocol for Senegalese sole and other *Solea* species.

Acknowledgements

The authors are grateful to projects financed by programme MAR2020, Direção Geral das Pescas e Agricultura (ReproF1 16-02-01-FMP-0059 and Algasole 16-02-01-FMP-0058); EBB (EAPA_501/2016), Interreg Atlantic area and Assemble plus JRA2 (H2020-INFRAIA-2016-2017). The work conducted received Portuguese national funds from FCT - Foundation for Science and Technology through project UID/Multi/04326/2019, and from the operational programmes CRESC Algarve 2020 and COMPETE 2020 through project EMBRC.PT ALG-01-0145-FEDER-022121.

References

Abascal, F.J., J. Cosson and C. Fauvel. 2007. Characterization of sperm motility in sea bass: the effect of heavy metals and physicochemical variables on sperm motility. J. Fish. Biol. 70: 509–522.

Agulleiro, M.J., A.P. Scott, N. Duncan, C.C. Mylonas and J. Cerdá. 2007. Treatment of GnRHa-implanted Senegalese sole (*Solea senegalensis*) with 11-ketoandrostenedione stimulates spermatogenesis and increases sperm motility. Comp. Biochem. Physiol. 147: 885–892.

Aitken, R.J., R. Bronson, T.B. Smith and G.N. De Iuliis. 2013. The source and significance of DNA damage in human spermatozoa; a commentary on diagnostic strategies and straw man fallacies. Mol. Hum. Reprod. 19: 475–485.

Alavi, S.M. and J. Cosson. 2006. Sperm motility in fishes. (II) Effects of ions and osmolality: A review. Cell Biol. Int. 30: 1–14.

Alavi, H.S.M., I.A.E. Butts, A. Hatef, I. Babiak, M. Mommens, E.A. Trippel et al. 2011. Sperm morphology, seminal plasma composition, ATP content and analysis of sperm motility and flagellar wave parameters in Atlantic halibut (*Hippoglossus hippoglossus* L.). Can. J. Zool. 89: 219–228.

Asturiano, J.F., L.A. Sorbera, M. Carrillo, S. Zanuy, J. Ramos, J.C. Navarro et al. 2001. Reproductive performance in male European sea bass (*Dicentrarchus labrax*, L.) fed two PUFA enriched experimental diets: a comparison with males fed a wet diet. Aquaculture 194: 173–190.

Ausió, J., R. González-Romero and C.L. Woodcock. 2014. Comparative structure of vertebrate sperm chromatin. J. Struct. Biol. 188: 142–155.

Baeza, R., I. Mazzeo, M.C. Vílchez, V. Gallego, D.S. Peñaranda, L. Pérez et al. 2015. Relationship between sperm quality parameters and the fatty acid composition of the muscle, liver and testis of European eel. Comp. Biochem. Physiol. A Mol. Integr. Physiol. 181: 79–86.

Beirão, J., E. Cabrita, F. Soares, M.P. Herraez and M.T. Dinis. 2008. Cellular damage in spermatozoa from wild-captured *Solea senegalensis* detected with comet analysis and annexin-V. J. Appl. Ichthyol. 24: 508–513.

Beirão, J., F. Soares, M.P. Herráez, M.T. Dinis and E. Cabrita. 2009. Sperm quality evaluation in *Solea senegalensis* during the reproductive season at cellular level. Theriogenology 72: 1251–61.

Beirão, J., S. Pérez-Cerezales, S. Martínez-Páramo and M.P. Herráez. 2010. Detection of early damage of sperm cell membrane in Gilthead seabream (*Sparus aurata*) with the nuclear stain YO-PRO 1. J. Appl. Ichthyol. 26: 794–796.

Beirão, J., F. Soares, M.P. Herráez, M.T. Dinis and E. Cabrita. 2011. Changes in *Solea senegalensis* sperm quality throughout the year. Anim. Reprod. Sci. 126: 122–129.

Beirão, J., L. Zilli, S. Vilella, E. Cabrita, C. Fernandez-Diez, C. Schiavone et al. 2012a. Fatty acid composition of the head membrane and flagella affects *Sparus aurata* sperm quality. J. Appl. Ichthyol. 28: 1017–1019.

Beirão, J., L. Zilli, S. Vilella, E. Cabrita, R. Schiavone and M.P. Herráez M.P. 2012b. Improving Sperm Cryopreservation with Antifreeze Proteins. Effect on Gilthead Seabream (*Sparus aurata*) Plasma Membrane Lipids. Biol Reprod. 59: 1–9.

Beirão, J., F. Soares, P. Pousão-Ferreira, P. Diogo, J. Dias, M.T. Dinis et al. 2015. The effect of enriched diets on *Solea senegalensis* sperm quality. Aquaculture 435: 187–194.

Bell, M.V., J.R. Dick and C. Buda. 1997. Molecular speciation of fish sperm phospholipids: Large amounts of dipolyunsaturated phosphatidylserine. Lipids 32: 1085–1091.

Bermejo, R., A. Kumar and M. Foiani. 2012. Preserving the genome by regulating chromatin association with the nuclear envelope. Trends Cell Biol. 22: 465–473.

Billard, R., J. Cosson, L.W. Crim and M. Suquet. 1995. Sperm physiology and quality. pp. 25–52. *In*: Bromage, N.R. and R.J. Robert [eds.]. Broodstock Management and Egg and Larval Quality. Blackwell Science Ltd, Oxford.

Billard, R., J. Cosson, F. Fierville, R. Brun, T. Rouault and P. Williot. 1999. Motility analysis and energetics of the Siberian sturgeon *Acipenser baerii* spermatozoa. J. Appl. Ichthyol. 15: 199–203.

Bromage, N., M. Bruce, N. Basavaraja, K. Rana, R. Shields, C. Young et al. 1994. Egg quality determinants in finfish: the role of overripening with special reference to the timing of stripping in the Atlantic halibut *Hippoglossus hippoglossus*. J. World Aq. Soc. 25: 13–21.

Bobe, J. and C. Labbé. 2010. Egg and sperm quality in fish. Gen. Comp. Endocrinol. 165: 535–548.

Boj, M., F. Chauvigné and J. Cerdà. 2015. Coordinated action of aquaporins regulates sperm motility in a marine Teleost. Biol. Reprod. 93: 1–11.

Butts, I.A., E.A. Trippel, A. Ciereszko, C. Soler, M. Slowińska, S.M.H. Alavi et al. 2011. Seminal plasma biochemistry and spermatozoa characteristics of Atlantic cod (*Gadus morhua* L.). Comp. Biochem. Physiol. A Mol. Integr. Physiol. 159: 16–24.

Cabrita, E., R. Alvarez, L. Anel, K.J. Rana and M.P. Herráez. 1998. Sublethal damage during cryopreservation of rainbow trout sperm. Cryobiology 37: 245–253.

Cabrita, E., V. Robles, S. Cuñado, J.C. Wallace, C. Sarasquete and M.P. Herráez. 2005. Evaluation of gilthead sea bream, *Sparus aurata*, sperm quality after cryopreservation in 5 ml macrotubes. Cryobiology 50: 273–284.

Cabrita, E., F. Soares and M.T. Dinis. 2006. Characterization of Senegalese sole, *Solea senegalensis*, male broodstock in terms of sperm production and quality. Aquaculture 261: 967–975.

Cabrita, E., V. Robles and M.P. Herráez. 2008a. Sperm quality assessment. pp. 16–24. *In*: Cabrita, E., V. Robles and M.P. Herráez [eds.]. Methods in Reproductive Aquaculture: Marine and Freshwater species. Biology series. CRCPress (Taylor and Francis group), London, UK.

Cabrita, E., F. Soares, C. Aragão, J. Beirão, W. Pinto, T. Petochi et al. 2008b. Milt quality in low and high cortisol responsive Senegalese sole males. EAS 2008, Krakovia, Poland.

Cabrita, E., V. Robles, C. Sarasquete and M.P. Herráez. 2011a. Sperm quality evaluation for broodstock improvement. pp. 146–161. *In*: Tiersch, T. and P.M. Mazik [eds.]. Cryopreservation of Aquatic Species, 2nd edition, World Aquaculture Society, Baton Rouge, Lousiana, USA.

Cabrita, E., F. Soares, J. Beirão, A. García-López, G. Martínez-Rodríguez and M.T. Dinis. 2011b. Endocrine and milt response of Senegalese sole, *Solea senegalensis*, males maintained in captivity. Theriogenology 75: 1–9.

Cabrita, E., S. Martínez-Páramo, P.J. Gavaia, M.F. Riesco, D.G. Valcarce, C. Sarasquete et al. 2014. Factors enhancing fish sperm quality and emerging tools for sperm analysis. Aquaculture 432: 389–401.

Cabrita, E., C. Oliveira, M.F. Riesco, M. Livramento, E. Mañanós, L.E.C. Conceição et al. 2015. Effect of hormonal treatments on Senegalese sole sperm quality. Proc. 5th Int. Workshop Biology of Fish Gametes. 7–11 September 2015, Ancona, Italy.

Cartón-García, F., M.F. Riesco, E. Cabrita, M.P. Herráez and V. Robles. 2013. Quantification of lesions in nuclear and mitochondrial genes of *Sparus aurata* cryopreserved sperm. Aquaculture 402: 106–112.

Celino, F.T., S. Yamaguchi, C. Miura, T. Ohta, Y. Tozawa, T. Iwai et al. 2011. Tolerance of spermatogonia to oxidative stress is due to high levels of Zn and Cu/Zn superoxide dismutase. PLoS One 6: e16938.

Chereguini, O., I. García de la Banda, M. Herrera, C. Martinez and M. De la Hera. 2003. Cryopreservation of turbot *Scophthalmus maximus* (L.) sperm: fertilization and hatching rates. Aquac. Res. 34: 739–747.

Christen, R., J.L. Gatti and R. Billard. 1987. Trout sperm motility. The transient movement of trout sperm is related to changes in concentration of ATP following the activation of the flagellar movement. Eur. J. Biochem. 166: 667–671.

Cosson, J., A.L. Groison, M. Suquet, C. Fauvel, C. Dreanno and R. Billard. 2008. Studying sperm motility in marine fish: an overview on the state of the art. J. Appl. Ichthyol. 24: 460–486.

Darszon, A., P. Labarca, T. Nishigaki and F. Espinosa. 1999. Ion Channels in Sperm Physiology. Physiol. Rev. 79: 481–510.

Dinis, M.T., L. Ribeiro, F. Soares and M.C. Sarasquete. 1999. A review on the cultivation potential of *Solea senegalensis* in Spain and Portugal. Aquaculture 176: 27–38.

Diogo, P., F. Soares, M.T. Dinis and E. Cabrita. 2008. Sperm motility in *Solea senegalensis*: relationship with water temperature, salinity and pH. Proc. 7th Flatfish Symp. 2–7 November 2008, Sesimbra, Portugal.

Diogo, P., F. Soares, J. Beirão, M.T. Dinis and E. Cabrita. 2010a. Parameters affecting sperm motility activation in *Solea senegalensis*: multivariate cluster analysis vs mean values analysis. Proc. Annual Meeting Aquaculture Europe. 5–8 October 2010, Porto, Portugal.

Diogo, P., F. Soares, M.T. Dinis and E. Cabrita. 2010b. Ovarian fluid influence on *Solea senegalensis* sperm motility. J. Appl. Ichthyol. 26: 690–695.

Dreanno, C., J. Cosson, M. Suquet, F. Seguin, F. Dorange and R. Billard. 1999a. Nucleotide content, oxydative phosphorylation, morphology, and fertilizing capacity of turbot (*Psetta maxima*) spermatozoa during the motility period. Mol. Reprod. Dev. 53: 230–243.

Dreanno, C., M. Suquet, C. Fauvel, J.R. Le Coz, G. Dorange, L. Quemener and R. Billard. 1999b. Effect of the aging process on the quality of seabass (*Dicentrarchus labrax*) semen. J. Appl. Ichthyol. 15: 176–180.

Dzyuba, V. and J. Cosson. 2014. Motility of fish spermatozoa: from external signaling to flagella response. Reproductive Biology 14: 165–175.

Fatehi, A.N., M.M. Bevers, E. Schoevers, B.A. Roelen, B. Colenbrander and B.M. Gadella. 2006. DNA damage in bovine sperm does not block fertilization and early embryonic development but induces apoptosis after the first cleavages. J. Androl. 27: 176–188.

Fauvel, C., M. Suquet and J. Cosson. 2010. Evaluation of fish sperm quality. J. Appl. Ichthyol. 26: 636–643.

Ferramosca, A., S. Pinto Provenzano, D.D. Montagna, L. Coppola and V. Zara. 2013. Oxidative stress negatively affects human sperm mitochondrial respiration. Urology 82: 78–83.

Figueroa, E., I. Valdebenito and J.G. Farias. 2014. Technologies used in the study of sperm function in cryopreserved fish spermatozoa. Aquac. Res. 47: 1691–1705.

Flajshans, M., K. Kohlmann and P. Ráb. 2007. Autotriploid tench *Tinca tinca* (L.) larvae obtained by fertilization of eggs previously subjected to postovulatory ageing *in vitro* and *in vivo*. J. Fish Biol. 71: 868–876.

Forné, I., M.J. Agulleiro, E. Asensio, J. Abián and J. Cerdà. 2009. 2-D DIGE analysis of Senegalese sole (*Solea senegalensis*) testis proteome in wild-caught and hormone-treated F1 fish. Proteomics 9: 2171–2181.

García-Herrero, S., N. Garrido, J.A. Martínez-Conejero, J. Remohí, A. Pellicer and M. Meseguer. 2011. Differential transcriptomic profile in spermatozoa achieving pregnancy or not via ICSI. Reprod. Biomed. Online. 22: 25–36.

Gibbons, I.R. 1968. The biochemistry of motility. Ann. Rev. Biochem. 37: 521–546.

González-Rojo, S., C. Fernández-Díez, S.M. Guerra, V. Robles and M.P. Herraez. 2014. Differential gene susceptibility to sperm DNA damage: analysis of developmental key genes in trout. PLoS One 9: e114161.

Guerra, S.M., D.G. Valcarce, E. Cabrita and V. Robles. 2013. Analysis of transcripts in gilthead seabream sperm and zebrafish testicular cells: mRNA profile as a predictor of gamete quality. Aquaculture 406: 28–33.

Guthrie, H.D., L.C. Woods, J.A. Long and G.R. Welch. 2008. Effects of osmolality on inner mitochondrial transmembrane potential and ATP content in spermatozoa recovered from the testes of striped bass (*Morone saxatilis*). Theriogenology 69: 1007–1012.

Guzmán, J., J. Ramos, C.C. Mylonas and E. Mañanós. 2007. Induction of spermiation in Senegalese sole (*Solea senegalensis*) by injection with human chorionic gonadotropin (hCG) or GnRHa implants. Proc. 6th Cong. AIEC. 10–13 September 2007. Cádiz, Spain.

Guzmán, J.M., B. Norberg, J. Ramos, C.C. Mylonas and E. Mañanós. 2008. Vitellogenin, steroid plasma levels and spawning performance of cultured female Senegalese sole (Solea senegalensis). Gen. Comp. Endocrinol. 156: 285–297.

Guzmán, J.M., J. Ramos, C.C. Mylonas and E. Mañanós. 2011a. Comparative effects of human chorionic gonadotropin (hCG) and gonadotropin-releasing hormone agonist (GnRHa) treatments on the stimulation of male Senegalese sole (*Solea senegalensis*) reproduction. Aquaculture 316: 121–128.

Guzmán, J.M., R. Cal, A. García-López, O. Chereguini, K. Kight, M. Olmedo et al. 2011b. Effects of in vivo treatment with the dopamine antagonist pimozide and gonadotropin-releasing hormone agonist (GnRHa) on the reproductive axis of Senegalese sole (*Solea senegalensis*). Comp. Biochem. Physiol. A. 158: 235–245.

Gwo, J.C., W.T. Yang, Y.T. Sheu and H.Y. Cheng. 2002. Spermatozoan morphology of four species of bivalve (*Heterodonta, Veneridae*) from Taiwan. Tissue and Cell 34: 39–43.

Halliwell, B. 1994. Free radicals, antioxidants, and human disease: curiosity, cause, or consequence? Lancet 344: 721–724.

Henrotte, E., V. Kaspar, M. Rodina, M. Psenicka, O. Linhart and P. Kestemont. 2010. Dietary n-3/n-6 ratio affects the biochemical composition of Eurasian perch (*Perca fluviatilis*) semen but not indicators of sperm quality. Aquac. Res. 41: 31–38.

Herráez, M.P., J. Ausió, A. Devaux, S. González-Rojo, C. Fernández-Díez, S. Bony et al. 2017. Paternal contribution to development: Spermatogenetic damage and repair in fish. Aquaculture 472: 45–59.

Horváth, A., W.R. Wayman, J.C. Dean, B. Urbanyi, T.R. Tiersch, S.D. Mims et al. 2008. Viability and fertilizing capacity of cryopreserved sperm from three North American Acipenseriform species: a retrospective study. J. Appl. Ichthyol. 24: 443–449.

Howell, B., L. Conceiçao, R. Prickett, P. Caňnavate and E. Mañanos. 2009. Sole farming: nearly there but not quite? A report of 4th workshop on the cultivation of soles. Aquaculture Europe 34: 24–27.

Howell, B., R. Prickett, P. Caňnavate, E. Mañanos, M.T. Dinis, L. Conceiçao et al. 2011. Sole farming: there or thereabouts! A report of the 5th workshop on the cultivation of soles. Aquaculture Europe 36: 42–45.

Izquierdo, M.S., H. Fernandez-Palacios and A.G.J. Tacon. 2001. Effect of broodstock nutrition on reproductive performance of fish. Aquaculture 197: 25–42.

Jamieson, B.G.M. and L.K.P. Leung. 1991. Introduction to fish spermatozoa and the micropyle. pp. 56-72. *In*: B.G.M. Jamieson [ed.]. Fish Evolution and Systematics: Evidence from spermatozoa. Cambridge University Press, Cambridge, UK.

Jeong, Y.J., M.K. Kim, H.J. Song, E.J. Kang, S.A. Ock, M. Kuma et al. 2009. Effect of alpha-tocopherol supplementation during boar semen cryopreservation on sperm characteristics and expression of apoptosis related genes. Cryobiology 58: 181–189.

Johnson, G.D., C. Lalancette, A.K. Linnemann, F. Leduc, G. Boissonneault and S.A. Krawetz. 2011. The sperm nucleus: chromatin, RNA, and the nuclear matrix. Reproduction 141: 21–36.

Jones, R. and D. Butler. 1988. Spermatozoon ultrastructure of *Platichthys flesus*. J. Ultrastruct. Mol. Struct. Res. 98: 71–82.

Kasinsky, H.E., J.M. Eirin-Lopez and J. Ausio. 2011. Protamines: structural complexity, evolution and chromatin patterning. Protein Pept. Lett. 18: 755–771.

King, L.S. and P. Agre. 1996. Pathophysiology of the aquaporin water channels. Annu. Rev. Physiol. 58: 619–648.

Kopeika, E. and J. Kopeika. 2007. Variability of sperm quality after cryopreservation in fish. pp. 347–396. *In*: Alavi, S.M.H., J. Cosson, K. Coward and G. Raffie [eds.]. Fish spermatology. Alpha Science Intl Ltd, Oxford, UK.

Kowaltowski, A.J., N.C. de Souza-Pinto, R.F. Castilho and A.E. Vercesi. 2009. Mitochondria and reactive oxygen species. Free Radic Biol Med. 47: 333–43.

Lahnsteiner, F. 2009. The role of free amino acids in semen of rainbow trout *Oncorhynchus* mykiss and carp *Cyprinus carpio*. J. Fish. Biol. 75: 816–33.

Lahnsteiner, F. 2010. A comparative study on the composition and importance of free amino acids in semen of gilthead sea bream, *Sparus aurata*, and perch, *Perca fluviatilis*. Fish. Physiol. Biochem. 36: 1297–305.

Lahnsteiner, F. and R.A. Patzner. 1997. Fine structure of spermatozoa of four littoral teleosts, *Symphodus ocellatus*, *Coris julis, Thalassoma pavo* and *Chromis chromis*. J. Submicrosc. Cytol. Pathol. 29: 477–485.

Lahnsteiner, F., F. Berger, T. Weismann and R. Patzner. 1996. The influence of various cryoprotectants on semen quality of the rainbow trout (*Oncorhynchus mykiss*) before and after cryopreservation. J. Appl. Ichthyol. 12: 99–106.

Lahnsteiner, F., B. Berger, T. Weismann and R.A. Patzner. 1998. Determination of semen quality of the rainbow trout, *Oncorhynchus mykiss*, by sperm motility, seminal plasma parameters, and spermatozoal metabolism. Aquaculture 163: 163–181.

Lahnsteiner, F., N. Mansour, M.A. McNiven and G.F. Richardson. 2009. Fatty acids of rainbow trout (*Oncorhynchus mykiss*) semen: Composition and effects on sperm functionality. Aquaculture 298: 118–124.

Lahnsteiner, F., N. Mansour and S. Caberlotto. 2010. Composition and metabolism of carbohydrates and lipids in *Sparus aurata* semen and its relation to viability expressed as sperm motility when activated. Comp. Biochem. Physiol. B. 157: 39–45.

Lalancette, C., D. Miller, Y. Li and S.A. Krawetz. 2008. Paternal contributions: New functional insights for spermatozoal RNA. J. Cell Biochem. 104: 1570–9.

Li, P., Z.H, Li,B. Dzyuba, M. Hulak, M. Rodina and O. Linhart. 2010. Evaluating the impacts of osmotic and oxidative stress on common carp (*Cyprinus carpio* L.) sperm caused by cryopreservation techniques. Biol. Reprod. 83: 852–8.

Liu, Q.H., J. Li, S.C. Zhang, F.H. Ding, X.Z. Xu, Z.Z. Xiao et al. 2006. An efficient methodology for cryopreservation of spermatozoa of red seabream, *Pagrus major*, with 2-mL cryovials. J. World. Aquacult. Soc. 37: 289–297.

Liu, X.F., X.Z. Liu, J.H. Lian, Y.G. Wang, F.L. Zhang, H. Yu et al. 2008. Large scale artificial reproduction and rearing of Senegal sole, *Solea senegalensis* Kaup. Mar. Fish. Res. 29(2): 10–29.

Lubzens E., N. Daube, I. Pekarsky, Y. Magnus, A. Cohen, F. Yusefovich et al. 1997. Carp (*Cyprinus carpio* L.) spermatozoa cryobanks—strategies in research and application. Aquaculture 155: 13–30.

Luger, K., A.W. Mader, R.K. Richmond, D.F Sargent and T.J. Richmond. 1997. Crystal structure of the nucleosome core particle at 2.8 A resolution. Nature 389: 251–260.

Mansour, N., M.A. McNiven and G.F. Richardson. 2006. The effect of dietary supplementation with blueberry, alpha-tocopherol or astaxanthin on oxidative stability of Arctic char (*Salvelinus alpinus*) semen. Theriogenology 66: 373–382.

Mañanós, E., N. Duncan and C.C. Mylonas. 2009. Reproduction and control of ovulation, spermiation and spawning in cultured fish. pp. 3–80. *In*: Cabrita, E., V. Robles and P. Herráez [eds.]. Methods in Reproductive Aquaculture: Marine and Freshwater Species. CRC Press, Taylor and Francis Group, Boca Raton, FL, USA.

Martínez-Pastor, F., E. Cabrita, F. Soares, L. Anel and M.T. Dinis. 2008. Multivariate cluster analysis to study motility activation of *Solea senegalensis* spermatozoa: a model for marine teleosts. Reproduction 135: 449–59.

Medina, A., C. Megina, E.J. Abascal and A. Calzada. 2000. The spermatozoon morphology of *Solea senegalensis* (Kaup, 1858) (*Teleostei, Pleuronectiformes*). J. Submicrosc. Cyto. Pathol. 32: 645–650.

Meseguer, M., M.J. de los Santos, C. Simón, A. Pellicer, J. Remohí and N. Garrido. 2006. Effect of sperm glutathione peroxidases 1 and 4 on embryo asymmetry and blastocyst quality in oocyte donation cycles. Fertil. Steril. 86: 1376–85.

Migaud, H., G. Bell, E. Cabrita, B. McAndrew, A. Davie, J. Bobe et al. 2013. Broodstock management and gamete quality in temperate fish. Rev. Aquac. 5: 194–223.

Miller, D. and G.C. Ostermeier. 2006. Spermatozoal RNA: Why is it there and what does it do? Gynecol. Obstet. Fertil. 34: 840–6.

Morais, S., C. Aragao, E. Cabrita, A. Estevez, M. Yúfera, L.M.P. Valente et al. 2016. New developments and biological insights into the farming of *Solea senegalensis* reinforcing its aquaculture potential. Rev. Aquac. 6: 1 37.

Morisawa, M., K. Suzuki and S. Morisawa. 1983. Effect of potassium and osmolarity on spermatozoa motility of salmonid fishes. J. Exp. Biol. 107: 105–113.

Muller, K., P. Muller, G. Pincemy, A. Kurz and C. Labbé. 2008. Characterization of sperm plasma membrane properties after cholesterol modification: Consequences for cryopreservation of rainbow trout spermatozoa. Biol. Reprod. 78: 390–399.

Nandi, S., P. Routray, S.D. Gupta, S.C. Rath, S. Dasgupta, P.K. Meher et al. 2007. Reproductive performance of carp, *Catla catla* (Ham.), reared on a formulated diet with PUFA supplementation. J. Appl. Ichthyol. 23: 684–691.

Nynca, J., G.J. Arnold, T. Fröhlich, K. Otte and A. Ciereszko. 2014. Proteomic identification of rainbow trout sperm proteins. Proteomics 14: 1569–1573.

Norambuena, F., S. Morais, A. Estévez, J.G. Bell, D.R. Tocher, J.C. Navarro et al. 2013. Dietary modulation of arachidonic acid metabolism in Senegalese sole (*Solea senegalensis*) broodstock reared in captivity. Aquaculture 372–375: 80–88.

Oliveira, C., I. Meiri, H. Rosenfeld, M. Custódio, M.T. Dinis and E. Cabrita. 2013. Daily rhythms of sperm quality and luteinizing hormone in plasma and in seminal fluid of gilthead seabream, *Sparus aurata*. Proc. 4th Int. Workshop Biology of Fish Gametes. September 2013, Faro, Portugal.

Ortega-Ferrusola, C., Y. Sotillo-Galan, E. Varela-Fernández, J.M. Gallardo-Bolaños, A. Muriel, L. González-Fernández et al. 2008. Detection of "Apoptosis-Like" changes during the cryopreservation process in equine sperm. J. Androl. 29: 213–221.

Ostermeier, G.C., D. Miller, J.D. Huntriss, M.P. Diamond and S.A. Krawetz. 2004. Reproductive biology: delivering spermatozoan RNA to the oocyte. Nature 429(6988): 154.

Ostermeier, G.C., R.J. Goodrich, J.S. Moldenhauer, M.P. Diamond and S.A. Krawetz. 2005. A suite of novel human spermatozoal RNAs. J. Androl. 26: 70–4.

Pérez, L., M.F. Riesco, M.C. Vílchez, J.F. Asturiano and E. Cabrita. 2015. Effect of extenders on Senegalese sole (*Solea senegalensis*) sperm. Proc. 5th Int. Workshop on Biology of Fish Gametes. 7–11 September 2015, Ancona, Italy.

Pérez, L., M.C. Vílchez, V. Gallego, M. Morini, D.S. Peñaranda and J.F. Asturiano. 2016. Role of calcium on the initiation of sperm motility in the European eel. Comp. Biochem. Physiol. A Mol. Integr. Physiol. 191: 98–106.

Pérez-Cerezales, S., S. Martínez-Páramo, E. Cabrita, F. Martínez-Pastor, P. de Paz and M.P. Herráez. 2009. Evaluation of oxidative DNA damage promoted by storage in sperm from sex-reversed rainbow trout. Theriogenology 71: 605–613.

Pérez-Cerezales, S., S. Martínez-Páramo, J. Beirao and M.P. Herráez. 2010. Fertilization capacity with rainbow trout DNA-damaged sperm and embryo developmental success. Reproduction 139(6): 989–997.

Pérez-Cerezales, S., A. Gutiérrez-Adan, S. Martínez-Páramo, J. Beirao and M.P. Herráez. 2011. Altered gene transcription and telomere length in trout embryo and larvae obtained with DNA cryodamaged sperm. Theriogenology 76: 1234–1245.

Perchec, G., C. Jeulin, J. Cosson, F. André and R. Billard. 1995. Relationship between sperm ATP content and motility of carp spermatozoa. J. Cell. Sci. 108: 747–753.

Pustowka, C., M.A. McNiven, G.F. Richardson and S.P Lall. 2000. Source of dietary lipid affects sperm plasma membrane integrity and fertility in rainbow trout *Oncorhynchus mykiss* (Walbaum) after cryopreservation. Aquac. Res. 30: 297–305.

Rasines, I., M. Gómez, I. Martín, C. Rodríguez, E. Mañanós and O. Chereguini. 2012. Artificial fertilization of Senegalese sole (*Solea senegalensis*): Hormone therapy administration methods, timing of ovulation and viability of eggs retained in the ovarian cavity. Aquaculture 326–329: 129–135.

Rasines, I., M. Gómez, I. Martín, C. Rodríguez, E. Mañanós and O. Chereguini. 2013. Artificial fertilisation of cultured Senegalese sole (Solea senegalensis): Effects of the time of day of hormonal treatment on inducing ovulation. Aquaculture 392–395: 94–97.

Riesco, M.F. and V. Robles. 2013. Cryopreservation Causes Genetic and Epigenetic Changes in Zebrafish Genital Ridges. PLoS One 8: e67614.

Riesco, M.F., C. Oliveira, F. Soares, P.J. Gavaia, M.T. Dinis and E. Cabrita. 2017. *Solea senegalensis* sperm cryopreservation: new insights on sperm quality. PLoS One 12(10): e0186542.

Rothfuss, O., T. Gasser and N. Patenge. 2010. Analysis of differential DNA damage in the mitochondrial genome employing a semi-long run real time PCR approach. Nucleic. Acids. Res. 38: e24.

Rurangwa, E., D.E. Kime, F. Ollevier and J.P. Nash. 2004. The measurement of sperm motility and factors affecting sperm quality in cultured fish. Aquaculture 234: 1–28.

Santos, J.S., G. O. Introíni, A.C.P. Veiga-Menoncello, A. Blasco, M. Rivera and S.M. Recco-Pimentel. 2016. Comparative sperm ultrastructure of twelve leptodactylid frog species with insights into their phylogenetic relationships. Micron. 91: 1–10.

Saperas, N., M. Chiva, M.T. Casas, J.L. Campos, J.M. Eirin-Lopez, L.J. Frehlick et al. 2006. A unique vertebrate histone H1-related protamine like protein results in an unusual sperm chromatin organization. FEBS J. 273: 4548–4561.

Serradeiro, R., M. Pinto, C. Oliveira, M.F., Riesco, I. Blanquet, J. Dias et al. 2015. Effect of dietary arachidonic acid supplementation on Senegalese sole sperm quality. Proc. EAS Annual Meeting, 20–23 October 2015. Rotterdam, The Netherlands.

Shaliutina, A., M. Hulak, B. Dzuyba and O. Linhart. 2012. Spermatozoa motility and variation in the seminal plasma proteome of Eurasian perch (*Perca fluviatilis*) during the reproductive season. Mol. Reprod. Dev. 79: 879–887.

Shaliutina, A., M. Hulak, I. Gazo, P. Linhartova and O. Linhart. 2013. Effect of short-term storage on quality parameters, DNA integrity, and oxidative stress in Russian (*Acipenser gueldenstaedtii*) and Siberian (*Acipenser baerii*) sturgeon sperm. Anim. Reprod. Sci. 139: 127–135.

Takai, H. and M. Morisawa. 1995. Change in intracellular K+ concentration caused by external osmolality change regulates sperm motility of marine and freshwater teleosts. J. Cell. Sci. 108: 1175–1181.

Thomson, L.K., S.D. Fleming, K. Barone, J.A. Zieschang and A.M. Clark. 2009. The effect of repeated freezing and thawing on human sperm DNA fragmentation. Fertil. Steril. 93: 1147–1156.

Valcarce, D.G. and V. Robles. 2016. Effect of captivity and cryopreservation on ROS production in *Solea senegalensis* spermatozoa. Reproduction 152: 439-46.

Valcarce, D.G., M.P. Herráez, O. Chereguini, C. Rodríguez and V. Robles. 2016. Selection of nonapoptotic sperm by magnetic-activated cell sorting in Senegalese sole (*Solea senegalensis*). Theriogenology 86(5): 1195–1202.

Verkman, A.S. 2002. Aquaporin water channels and endothelial cell function. J. Anat. 200: 617–627.

Walter, C.A., G.W. Intano, J.R. McCarrey, C.A. McMahan and R.B. Walter. 1998. Mutation frequency declines during spermatogenesis in young mice but increases in old mice. Proc. Natl. Acad. Sci. USA. 95: 10015–9.

Wassall, S. and W. Stillwell. 2009. Polyunsaturated fatty acid-cholesterol interactions: domain formation in membranes. Biochim. Biophys. Acta Biomembr. 1788: 24–32.

Waterhouse, K.E., P.O. Hofmo, A. Tverdal and R.R. Miller. 2006. Within and between breed differences in freezing tolerance and plasma membrane fatty acid composition of boar sperm. Reproduction 131: 887–894.

Watson, C.E. and P.L. Davies. 1998. The high molecular weight chromatin proteins of winter flounder sperm are related to an extreme histone H1 variant. J. Biol. Chem. 273: 6157–6162.

Weinbauer, G.F., C. M. Luetjens, M. Simoni, and E. Nieschlag. 2010. Physiology of testicular function. pp. 11–59. *In*: Nieschlag, E., H.M. Behre and S. Nieschlag [eds.]. Andrology: Male Reproductive Health and Dysfunction. Springer, Netherlands.

Yavaş, İ., Y. Bozkurt and C. Yildiz. 2014. Cryopreservation of scaly carp (*Cyprinus carpio*) sperm: effect of different cryoprotectant concentrations on post-thaw motility, fertilization and hatching success of embryos. Aquac. Int. 22: 141–148.

Zilli, L., R. Schiavone, F. Chauvigné, J. Cerdà, C. Storelli and S. Vilella. 2009. Evidence for the Involvement of aquaporins in sperm motility activation of the teleost gilthead sea bream (*Sparus aurata*). Biol. Reprod. 81: 880–888.

B-1.5

Mating Behaviour

Neil Duncan,[1,*] **Ignacio Carazo,**[1] **Olvido Chereguini**[2]
and Evaristo Mañanós[3]

1. Introduction

The study of mating behaviour of the Senegalese sole (*Solea senegalensis*) was undertaken to investigate the total reproductive failure observed in hatchery Senegalese sole. The term hatchery Senegalese sole is used here to refer to any sole that was hatched and reared in captivity as opposed to wild fish that were captured from the wild and acclimatised to captivity. The reproductive failure observed in hatchery Senegalese sole is characterised by the spawning of unviable eggs that do not develop to hatch (Guzman et al. 2009) as the eggs were not fertilised (Carazo 2013). In comparison, wild Senegalese sole held in captivity spawn fertilised eggs that have sufficient quality and quantity for commercial aquaculture production (Anguis and Cañavate 2005, Martin et al. 2014). Guzman et al. (2009) described the reproductive output of hatchery stocks, which only spawned unviable eggs and annual total fecundities ranged from 3,000 to 84,000 eggs kg[-1] of female body weight. Anguis and Cañavate (2005) described the reproductive output of wild captive stocks that had mean annual rates of hatching ranging from 56.5 ± 25% to 69.7 ± 24% and annual total fecundities that ranged from 1,150,000 to 1,650,000 eggs kg[-1] body weight. The comparison of these studies is perhaps compromised as the studies were in different centres with different holding conditions. However, when hatchery and wild stocks are held under the same conditions the reproductive failure of hatchery stocks persists. For example the reproductive output of hatchery and wild stocks held in identical conditions has been compared in two research centres. In IRTA (Sant Carles de la Rapita, Tarragona, Spain), the hatchery stocks only spawned unviable eggs and annual total fecundities ranged from 102,450 to 231,700 eggs kg[-1] of female body weight. Compared to wild captive

---

[1] IRTA Sant Carles de la Rapita, AP200, 43540 Sant Carles de la Rapita, Tarragona, Spain.
[2] Spanish Institute of Oceanography, Santander Oceanographic Centre, Promontorio de San Martín, s/n. Apdo. 240, 39080 Santander, Spain.
[3] Institute of Aquaculture of Torre la Sal, Spanish Council for Scientific Research (CSIC), Torre la Sal s/n, 12595-Cabanes, Castellón, Spain.
* Corresponding author: neil.duncan@irta.cat

stocks that had mean annual rates of hatching of 30 ± 31 to 36 ± 27% and annual total fecundities ranged from 351,400 to 724,150 eggs kg⁻¹ body weight. In IEO (Santander, Cantabria, Spain) the hatchery stocks only spawned unviable eggs and had annual total fecundities that ranged from 99,560 to 413,790 eggs kg⁻¹ of female body weight, whereas wild captive stocks that had mean annual fertilization rates hatching rates of 49 ± 5 to 65 ± 3%, and annual total fecundities ranged from 709,280 to 1,289, 450 eggs kg⁻¹ body weight.

However, the hatchery stocks do mature to produce viable gametes. The gametes were stripped from hatchery breeders and successfully fertilised *in vitro* (Rasines et al. 2012, 2013). Large volumes of ova were obtained from females by inducing ovulation with an agonist of gonadotropin-releasing hormone (GnRHa) and the ova were fertilised with small amounts of sperm obtained by pressure to the testes of hatchery males (Rasines et al. 2012, 2013). A reproductive dysfunction where breeders mature to the end of the reproductive cycle, but do not spawn fertilised eggs is not uncommon in captive fish and has been attributed to environmental or social constraints imposed by the culture system (Zohar and Mylonas 2001). However, the situation described for Senegalese sole indicates that the environment of adult fish is not limiting as wild fish spawned viable eggs under identical conditions to hatchery fish that did not spawn viable eggs. These observations perhaps focus the cause of the dysfunction on social or behavioural aspects. Discussions have given rise to a general hypothesis that developmental or social aspects prior to puberty have caused the reproductive failure observed in hatchery breeders (Howell et al. 2011, Rasines et al. 2012, Norambuena et al. 2012).

Therefore, the aim of the study of mating behaviour and this chapter was: First to describe the successful mating behaviour of wild fish held in captivity. Secondly, to describe the unsuccessful mating behaviour of hatchery fish. Thirdly, experiments were completed to further understand the behavioural dysfunction and to attempt to develop solutions for the dysfunction. The experiments examined the use of hormones to induce successful mating behaviour and the effect of maintaining wild fish with hatchery fish. Lastly, the chapter discusses the cause of the reproductive behavioural dysfunction and future research aims.

2. General behaviour

Senegalese sole held in captivity appear to be a sedentary and social fish. The sole are most commonly observed to remain stationary on the bottom of the tank and buried in sand when sand is provided in the tank. The stationary fish are observed positioned as isolated individuals or in groups with the fish resting on each other. Breeders have been observed to remain stationary for approximately 70% of the time during a 24 hour period and exhibit a regular circadian pattern of activity with periods of low and increased activity. The pattern of activity tends to be low during the day with increased activity during the night. The period of increased activity can extend from late afternoon through to lights on in the morning. Differences in the activity pattern have been observed between juveniles and breeders. Juveniles were active during the night, particularly the first part of the night from lights off until midnight and were classified as nocturnal (Bayarri et al. 2004). In comparison breeders had increased activity from the afternoon into the early part of the night and low activity from midnight through to midday (Carazo 2013, Carazo et al. 2013). Carazo et al. (2013) suggested this difference was caused by the habitation of breeders to husbandry care and in particular feeding during the day.

The behaviours associated with these patterns in activity can be grouped into four broad groups, staying quiet on the bottom, fish resting on each other, burying and swimming (Carazo 2013). As mentioned above a large part of the day is spent staying quiet on the bottom of the tank in a kind of resting period. This staying quiet on the bottom may be a situation where fish rest on each other with little or no movement. However, the behaviour of resting on another sole appears to have a strong social aspect and has been observed to be a behaviour associated with high activity. Particularly the upper fish of the resting pair was observed to be active making many small movements and interactions that differentiate this behaviour of staying quiet. This resting on each other provides a situation for chemical communication with the olfactory system. Studies have demonstrated the ability of Senegalese sole to detect conspecifics and the sole differentiated between sex and maturity status of the fish emitting the chemical signal (Carazo 2013, Fatsini et al. 2017). Like most flatfish the Senegalese sole has an upper and lower olfactory rosette. The upper olfactory gave a higher electro-olfactogram (EOG) response with conspecific-derived odorants compared to the lower olfactory rosette (Velez et al. 2007) and Carazo (2013) and Fatsini et al. (2017) demonstrated that sole gave different olfactory responses to urine from mature and immature sole and male and female sole. Therefore, the sole that has a fish resting on it is in a good position to receive chemical signals whilst the fish on top is in a good position to emit signals. The burying behaviour is a distinct behaviour of flatfish and in the sole consists of raising the head and sending a wave down the body and particularly the lateral fins to raise substrate up over the fish and achieve burying the fish (Kruuk 1963, Gibson 2005). Lastly swimming is an obvious behaviour used to move through the water environment. The sole swim by sending waves down the whole body and also by gliding. The sole most often swim close to the bottom, but also swim through the water column.

3. Successful mating behaviour

Successful mating behaviour has only been described in wild Senegalese sole acclimated to captivity (Carazo 2013, Carazo et al. 2016). The successful mating behaviour was registered as an increase in activity of the fish. The period of increased activity was from 17:00 to 24:00 and was longer than the period of activity observed in the same period without mating behaviour (Fig. 1). This increase in activity associated with mating behaviour was approximately three times higher than the period of activity observed without mating behaviour. Different photoperiods did not appear to affect the daily timing of the increase in activity. The timing (17:00 to 24:00) of this period was similar under different photoperiods either a continuous light-dark (LD) of 16:8 (lights on at 08:00 and lights off at 24:00) or a natural photoperiod with close to LD12:12 (lights on at ~ 07:00 and lights off at ~ 19:00) at the time of spawning. Therefore, the period of increased activity was observed both late in the day before the lights were switched off (LD16:8) and extending from before dusk to after dusk under a natural photoperiod. All spawning was observed during this period of increased activity. These observations were in agreement with a study that found eggs were collected in the same period (21:00–01:00) from a tank of breeders maintain in natural conditions (Oliveira et al. 2009).

The mating behaviour began with mate selection, which was the principal cause of the increase in activity. Mate selection was clearly evident as all spawning was in pairs, males displayed to females, males solicit that females swim from the bottom of the tank to spawn and females either accepted to spawn with the male or reject the male by swimming away. The strict paired spawning implies that the male and female had to select each other to spawn together. Paired spawning was also confirmed with parental analysis of progeny using microsatellites. The paternity of progeny demonstrated that the majority of spawns were between a single pair (one family) and few contained more than one family. In

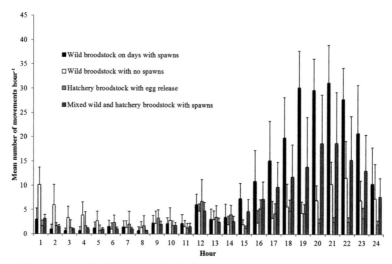

Fig. 1. Activity of Senegalese sole (*Solea senegalensis*) during a 24 hour period. Average (n = 5 days) number times a breeder moved over a fixed point in the tank for a wild broodstock on days with spawns (black), a wild broodstock with no spawns (white), a hatchery broodstock with egg release (light grey) and mixed wild and hatchery broodstock with spawns (dark grey). The light period was 08:00 to 24:00. The broodstocks were fed at ~ 11:00. Figure modified from Carazo et al. (2016).

addition to single pairs providing an entire spawn, over the spawning season certain pairs of breeders in the broodstock dominate spawning to produce most of the eggs and fidelity may exist between pairs (Porta et al. 2006a, Martin et al. 2014).

The process or criteria used by the sole to select a mate is not clear, but is probably related to the displaying and contact between fish that was observed in association with spawning. The two principal kinds of displaying were swimming in a procession and the male resting on the female. The swimming in a procession is a very distinctive behaviour, which has been termed the "following" behaviour (Fig. 2) (Carazo 2013, Carazo et al. 2016). The "following" behaviour is when two or more fish swim in a procession and the fish that are following copy closely the turns and direction of the lead fish. The behaviour usually involves males, but females have also been observed to participate. This "following" behaviour is most common during periods with spawning and accounts for a large part of the 3 fold increase in activity that is associated with spawning (Fig. 1). This behaviour would appear to give the males the opportunity to demonstrate some kind of dominance or swimming ability. The females remain mostly still and quiet on the bottom. In between the "following" behaviour the fish return to rest on the bottom and the male often rests on a female. These behaviours take up the entire period of increased activity that often lasts from 17:00 to 24:00. When the behaviour of a successfully spawning pair was traced back in a video recording, the male was observed to be involved in "following" behaviours for up to 1–2 hours before spawning (Fig. 2).

As mentioned the males also rest on the female (Fig. 2). The female may accept or reject a resting male. During the period of mating behaviour the male when resting on a female will move over the back and head of the female and appears to be both communicating with the female and attempting to encourage that the female swims. The male has been observed to use the burying behaviour to attempt to disturb the female. The males have also been observed to protect the female from other males. For example the male has been observed to push a second approaching male away from the female with its tail whist

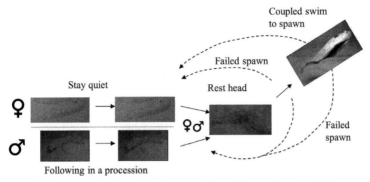

Fig. 2. The sequence of behaviours observed during courtship of Senegalese sole (*Solea senegalensis*). Initially females remain mainly still and quiet whilst males display by swimming in a procession. Males then rest on females to gain acceptance to make a synchronised coupled swim to the surface. The coupled swim may fail and the couple break apart or the couple swim to the surface and successfully release gametes to complete spawning. Figure modified from Carazo et al. (2016).

covering the female's eyes to isolate the female from the approaching male. This behaviour of shielding a female from another male was termed the "guardian" behaviour (Carazo 2013, Carazo et al. 2016). The aim of these behaviours on the part of the male is to encourage the female to swim from the bottom to start an upward coupled swim to spawn.

The female displays various behaviours in response to this attention from a male resting on its back. The behaviours appear to be aimed at escaping, avoiding or accepting the male. To escape the female swims away from the male at velocity to avoid further contact with the male. Avoidance was achieved by either staying quiet and firm on the bottom of the tank to essentially ignore the male or by using the burying behaviour to displace the male resting on its back. The female would accept the male by swimming from the bottom of the tank and allowing the male to immediately place himself underneath so that the dorsal side of the male was closely pressed to the ventral side of the female. The male and female pressing together would then make a coupled swim in synchrony to the surface to release gametes. This coupled swim would last 1–3 minutes and gamete release a few seconds from when gametes were visible until the couple broke apart and returned to the bottom of the tank. All pairs that spawned were observed to participate in these resting behaviours, for a period up to 15 minutes, immediately before spawning.

However, coupled swims that culminated in the successful release of gametes were less frequent than unsuccessful coupled swims where the pair broke apart before releasing gametes (Fig. 2). The unsuccessful coupled swims were 5.6 times more frequent than successful coupled swims (Carazo 2013). The most common unsuccessful coupled swim was when the female essentially rejected the male's attempts to swim under her. Others were when the female appeared to accept the male and the coupled swim began, but was interrupted by another sole often a male or by a physical object such as the side of the tank or an airline.

This description of the mating behaviour of Senegalese sole has many similarities with the mating behaviour observed in other flatfish species. Mating behaviour has been described to varying degrees for 9 flatfish species which are: winter flounder (*Pseudopleuronectes americanus*) (Breder 1922, Stoner et al. 1999); European plaice (*Pleuronectes platessa*) (Forster 1953); kobe flounder (*Crossorhombus kobensis*) (Moyer et al. 1985); common sole (*Solea solea*) (Baynes et al. 1994); eyed flounder (*Bothus ocellatus*) (Konstatinou and Shen 1995); large scale flounder (*Engyprosopon grandisquama*) (Manabe et al. 2000); wide-eyed flounder (*Bothus podas*) (Carvalho et al. 2003); bastard halibut (*Tarphops oligolepis*) (Manabe and Shinomiya 2001) and greenback flounder (*Rhombosolea tapirina*) (Pankhurst and Fitzgibbon 2006). The males of these flatfish species, also displayed to the females to gain acceptance to spawn together. Male displaying included touching or resting on the female, arching the body, raising the head, waving the pectoral fin and making short upward swims (Moyer et al. 1985, Manabe et al. 2000, Manabe and Shinomiya 2001, Carvalho et al. 2003, Pankhurst and Fitzgibbon 2006). One species the winter flounder was also observed to engage in the "following" behaviour where fish swam in a procession (Stoner et al. 1999). All species spawned as pairs during an upward coupled swim with the exception being winter flounder that spawned both as a pair and a group of one female with two or more males (Stoner et al. 1999). The males of the bastard halibut (Manabe and Shinomiya 2001) and four Bothidea species (Moyer et al. 1985, Konstatinou and Shen 1995, Manabe et al. 2000, Carvalho et al. 2003) were described to defend a territory in the natural environment. The territory had importance for reproduction and either contained a harem of females (Moyer et al. 1985, Konstatinou and Shen 1995, Carvalho et al. 2003) or females visited for courtship and spawning (Manabe et al. 2000).

Altogether this description of mating behaviour of Senegalese sole highlights the complexity of the mate selection and spawning. The males in particular engage in swimming processions and close contact with the spawning female to achieve acceptance to make a coupled swim. This high activity from the males is evident as the males have twice the activity of females (Fig. 3). The coupled swim requires that the two sole swim in coordination holding the genital pores closely together to release the gametes at the same time and in very close proximity. The male seeks to initiate the coupled swim and appears to have a guiding role pushing from underneath to maintain the two fish closely together whilst rising through the water column to the surface. A number of important points can be taken from this description of mating behaviour, (a) the complete synchronised coupled swim to the surface appears to be essential for the fertilisation and the viability of the eggs, (b) the mating behaviour and particularly the swimming in a procession represents a clear increase in activity of the breeders, (c) the behaviours that the male fish must execute to achieve successful spawning are complex and varied.

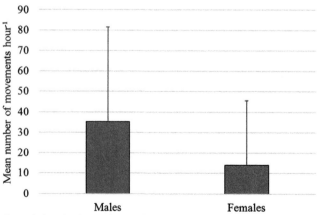

Fig. 3. Activity of male and female Senegalese sole (*Solea senegalensis*) during a 24 hour period. Average (n = 5 days) number times a male or female moved over a fixed point in the tank on days with spawns.

4. Unsuccessful mating behaviour

Unsuccessful mating behaviour was studied in a broodstock of hatchery Senegalese sole that infrequently liberated eggs that were not viable when collected. The broodstock consisted of 8 male and 8 female breeders of a first generation reared in captivity. The behaviour was video recorded over a period of 19 days in which four liberations of eggs were collected. The behaviour was studied during the four 24 h periods prior to eggs being collected and compared to the behaviour on 5 randomly selected 24 h periods after which no eggs were collected. The activity pattern over 24 h and the level of activity were similar during the periods before egg collection and before no eggs were collected. In both periods, the hatchery fish had an increase in activity at feeding time (11:00) and late afternoon–evening. A similar pattern of activity was observed in wild fish held in captivity. However, the levels of activity in the wild fish were higher particularly in the late afternoon–evening period prior to egg collection. In the wild fish, the late afternoon–evening increase in activity prior to egg collection was associated with mating behaviour and spawning. The increase was 3 times the levels observed in similar periods that were not followed with egg collection. This was not the situation in hatchery fish that had similar levels of activity prior to egg collection and prior to no egg collection. The increase in activity of the hatchery fish during late afternoon–evening was 6 times lower compared to wild fish prior to egg collection (Fig. 1). There appeared to be no increase in activity associated with egg release and associated mating behaviour in hatchery fish as was observed in wild fish.

However, although hatchery fish had similar levels of activity prior to egg collection and prior to no egg collection there was a difference in the types of behaviours observed. The different behaviours were counted during the peak half hour of activity during the afternoon–evening period. The main difference was an increase in the behaviour of fish resting on another (rest head) during the period prior to egg collection compared to prior to no egg collection. The act of fish resting on another fish has close association with spawning and was observed immediately before all spawning events in wild fish. However, the "rest head" behaviour is also a common behaviour outside of the period of mating behaviour. When the behaviours before egg collection were compared between wild and hatchery fish it was observed that hatchery fish did not make two behaviours that were associated with mating in wild fish. Hatchery fish did not engage in swimming in procession or "following" behaviours during the peak half hour of activity and no synchronised coupled swims were observed during the entire 24 h period prior to egg collection.

Therefore, the important behaviours or points highlighted in successful mating behaviour were not completed during unsuccessful mating behaviour. A number of important points can be taken from this description of unsuccessful mating behaviour, (a) the synchronised coupled swim was not observed and the eggs were not fertilised, (b) the mating behaviour and particularly the swimming in a procession was not observed and there was no increase in activity, (c) the complex behaviours that the male fish must execute to achieve successful spawning such as displaying by swimming in a procession and making the synchronised coupled swim to the surface were not observed during the peak of activity.

5. Application of hormones to induce mating behaviour

The use of hormones to induce spawning is a valuable therapy to overcome reproductive dysfunctions, which has been extensively applied and reviewed (Zohar and Mylonas 2001, Mañanós et al. 2008). The hormones used are applied to induce the final stages of maturation and in many examples also induce mating behaviour with successful spontaneous spawning. The two principal hormones used are GnRHa and human chorionic gonadotropin (hCG). In addition to these two hormones, pheromones have been used to induce mating behaviour in a few species, particularly carps (Sorensen and Stacey 2004). Here three experiments are described that examined the effect GnRHa, hCG and the pheromone prostaglandin F2α (PGF) on egg liberation and mating behaviour in Senegalese sole (Carazo 2013).

In recent decades GnRHa has been extensively used to induce spawning particularly for marine fish species. The GnRHa is applied to principally induce final oocyte maturation, ovulation and spawning in females and spermiation and spawning in males. An example would be the use of GnRHa to induce the spawning of meagre (*Argyrosomus regius*) (Fernandez-Palacios et al. 2014). Meagre in advanced stages of vitellogenesis (large vitellogenic oocytes) and with easily expressed sperm were treated with a single injection of GnRHa. The optimal treatment of 15 µg kg^{-1} of GnRHa induced spawning and high numbers of good quality eggs that were collected from surface flow egg collectors connected to the spawning tanks. Therefore, the GnRHa treatment induced the final stages of maturation and the mating behaviour for the breeders to successfully fertilise the eggs that were collected. These kinds of GnRHa treatments have been applied to hatchery Senegalese sole to increase the number of eggs collected from treated groups compared to control groups (Agulleiro et al. 2006, Guzman et al. 2009, 2011a). However, all eggs collected were not viable indicating that the correct mating behaviour was not induced. The mating behaviour was examined in hatchery Senegalese sole that were treated with GnRHa (Carazo 2013). Two groups (five males and five females per group) of hatchery sole that were in advanced stages of maturation were set up. A treatment of 5 µg kg^{-1} of GnRHa was applied every three days (similar treatment used by Agulleiro et al. 2006) and the behaviour was studied. A total of 5 spawns with a total volume of eggs of 670 mL were collected from the two tanks. However, no fertilised eggs were obtained and none of the behaviours associated with mating behaviour were observed. There were no synchronised coupled swims to the surface and no swimming in procession by the males.

The use of hCG has a longer history than GnRHa as an effective hormone to induce both the final stages of maturation and mating behaviour to obtain fertilised eggs (Zohar and Mylonas 2001, Mañanós et al. 2008). In Senegalese sole a single fertilised spawn was obtained when females were treated with a GnRHa implant and males were treated with a weekly dose of 1000 IU kg^{-1} of hCG (Guzman et al. 2011b). Interestingly the females were all treated the same with a GnRHa implant (both control and experimental groups) and the treatment of males with hCG increased the number of spawns and the total number of eggs compared to controls where males did not received any treatment. The total number of spawns from the hCG experimental group was 19 unfertilised spawns and one spawn with fertilised eggs. Therefore, a similar experiment was conducted to examine the mating behaviour when male hatchery Senegalese sole were treated with hCG and females treated with GnRHa implants (Carazo 2013). Four groups (three males and two females per group) of hatchery sole that were in advanced stages of maturation were set up. An implant of approximately 50 µg kg^{-1} of GnRHa was applied to all females. The males in two groups were treated with weekly injections of 1000 IU kg^{-1} of hCG and the males in two other groups were not treated (control groups). The treatment of males with hCG increased both the number of spawns and the number of eggs per spawn compared to control groups as was also observed by Guzman et al. (2011b). However, no fertilised eggs were obtained and none of the behaviours associated with mating behaviour were observed. There were no synchronised coupled swims to the surface and no swimming in procession by the males.

Pheromones are generally accepted as the main control of mating behaviour in fish. However, the correct identification of a chemical acting as a pheromone combined with the demonstration of emission of the chemical by one individual (emitter) and reception of the chemical by a different individual (receptor) to stimulate behaviour has been achieved in very few species. Sorensen and Stacey (2004) reviewed the actions of pheromones in reproduction and spawning of the goldfish (*Carassius auratus*). The female goldfish emitted the prostaglandin $F_{2\alpha}$ (PGF) during ovulation. The emitted PGF was received by the male to stimulate male spawning behaviour and the successful spawning between the male and female. The model of female emitter to stimulate mating behaviour in the receptor male and achieve successful spawning appears to apply to the behaviour observed in Senegalese sole. Therefore, an experiment was conducted to examine the mating behaviour when hatchery Senegalese sole were treated with PGF (Carazo 2013). Six groups (three males and two females per group) of hatchery sole that were in advanced stages of maturation were set up. Two control groups were not treated with any hormone or pheromone. The sole in the four experimental groups were treated with an injection of 5 µg kg^{-1} of GnRHa every three days. Once eggs were collected from the tank the sole were treated with PGF (Prosolvin, Luprostiol an analogue of prostaglandin $F_{2\alpha}$) that was applied as an injection of 100 µg kg^{-1} every two days. No spawns were obtained from one of the experimental tanks and the control tanks. A total of nine spawns were obtained from the three remaining experimental tanks and six of these spawns were after the PGF treatment was initiated. A total 485 mL of eggs were collected of which 190 mL were collected after the PGF treatment was initiated. However, no fertilised eggs were obtained and none of the behaviours associated with mating behaviour were observed. There were no synchronised coupled swims to the surface and no swimming in procession by the males.

Altogether the experiments with hormones and pheromones were not successful in stimulating mating behaviour. All hormone and pheromone treatments increased the numbers of eggs liberated by the females through the action of the hormones on ovulation in the females. However, the hormones reduced the activity of the fish compared to the control groups and did not induce the mating behaviours. The situation was the same or similar as in hatchery fish that did not receive hormone or pheromone stimulation. Therefore, the same points can be repeated in relation to the doses and experimental conditions used in these experiments, (a) The synchronised coupled swim was not induced and, therefore, the eggs were not fertilised (b) the mating behaviour and particularly the swimming in a procession were not induced and there was no increase in activity (c) the complex behaviours that the male fish must execute to achieve successful spawning were not induced by GnRHa, hCG or PGF.

6. Holding wild and hatchery fish together

To differentiate between the behavioural capacities of hatchery males and females, experiments were undertaken that mixed the sexes by origin of wild or hatchery. Two experimental groups were set up a year before the spawning season in which behaviour was examined. One group was a mixture of three wild females with five hatchery males. The wild females had been previously held with wild males and the group of breeders had produced viable spawns. The hatchery males were two 1st generation males and three 2nd generation males that were hatched and reared in captivity. The fish were all adults with reproductive capacity and the mean weight of the fish in the group was 1.26 ± 0.25 Kg.

The second group was a mixture of ten hatchery females with seven wild males. The hatchery females were four 1st generation females and six 2nd generation females that were hatched and reared in captivity. The wild males had been previously held with wild females and the group of breeders had produced viable spawns. The fish were all adults with reproductive capacity and the mean weight of the fish in the group was 1.22 ± 0.29 Kg. The two groups were video recorded and the behaviour was analysed during one month from April to May. The group of wild females mixed with hatchery males, produced three liberations of unfertilised eggs during the month of observation. Two liberations of eggs were of small volumes (< 25 mL) and one liberation was of 144 mL. The behaviour was similar to that observed in a stock of all hatchery fish. There was no increase in activity and the behaviours, synchronised coupled swims and swimming in procession, associated with mating behaviour were not observed.

However, the group of hatchery females mixed with wild males did produce fertilised spawns. In total of six liberations of eggs were collected of which three were fertilised spawns and three were unfertilised eggs. The activity and behaviour was similar to that observed in stocks of just wild fish. The spawning of fertilised eggs was associated with an increase in activity in the afternoon–evening period prior to egg collection. During the afternoon–evening periods of increased activity the behaviours associated with mating behaviour were observed. The fish were observed swimming in procession and this contributed to the increase in activity (Fig. 1). A total of two synchronised coupled swims that terminated at the surface with the liberation of gametes were observed to demonstrate the full sequence of mating behaviour was completed with the successful fertilisation of gametes. Therefore, the hatchery females had the capacity to complete the mating behaviour to achieve the fertilisation of released eggs with wild males. However, the hatchery males did not complete the mating behaviour with wild females and eggs were not fertilised and were not viable. This is perhaps in agreement with the observation that the males have a more complex and active role in the mating behaviour. The hatchery males did not have the capacity to complete the complex sequence of behaviours that represent the male role in the mating behaviour. The hatchery males did not initiate the mating behaviour with an increase in activity that consisted of displays of swimming in a procession and did not solicit and make coupled swims with the females. These observations demonstrate that a male behavioural reproductive dysfunction is responsible for the complete failure of hatchery breeders to spawn viable eggs.

7. Discussion

These studies of mating behaviour have identified that a reproductive behavioural dysfunction in hatchery males is responsible for the failure to produce viable eggs from stocks hatched and reared in captivity. This dysfunction was shown to be an inability of hatchery males to execute the courtship and spawning in an environment that provided the correct stimulus for courtship and spawning by wild males. The spawning environment with wild females provided for the hatchery males both the physical (light, temperature, nutrition, etc.) and social (cues from females) stimulus that enabled wild males to spawn successfully. Therefore, the cause of the reproductive behavioural dysfunction in hatchery males must be related to: (a) the hatchery males are prevented from participating and/or (b) the hatchery males do not have the capacity (or will/sex drive?) to participate.

That the hatchery males are prevented from participating in courtship and spawning, appears unlikely as all hatchery males do not participate. However, during mate selection and courtship of wild fish, some wild males appeared to be excluded by the behaviour of both males and females that successfully spawned. This can be concluded as spawning was only in pairs and the courtship included behaviours where males sort selection by females that either selected or rejected the male. The courtship began with males displaying to the females to obtain acceptance to rest on the female. Once resting on the female, the males defended or prevented other males from approaching and resting on the female and attempted to gain acceptance to make the paired couple swim to the surface to spawn. Failed coupled swims to spawn were more frequent than successful coupled swims indicating that females avoided inappropriate males. Microsatellite studies to identify the paternity of larvae grown from spawns from wild fish confirmed that certain males dominate spawning whilst the remaining males did not participate and were excluded (Porta et al. 2006a, Martin et al. 2014). Martin et al. (2014) examined four years of spawning from three tanks of wild breeders and found that the percentage of males that did not participate in spawning per year varied found 40 to 90%. In addition to this, although not published, it is not uncommon that a tank of wild broodstock do not spawn fertilised eggs (personal observation). Such tanks of wild fish have the same environment as wild fish that do spawn viable eggs. Therefore, it would appear that in wild fish it is common that males are prevented from spawning and that this can extend to the point were no breeders spawn fertilised eggs as is observed in hatchery breeders. This indicates that the difference between wild and hatchery stocks is more subtle than it appears, both stocks exclude males from spawning which in hatchery stocks extends to all males a situation that is less frequent in wild stocks.

Mate selection forms part of the mating system, which will also have implications for participation in spawning. Many different mating systems exist in animals and it is common that males are excluded and must compete for the opportunity to mate with females (Krasnec et al. 2012). In general, terms the non-participation of certain individuals is biased to males as females select the same dominant male that is perceived to have greater fitness. In addition, non-participating individuals (males and females) can have a role in improving the survival of offspring that have common genes. A harem type mating system where a dominant male mates with several females has been described in flatfish. As mentioned in the introduction, the males of the bastard halibut (Manabe and Shinomiya 2001) and four Bothidea species (Moyer et al. 1985, Konstatinou and Shen 1995, Manabe et al. 2000, Carvalho et al. 2003) were described to defend a territory in the natural environment. The territory either contained a harem of females (Moyer et al. 1985, Konstatinou and Shen 1995, Carvalho et al. 2003) or females visited for courtship and spawning (Manabe et al. 2000). However, a harem type mating system does not appear to exist in tanks of wild Senegalese sole breeders. In a harem mating system few males would be expected to spawn with many more females. Martin et al. (2014), observed that the percentage of females that did not participate in spawning was similar to males indicating that females were not in a harem associated to a dominant male(s). The spawning was dominated by pairs that showed fidelity to each other both within a year and from one year to the next. Dominant pairs provided 61.7% of the spawns and six pairs exhibited fidelity over a three year period. Therefore, the mating system of wild fish held in captivity appears to be monogamy with fidelity and a high percentage of individuals (generally >50%) did not find a partner even though potential partners were available. A harem mating system cannot be discounted, but the apparent monogamy mating system would logically aim that all or a high percentage of fish with reproductive potential find a partner and reproduce. However, the percentage of breeders that find a partner amongst wild captive breeders is low and can be zero and is almost always zero in hatchery breeders. The low participation may be related to social interactions in limited space or that many breeders do not have adequate traits to be selected and form a pair. The low or no participation also highlights that the selection process is complex and the traits that enable a spawning pair to form are unknown. In the case of wild male breeders it is assumed that all breeders placed in a tank have the capacity to reproduce. Therefore, placing wild breeders in a tank to provide a limited number of possible mates in a confined space does not appear to be an adequate approach to have high participation amongst breeders. The same approach applied to hatchery males resulted in a total failure to form pairs even when wild females are present. It would appear probable that the same cause of failed spawning in wild males affects hatchery males, but it is also evident that additional causes may further hinder the participation of hatchery males. In the case of hatchery male breeders it cannot be assumed that the all breeders placed in a tank have the capacity to reproduce or have passed through the correct development to complete reproductive behaviour and spawning. Therefore, it should be considered that hatchery males do not have the capacity to complete the courtship and spawn due to some aspect of development. The incorrect development of courtship and spawning could be related to genetic changes between generations, organogenesis/organ development and cognition of behaviour and social interaction.

A number of genetically related aspects have been considered including the role of siblings in hatchery broodstocks, genetic selection due to domestication over a generation and males that are genetically females (neo-males). Porta et al. (2006b) found a significant decline in genetic variation from a captive wild broodstock to the first generation of progeny. The reduction in genetic variation was due to the low participation of breeders in the spawning resulting in the first generation consisting of a few families. Therefore, an unselected hatchery broodstock will contain many full and half sibling individuals from a few families and this may affect the breeding potential of the broodstock. However, setting up hatchery broodstocks that did not contain full and half sibling relationships between males and females did not provide fertilised spawns (personal observation, Carazo 2013). A second genetic consideration was that the captive system that enabled only a few individuals to reproduce exerted a genetic selection that combined with genetically selective mortality in the rearing system favoured males with low reproductive fitness being chosen as breeders. This genetic selection would need to favour males with low reproductive fitness whilst female's reproductive fitness was not affected as drastically. This scenario was considered in relation to behavioural syndromes or stress coping styles. Stress coping styles describe the physiological and behavioural response of an individual in a stressful situation and have been categorised to extend from proactive fish that have a fight or flight response to reactive fish that freeze or hide in response to a stressor (Koolhaas et al. 1999). However, hatchery Senegalese sole appeared to have the same range of stress coping styles as wild breeders and stress coping style did not appear to be related to reproductive fitness (Ibarra-Zatarain 2015). Although the effects of genetic selection cannot be discounted it does appear unlikely that such a drastic reduction in reproductive fitness over one generation can affect males whilst females were unaffected. Lastly it has been considered that neo-males may be selected as breeders. Neo-males are fish that are genetically females and that have matured as a male in response to the rearing environment during sex determination. The rearing environment has been shown to change sex ratios in Senegalese sole that were as high as 80% males (Blanco-Vives et al. 2011). However, a ratio of 80% males would include approximately 50% normal males and 30% neo-males. Therefore, although neo-males is another factor that could reduce the reproductive fitness of hatchery males it does appear unlikely that all hatchery males that were selected as breeders in all hatchery broodstocks were neo-males.

Incorrect organogenesis or organ development in hatchery males would affect the reproductive fitness of males. The organs implicated in the control of courtship and spawning are the brain-pituitary-gonad axis and the olfactory system. The importance of these organs in the correct execution of courtship and spawning has been demonstrated in the goldfish (Sorensen and Stacey 2004) and the importance of these organs in the courtship and spawning of Senegalese sole will be considered. It is obvious that a fully developed testis with spermiation would be a requirement for mate selection and spawning. As indicated in the introduction, the hatchery stocks mature to produce viable gametes that have been stripped and successfully fertilised *in vitro* (Rasines et al. 2012, 2013). Large volumes of ova were obtained from females by inducing ovulation with GnRHa (Rasines et al. 2012, 2013) and it would appear that female gamete production is not a constraint in the reproduction of hatchery stocks. Hatchery males also produce viable gametes, however, the sperm quality and quantity from hatchery males was slightly inferior compared to wild males (Cabrita et al. 2006). The volume of sperm collected from hatchery males ranged from 5 to 20 µl compared to 10 to 80 µl from wild males. Cell density and sperm production (total spermatozoa (spz) per stripping), respectively, ranged from 0.7 to 1.2×10^9 spz mL^{-1} and 20×10^6 from hatchery males compared to 1–2 $\times 10^9$ spz mL^{-1} and 40–60 $\times 10^6$ spermatozoa from wild males (Cabrita et al. 2006). Therefore, sperm production is low in both hatchery and wild males and hatchery males have a slightly lower production. The observation that hatchery males produce viable sperm strongly suggests that hatchery males can participate in courtship and spawning, however, the lower sperm production may signal to females that hatchery males have lower reproductive fitness and should be avoided for courtship and spawning.

Signalling and the olfactory system has been shown to play a central role in courtship and spawning in the few fish species that have been studied (Sorensen and Stacey 2004, Yambe et al. 2006, Keller-Costa et al. 2014). The olfactory system of the Senegalese sole has been demonstrated to respond to conspecific fluids (Velez et al. 2007, Carazo 2013). Velez et al. (2007) used an electro-olfactogram (EOG) to demonstrate that sole detected bile fluid, intestinal fluid and mucus. The upper olfactory epithelium (in the upper nostril) was significantly more sensitive to the fluids compared to the lower olfactory epithelium (in the lower nostril). Continuing from this study, Carazo (2013) and Fatsini et al. (2016) obtained EOG responses in juveniles and adults to both intestinal fluids and urine from adult sole. The EOG responses were significantly higher to fluids (both intestinal and urine) from mature compared to immature adult sole. Therefore, conspecific communication through the olfactory system in sole may have a role in the control of reproductive behaviour as has been observed in other fish species. No differences were observed in the structure of the olfactory system between hatchery and wild males (Fatsini et al. 2017) suggesting that structurally hatchery males have the same capacity to communicate as wild males. However, the same study compared the transcriptome of the olfactory rosettes from hatchery and wild males to find that over 2000 transcripts were differentially expressed between cultured and wild males (Fatsini et al. 2017). Therefore, further research is required to determine the role of the olfactory system in sole reproduction and to answer two related questions: (a) Does the olfactory system function correctly in hatchery males? (b) Do hatchery males communicate through the olfactory system to receive and signal with the correct stimuli to attract females?

Lastly the cognition of behaviour and social interaction during development will be considered. The development of behaviour is controlled by the interaction of genes and environment (Breed and Sanchez 2010). Genes capture instinctive or innate behaviours that are required for the survival of the species and provide a structure to adjust and learn behaviours in response to the changing environment. Examples of innate behaviours are a cockroach avoiding light or a dog circling before sleeping as if to make a bed in vegetation (Breed and Sanchez 2010). Examples of behaviours that are learnt or imprinted from the environment are recognition of their mother by young geese or leaning what food to eat or avoid (Breed and Sanchez 2010). It appears unlikely that hatchery males would lose innate behaviours required for reproduction due to genetic selection over one generation (see previous discussion). However, perhaps the captive environment does not provide the correct stimulus to learn behaviours required for reproduction. The captive environment does not provide any opportunities to learn reproductive behaviours as common farming practice in aquaculture is to rear each cohort in isolation for other cohorts to ensure there is no disease transfer between generations. Combined with this the male sole must execute a complex sequence of behaviours that include displaying by swimming in a procession, resting on the female, soliciting the female to swim and guiding the synchronised coupled swim to the surface to spawn. Therefore, it is possible that aspects of this complex behavioural sequence need to be learnt or imprinted to achieve the correct execution by hatchery males. This has been partially demonstrated as holding hatchery males with wild pairs that spawned successfully was observed to increase participation of hatchery males in reproductive behaviour such as the swimming in procession and a few hatchery males were observed to participate in a few isolated spawns (Duncan et al. 2014, Fatsini 2017). However, further work is need to determine the role of cognition and imprinting in the reproductive behaviour of male Senegalese sole.

Conclusions

A reproductive behavioural dysfunction in hatchery males is responsible for the failure to produce viable eggs from stocks hatched and reared in captivity. Two areas that should be carefully considered to determine both the cause of and solutions for the behavioural dysfunction are (a) organogenesis and organ development of the testis or olfactory system in particular with reference to the control of reproductive behaviour (b) the role of cognition and imprinting in the development of reproductive behaviour.

Acknowledgments

The authors gratefully thank students and staff that have made these time consuming behavioural studies possible. The studies were supported by two funding agencies; (1) Ministerio de Agricultura, Pesca, Alimentación y Medio Ambiente -JACUMAR Sole Projects coordinated nationally by J. Pedro Cañavate, in Catalonia by ND, in Cantabria by OC and in Valencia by EM and (2) Instituto Nacional de Investigación y Tecnología Agraria y Alimentarías (INIA)-FEDER projects RTA2005-00113, RTA2011-00050 and RTA2014-00048 coordinated by ND.

References

Agulleiro, M.J., V. Anguis, J.P. Cañavate, G. Martínez-Rodríguez, C.C. Mylonas and J. Cerdá. 2006. Induction of spawning of captive reared Senegalese sole (*Solea Senegalensis*) using different delivery systems for gonadotropin-releasing hormone agonist. Aquaculture 257: 511–524.

Anguis, V. and J.P. Cañavate. 2005. Spawning of captive Senegal sole (*Solea senegalensis*) under a naturally fluctuating temperature regime. Aquaculture 243: 133–145.

Bayarri, M J., J.A. Muñoz-Cueto, J.F. López-Olmeda, L.M. Vera, M.A. Rol de Lama, J.A. Madrid and F.J. Sánchez-Vázquez. 2004. Daily locomotor activity and melatonin rhythms in Senegal Sole (*Solea senegalensis*). Physiology & Behavior 81: 577–583.

Baynes, S.M., B.R. Howell, T.W. Beard and J.D. Hallam. 1994. A description of spawning behaviour of captive Dover sole, *Solea solea* (L.). Netherlands Journal of Sea Research 32: 271–275.

Blanco-Vives, B., L.M. Vera, E. Mañanós, J. Ramos, M.J. Bayarri and F.J. Sánchez-Vázquez. 2011. Exposure of larvae to daily thermocycles affects gonad development, sex ratio and sexual steroids in *Solea senegalensis*, Kaup. Journal of Experimental Zoology 315: 162–169.

Breder, C.M. 1922. Description of the spawning habits of *Pseudopleuronectes americanus* in captivity. Copeia. 102: 3–4.

Breed, M. and L. Sanchez. 2010. Both Environment and Genetic Makeup Influence Behavior. Nature Education Knowledge 3(10): 68.

Cabrita, E., F. Soares and M.T. Dinis. 2006. Characterization of Senegalese sole, *Solea senegalensis*, male broodstock in terms of sperm production and quality. Aquaculture 261: 967–975.

Carazo, I., O. Chereguini, I. Martín, F. Huntingford and N. Duncan. 2016. Reproductive ethogram and mate selection in captive wild Senegalese sole (*Solea senegalensis*). Spanish Journal of Agricultural Research 14(4) e0401. http://dx.doi.org/10.5424/sjar/2016144-9108.

Carazo, I., F. Norambuena, C. Oliveira, F.J. Sanchez-Vazquez and N.J. Duncan. 2013. The effect of night illumination, red and infrared light, on locomotor activity, behaviour and melatonin of Senegalese sole (*Solea senegalensis*) broodstock. Physiology & Behavior. 118: 201–207.

Carazo, I. 2013. Reproductive behaviour and physiology of Senegalese sole, (*Solea senegalensis*) broodstock in captivity. PhD Thesis, University of Barcelona, Barcelona, Spain, pp. 326, http://hdl.handle.net/10803/107945.

Carvalho, N., P. Alfonso and R. Serrao Santos. 2003. The haremic mating system and mate choice in the wide-eyed flounder, *Bothus podas*. Environmental Biology of Fishes. 66: 249–258.

Duncan, N., I. Carazo, E. Fatsini, J. Napuchi, Z. Ibarra, I. Martin and O. Chereguini. 2014. Reproductive behaviour of the Senegalese sole (*Solea senegalensis*): a review with indications of possible behavioural solutions to the G1 reproductive dysfunction. Book of Abstracts, Aquaculture Europe 2014, San Sebastian, Spain.

Fatsini, E. 2017. Reproduction, olfaction and dominance behaviour in Senegalese sole (*Solea senegalensis*) PhD thesis. Universitat Politècnica de València. doi:10.4995/Thesis/10251/81550 pp. 204. http://hdl.handle.net/10251/81550.

Fatsini, E., R. Bautista, M. Manchado and N.J. Duncan. 2016. Transcriptomic profiles of the upper olfactory rosette in cultured and wild Senegalese sole (Solea senegalensis) males. Comparative Biochemistry and Physiology Part D: Genomics and Proteomics 20: 125–135.

Fatsini, E., I. Carazo, F. Chauvigné, M. Manchado, J. Cerdà, P.C. Hubbard and N.J. Duncan. 2017. Olfactory sensitivity of the marine flatfish *Solea senegalensis* to conspecific body fluids. Journal of Experimental Biology, 220: 2057–2065.

Fernandez-Palacios, H., D. Schuchardt, J. Roo, M. Izquierdo, C. Hernandez-Cruz and N. Duncan. 2014. Dose-dependent effect of a single GnRHa injection on the spawning of meagre (*Argyrosomus regius*) broodstock reared in captivity. Spanish Journal of Agricultural Research 12: 1038-1048 doi:http://dx.doi.org/10.5424/sjar/2014124-6276.

Forster, G.R. 1953. The spawning behavior of plaice. Journal of the Marine Biological Association of the United Kingdom 32: 319. http://dx.doi.org/10.1017/S0025315400014569.

Gibson, R.N. 2005. The behaviour of flatfishes. pp. 213–239. *In*: Gibson, R.N. [ed.]. Flatfishes Biology and Exploitation. Fish and Aquatic Resources Series 9. 1st ed. Oxford UK: Blackwell Science Ltd.

Guzmán, J.M., R. Cal, A. García-López, O. Chereguini, K. Kight, M. Olmedo, C. Sarasquete, C.C. Mylonas, J.B. Peleteiro, Y. Zohar and E. Mañanós. 2011a. Effects of in vivo treatment with the dopamine antagonist pimozide and gonadotropin-releasing hormone agonist (GnRHa) on the reproductive axis of Senegalese sole (*Solea senegalensis*). Comparative Biochemistry and Physiology - A. 158: 235–245.

Guzmán, J.M., J. Ramos, C.C. Mylonas and E. Mañanós. 2009. Spawning performance and plasma levels of GnRHa and sex steroids in cultured female Senegalese sole (*Solea senegalensis*) treated with different GnRHa-delivery systems. Aquaculture 291: 200–209.

Guzmán, J.M., J. Ramos, C.C. Mylonas and E. Mañanós. 2011b. Comparative effects of human chorionic gonadotropin (hCG) and gonadotropin-releasing hormone agonist (GnRHa) treatments on the stimulation of male Senegalese sole (*Solea senegalensis*) reproduction. Aquaculture 316: 121–128.

Howell, B., R. Pricket, P. Canavate, E. Mananos, M.T. Dinis, L. Conceicao and L.M.P. Valente. 2011. Sole farming: there or thereabouts! A report of the 5th workshop on the cultivation of soles. Aquaculture Europe 36: 42–45.

Ibarra-Zatarain, Z. 2015. The role of stress coping style in reproduction and other biological aspects in the aquaculture species, Senegalese sole (*Solea senegalensis*) and gilthead seabream (*Sparus aurata*). PhD Thesis, Autonomous University of Barcelona, Barcelona, Spain, pp. 233, http://www.tdx.cat/handle/10803/322809.

Keller-Costa, T., P.C. Hubbard, C. Paetz, Y. Nakamura, J.P. da Silva, A. Rato, E.N. Barata, B. Schneider and A.V.M. Canario. 2014. Identity of a Tilapia Pheromone Released by Dominant Males that Primes Females for Reproduction. Current Biology 24: 2130–2135.

Konstantinou, H. and D.C. Shen. 1995. The social and reproductive behavior of the eyed flounder, *Bothus ocellatus*, with notes on the spawning of *Bothus lunatus* and *Bothus ellipticus*. Environmental Biology of Fishes. 44: 311–324.

Koolhaas, J.M., S.M. Korte, S.F. De Boer, B.J. Van Der Vegt, C.G. Van Reenen, H. Hopster, I.C. De Jong, M.A. Ruis and H.J. 1999. Coping styles in animals: current status in behavior and stress-physiology. Neurosci. Biobehav. Rev. 23: 925–935.

Krasnec, M.O., C.N. Cook and M.D. Breed. 2012. Mating Systems in Sexual Animals. Nature Education Knowledge 3(10): 72.

Kruuk, H. 1963. Diurnal periodicity in the activity of the common sole, *Solea vulgaris* quensel. Netherlands Journal of Sea Research. 2: 1–28.

Manabe, H., M. Ide and A. Shinomiya. 2000. Mating system of the lefteye flounder, *Engyprosopon grandisquama*. Ichthyological Research 47: 69–74.

Manabe, H. and A. Shinomiya. 2001. Two spawning seasons and mating system of the bastard halibut, *Tarphops oligolepis*. Ichthyological Research 48: 421–424.

Mañanós, E., N. Duncan and C.C. Mylonas. 2008. Reproduction and control of ovulation, spermiation and spawning in cultured fish. pp. 3–80. *In*: Cabrita, E., V. Robles and M.P. Herráez [eds.]. Methods in Reproductive Aquaculture: Marine and Freshwater Species. CRC Press, Taylor and Francis Group, Boca Raton.

Martin, I., I. Rasines, M. Gomez, C. Rodrıguez, P. Martınez and O. Chereguini. 2014. Evolution of egg production and parental contribution in Senegalese sole, *Solea senegalensis*, during four consecutive spawning seasons. Aquaculture 424-425: 45–52.

Moyer, J.T., Y. Yogo, M.J. Zaiser and H. Tsukahara. 1985. Spawning behavior and social organization of the flounder *Crossorhombus kobensis* (Bothidae) at Miyake-jima. Japanese Journal of Ichthyology 32: 363–367.

Norambuena, F., A. Estevez, G. Bell, I. Carazo and N. Duncan. 2012. Proximate and fatty acid composition in muscle, liver and gonads of wild versus cultured broodstock of Senegalese sole (*Solea senegalensis*). Aquaculture 356–357: 176–185.

Oliveira, C., M.T. Dinis, F. Soares, E. Cabrita, P. Pousão-Ferreira and F.J. Sanchez-Vazquez. 2009. Lunar and daily spawning rhythms of Senegal sole *Solea senegalensis*. Journal of Fish Biology 75: 61–74.

Pankhurst, N.W. and Q.P. Fitzgibbon. 2006. Characteristics of spawning behaviour in cultured greenback flounder *Rhombosolea tapirina*. Aquaculture 253: 279–289.

Porta, J., J.M. Porta, G. Martínez-Rodriguez and M.C. Alvarez. 2006a. Development of a microsatellite multiplex PCR for Senegalese sole (*Solea senegalensis*) and its application to broodstock management. Aquaculture 256: 159–166.

Porta, J., J.M. Porta, G. Martınez-Rodriguez and M.C. Alvarez. 2006b. Genetic structure and genetic relatedness of a hatchery stock of Senegal sole (*Solea senegalensis*) inferred by microsatellites. Aquaculture 251: 46–55.

Rasines, I., M. Gómez, I. Martín, C. Rodríguez, E. Mañanos and O. Chereguini. 2012. Artificial fertilization of Senegalese sole (*Solea senegalensis*): Hormone therapy administration methods, timing of ovulation and viability of eggs retained in the ovarian cavity. Aquaculture 326-329: 129–135.

Rasines, I., M. Gomez, I. Martin, C. Rodriguez, E. Mañanós and O. Chereguini. 2013. Artificial fertilization of cultured Senegalese sole (*Solea senegalensis*): effects of the time of day of hormonal treatment on inducing ovulation. Aquaculture 392-395: 94–97.

Sorensen, P.W. and N.E. Stacey. 2004. Brief review of fish pheromones and discussion of their possible uses in the control of non-indigenous teleost fishes. New Zeal. J. Mar. Fresh. 38: 399–417.

Stoner, A.W., A.J. Bejda, J.P. Manderson, B.A. Phelan, L.L. Stehlik and J.P. Pessutti. 1999. Behavior of winter flounder, *Pseudopleuronectes americanus*, during the reproductive season: laboratory and field observations on spawning, feeding, and locomotion. Fishery Bulletin 97: 999–1016.

Velez, Z., P.C. Hubbard, E.N. Barata and A.V.M. Canário. 2007. Differential detection of conspecific-derived odorants by the two olfactory epithelia of the Senegalese sole (*Solea senegalensis*). General and Comparative Endocrinology 153: 418–425.

Yambe, H., S. Kitamura, M. Kamio, M. Yamada, S. Matsunaga, N. Fusetani and F. Yamazaki. 2006. L-Kynurenine, an amino acid identified as a sex pheromone in the urine of ovulated female masu salmon. Proc. Natl. Acad. Sci. 103: 15370–15374.

Zohar, Y. and C.C. Mylonas. 2001. Endocrine manipulations of spawning in cultured fish: from hormones to genes. Aquaculture 197: 99–136.

B-2.1

The Biological Clock of Sole
From Early Stages to Adults

Águeda Jimena Martín-Robles,[1,2,*] *Carlos Pendón-Meléndez*[1]
and *José Antonio Muñoz-Cueto*[2]

1. Introduction to circadian rhythms and the molecular clock

Millions of years of evolution under a cyclic and predictable environment has had deep influence in the life of most if not all-living organisms and has favoured the development of internal timekeeping mechanisms or biological clocks. Animals have been subjected to diverse cyclic events driven by geophysical cycles, such as the daily and annual variations of light and darkness provoked by the rotation of the Earth around its axis and around the sun, and tidal and lunar cycles caused by the rotation of the moon around the Earth. In addition, animals are exposed to daily changes in other factors such as temperature or food availability. Only animals lived in caves or in the depth of the oceans, have avoided some of these environmental fluctuations (Cymborowski 2010).

Biological clocks are internal systems to keep time and represent an adaptive advantage as may allow organisms the anticipation of environmental periodic variations, being prepared in advance for forthcoming changes, thus optimizing biochemical, physiological and behavioural processes. This temporal organization is reflected by the existence of biological rhythms, internal cycles in a wide range of processes and variables in living organisms that occur at regular and predictable time intervals. In fact, rhythmicity is one of the basic features of living organisms.

Several types of biological rhythms have been described with different periodicities (from seconds to years) but the most ubiquitous rhythms in nature are circadian rhythms.

[1] Biochemistry and Molecular Biology, Department of Biomedicine, Biotechnology and Public Health. Faculty of Sciences, Marine Campus of International Excellence (CEIMAR) and Agrifood Campus of International Excellence (ceiA3), University of Cádiz, 11519 Puerto Real, Cádiz, Spain.
[2] Department of Biology, Faculty of Marine and Environmental Sciences and Institute of Marine Research (INMAR). Marine Campus of International Excellence (CEIMAR) and Agrifood Campus of International Excellence (ceiA3), University of Cádiz, 11519 Puerto Real, Cádiz, Spain.
* Corresponding author: agueda.jimena@uca.es

The term *circadian* derives from the Greek words *circa* (about) and *dian* (day) and includes oscillations of biological processes with a period of approximately one day or 24 h, i.e., rhythms that are related with daily environmental cycles. Circadian rhythms are involved in many aspects of metabolism, physiology, and behaviour and are a shared characteristic of all living organisms from bacteria to mammals. One of the main characteristics of circadian rhythms is that they keep on under constant environmental conditions, in absence of temporal cues. Therefore, they are not passive responses to the environment; instead, they arise within the organism itself, driven by self-sustained endogenous oscillators. Under these non-cycling conditions, the rhythm free runs with its natural internal period, which closely approximates 24 h, although normally exhibits a slight difference with the period of the geophysical cycle (longer or shorter depending on species). In addition, circadian rhythms exhibit other features. They are entrained to environmental cycles and temperature compensated. The resetting of the clock so that the organisms remain in phase with their surroundings (internal time matches local time) is accomplished by external synchronizing cues called *zeitgebers*, meaning time givers in German. Some of the most potent external time cues to reset the circadian phase of the internal clocks are changes in light (Pando et al. 2001), temperature (Rensing and Ruoff 2002) and food availability (Mistlberger and Antle 2011). Circadian rhythms have also developed temperature compensation mechanisms that remain less understood (Ki et al. 2015). Whereas the rate of most biochemical reactions is temperature-dependent, the period of the circadian rhythms is roughly consistent over a range of temperatures.

Therefore, circadian rhythms are endogenously generated but respond to changes in the environment. They are adjusted to the geophysical time, being the rhythm daily synchronized with the environmental cycle and allowing internal processes occur at proper time.

But how do organisms generate endogenous rhythmicity? The answer relies on autoregulatory feedback loops that are at the heart of oscillatory processes and constitute the molecular bases of circadian rhythms. Endogenous oscillations are ultimately generated by a self-sustainable molecular clock that comprises interlocked transcriptional-translational feedback loops involving a set of so-called clock genes (Fig. 1). Molecular clocks exist at all levels of organization from cells, through tissues to organs, and transmit their temporal signals to the organism by driving rhythms of the clock controlled gene expression. At the core of the vertebrate circadian clockwork model, the positive transcriptional loop is constituted by *Clock* (Circadian Locomotor Output Cycles Kaput) and *Bmal1* (Brain and Muscle ARNT-like protein) genes, whereas the negative loop is mainly formed by *Per* (Period, *Per1*, 2 and 3) and *Cry* (Cryptochrome, *Cry1* and 2) clock genes (Fig. 1). CLOCK and BMAL proteins act as transcription factors that heterodimerize and induce the transcription of *Per* and *Cry* genes by binding to a specific promoter region. After translation, PER and CRY proteins interact in the cytoplasm and, after nuclear entry, negatively regulate their own transcription by repressing the transcriptional activation driven by CLOCK:BMAL complexes (Reppert and Weaver 2001, Partch et al. 2014) (Fig. 1). An additional loop comprised by *Rev-erbα* nuclear receptor and *ROR* (Retinoic acid related Orphan Receptor) genes confers stability and robustness to the system, by directing rhythmic *Bmal1* expression (Emery and Reppert 2004, Guillaumond et al. 2005). Moreover, post-transcriptional and post-translational modifications, including phosphorylation of clock proteins mediated by *Casein Kinase 1δ* and ε (CK1δ/ε), may impact upon the workings of the clock by modulating the period of the circadian oscillator, allowing them to run with an approximately 24 h period (Lee et al. 2001, 2004, 2009, 2011).

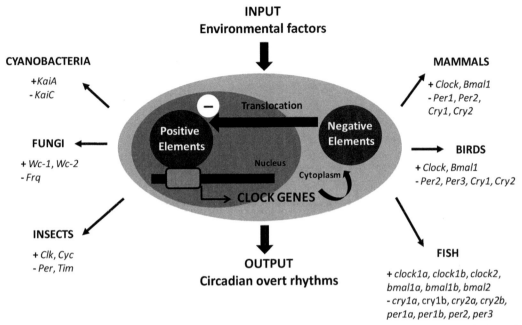

Fig. 1. Conservation of the molecular clock mechanism along the phylogenetic scale. The key positive (indicated by "+" symbols) and negative (indicated by "−" symbols) elements that form the interconnected feedback loops of the core oscillator are shown in diverse models such as Cyanobacteria, Fungi, Insects, Fish, Birds and Mammals. Positive elements activate the expression of genes encoding negative elements that, in turn, inhibit the activities of the positive elements. Phosphorylation of negative elements in the cytoplasm leads to their eventual degradation, allowing the restoration of the cycle. The same basic mechanism is present in different organisms, although the components vary. Some rhythmic biological outputs driven by this mechanism are also shown. These oscillators are daily adjusted by environmental cues to remain in phase with their surroundings and control rhythmic physiology and behaviour. Modified from Bell-Pedersen et al. 2005.

This autoregulatory mechanism arises cell-autonomously even in multicellular organisms and results in a cyclic, self-sustained expression of clock genes with an approximately 24 h period. Although the anatomical location of the main circadian oscillators and their degree of autonomy may differ through different organisms, this mechanism is known to be highly conserved along the phylogenetic scale and has been implicated in all circadian clocks investigated to date in such a different model systems as Cyanobacteria (prokaryotes), *Neurospora*, *Drosophila* and mouse (eukaryotes) (Bell-Pedersen et al. 2005, Pegoraro and Tauber 2011, Crane and Young 2014). Chronobiological studies have also acquired a growing interest relevant for human beings as when the molecular clock is desynchronized, may precipitate symptoms of psychiatric diseases and mutations in particular clock genes led to sleep wake disorders (Toh 2008, Parekh and McClung 2016).

2. Fish as models for the study of molecular clocks

Over the last two decades, the field of the molecular bases of circadian rhythms and their synchronisation to external/internal cues has expanded considerably. After describing the key molecular components of the circadian clock in mammals, homologs of clock genes have been discovered and their expression has been analysed in a wide range of species including fish, that have ultimately proven to be valuable complementary models for studying key aspects of clock biology (Idda et al. 2012). Due to its phylogenetic position, fish represent an excellent alternative model to other vertebrate groups to study the evolution of circadian clocks. They are one of the most successful groups of vertebrates, comprising more than 25,000 species that inhabit in a variety of ecological niches from tropical to extreme latitudes and from continental lagoons to deep ocean floors (Nelson et al. 2016). They also show enormous diversity in the development of feeding and reproductive strategies and activity rhythms, some species presenting the ability to display either diurnal or nocturnal behaviour that reflect the existence of a complex and flexible circadian organization (Falcón et al. 2010, Kulczykowska et al. 2010, Migaud et al. 2010). Moreover, unusual species inhabiting extreme environments have provided surprising results (Cavallari et al. 2011, Beale et al. 2013). Altogether, these evidences reinforce fish as one of the most fascinating models to study the plasticity of the circadian clocks and how they adapt to diverse environmental conditions during evolution. In fact, studies on fish circadian clocks have significantly contributed to our understanding of the basic mechanisms driving endogenous oscillations.

One of the features of the fish molecular clocks is the presence of additional copies of many clock genes compared to tetrapods resulting from genome duplication events in the teleost lineage (Postlethwait et al. 1998, Meyer and Van de Peer 2005). The existence of extra copies of clock genes may indicate certain degree of redundancy within the fish molecular clock, with functions being shared by the extra genes. However, it is known that duplication events are an important mechanism for the evolution of diversity and the origin of novel gene functions, and they have been proposed to be one of the reasons for the evolutionary success and extraordinary biological diversity of teleost fish (Meyer and Van de Peer 2005). Subsequent differential gene loss after the duplication resulted in retention of different clock duplicates in different teleost fish, that further suggest retained duplicates do have specific, non redundant roles, reflecting an increased complexity not seen in other models (Wang 2008a, 2008b, 2009, Idda et al. 2012, Whitmore 2010). This divergent resolution of duplicated genes could have contributed to the evolution of distinct fish circadian clock mechanisms.

Thus, regarding the positive elements of the clock mechanism, three *clock* genes (*clock1a, 1b* and *2*) and three *bmal* genes (*bmal1a, 1b* and *2*) have been identified in zebrafish, *Danio rerio* (Fig. 1), and have shown to interact in various heterodimeric combinations that display different transactivation properties (Hirayama et al. 2003, Ishikawa et al. 2002, Wang 2008b, 2009). In most tissues, *clock* and *bmal* genes exhibit robust mRNA expression rhythms (Whitmore et al. 1998a, Cermakian et al. 2000). In addition, four *period* genes, two *per1* homologs (*per1a* and *per1b*) together with single *per2* and *per3*, and six *cry* genes (*cry1a, 1b, 2a, 2b, 3* and *4*) form the negative loop of the clock mechanism in zebrafish (Wang 2008a, Kobayashi et al. 2000). Both *per2* and *cry1a* are light-driven genes, and function in combination as an element of the light input pathway for the photic entrainment of the clock, while the remaining *per* and *cry* genes are predominantly clock-regulated (Hirayama et al. 2003, Tamai et al. 2007, Vatine et al. 2009). Interestingly, not all *cry* genes have been shown to repress Clock:Bmal directed transcriptional activation. Phylogenetic analysis of *cry* genes has shown that *cry1a, 1b, 2a* and *2b* display high homology with mammalian *Cryptochrome* genes and also repress Clock and Bmal activation, whereas *cry3* and *cry4* resemble *Drosophila* sequences and do not repress Clock and Bmal, being *cry4* the most divergent gene from vertebrate *Crys* (Kobayashi et al. 2000). Temporal expression patterns of *per* and *cry* genes vary significantly, although most of them display high amplitude mRNA rhythms (Pando et al. 2001, Kobayashi et al. 2000). Together, combined evidence further supports the hypothesis of divergent and specialized functions for the fish *per* and *cry* genes.

Although initially the zebrafish got all the attention on research of clock biology and has emerged as a useful model to study key aspects in the circadian system of vertebrates (Vatine et al. 2011), it is becoming obvious that zebrafish cannot really represent all of this large and diverse group of vertebrates. Fortunately, in recent years quite a few teleost species have been added to the list of organisms where the molecular clock has been characterized, being of an immense comparative value. This include diverse species such as goldfish *Carassius auratus* (Velarde et al. 2009), rainbow trout *Oncorhynchus mykiss* (López-Patiño et al. 2011), medaka *Oryzias latipes* (Cuesta et al. 2014), some reef fishes such as *Siganus guttatus* and *Halichoeres trimaculatus* (Park et al. 2007, Hur et al. 2012), Atlantic cod (*Gadus morhua*) (Lazado et al. 2014), Nile tilapia *Oreochromis niloticus* (Costa et al. 2016) and two blind cavefish species, *Phreatichthys andruzzii* and *Astyanax mexicanus* (Cavallari et al. 2011, Beale et al. 2013). Some species studied are particularly attractive owing to their high commercial interest for marine aquaculture and the potential implications of circadian research on aquaculture practice, such as Atlantic salmon *Salmo salar* (Davie et al. 2009), European sea bass *Dicentrarchus labrax* (Sánchez et al. 2010, Del Pozo et al. 2012, Herrero and Lepesant 2014), gilthead seabream *Sparus aurata* (Vera et al. 2013, Mata-Sotres et al. 2015), and the flatfish species under study in this book, the Senegalese sole *Solea senegalensis* (Kaup, 1858) (Martín-Robles et al. 2011, 2012a).

3. Senegalese sole as a model for chronobiology

As it has been mentioned in other chapters, the Senegalese sole is a member of Acanthopterygii, belonging to the highly evolved Pleuronectiform order of euteleosts. It is considered a promising important species for diversification of marine aquaculture in Southern Europe as a result of market saturation with gilthead seabream, European seabass and turbot. However, as has been reviewed in this volume, Senegalese sole intensive production is still slowly growing, and the consolidation of sole farming as a cultivation of strategic interest for aquaculture industry is being delayed. This has been explained mostly by serious disease problems, high mortality at weaning, variable growth and reproductive problems, as stable and viable embryos are only obtained from wild-caught broodstocks (Anguis and Cañavate 2005). Many of the problems in the viability of farming for new species are derived from extrapolation of rearing conditions from other well established species in relation to photoperiod, temperature and feeding protocols. A large and sustained research investment, together with technical improvements, have led to a better understanding of the particular requirements of this species, and favoured an important progress in productivity, giving a new impetus to the cultivation of Senegalese sole towards a competitive and sustainable industry (Morais et al. 2016).

From a chronobiological point of view, information about its biological rhythms, the organization of the circadian system and the effects of environmental factors is considered crucial for the establishment of sustainable and economically relevant Senegalese sole farming. In this sense, the Senegalese sole has acquired a great interest for chronobiologists that were firstly motivated for its high market value and the potential contribution of research to sole aquaculture. Afterwards, this species has emerged as an attractive teleost model alternative to the zebrafish to enlarge our knowledge on fish circadian system due to its different habits. The metamorphic process that sole experiences during ontogeny has been an important incentive that has further motivated the selection of this species as a model study for chronobiology. Moreover, the fact that the Pleuronectiform order is one of most evolved teleost orders from an evolutionary perspective makes Senegalese sole a species with both phylogenetic and comparative interest.

The attention of chonobiologists originally turned to sole years ago, when monitoring systems based on infrared photocells attached to the tanks were used to record activity rhythms of sole, revealing its nocturnal locomotor activity pattern (Bayarri et al. 2004). From then, remarkable progress, mainly focused in describing different clock outputs and the development of the photoreceptive structures, have been made during the last few years. Thus, recent studies have shown that adult sole specimens also display daily and/ or seasonal rhythms in feeding behaviour, sexual steroids, reproduction and spawning, as well as in endocrine factors involved in growth and stress response, all of them strongly influenced by photoperiod and temperature conditions (Anguis and Cañavate 2005, García-López et al. 2007, Guzmán et al. 2008, Boluda-Navarro et al. 2009, Oliveira et al. 2007, 2009, López-Olmeda et al. 2013, 2016 Carazo et al. 2013). Daily and seasonal rhythms in plasma melatonin (Bayarri et al. 2004, Vera et al. 2007) and melatonin receptors (Oliveira et al. 2008, Confente et al. 2010, Lan-Chow-Wing et al. 2014) have also been documented in sole, showing the usual melatonin profile with lower levels during the day and higher levels during the night. Several studies performed in sole larvae showed that these rhythms are established at early developmental stages, as revealed by the existence of daily variations on the expression of the melatonin biosynthesis enzyme *arylalkylamine N-acetyltransferase 2 (aanat2)* from 2 days post-fertilization (Isorna et al. 2009a). Besides, hatching rhythms have been described and locomotor activity rhythms have been reported to arise from few days post-hatching (Blanco-Vives et al. 2011a, 2012), being larval development tightly influenced by both light (photoperiod and spectrum) and temperature conditions (Yúfera et al. 1999, Parra and Yúfera 2001, Cañavate et al. 2006, Blanco-Vives et al. 2010, 2011b). The ontogeny of the photoreceptive structures present in the animal, the pineal organ and the retina, has been described as well. Both structures are differentiated and acquire their photoreceptive capacity very early during development, although the first opsin-immunoreactive cells are detected earlier in the pineal gland than in the retina (El M'Rabet et al. 2008, Isorna et al. 2009a, 2011, Bejarano-Escobar et al. 2010). In addition, what is really fascinating of this species is that in the course of their development, Senegalese sole undergoes a real metamorphic process that encompasses a light-dependent switch from diurnal to nocturnal in locomotor activity and feeding behaviour, which occur just before the onset of this process (Cañavate et al. 2006, Blanco-Vives et al. 2012). Deciphering the underlying mechanisms of this chronotype switch will be challenging, but from a chronobiological perspective quite fascinating. All these evidences confirm the marked rhythmic character of this species and the presence of a functional molecular clock from early ontogenetic stages. However, until very recently the genetic bases responsible for the generation of these circadian outputs have been unexplored.

In this chapter, we will review studies on the characterization of the molecular clock in Senegalese sole, focusing on the core clock components available so far in this species and their expression profiles during different stages of its life cycle, from few hours after fertilization to early development, metamorphosis and in adult specimens. We will also put our attention on the effects of several lighting conditions in the onset of molecular clock activity and gene expression during crucial developmental phases such as the first days of development before the exogenous feeding and during the metamorphic stages, emphasizing the relevant knowledge achieved in our laboratories.

4. Components of the Senegalese sole molecular clock

The Senegalese sole circadian clock has a number of key molecular components similar to those described from other vertebrates. Some of the core clock genes from the negative and positive feedback loops have been cloned and its central and peripheral expression have been characterized in our laboratories. The first isolated clock gene was *per3* (Martín-Robles et al. 2011). A gene walking based strategy and 5′-3′ rapid amplification of cDNA ends (RACE) allowed us to obtain the full-length coding region of sole *per3*, which generates a putative Per3 protein of 1267 amino acids. Later on, partial sequences of *per1*, *per2* and *clock* were obtained, from which 238, 166, and 204 partial amino acid sequences were deduced (Martín-Robles et al. 2012a). Highly conserved protein domains and important sequence motifs previously described to be essential for the molecular clockwork were identified among sole sequences by *in silico* structural analysis. In the case of Per3, these included two tandemly organized conserved Per-Arnt-Sim (PAS) binding domains (PAS A-PAS B) in the N-terminal part of the protein and a cytoplasmic localization domain (CLD) following the PAS domains. In addition, similar to other vertebrate PER proteins, some others key regions were found such as a putative nuclear localization signal (NLS), a nuclear export signal (NES), two putative casein kinase 1ε phosphorylation sites (CK1ε site) and a carboxy-terminal serine/threonine-glycine (SG) repeat region.

In the case of sole Per1 and Per2 partial proteins, conserved PAS binging domains were also identified and a basic helix-loop-helix (bHLH) DNA binding domain was recognized in sole partial Clock. All these functional regions are shared between transcription factors regulating the circadian clock and are required for its proper functioning in different species (Hirayama and Sassone-Corsi 2005). On one hand, PAS domains function as dimerization domains required for protein associations to control their post-translational regulation and subcellular distribution (Ponting and Aravind 1997, Whitmore et al. 1998a,b), and the CLD domain confers cytoplasmic localization to the monomeric forms of Per and seems to contribute to proper structure of the PAS domain and protein folding (Yagita et al. 2000, 2002, Vielhaber et al. 2001). On the other hand, NLS and NES sequence motifs also determine the cellular trafficking of clock proteins allowing their active transport between the cytoplasm and the nucleus (Kaffman and O'Shea 1999, Yagita et al. 2000, 2002) and CK1ε phosphorylation sites, serve as key regulators of Per activity and contribute to establish and maintain the 24 h approximated period of circadian rhythms by promoting degradation of clock proteins (Lee et al. 2004). The presence of NLS, CKIε phosphorylation sites and NES in sole Per3 suggests that this protein could be shuttling continuously between the cytoplasm and the nucleus, which is a fundamental feature of the oscillation process.

All predicted amino acid sequences showed high identity when compared with their respective teleost clock proteins and lower identity with other vertebrate sequences. Phylogenetic analysis grouped sole Per1, Per2, Per3 and Clock amino acid sequences within clearly separate branches together with their corresponding teleost clock proteins, showing higher divergence in relation to amphibian, avian, and mammalian sequences. The Senegalese sole Per1 identified clusters together with sea bass Per1, medaka and zebrafish Per1b, sole Per2 with medaka Per2b and zebrafish Per2, sole Per3 with medaka Per3 and the sole Clock is more closely related to zebrasfish Clock1a. The presence of two products in northern blot analysis performed with a *per3* probe could suggests that two *per3* exists in sole, although a single *per3* orthologous to mammalian Per3 has been preserved in other fish species such as zebrafish, Japanese pufferfish (*Takifugu rubripes*), green pufferfish

(*Tetraodon nigroviridis*) and medaka (Wang 2008a). Evolution of clock genes reflects the existence of several gene duplication events in the teleost lineage and additional copies are present in fish compared to tetrapods. Alternatively, two different spliced variants generated by alternative splicing of the same *per3* gene could be present in sole, at least in some tissues. It has been hypothesized that during the third-round genome duplication that occurred in the teleost lineage, *per3* gave rise to *per3a* and *per3b*, but *per3b* was quickly lost (Wang 2008a). Further studies are required to corroborate that these genes are the only *per1*, *per2*, *per3* and *clock* existing in sole, as differential gene loss after the duplication resulted in retention of different duplicates in different teleost species. The identification of clock genes in Senegalese sole enabled us to start with the characterization of the molecular clock mechanism in this species, from early developmental stages to adult specimens.

5. The onset of the circadian clock during sole development

One of the key questions concerning the molecular clock is when and how the oscillatory mechanism first emerges during development and the significance of exposure to environmental factors at early stages. It is held that proper development of the circadian system during ontogeny and entrainment to natural cycling conditions appear essential for the future survival of an organism (Vallone et al. 2007). Fish have a particular advantage over studies in other vertebrate groups when examining processes involved in early embryo development, as they produce large number of transparent eggs and embryos that develop outside the female (Whitmore 2010). However, the main limitation of temperate fish species is eggs availability because they usually spawn once or twice a year, as the maturation and spawning is dependent upon seasonal changes in light and temperature. Senegalese sole is a long-day breeder that shows two spawning periods induced by fluctuating water temperatures. The main reproductive season occurs during spring, and secondary spawning events have also been reported in autumn (Anguis and Cañavate 2005, Vera et al. 2007). A very marked nocturnal spawning rhythm has been observed in this species, showing a preference to spawn during the first half of the night (Oliveira et al. 2009). Therefore, sole fertilized eggs can be collected during the first hours of the night and transferred to tanks in constant darkness, ensuring that the eggs do not receive any light before the experimental treatments.

Circadian clock ontogeny has been investigated in a limited number of teleost fish species such as zebrafish, rainbow trout and medaka, where molecular clocks are present and functional from early embryonic stages and require light signals for accurate entrainment (Dekens and Whitmore 2008, Davie et al. 2011, Cuesta et al. 2014). Moreover, these embryonic clocks are affected by different light spectrum, and daily rhythms of clock gene expression appeared earlier in larvae reared under light-dark cycles of blue wavelengths than in those reared under white or red wavelengths (Di Rosa et al. 2015). The only species with commercial interest for marine aquaculture where the onset of the molecular clock has been characterized is the Senegalese sole. One study reported clock gene expression during gilthead seabream ontogeny under different photoperiods, but from developing larvae of 10 days post-hatching (dph) (Mata-Sotres et al. 2015).

It is a common practice in larviculture to set-up a 24 h illumination protocol during the first weeks of larval rearing in order to increase food ingestion, accelerating growth and increasing farming production. However, photoperiod manipulations could seriously compromise the establishment, organization, and function of the molecular clock during ontogeny. In Senegalese sole, studies of circadian outputs have provided evidence for the existence of a functional molecular clock during early ontogenetic stages and pointed to differences in the establishment of the circadian system according to the environmental conditions in which sole develops (Blanco-Vives et al. 2011a, 2012, Isorna et al. 2009a, 2011). Moreover, early development of sole is tightly influenced by lighting conditions (Blanco-Vives et al. 2010, 2011b). But when does the molecular clock start ticking during sole embryo development? To address this question, we decided to explore the onset of rhythmic clock gene expression during the first several days of sole development, from 0 days post-fertilization (dpf) until the end of the yolk-sack stage and beginning of exogenous feeding at 4 dpf, avoiding external food supply and the possible synchronizing role of feeding-related cues. During the spring reproductive season, sole fertilized eggs were collected before dawn and, in order to determine the effect of photoperiod on the establishment and maturation of the clock, they were maintained under constant temperature and 14 h: 10 h light-dark (LD 14:10) cycles, 24 h constant light (LL) and 24 h constant dark (DD) (Martín-Robles et al. 2012b). This represented an important piece of work from both a commercial and comparative point of view as we provided evidence, for the first time in a highly evolved marine fish, that an embryonic molecular clock is present from the first day post-fertilization. Our findings also reinforced earlier studies performed in very distant species such as zebrafish, medaka and rainbow trout, suggesting conserved mechanisms in the emergence of clock mechanisms during evolution of teleost fish.

Rhythms of clock gene expression, as well as of some clock outputs, such as melatonin synthesis and *aanat2* expression in the pineal gland, are some of the earliest detectable circadian rhythms in developing zebrafish and medaka embryos (Gothilf et al. 1999, Kazimi and Cahill 1999, Ziv et al. 2005, Cuesta et al. 2014, Di Rosa et al. 2015). In Senegalese sole, rhythmic expression of *per1*, *per2*, *per3* and *clock* genes was detected as early as 0–1 dpf under all lighting conditions tested (Martín-Robles et al. 2012b, Fig. 2), supporting the most recent accepted hypothesis about the origin of circadian clocks in fish. This hypothesis maintains that circadian oscillation undergoes a natural, light-independent, ontogenetic activation during the first hours of development, i.e., the process of starting the clock is genetically determined (Dekens and Whitmore 2008). However, the environment is crucial to ensure the synchronization of the mechanism. The four clock genes peaked during daytime at 0–1 dpf, and a phase adjustment of the rhythms was observed in the following days depending on the photoperiod applied (Fig. 2). Under LD conditions, rhythms in *per1* and *per3* were gradually phase-advanced each day to reach their characteristic profile of negative elements of the clock mechanism at 3–4 dpf, with the acrophases (peak time) placed anticipating lights on (Fig. 2). At 2–3 dpf, *per3* daily rhythm was transitorily lost, but robust rhythmic oscillation was observed again by the next day (3-4 dpf). A progressive increment in overall expression levels and cycling amplitudes was also detected throughout development (Fig. 2). *Per2* was actively and rhythmically expressed from 0–1 to 3–4 dpf as well, with the acrophase of the rhythm always maintained during the light phase, few hours after lights on (Martín-Robles et al. 2012b, Fig. 2), which reinforces that *per2* is a light-inducible gene (Vatine et al. 2009). Interestingly, amplitudes and mesors were particularly increased during two specific moments of sole development, from 0-1 to 1-2 and from 2–3 to 3–4 dpf (see scales in the graphs, Fig. 2). In the case of *clock*, daily expression rhythms were gradually phase-delayed and although they were transiently lost at 1–2 dpf, by 2–3 dpf rhythmicity was recovered and maintained with robust cyclic oscillations at 3–4 dpf, peaking during the end of the light phase (Martín-Robles et al. 2012b, Fig. 2), which is consistent with its putative role as positive element of the clock in mammals and fish (Pando and Sassone-Corsi 2002). The transient absence of rhythmicity of some clock components, such as *clock* and *per3*, at particular ontogenetic stages (1–2 and 2–3 dpf, respectively) could be related to their resetting and suggests the differential regulation of

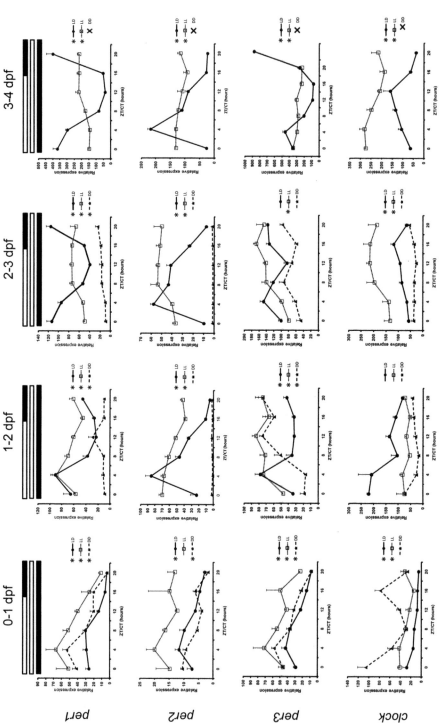

Fig. 2. Developmental onset of the circadian clock in the Senegalese sole. Relative mRNA expression of *per1*, *per2*, *per3* and *clock* during early development of Senegalese sole, from 0 to 4 days post fertilization (dpf) is shown. Fertilized eggs and larvae were maintained under LD (14h L:10 h D), LL and DD conditions by means of a photoperiod control system placed in each tank. Samples were taken every 4 h and relative expression was quantified by Real-Time quantitative PCR (RT-qPCR). Each value represents the mean ± SEM (n = 5). The hours in the x-axis represent zeitgeber time (ZT) in LD conditions and circadian time (CT) in LL and DD conditions. Photoperiod conditions are indicated by white and black bars above each graph, white bars representing light periods and black bars representing periods of darkness. The asterisks in the legends indicate significant daily rhythms by cosinor analysis (p < 0.05) and the cross in DD conditions at 3–4 dpf denotes mortality of the larvae under this condition. Modified from Martín-Robles et al. 2012b.

core circadian clock elements during sole development. In any case, clock gene expression patterns and acrophases at 3–4 dpf already resembled adult neural profiles (Martín-Robles et al. 2011, 2012a).

When compared to other fish species, the expression pattern of *per1* in sole larvae under LD conditions resembles that described in zebrafish, peaking during the early light phase in the first 2 days of development and just prior to lights on as the embryo develops (Dekens and Whitmore 2008, Vallone et al. 2004). In rainbow trout, peak expression of *per1* also occurred just before lights on in embryos raised in LD conditions, but in contrast to zebrafish and sole, no significant variations were detected on the first day of development, which might be related to differences in embryonic development between species (Davie et al. 2011). Moreover, sole *per1* profile is comparable to medaka *per1b* expression in developing larvae. In this species, *per1b* rhythmicity was detected at 2–3 dpf, although previous developmental stages were not tested (Cuesta et al. 2014).

Early expression of sole *per2* under LD cycles was also similar to that reported in zebrafish, where this gene is rhythmically expressed on the first day of development and peaks during the light phase (Dekens and Whitmore 2008, Lahiri et al. 2005, Ziv et al. 2005, Di Rosa et al. 2015). However, we found rhythmic *per3* expression from 0 to 2 dpf and at 3–4 dpf, while exposure of the zebrafish embryos to five-six solid LD cycles is needed for the establishment of robust *per3* significant oscillations (Kaneko and Cahill, 2005), although other studies detected *per3* rhythms earlier during zebrafish development (Delaunay et al. 2000, 2003). Results obtained from medaka larvae on *per3* expression also suggest earlier oscillation of this gene as statistical differences between peak and through daily points were observed from 2–3 dpf (Cuesta et al. 2014). Sole *per3* profile at 3–4 dpf was comparable to that documented in gilthead seabream larvae (Mata-Sotres et al. 2015).

Regarding the onset of *clock* rhythmicity, these profiles were similar to those found in rainbow trout, in which *clock* is rhythmically expressed during the first day of development, although the acrophase was different from sole because, similar to *per1*, it was placed just before lights on (Davie et al. 2011). *Clock* expression profile at 3–4 dpf was comparable with the expression pattern found in gilthead seabream larvae as well (Mata-Sotres et al. 2015). By contrast, the expression of the transcriptional activator in Senegalese sole was distinct to that reported in developing zebrafish and medaka embryos subjected to LD cycles, where rhythmic *clock* gene expression emerged later during development, being constant until the fourth day of development in zebrafish and until 12–14 dpf in medaka larvae (Dekens and Whitmore 2008, Cuesta et al. 2014, Di Rosa et al. 2015). This is also seen in *bmal1* expression, suggesting that genes from the positive loop of the molecular clock could need longer time to start oscillating in these species and that different components of the molecular clock mechanism mature at slightly different times during early development, although it is unknown which mechanisms are responsible for delaying the appearance of *clock1* and *bmal1* rhythms compared with *pers* (Cuesta et al. 2014, Di Rosa et al. 2015). Interestingly, this later emergence of *clock* and *bmal* rhythms has been recently related with both the developmental stage and the time of hatching (Cuesta et al. 2014). Therefore, it seems that rhythmicity of some clock genes in sole larvae is established even earlier than in zebrafish and medaka, the maturation of the circadian clock occurring over a less extended period of time, although a robust circadian oscillator is only present at the 4th day of development in both zebrafish and sole. The role of the developmental stage and hatching time as factors contributing to the time emergence of clock gene rhythmicity will require further research in sole.

The pineal organ of sole contains immunoreactive opsins from 2 dpf, whereas retinal photopigments are detected later at 3 dpf (El M'Rabet et al. 2008, Bejarano-Escobar et al.

2010). Thus, sole embryos are light sensitive and exhibit a functional circadian clock before the differentiation of any specialized photoreceptive structures, as it has been shown in zebrafish and rainbow trout (Davie et al. 2011, Dekens and Whitmore 2008), which predicts that non-pineal, non-retinal photoreceptors are present at these early developmental stages. Interestingly, it has been recently shown that an extensive range of photopigments are widely expressed in zebrafish (Davies et al. 2015). Overall expression levels and cycling amplitudes increased with age in sole, suggesting that more cells start to express clock genes with high-amplitude oscillations. Whether the appearance of functional photoreceptive organs at this developmental stage contributes to the significant raise in the amplitude of all clock genes studied and in resetting the phase of these rhythms remains to be investigated in sole. In this sense, it was interesting that both mesor and amplitude of *per2* were increased from 0–1 to 1–2 dpf and from 2–3 to 3–4 dpf, which roughly corresponds to the onset of photosensitivity in the pineal gland and the retina of sole, respectively (Martín-Robles et al. 2012b).

The gradual entrainment described above under cycling conditions of light and dark, permitting to develop a mature circadian oscillator, was markedly affected by both LL and DD (Fig. 2). In developing sole embryos reared under LL conditions, rhythmic oscillations in the expression of the four clock genes persisted during the experiment, but an important reduction of rhythm amplitudes was manifest, lacking the robustness found under light-dark conditions (Martín-Robles et al. 2012b, Fig. 2). Moreover, the expression patterns were notably different than those observed under LD cycles, being acrophases displaced several hours, so that the characteristic neural profile of the four clock genes was not accomplished. On the other hand, the effects of constant darkness (DD) in the establishment and organization of the molecular clock mechanism and embryo survival were quite dramatic. Rhythmicity was rapidly lost from 0–1 dpf onwards, being *per2* and *clock* daily rhythms the most drastically affected (Fig. 2). Consistent with previous studies, prolonged DD exposure also compromised the future survival of developing sole and led to an elevated mortality rate of larvae compared to LD or LL photoperiods. Under DD, Senegalese sole showed the lowest hatching rate and the larvae that were able to survive showed the poorest performance and never achieved the metamorphic process (Blanco-Vives et al. 2010, 2011a, 2012). This mortality begins before the onset of the external feeding indicating that this phenomenon is not only a consequence of a reduced ingestion rate under DD conditions. This is also evident in other fish species such as zebrafish and European sea bass, where photoperiod effects on larval performance, survival and the occurrence of malformations have been also examined (Villamizar et al. 2009, 2014). Therefore, high-amplitude rhythms in the four clock genes and similar in phase to those observed in adult neural tissues (Martín-Robles et al. 2011, 2012a) were only achieved under LD conditions, at 3–4 dpf.

This oscillator may sustain some early clock outputs mentioned before in sole, such as hatching, expression of *aanat* genes, and locomotor activity (Blanco-Vives et al. 2011a, 2012, Isorna et al. 2009a, 2011). Interestingly, Senegalese sole larvae showed a marked rhythmic activity pattern only under LD cycles and from 3 days post hatching (dph) onwards, i.e., from 4 dpf (Blanco-Vives et al. 2012). In addition, the onset of external feeding in sole also occurs at 4 dpf (Cañavate et al. 2006, Blanco-Vives et al. 2012). This timing roughly coincides with the complete organization of molecular clock rhythms in this species, at least for *per1*, *per2*, *per3*, and *clock* genes. The general reduction of amplitudes or loss of circadian rhythmicity under exposure to constant conditions could be at the base of the absence of clear daily locomotor activity rhythms observed in sole maintained under these constant regimes (Blanco-Vives et al. 2012). However, some of these events such as hatching and rhythmic expression of *aanat* genes take place before 3–4 dpf and the complete organization of the molecular loops at the whole embryo level. We cannot discard that particular neural areas in the embryo show a complete organization of the molecular loops before 3–4 dpf, that we were not able to detect at the whole embryo level. These neural areas could be directing some early rhythms in sole. In this sense, the pineal organ has been shown to contain a circadian clock in fish (Cahill 1996, Menaker et al. 1997, Falcón et al. 2010). It is the first photoreceptive structure differentiated during sole development and has been proposed to mediate the light effects on the timing of hatching in sole (Isorna et al. 2009a, Blanco-Vives et al. 2011). *aanat2* expression is detected as early as 12 h post fertilization (0 dpf), and this expression increases between 0 and 2–4 dpf, showing day-night variations from 2 dpf, with lower expression during the day and higher expression at night (Isorna et al. 2009a). This could suggest that the sole pineal clock is already functional at 2 dpf. Alternatively, the appearance of these early clock outputs would not require the complete organization of the molecular clock mechanism.

It is known that changes in environmental cues, such as light or temperature, during the first day of development are required for the appearance of circadian clock outputs. Zebrafish embryos raised in constant darkness fail to show robust locomotor activity rhythms, as well as circadian rhythms in cell cycle timing and melatonin synthesis (Kazimi and Cahill 1999, Hurd and Cahill 2002, Dekens et al. 2003, Di Rosa et al. 2005). A reduction in the number of LD cycles results in a corresponding decrease in the amplitude of the subsequent free-running rhythms of behaviour (Hurd and Cahill 2002). However, a different effect of LL and DD conditions was found on the sole molecular clock. As mentioned before, rhythmicity on DD conditions was rapidly lost in most clock genes analysed, while rhythmic oscillation with lower amplitude was detected from 0-1 dpf under continuous illumination (Martín-Robles et al. 2012b). Despite the same behavioural outcome, different mechanisms might explain these results in DD and LL. On one hand, it has been recently demonstrated that under constant darkness, individual clocks are functioning at single cell level, but the oscillations are asynchronous so rhythms cannot be detected at the whole animal level (Dekens and Whitmore 2008). The effect of exposure to environmental cue such as a light-dark cycle is to reset single cell clocks, leading to synchronized oscillations within the embryo. Asynchrony of cellular oscillations could explain the lack of rhythmicity in zebrafish clock outputs maintained under DD conditions (Dekens and Whitmore 2008, Hurd and Cahill 2002). Whether the loss of rhythmicity in the subsequent days is due to the loss of gene expression rhythms or to the desynchronization between individual cellular clocks remains to be elucidated in sole. On the other hand, prolonged light treatment has been demonstrated to stop or dramatically reduce the amplitude of the circadian oscillator, through a sustained increase in *cry1a* levels that act as a strong light-induced repressor of clock function (Tamai et al. 2007, Vallone et al. 2004). Accordingly, in our LL experiment, significant daily rhythms were maintained, but amplitudes of all clock genes studied were much lower than in the LD conditions. The pronounced attenuation or loss of rhythms observed in constant conditions supports the significance of early exposure to LD transitions to entrain the embryonic circadian clock and maintain strong circadian rhythms.

As a whole, this work revealed the onset of the Senegalese sole circadian clock at the first day of development. This embryonic clock required LD exposure to gradually entrain clock gene expression, which reflects the successive appearance of sole key clock outputs. In addition, our results confirmed the significance of cycling LD conditions when raising fish embryos and larvae, as daily rhythms of clock gene expression were strongly attenuated and/or loss when constant light or dark conditions were applied.

6. The molecular clock during Senegalese sole metamorphosis

In the course of their development, flatfish species as Senegalese sole undergo a real metamorphic process that represents a dramatic transition from symmetric pelagic larva to an asymmetric benthic juvenile and involves tissue differentiation, biochemical, molecular, and physiological changes (Fernández-Díaz et al. 2001, Parra and Yúfera 2001, Power et al. 2001, Isorna et al. 2009b, Bejarano-Escobar et al. 2010). Metamorphosis drastically modifies the anatomy of Pleuronectiform species, implying eye migration towards the upper-pigmented side of the body (Policansky 1982, Fernández-Díaz et al. 2001) and craniofacial remodelling, which determines their characteristic juvenile and adult asymmetry (Brewster 1987, Wagemans et al. 1998, Rodríguez-Gómez et al. 2000a, 2000b, Okada et al. 2001, Schreiber 2006). This transformation is accompanied by changes in their habitats, where they come across markedly different environmental light conditions as pre-metamorphic pelagic larvae swimming in the water column develop into benthic juveniles living on the bottom of the sea. In fact, adaptive mechanisms intended to enhance light sensitivity have been described in Senegalese sole photoreceptive structures such as the pineal gland (Confente et al. 2008).

Of particular interest in this species is that a switch from diurnal to nocturnal in locomotor activity rhythms and feeding behaviour precedes this rearrangement of the body plan (Cañavate et al. 2006, Blanco-Vives et al. 2012). Studies on animals that show the ability to switch their chronotypes are of special interest in deciphering the mechanisms responsible for temporal preference (Sánchez-Vázquez et al. 1995, 1996, Reebs 2002, Vivanco et al. 2010). Once we had established the emergence of the sole circadian clock during early development, and evidenced that molecular oscillations were fully organized at 4 dpf, we considered exploring whether the mechanism was maintained during this crucial stage and whether changes in the molecular clock reflected the switch in sole chronotype. This was not found to be the case. Moreover, we investigated the effects of transitory constant light or dark conditions in *per1*, *per2*, *per3* and *clock* daily rhythms. For this purpose, developing Senegalese sole were reared from the day of fertilization (0 dpf) to 24 dpf (when metamorphosis was completed) under three different experimental conditions. In the first experimental group, larvae were maintained under LD 14:10 from 0 to 24 dpf (LD group). The two remaining groups were reared under LD from 0 to 9 dpf, followed by transient constant light (LL-LD group) or dark (DD-LD group) conditions from 10 to 14 dpf and then they were returned to LD conditions from 15 to 24 dpf. Diel clock gene expression was determined during pre-, early-, middle-, and post-metamorphic stages. This represented the first study to describe molecular clock rhythms during crucial developmental phases such as metamorphosis in vertebrates (Martín-Robles et al. 2013).

The four components of the sole clock analysed showed robust expression rhythms along the metamorphic process, although with declining amplitudes and expression levels (Fig. 3). They displayed the usual profiles described for the negative and positive elements of the fish circadian system, where acrophases of *per1* and per3 were placed at dawn, *per2* during the first half of the day and *clock* at the second half of the light phase (Whitmore et al. 1998a, Zhdanova et al. 2008, Park et al. 2007, Velarde et al. 2009, Sánchez et al. 2010, López-Patiño et al. 2011). These rhythms could sustain the clear rhythmic pattern of locomotor activity observed in sole during the metamorphic process (Blanco-Vives et al. 2012). All phases were maintained during pre-, early-, middle-, and post-metamorphic stages and were clearly similar to those found earlier at 4 dpf and in adult neural tissues (Martín-Robles et al. 2011, 2012a, 2012b), indicating that molecular clock oscillations are established very early during development, are preserved during sole metamorphosis and

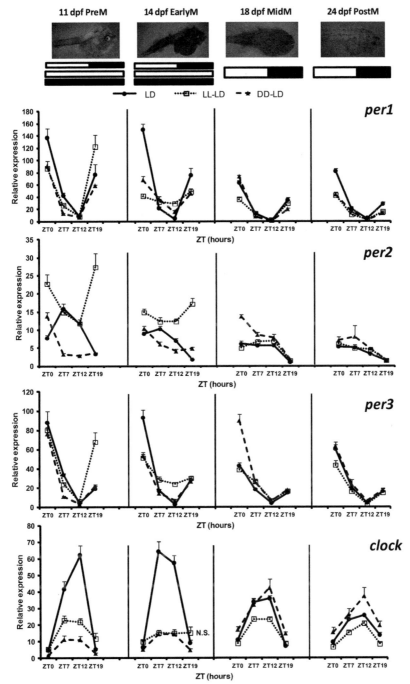

Fig. 3. Relative expression of molecular clock components throughout Senegalese sole metamorphosis. The LD group was maintained under LD (14h L:10h D) all over the experiment, and transient constant conditions of light or dark were applied from 10 to 14 dpf in the LL-LD and DD-LD groups, respectively. Larvae and post-larvae were sampled at ZT/CT 0, 7, 12 and 19 during four developmental stages of the metamorphic process, coinciding with pre- (11 dpf), early- (14 dpf), middle- (18), and post- (24 dpf) metamorphosis. The bars above the graphs indicate the photoperiod conditions within a given 24 h period, white bars representing light phases and black bars representing phases of darkness. Relative expression was quantified by RT-qPCR. Each value represents the mean ± SEM (n = 5). All profiles were significantly rhythmic (cosinor analysis p < 0.05) with the exception of *clock* during 14 dpf in the LL-LD group. N.S: non-significant rhythm. Modified from Martín-Robles et al. 2013.

retained until adult life. Besides, there is no reversal in the phase of clock gene rhythms between pre- and post-metamorphic animals that would be coincident with the switch from diurnal to nocturnal locomotor activity. This fact provides evidence suggesting that the key for diurnality/nocturnality does not rely upon changes on the molecular clock, at least in the genes currently studied. However, specialized central pacemakers could dictate the phase of locomotor activity, which cannot be visualized in whole animal RNA extracts. Otherwise, it could arise from mechanisms outside the core clock that could operate upstream, downstream from the neural oscillators, or both, i.e., in the light input pathway or in the neural/humoral output connecting to the physiological and behavioural rhythmic processes (Mrosovsky and Hattar 2005).

An important decrease in clock gene amplitudes and mesors was detected as Senegalese sole metamorphosis proceeds, which is particularly evident from middle-metamorphosis for *per1*, *per3* and *clock* and from early-metamorphosis for sole *per2* (see graphs profiles, Fig. 3). It is unknown whether the metamorphic process itself may have an effect on expression levels of clock genes or this decline is associated with changes in tissue-dependent zeitgebers and reflects a reorganization of the circadian clock in different tissues. We have observed this pattern of expression in many other genes in sole, with peaks during early development at 4 dpf and a drop later during metamorphosis (Isorna et al. 2009a, 2011, Lan-Chow-Wing et al. 2014). These changes could be mediated by the thyroid system, which is up regulated (hormone levels, receptor expression, deiodinase expression and activity) during sole metamorphosis and promotes the morphological, molecular, and physiological changes that take place during metamorphosis in fish (Power et al. 2001, Marchand et al. 2004, Isorna et al. 2009b). Alternatively, the phase of the rhythms in different tissues could vary throughout sole development resulting from synchronization to different tissue-dependent zeitgebers, which could have led to a decrease in amplitudes when whole animals are used. The shift from diurnal to nocturnal preference of sole feeding behaviour could account for these possible phase differences. In fact, peripheral (non-neural) tissues can be entrained by food-related cues in fish, affecting the phases of clock genes in these organs (López-Olmeda et al. 2010, Feliciano et al. 2011, Nisembaum et al. 2012). In sole adult (nocturnal) specimens, clock gene rhythms in central tissues are different in phase in relation to peripheral tissues such as liver, also suggesting different entraining cues (light vs. feeding time) (Martín-Robles et al. 2011, 2012a). Further research at tissue or organ level is needed in sole metamorphic larvae as well as studies in other metamorphic species.

During exposure to transient constant light and dark conditions, *per1*, *per2*, *per3* and *clock* all have distinct responses, which predominantly affected *per1*, *per3* and *clock* amplitudes and *per2* phase (Martín-Robles et al. 2013), supporting that the three *per* genes have different regulatory mechanisms (Fig. 3). Whereas *per1* and *per3* are predominantly clock-regulated, endogenously driven, *per2* expression is regulated by light (Shearman et al. 1997, Zylka et al. 1998, Zhuang et al. 2000, Pando et al. 2001, Okabayashi et al. 2003, Besharse et al. 2004, Ziv et al. 2005). Upon release into DD and after 5 days in these conditions (DD-LD group), rhythmic expression consistent with the previous LD cycles persisted in *per1*, *per3*, and *clock*, demonstrating its circadian nature (Whitmore et al. 2000, Pando et al. 2001). Amplitudes were reduced but the phase of the rhythms was maintained, i.e., acrophases did not vary considerably in relation to the LD group. In contrast, *per2* daily expression was revealed as rhythmic but 6-h phase advanced in relation to the LD group, peaking during the subjective dawn (Martín-Robles et al. 2013). Under transient LL conditions (LL-LD group), clock gene expression rhythms were much more affected, supporting that sustained light treatment markedly reduces clock function (Tamai et al. 2007). Amplitudes of *per1* and *per3* were significantly reduced after 5 days of continuous light, but their phases were almost unaltered. *Clock* daily rhythm was lost at 14 dpf, after 5 days in constant light, that could be associated to the decrease in the amplitude of *pers* oscillations. On the contrary, *per2* phase was clearly disrupted, with peak expression falling into the subjective night. *Per2* was also up regulated under these conditions. Although *per2* phase was remarkably affected by lighting conditions, a rhythmic expression was maintained either after 5 days under DD and LL transient regimes (Fig. 3). As an element of the light input pathway, and together with *cry1a*, *per2* has been proposed to play key roles in circadian entrainment in fish (Tamai et al. 2004, 2007, Vatine et al. 2009). This evidence could suggest a clock-regulated control of Senegalese sole *per2* in the absence of cycling conditions. In contrast, in zebrafish cell lines and transgenic embryos, *per2* expression ceases immediately to oscillate and remains constant following transfer to DD (Pando et al. 2001, Vatine et al. 2009). However, under LL conditions, oscillations in *per2* are maintained in the transgenic embryos, suggesting also both light- and clock-regulated transcriptional control of *per2* in this species (Vatine et al. 2009). Whether these differences in *per2* rhythmic expression under constant conditions represent real species-specific differences or a consequence of the different experimental approaches and sensitivity of the techniques used remain to be elucidated. Additionally, other environmental cues could be sustaining *per2* rhythms under constant conditions in sole.

Perhaps the most surprising finding of this study was the effects of transient constant light conditions on the molecular clock. As in our experiment carried out during sole early development, LL conditions provoked a reduction in the amplitudes of the rhythms in all clock genes. However, although this time we reinstated the cycling conditions, a long-lasting effect was observed. Three day after the restoration of the LD conditions, *per1* and *clock* amplitudes were not as robust as those observed for the LD group, and even after 9 days under LD cycles (24 dpf) *per1* amplitude was still lower than in the LD group (Fig. 3). In addition, significant differences were found in the last sampling day at 24 dpf between different photoperiods in both *per1* and *per3*. It is unknown how many light-dark cycles need the sole clock to develop high amplitude rhythms after constant lighting conditions and what is the effect of a low-amplitude running clock at the animal level. However, this evidence could have a practical interest because transient constant light conditions are commonly used in sole aquaculture to accelerate growth and development. Our data highlight that constant photoperiod regimes, especially continuous illumination, even applied for a short period of time have a prolonged effect on the molecular clock of sole. Consequently, outputs directly regulated by the clock might be altered and this fact should be taken into account to improve aquaculture practice and welfare in this species. The present results, together with recent studies performed in this species (Blanco-Vives et al. 2010, 2012), emphasize the high sensitivity of the sole circadian system to light cues and also underline the important role of maintaining LD cycles for the proper development of sole larvae.

To summarize the work done, our results confirmed that clock gene rhythms were maintained throughout Senegalese sole metamorphosis with similar phase, providing evidence that molecular rhythms are established early during development and preserved during the diurnal-nocturnal switching period. However, a general decline of amplitudes and expression levels was observed during this hallmark process that could be related to changes in tissue-specific clock gene expression patterns. In addition, our results revealed the differential sensitivity of sole clock components to transitory light cues and reinforced the role of *per2* in light entrainment. These findings also point out that the rearing of

Senegalese sole larvae in aquaculture facilities should consider the prolonged effect that transient light changes applied during this process have in the molecular clock machinery.

7. Characterization of the molecular clock in central and peripheral tissues of adult Senegalese sole

In the last part of this chapter, we will end up describing recent knowledge on the organization of the molecular clock in adult sole. For technical reasons, in developmental studies we quantified clock gene expression in whole embryo and/or larvae body because of the difficulty involved in precisely dissecting particular tissues during early ontogenetic stages. The use of sole adult specimens allowed us to dissect central and peripheral (non-neural) tissues and studied the temporal expression patterns of clock genes under light-dark cycles (Martín-Robles et al. 2011, 2012a). In a preliminary study, we analysed the tissue distribution of sole clock genes in several central regions (telencephalon, diencephalon, mesencephalon, metencephalon, rhombencephalon and retina) and peripheral tissues (gills, heart, liver, kidney, intestine and ovary). We found a ubiquitous spatial expression pattern, as clock genes transcripts were detected in almost all neural and peripheral tissues tested. This was consistent with all other species in which clock gene expression have been examined, from insects to mammals, demonstrating that they are not restricted to central pacemaker structures but are widely expressed even outside the brain, reinforcing that sole may also have multiple clocks dispersed through the body (Shearman et al. 1997, Whitmore et al. 1998a).

The presence of clock genes in many tissues turned radically around classical view of circadian oscillators, as for a long time it was believed that circadian rhythms were generated in specialized neural structures such as the suprachiasmatic nucleus (SCN) of mammals, the pineal gland of amphibians and fish, or the optic lobe of insects. Various tissues all over the body of *Drosophila*, fish and mammals have been shown to express clock genes rhythmically (Plautz et al. 1997, Yamazaki et al. 2000, Yoo et al. 2004, Velarde et al. 2009, Kaneko et al. 2006). However, the oscillatory properties differ among tissues and the SCN acts as a central pacemaker, which is mainly reset by light and orchestrates all other oscillators to coordinate daily temporal organization of physiology and behaviour (Reppert and Weaver 2001). This hierarchic organization of the circadian system has not been found to date in fish, and the location of a central pacemaker or even its presence is controversial. Although the SCN is an anatomically well-defined structure in the fish brain, its potential function as a central pacemaker in the circadian system is unknown. A recent study showed that multiple adult brain regions contained endogenous oscillators that were directly light responsive, revealing a high degree of pacemaker complexity within fish brain. Some of these brain regions include the teleost equivalent of the SCN, as well as numerous hypothalamic nuclei, the periventricular grey zone of the optic tectum and granular cells of the rhombencephalon (Moore and Whitmore 2014).

We approached the study on the temporal expression pattern of sole clock genes in neural areas and tissues where we found the most remarkable expression of clock genes, i.e., the retina, mesencephalic optic tectum/tegmentum, diencephalon, cerebellum/rhombencephalon and, regarding peripheral tissues, the liver (Figs. 4 and 5). We also selected these neural regions because they have been implicated in the processing of visual/light information, they express melatonin receptors and exhibits significant day/night differences in the density of melatonin binding sites in sole (Oliveira et al. 2008, Confente et al. 2010). Moreover, the diencephalon harbours the teleost equivalent of the SCN and it has been considered that the teleost cerebellum serves visual, motor learning

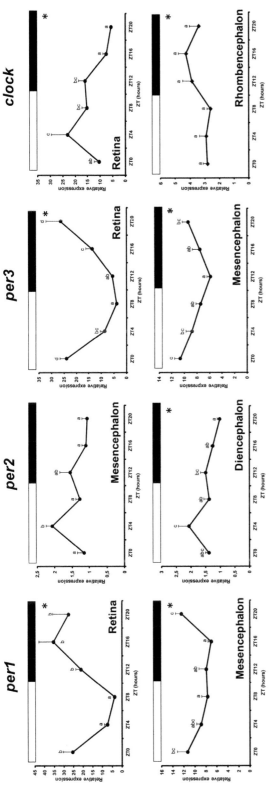

Fig. 4. Daily profiles of molecular clock components in central tissues of adult Senegalese sole. Relative expression was analysed by RT-qPCR in retina, diencephalon, mesencephalon and cerebellum/rhombencephalon. Only genes and tissues showing significant daily variations revealed by one-way ANOVA and/or rhythmic profiles (cosinor analysis p < 0.05) are displayed, being the later indicated by an asterisk. Animals were reared in indoor facilities receiving natural environmental light (12 h light:12 h darkness). Samples were taken every 4 h and each value represents the mean ± SEM of six different specimens (n = 6). The bars above each graph indicate the daily photoperiod conditions. White bars represent the light phase and black bars represent phases of darkness. Different letters indicate statistically significant differences between mean values (p < 0.05). Mesencephalon: optic tectum and tegmentum. Rhombencephalon: cerebellum, vestibulolateral lobe and medulla oblongata. Modified from Martín-Robles et al. 2011, 2012a.

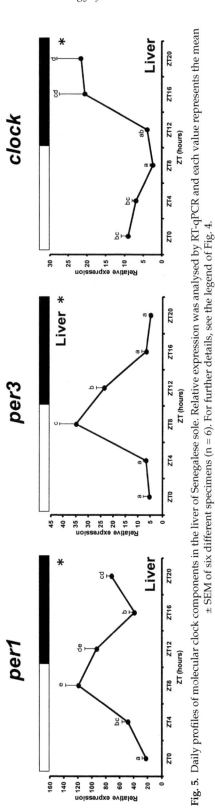

Fig. 5. Daily profiles of molecular clock components in the liver of Senegalese sole. Relative expression was analysed by RT-qPCR and each value represents the mean ± SEM of six different specimens (n = 6). For further details, see the legend of Fig. 4.

and coordination functions (Finger 1983, Wullimann 1998). The retina of most vertebrate species is also a remarkably rhythmic tissue containing a circadian clock that regulates daily retinal physiology and photoreceptor function (Iuvone et al. 2005). On the other hand, an oscillator entrained by feeding cues has been reported in the liver of mammals and fish (Stokkan et al. 2001, López-Olmeda et al. 2010, Cavallari et al. 2011, Feliciano et al. 2011, Costa et al. 2016). Diel expression profiles of the four clock genes analysed in adult sole were quite similar to those recently described in several fish species, but presented more divergence with those documented in amphibian or mammals (Kamphuis et al. 2005, Peirson et al. 2006, Zhuang et al. 2000). Our results indicated differences in phase depending on the tissue analysed, being the retina and the liver the most rhythmic tissues. In sole retina, *per1*, *per3* and *clock* displayed significant cyclic oscillations over a 24 h cycle, as well as *per1* and *per3* in the mesencephalic optic tectum/tegmentum (Martín-Robles et al. 2011, 2012a). Mesencephalic *per1* and *per3* expression resembled *per1* and *per3* daily profiles observed in the retina, although presenting a 2–3 h phase delay in relation to the photoreceptive organ (Fig. 4). In the diencephalon, only *per2* transcripts displayed daily rhythms, its expression being higher during daytime, and in the case of the cerebellum/rhombencephalon, rhythmic *clock* expression was detected (Fig. 4). Concerning the liver, this tissue showed daily rhythms in *per1*, *per3* and *clock* expression (Fig. 5).

The expression patterns of *pers* and *clock* genes in neural tissues are consistent with their roles as negative and positive elements of the molecular clock mechanism, respectively. Highest mRNA levels of *per1* preceded those of *per3* and both genes anticipated dawn, as were up regulated before sunrise, showing their peaks during the end of the night. In turn, *per2* acrophase was found during the first half of the day and *clock* displayed an inverted patter in relation to *per1* and *per3*, peaking during daytime (Fig. 4). Although the phase of these rhythms was quite similar in visually-related neural structures such as the retina and the mesencephalic optic tectum, there was a significant difference in their amplitudes, being considerably higher in the former than in the latter. These differences may be probably caused by the endogenous properties of their oscillators, differences in photosensitivity and/or phototransduction, in the coupling among cells or in periods, phases and/or amplitudes of cells present. These patterns were already seen in sole at 3–4 dpf (Martín-Robles et al. 2012b), and were comparable to the expression profiles characterized in the retina or brain of several fish species such as the golden rabbitfish, goldfish, rainbow trout, zebrafish, European seabass or Nile tilapia (Park et al. 2007, Velarde et al. 2009, López-Olmeda et al. 2010, Sánchez et al. 2010, López-Patiño et al. 2011, Costa et al. 2016).

Despite the expression pattern of the analysed clock genes is well conserved, we found some peculiarities concerning retinal rhythms in sole. One of them is that surprisingly, *per2* expression was arrhythmic in sole retina, although a daily rhythm of this gene has been described in the retina of most species studied including teleost fish (Yoshimura et al. 2000, Zhuang et al. 2000, Kamphuis et al. 2005, Chaurasia et al. 2006, Velarde et al. 2009). As we have seen along our experiments, transcript levels of *per2* change rapidly in response to variations in light conditions and, as mentioned before, a role of this clock gene in phase resetting by light has been demonstrated (Pando et al. 2001). Sole retina did not show evident daily rhythms in melatonin secretion either, but differences in melatonin content were found between both eyes (Bayarri et al. 2004). The lack of *per2* rhythmicity in sole retina could also be associated to differences in clock gene expression between left and right eyes or, alternatively, to shifted phases of *per2* expression in different retinal cell types (Tosini et al. 2008). In addition, it could constitute a specific feature of the retina of sole associated with seasonal variations in *per2* retinal rhythms and/or to the benthic habits of sole, which determines differences in light exposition (light intensity and/or spectrum) in relation to pelagic species. The other peculiarity of sole retina was observed in *clock* expression. Its phase was slightly advanced in comparison with other teleost species, peaking during midday, while minimum expression values were reached at night (Fig. 4). In the rainbow trout and zebrafish, retinal *clock* showed a major peak of expression at the end of the light phase and the onset of the night (Whitmore et al. 1998a, Zhdanova et al. 2008, López-Patiño et al. 2011). Moreover, *clock* is constitutively expressed in the retina of quail (*Coturnix japonica*) and *Xenopus* (Yoshimura et al. 2000, Zhu et al. 2000), but contradictory results have been obtained in both avian and mammalian species depending on the techniques used and the species studied (Iuvone et al. 2005, Helfer et al. 2006). The regulatory mechanisms responsible for these differences in *clock* mRNA levels in the retina and their significance on the circadian clock function are unknown. Taken together, our results pointed to interspecies differences in circadian organization of the vertebrate retina, as it has been suggested before (Iuvone et al. 2010).

It is interesting to note that the sole diencephalon includes the homologous to the mammalian SCN of fish (Springer and Mednick 1984, Yáñez et al. 2009), and this nucleus represents the central pacemaker in mammals and shows high amplitude rhythms of clock gene expression (Klein et al. 1991). However, we only detected significant daily rhythms in *per2* expression, with maximum values during daytime. There is the possibility of different phase relationships of gene expression in certain diencephalic nuclei, which would have been missed when using the entire diencephalon. Therefore, it is also possible that a pacemaker structure exists in the sole hypothalamus, but it remains masked when this region is processed as a whole. As we indicated before, a brain area with similar pacemaker properties as mammalian SCN has not been yet found in fish and whether the circadian system in fish functions in a hierarchical manner is still a matter of controversy. In this sense, recent elegant studies performed in fish mapping in detail clock gene expression in brain, demonstrated that multiple brain nuclei express clock genes and show robust day-night variations, including the teleostean SCN, although with no differences in labelling compared to other brain regions (Moore and Whitmore 2014, Sánchez-Bretaño et al. 2015a). Consequently, it has been suggested that the fish circadian system is not dominated by a central structure, and that the complexity on the distribution of potential circadian oscillators in the brain is much higher than initially thought. The brain areas analysed in sole represent quite large neural regions and so, a more precise anatomical approach appears necessary to interpret the presence or the lack of rhythmicity of some clock genes and to explore in more detail whether discreet areas of the brain might contain circadian clocks in sole. Nowadays, detailed reports on clock genes-expressing cells in fish brain are very scarce, and precise information is limited to rainbow trout, flounder, zebrafish and goldfish (Mazurais et al. 2000, Watanabe et al. 2012, Weger et al. 2013, Moore and Whitmore 2014, Sánchez-Bretaño et al. 2015a). Studies on clock outputs carried out in animals with lesions in certain brain nuclei could definitively constitute a valuable approach to identify putative pacemaker areas similar to the mammalian SCN. Furthermore, whether diencephalic expression of *per2* is present in putative deep brain photoreceptors or in cells receiving direct retinal/pineal projections remains to be investigated.

In our preliminary tissue distribution study, we detected an intense signal in sole hindbrain and temporal expression of clock genes in cerebellum/rhombencephalon revealed a 24 h-rhythm in *clock* mRNA levels. The cerebellum represents one of the brain regions where clock genes rhythms have been previously documented in mammals and fish (Akiyama et al. 1999, Namihira et al. 1999, Mazurais et al. 2000, Farnell et al. 2008, Weger et al. 2013, Moore and Whitmore 2014, Sánchez-Bretaño et al. 2015a). Moreover, it has been reported that clock gene oscillations in mammalian cerebellum can be shifted in response to restricted feeding, supporting the presence of a circadian oscillator sensitive to feeding cues in this area. The cerebellum has also been related with the development of food anticipatory activity, a pronounced increase in locomotor activity prior to food access, which is markedly reduced when depleting Purkinje cells (Mendoza et al. 2010). In this regard, it is interesting to note that the phase of the rhythm of sole *clock* in the cerebellum differed with the pattern found in other neural tissues, resembling the profile observed in a peripheral tissue such as the liver, with the acrophase placed during the first half of the night. This evidence could suggest for the first time in fish, a role of the cerebellum as an oscillator sensitive to food-related cues.

Together with the retina, the liver is one of the tissues that exhibited more robust rhythms in sole, showing significant cyclic oscillations for *per1, per3* and *clock* (Fig. 5). In all cases, an antiphase expression profile was revealed in sole liver in relation to neural tissues, with a 10-12 h phase shift compared to mesencephalic and retinal rhythms. The presence of circadian oscillators outside the central nervous system and neural structures is a common feature of both invertebrates and vertebrates (Plautz et al. 1997, Whitmore et al. 1998a, Yamazaki et al. 2000). The liver plays a key role in the regulation of metabolism and is one of the most studied peripheral clocks, being highly sensitive to different internal and external cues such as feeding time. Rhythmic clock gene expression has already been reported in the liver of mammals and fish (Yamamoto et al. 2004, Yoo et al. 2004, Kaneko et al. 2006, Velarde et al. 2009, López-Olmeda et al. 2010, Costa et al. 2016). The phase shift difference found in sole suggests an uncoupling of the peripheral clock from neural tissues and the entrainment of the liver oscillator to cues different from light, highlighting a role of feeding time. Furthermore, these results suggest that the liver of sole is not directly light responsive, contrary to that reported in zebrafish (Whitmore et al. 2000). The situation in Senegalese sole appears similar to that found in nocturnal mammals. Regardless of lighting conditions, daytime restricted feeding of nocturnal rodents completely inverted the phase of circadian gene expression in many tissues including liver, where food-induced phase resetting proceeds faster than in other peripheral tissues (Damiola et al. 2000, Stokkan et al. 2001). Feeding entrainment of peripheral tissues has also been described in fish species such as zebrafish, goldfish, gilthead sea bream or Nile tilapia, in which clock gene expression in liver differed depending on the feeding time (López-Olmeda et al. 2010, Feliciano et al. 2011, Vera et al. 2013, Costa et al. 2016). In addition to the phase shift, a higher mesor and amplitude of *per1* was found in sole liver compared to the retina and the optic tectum but also compared to *per2* and *clock*, suggesting that *per1* could play an important role in liver feeding entrainment, being more susceptible to feeding cues. This result was consistent with that reported in goldfish, in which the negative elements *per1* and *cry3* also showed higher amplitude rhythms in liver compared to the brain, pointing to the liver as more sensitive to feeding entrainment (Feliciano et al. 2011). This was further confirmed in recent studies demonstrating that genes from the negative loop of the clock are more sensitive to both environmental and endogenous regulation than other clock genes (Sánchez-Bretaño et al. 2015b, Sánchez-Bretaño et al. 2016). On-going studies are being carried out to clarify how different light regimes (LD versus constant conditions) and periodically restricted food access (daytime, night-time or random restricted feeding) are able to entrain the brain and liver oscillators in sole.

Conclusions and future perspectives

As it has been reflected in this chapter, a significant progress in characterizing the Senegalese sole circadian clock has been made in the last few years. However, much work remains to be done focused in the examination of the core clock mechanism itself, the input pathways to the clock, and the downstream events or clock-regulated output processes. Regarding the clock mechanism, and although a number of clock components are available so far, it is becoming necessary to isolate other key clock elements such as *bmal* and *crys* to complete the clock oscillation puzzle in sole. The sole clock appears highly sensitive to important entraining cues such as light, especially during early ontogenetic stages, and discovering the molecules responsible for this entrainment is being currently one key aspect in sole circadian biology. In line with this, the light-dependent switch from diurnal to nocturnal habits seen in this species has lead us to start the search for determining the identity of the responsible photopigment/s. Moreover, research related with the role of other entraining cues such as feeding time and temperature and with some clock-regulated outputs such as the cell cycle and/or metabolism is still scarce. Clearly, we are technically limited in relation to zebrafish and other model organisms, as the last key technologies to perform functional analysis has not yet been developed in marine fish such as the Senegalese sole. Also, we lack stable cell lines to use them as sole-based *in vitro* systems to look at light-dependent signalling, oscillation and entrainment. In spite of the difficulties that we face, molecular and genetic tools are being released in sole (see last chapter of this book). Thus, future efforts will be directed to fill gaps in chronobiological knowledge in this exciting species and to advance in the understanding of the great diversity in the organisation and functioning of the teleost circadian system.

Acknowledgments

These works were supported by grants from the Spanish Ministry of Science and Innovation (MICINN, MICINN, AGL2007-66507-C02-01; AGL2010-22139-C03-03).

References

Akiyama, M., T. Kirihara, S. Takahashi, Y. Minami, Y. Yoshinobu, T. Moriya et al. 1999. Modulation of mPer1 gene expression by anxiolytic drugs in mouse cerebellum. Br. J. Pharmacol. 128: 1616–1622.

Anguis, V. and J.P. Cañavate. 2005. Spawning of captive Senegal sole (*Solea senegalensis*) under a naturally fluctuating temperature regime. Aquaculture 243: 133–145.

Bayarri, M.J., J.A. Muñoz-Cueto, J.F. López-Olmeda, L.M. Vera, M.A. Rol de Lama, J.A. Madrid et al. 2004. Daily locomotor activity and melatonin rhythms in Senegal sole (*Solea senegalensis*). Physiol. Behav. 81: 577–583.

Beale, A., C. Guibal, T.K. Tamai, L. Klotz, S. Cowen, E. Peyric et al. 2013. Circadian rhythms in Mexican blind cavefish *Astyanax mexicanus* in the lab and in the field. Nat. Commun. 4: 2769.

Bejarano-Escobar, R., M. Blasco, W.J. Degrip, J.A.- Oyola-Velasco, G. Martín-Partido and J. Francisco-Morcillo. 2010. Eye development and retinal differentiation in an altricial fish species, the Senegalese sole (*Solea senegalensis*, Kaup 1858). J. Exp. Zool. B Mol. Dev. Evol. 314: 580–605.

Bell-Pedersen, D., V.M. Cassone, D.J. Earnest, S.S. Golden, P.E. Hardin, T.L. Thomas and M.J. Zoran. 2005. Circadian rhythms from multiple oscillators: lessons from diverse organisms. Nat. Rev. Genet. 6: 544–556.

Besharse, J.C., M. Zhuang, K. Freeman and J. Fogerty. 2004. Regulation of photoreceptor Per1 and Per2 by light, dopamine and a circadian clock. Eur. J. Neurosci. 20: 167–174.

Blanco-Vives, B., N. Villamizar, J. Ramos, M.J. Bayarri, O. Chereguini and F.J. Sánchez-Vázquez. 2010. Effect of daily thermo- and photo-cycles of different light spectrum on the development of Senegal sole (*Solea senegalensis*) larvae. Aquaculture 306: 137–145.

Blanco-Vives, B., M. Aliaga-Guerrero, J.P. Cañavate, J.A. Muñoz-Cueto and F.J. Sánchez-Vázquez. 2011a. Does lighting manipulation during incubation affect hatching rhythms and early development of sole? Chronobiol. Int. 28: 300–306.

Blanco-Vives, B., L.M. Vera, J. Ramos, M.J. Bayarri, E. Mañanos and F.J. Sánchez-Vázquez. 2011b. Exposure of larvae to daily thermocycles affects gonad development, sex ratio, and sexual steroids in *Solea senegalensis*, Kaup. J. Exp. Zool. A Ecol. Genet. Physiol. 315: 162–169.

Blanco-Vives, B., M. Aliaga-Guerrero, J.P. Cañavate, G. García-Mateos, A.J. Martín-Robles, P. Herrera-Perez et al. 2012. Metamorphosis induces a light-dependent switch in Senegalese sole (*Solea senegalensis*) from diurnal to nocturnal behavior. J. Biol. Rhythms. 27: 135–144.

Boluda-Navarro, D., V.C. Rubio, R.K. Luz, J.A. Madrid and F.J. Sánchez-Vázquez. 2009. Daily feeding rhythms of Senegalese sole under laboratory and farming conditions using self-feeding systems. Aquaculture. 291: 130–135.

Brewster, B. 1987. Eye migration and cranial development during flatfish metamorphosis: a reappraisal (Teleostei:Pleuronectiformes). J. Fish Biol. 31: 805–834.

Cahill, G.M. 1996. Circadian regulation of melatonin production in cultured zebrafish pineal and retina. Brain Res. 708: 177–181.

Cañavate, J.P., R. Zerolo and C. Fernández-Díaz. 2006. Feeding and development of Senegal sole (*Solea senegalensis*) larvae reared in different photoperiods. Aquaculture. 258: 368–377.

Carazo, I., F. Norambuena, C. Oliveira, F.J. Sánchez-Vázquez and N.J. Duncan. 2013. The effect of night illumination, red and infrared light, on locomotor activity, behaviour and melatonin of Senegalese sole (*Solea senegalensis*) broodstock. Physiol. Behav. 118: 201–207.

Cavallari, N., E. Frigato, D. Vallone, N. Fröhlich, J.F. López-Olmeda, A. Foa et al. 2011. A blind circadian clock in cavefish reveals that opsins mediate peripheral clock photoreception. PLOS Biol. 9: e1001142.

Cermakian, N., D. Whitmore, N.S. Foulkes and P. Sassone-Corsi. 2000. Asynchronous oscillations of two zebrafish CLOCK partners reveal differential clock control and function. Proc. Natl. Acad. Sci. USA 97: 4339–4344.

Chaurasia, S.S., N. Pozdeyev, R. Haque, A. Visser, T.N. Ivanova and P.M. Iuvone. 2006. Circadian clockwork machinery in neural retina: evidence for the presence of functional clock components in photoreceptor-enriched chick retinal cell cultures. Mol. Vis. 12: 215–223.

Confente, F., A. El M'Rabet, A. Ouarour, P. Voisin, W.J. De Grip, M.C. Rendón et al. 2008. The pineal complex of Senegalese sole (*Solea senegalensis*): anatomical, histological and immunohistochemical study. Aquaculture. 285: 207–215.

Confente, F., M.C. Rendón, L. Besseau, J. Falcón and J.A. Muñoz-Cueto. 2010. Melatonin receptors in a pleuronectiform species, *Solea senegalensis*: Cloning, tissue expression, day-night and seasonal variations. Gen. Comp. Endocrinol. 167: 202–214.

Costa, L.S., I. Serrano, F.J. Sánchez-Vázquez, J.F. López-Olmeda. 2016. Circadian rhythms of clock gene expression in Nile tilapia (*Oreochromis niloticus*) central and peripheral tissues: influence of different lighting and feeding conditions. J. Comp. Physiol. 186: 775–785.

Crane, B.R. and M.W Young. 2014. Interactive features of proteins composing eukaryotic circadian clocks. Annu. Rev. Biochem. 83: 191–219.

Cuesta, I.H., K. Lahiri, J.F. López-Olmeda, F. Loosli, N.S Foulkes and D. Vallone. 2014. Diferential maturation of rhyhtmic clock gene expression during early development in medaka (*Oryzias latipes*). Chronobiol. Int. 31: 468–478.

Cymborowski, B. 2010. Introduction to circadian rhythms. pp. 155–184. *In*: Kulczykowska, E., W. Popek and B.G. Kapoor [eds.]. Biological clock in fish. Science Publishers, Enfield, New Hampshire, USA.

Damiola, F., N. Le Minh, N. Preitner, B. Kornmann, F. Fleury-Olela and U. Schibler. 2000. Restricted feeding uncouples circadian oscillators in peripheral tissues from the central pacemaker in the suprachiasmatic nucleus. Genes Dev. 14: 2950–2961.

Davie, A., M. Minghetti and H. Migaud. 2009. Seasonal variations in clock-gene expression in Atlantic salmon (*Salmo salar*). Chronobiol. Int. 26: 379–395.

Davie, A., J.A. Sánchez, L.M. Vera, F.J. Sánchez-Vázquez and H. Migaud. 2011. Ontogeny of the circadian system during embryogenesis in rainbow trout (*Oncorhynchus mykyss*) and the effect of prolonged exposure to continuous illumination on daily rhythms of per1, clock, and aanat2 expression. Chronobiol. Int. 28: 177–186.

Davies, W.I., T.K. Tamai, L. Zheng, J.K. Fu, R.G. Foster, D. Whitmore et al. 2015. An extended family of novel vertebrate photopigments is widely expressed and displays a diversity of function. Genome Res. 25: 1666–1679.

Dekens, M.P., C. Santoriello, D. Vallone, G. Grassi, D. Whitmore and N.S. Foulkes. 2003. Light regulates the cell cycle in zebrafish. Curr Biol. 13: 2051–2057.

Dekens, M.P. and D. Whitmore. 2008. Autonomous onset of the circadian clock in the zebrafish embryo. EMBO J. 27: 2757–2765.

Del Pozo, A., A. Montoya, L.M. Vera and F.J. Sánchez-Vázquez. 2012. Daily rhythms of clock gene expression, glycaemia and digestive physiology in diurnal/nocturnal European seabass. Physiol. Behav. 106: 446–450.

Delaunay, F., C. Thisse, O. Marchand, V. Laudet and B. Thisse. 2000. An inherited functional circadian clock in zebrafish embryos. Science 289: 297–300.

Delaunay, F., C. Thisse, B. Thisse and V. Laudet V. 2003. Differential regulation of Period 2 and Period 3 expression during development of the zebrafish circadian clock. Gene Expr. Patterns 3: 319–324.

Di Rosa, V., E. Frigato, J.F. López-Olmeda, F.J. Sánchez-Vázquez and C. Bertolucci. 2015. The light wavelength affects the ontogeny of clock gene expresion and activity rhythms in zebrafish larvae. PLoS One 10: e0132235.

El M'Rabet, A., F. Confente, A. Ouarour and J.A. Muñoz-Cueto. 2008. Ontogenia del órgano pineal del lenguado, *Solea senegalensis*. pp 179–181. *In*: Muñoz-Cueto, J.A., J.M. Mancera and G. Martínez-Rodríguez [eds.]. Avances en Endocrinología Comparada Vol 4. Serv. Public. Universidad de Cádiz, Cádiz, España.

Emery, P. and S.M. Reppert. 2004. A rhythmic Ror. Neuron. 43: 443–446.

Falcón, J., H. Migaud, J.A. Muñoz-Cueto and M. Carrillo. 2010. Current knowledge on the melatonin system in teleost fish. Gen. Comp. Endocrinol. 165: 469–482.

Farnell, Y.Z., G.C. Allen, S.S. Nahm, N. Neuendorff, J.R. West, W.J. Chen et al. 2008. Neonatal alcohol exposure differentially alters clock gene oscillations within the suprachiasmatic nucleus, cerebellum, and liver of adult rats. Alcohol Clin. Exp. Res. 32: 44–552.

Feliciano, A., Y. Vivas, N. de Pedro, M.J. Delgado, E. Velarde and E. Isorna E. 2011. Feeding time synchronizes clock gene rhythmic expression in brain and liver of goldfish (*Carassius auratus*). J. Biol. Rhythms. 26: 24–33.

Fernández-Díaz, C., M. Yúfera, J.P. Cañavate, F.J. Moyano, F.J. Alarcón and M. Díaz. 2001. Growth and physiological changes during metamorphosis of Senegal sole reared in the laboratory. J. Fish Biol. 58: 1086–1097.

Finger, T.E. 1983. Organization of the teleost cerebellum. pp. 261–284. *In*: Davis, R.E. and R.G. Northcutt [eds.]. Fish neurobiology: Higher Brain Areas and Functions, 2. University of Michigan Press, Ann Arbor, Michigan, USA.

García-López, A., E. Couto, A.V. Canario, C. Sarasquete and G. Martínez-Rodríguez. 2007. Ovarian development and plasma sex steroid levels in cultured female Senegalese sole *Solea senegalensis*. Comp. Biochem. Physiol. A Mol. Integr. Physiol. 146: 342–354.

Gothilf, Y., S.L. Coon, R. Toyama, A. Chitnis, M.A. Namboodiri and D.C. Klein. 1999. Zebrafish serotonin N-acetyltransferase-2: marker for development of pineal photoreceptors and circadian clock function. Endocrinology. 140: 4895–4903.

Guillaumond, F., H. Dardente, V. Giguère and N. Cermakian. 2005. Differential control of *Bmal1* circadian transcription by REV-ERB and ROR nuclear receptors. J. Biol. Rhythms 20: 391–403.

Guzmán, J.M., B. Norberg, J. Ramos, C.C. Mylona and E.L Mañanós. 2008. Vitellogenin, steroid plasma levels and spawning performance of cultured female Senegalese sole (*Solea senegalensis*). Gen. Comp. Endocrinol. 156: 285–297.

Helfer, G., A.E. Fidler, D. Vallone, N.S. Foulkes and R. Brandstaetter. 2006. Molecular analysis of clock gene expression in the avian brain. Chronobiol. Int 23: 113–127.

Herrero, M.J. and J.M. Lepesant. 2014. Daily and seasonal expression of clock genes in the pituitary of the European sea bass (*Dicentrarchus labrax*). Gen. Comp. Endocrinol. 208: 30–38.

Hirayama, J., I. Fukuda, T. Ishikawa, Y. Kobayashi and T. Todo. 2003. New role of zCRY and zPER2 as regulators of sub-cellular distributions of zCLOCK and zBMAL proteins. Nucleic Acids Res. 31: 935–943.

Hirayama, J. and P. Sassone-Corsi. 2005. Structural and functional features of transcription factors controlling the circadian clock. Curr. Opin. Genet. Dev. 15: 548–556.

Hur, S.P., Y. Takeuchi, H. Itoh, M. Uchimura, K. Takahashi, H.C. Kang et al. 2012. Fish sleeping under sandy bottom: interplay of melatonin and clock genes. Gen. Comp. Endocrinol. 177: 37–45.

Hurd, M.W. and G.M. Cahill. 2002. Entraining signals initiate behavioral circadian rhythmicity in larval zebrafish. J. Biol. Rhythms. 17: 307–314.

Idda, M.L., C. Bertolucci, D. Vallone, Y. Gothilf, F.J. Sánchez-Vázquez and N.S. Foulkes. 2012. Circadian clocks: lessons from fish. Prog. Brain Res. 199: 41–57.

Ishikawa, T., J. Hirayama, Y. Kobayashi and T. Todo. 2002. Zebrafish CRY represses transcription mediated by CLOCK-BMAL heterodimer without inhibiting its binding to DNA. Genes Cells. 7: 1073–1086.

Isorna, E., A. El M'rabet, F. Confente, J. Falcón and J.A. Muñoz-Cueto. 2009a. Cloning and expression of arylalkylamine N-acetyltranferase-2 during early development and metamorphosis in the sole *Solea senegalensis*. Gen. Comp. Endocrinol. 161: 97–102.

Isorna, E., M.J. Obregón, R.M. Calvo, R. Vázquez, C. Pendón, J. Falcón et al. 2009b. Iodothyronine deiodinases and thyroid hormone receptors regulation during flatfish (*Solea senegalensis*) metamorphosis. J. Exp. Zool. B. 312: 231–246.

Isorna, E., M. Aliaga-Guerrero, A. El M'Rabet, A. Servili, J. Falcón and J.A. Muñoz-Cueto. 2011. Identification of two arylalkylamine N-acetyltranferase 1 genes with different developmental expression profiles in the flatfish *Solea senegalensis*. J. Pineal Res. 51: 434–444.

Iuvone, P.M., G. Tosini, N. Pozdeyev, R. Haque, D.C. Klein and S.S. Chaurasia. 2005. Circadian clocks, clock networks, arylalkylamine N-acetyltransferase, and melatonin in the retina. Prog. Retin. Eye Res. 24: 433–456.

Iuvone, P.M., E. Velarde, M.J. Delgado, A.L. Alonso-Gómez and R. Haque. 2010. Circadian clocks in retina of goldfish. pp. 251–264. *In*: Kulczykowska, E., W. Popek and B.G. Kapoor [eds.]. Biological clock in fish. Science Publishers, Enfield, USA.

Kaffman, A. and E.K O'Shea. 1999. Regulation of nuclear localization: a key to a door. Annu. Rev. Cell. Dev. Biol. 15: 291–339.

Kamphuis, W., C. Cailotto, F. Dijk, A. Bergen and R.M. Buijs. 2005. Circadian expression of clock genes and clock-controlled genes in the rat retina. Biochem. Biophys. Res. Commun. 330: 18–26.

Kaneko, M. and Cahill G.M. 2005. Light-dependent development of circadian gene expression in transgenic zebrafish. PLoS Biol. 3: e34.

Kaneko, M., N. Hernandez-Borsetti and G.M. Cahill. 2006. Diversity of zebrafish peripheral oscillators revealed by luciferase reporting. Proc. Natl. Acad. Sci. USA 103: 14614–14619.

Kazimi, N. and G.M. Cahill. 1999. Development of a circadian melatonin rhythm in embryonic zebrafish. Brain Res. Dev. Brain Res. 117: 47–52.

Ki, Y., H. Ri, H. Lee, E. Yoo, J. Choe and C. Lim. 2015. Warming up your tick-tock: temperature-dependent regulation of circadian clocks. Neuroscientist. 21: 503–518.

Klein, D.C., R.Y. Moore and S.M. Reppert. 1991. Suprachiasmatic Nucleus: The Mind's Clock. Oxford University Press, New York.

Kobayashi, Y., T. Ishikawa, J. Hirayama, H. Daiyasu, S. Kanai, H. Toh et al. 2000. Molecular analysis of zebrafish photolyase/cryptochrome family: two types of cryptochromes present in zebrafish. Genes Cells. 5: 725–738.

Kulczykowska, E., W. Popek and B.G. Kapoor. 2010. Biological clock in fish. Science Publishers, Enfield, New Hampshire, USA.

Lahiri, K., D. Vallone, S.B. Gondi, C. Santoriello, T. Dickmeis and N.S. Foulkes. 2005. Temperature regulates transcription in the zebrafish circadian clock. PLoS Biol. 3: e351.

Lan-Chow-Wing, O., F. Confente, P. Herrera-Pérez, E. Isorna, O. Chereguini, M.C. Rendón et al. 2014. Distinct expression profiles of three melatonin receptors during early development and metamorphosis in the flatfish *Solea senegalensis*. Int. J. Mol. Sci. 15: 20789–20799.

Lazado, C.C., H.P. Kumaratunga, K. Nagasawa, I. Babiak, A. Giannetto and J.M. Fernandes. 2014. Daily rhythmicity of clock gene transcripts in atlantic cod fast skeletal muscle. PLoS One. 9: e99172.

Lee, C., J.P. Etchegaray, F.R. Cagampang, A.S. Loudon and S.M. Reppert. 2001. Posttranslational mechanisms regulate the mammalian circadian clock. Cell. 107: 855–867.

Lee, C., D.R. Weaver and S.M. Reppert. 2004. Direct association between mouse PERIOD and CKIε is critical for a functioning circadian clock. Mol. Cell Biol. 24: 584–594.

Lee, H., R. Chen, Y. Lee, S. Yoo and C. Lee. 2009. Essential roles of CKIdelta and CKIepsilon in the mammalian circadian clock. Proc. Natl. Acad. Sci. USA 106: 21359–21364.

Lee, H.M., R. Chen, H. Kim, J.P. Etchegaray, D.R. Weaver and C. Lee. 2011. The period of the circadian oscillator is primarily determined by the balance between casein kinase 1 and protein phosphatase 1. Proc. Natl. Acad. Sci. USA. 108: 16451–16456.

López-Olmeda, J.F., E.V. Tartaglione, H.O. De la Iglesia and F.J. Sánchez-Vázquez. 2010. Feeding entrainment of food-anticipatory activity and per1 expression in the brain and liver of zebrafish under different lighting and feeding conditions. Chronobiol. Int. 27: 1380–1400.

López-Olmeda, J.F., B. Blanco-Vives, I.M. Pujante, Y.S. Wunderink, J.M. Mancera and F.J. Sánchez-Vázquez. 2013. Daily rhythms in the hypothalamus-pituitary-interrenal axis and acute stress responses in a teleost flatfish, *Solea senegalensis*. Chronobiol. Int. 30: 530–539.

López-Olmeda, J.F., I.M. Pujante, L.S. Costa, A. Galal-Khallaf, J.M. Mancera and F.J. Sánchez-Vázquez. 2016. Daily rhythms in the somatotropic axis of Senegalese sole (*Solea senegalensis*): The time of day influences the response to GH administration. Chronobiol. Int. 33: 257–267.

López-Patiño, M.A., A. Rodríguez-Illamola, M. Conde-Sieira, J.L. Soengas and J.M. Miguez. 2011. Daily rhythmic expression patterns of clock1a, bmall, and per1 genes in retina and hypothalamus of the rainbow trout, *Oncorhynchus mykiss*. Chronobiol. Int. 28: 381–389.

Marchand, O., M. Duffraisse, G. Trigueneaux, R. Safi and V. Laudet. 2004. Molecular cloning and developmental expression patterns of thyroid hormone receptors and T3 target genes in the turbot (*Scophtalmus maximus*) during post-embryonic development. 2004. Gen. Comp. Endocrinol. 135: 345–357.

Martín-Robles, A.J., E. Isorna, D. Whitmore, J.A. Muñoz-Cueto and C. Pendon. 2011. The clock gene Period3 in the nocturnal flatfish *Solea senegalensis*: Molecular cloning, tissue expression and daily rhythms in central areas. Comp. Biochem. Physiol. A Mol. Integr. Physiol. 159: 7–15.

Martín-Robles, A.J., D. Whitmore, F.J. Sánchez-Vázquez, C. Pendon and J.A. Muñoz-Cueto. 2012a. Cloning, tissue expression pattern and daily rhythms of Period1, Period2, and Clock transcripts in the flatfish Senegalese sole, *Solea senegalensis*. J. Comp. Physiol. B. 182: 673–685.

Martín-Robles, A.J., M. Aliaga-Guerrero, D. Whitmore, C. Pendón and J.A. Muñoz-Cueto. 2012b. The circadian clock machinery during early development of Senegalese sole (*Solea senegalensis*): effects of constant light and dark conditions. Chronobiol. Int. 29: 1195–1205.

Martín-Robles, A.J., D. Whitmore, C. Pendón and J.A. Muñoz-Cueto. 2013. Differential effects of transient constant light-dark conditions on daily rhythms of Period and Clock transcripts during Senegalese sole metamorphosis. Chronobiol. Int. 30: 699–710.

Mata-Sotres, J.A., G. Martínez-Rodríguez, J. Pérez-Sánchez, F.J. Sánchez-Vázquez and M. Yúfera. 2015. Daily rhythms of clock gene expression and feeding behavior during the larval development in gilthead seabream, *Sparus aurata*. Chronobiol. Int. 32: 1061–1074.

Mazurais, D., G. Le Dréan, I. Brierley, I. Anglade, N. Bromage, L. Williams et al. 2000. Expression of clock gene in the brain of rainbow trout: comparison with the distribution of melatonin receptors. J. Comp. Neurol. 422: 612–620.

Menaker, M., L.F. Moreira and G. Tosini. 1997. Evolution of circadian organization in vertebrates. Braz. J. Med. Biol. Res. 30: 305–313.

Mendoza, J., P. Pevet, M.P. Felder-Schmittbuhl, Y. Bailly and E. Challet. 2010. The cerebellum harbors a circadian oscillator involved in food anticipation. J. Neurosci. 30: 1894–1904.

Meyer, A. and Y. Van de Peer. 2005. From 2R to 3R: evidence for a fish-specific genome duplication (FSGD). Bioessays. 27: 937–945.

Migaud, H., A. Davie and J.F. Taylor. 2010. Current knowledge on the photoneuroendocrine regulation of reproduction in temperate fish species. J. Fish Biol. 76: 27–68.

Mistlberger, R.E. and M.C. Antle. 2011. Entrainment of circadian clocks in mammals by arousal and food. Essays Biochem. 49: 119–136.

Moore, H.A. and D. Whitmore. 2014. Circadian rhythmicity and light light sensitivity of the zebrafish brain. PLoS One. 9: e86176.

Morais, S., C. Aragao, E. Cabrita, L.E.C. Conceicao, M. Constenia, B. Costas et al. 2016. New developments and biological insights into the farming of *Solea senegalensis* reinforcing its aquaculture potential. Rev. Aquaculture. 8: 227–263.

Mrosovsky, N. and S. Hattar. 2005. Diurnal mice (*Mus musculus*) and other examples of temporal niche switching. J. Comp. Physiol. A Neuroethol. Sens. Neural Behav. Physiol. 191: 1011–1024.

Namihira, M., S. Honma, H. Abe, Y. Tanahashi, M. Ikeda and K. Honma. 1999. Daily variation and light responsiveness of mammalian clock gene, Clock and BMAL1, transcripts in the pineal body and different areas of brain in rats. Neurosci. Lett. 267: 69–72.

Nelson, J.S., T. Grande and M.V.H. Wilson. 2016. Fishes of the World, Fifth Edition. John Wiley and Sons, United States.

Nisembaum, L.G., E. Velarde, A.B. Tinoco, C. Azpeleta, N. De Pedro, A.L. Alonso-Gómez et al. 2012. Light-dark cycle and feeding time differentially entrains the gut molecular clock of the goldfish (*Carassius auratus*). Chronobiol. Int. 29: 665–673.

Okabayashi, N., S. Yasuo, M. Watanabe, T. Namikawa, S.Ebihara and T. Yoshimura. 2003. Ontogeny of circadian clock gene expression in the pineal and the suprachiasmatic nucleus of chick embryo. Brain Res. 990: 231–234.

Okada, N., Y. Takagi, T. Seikai, M. Tanaka and M. Tagawa. 2001. Asymmetrical development of bones and soft tissues during eye migration of metamorphosing Japanese flounder, *Paralichthys olivaceus*. Cell Tissue Res. 304: 59–66.

Oliveira, C., A. Ortega, J.F. López-Olmeda, L.-M. Vera and F.J. Sánchez-Vázquez. 2007. Influence of constant light and darkness, light intensity, and light spectrum on plasma melatonin rhythms in Senegal sole. Chronobiol. Int. 24: 615–627.

Oliveira, C., J.F. López-Olmeda, M.J. Delgado, A.L. Alonso-Gómez and F.J. Sánchez-Vázquez. 2008. Melatonin binding sites in Senegal sole: day/night changes in density and location in different regions of the brain. Chronobiol. Int. 25: 645–652.

Oliveira, C., M.T. Dinis, F. Soares, E. Cabrita, P. Pousao-Ferreira, F.J. Sánchez-Vázquez. 2009. Lunar and daily spawning rhythms of Senegal sole *Solea senegalensis*. J. Fish Biol. 75: 61–74.

Pando, M.P., A.B. Pinchak, N. Cermakian and P. Sassone-Corsi. 2001. A cell-based system that recapitulates the dynamic light-dependent regulation of the vertebrate clock. Proc. Natl. Acad. Sci. USA 98: 10178–10183.

Pando, M.P. and P. Sassone-Corsi. 2002. Unraveling the mechanisms of the vertebrate circadian clock: zebrafish may light the way. Bioessays 24: 419–426.

Parekh, P.K. and C.A. McClung. 2016. Circadian Mechanisms Underlying Reward-Related Neurophysiology and Synaptic Plasticity. Front. Psychiatry 6: 187.

Park, J.G., Y.J. Park, N. Sugama, S.J. Kim and A. Takemura. 2007. Molecular cloning and daily variations of the Period gene in a reef fish *Siganus guttatus*. J. Comp. Physiol. A Neuroethol. Sens. Neural. Behav. Physiol. 193: 403–411.

Parra, G. and M. Yúfera. 2001. Comparative energetics during early development of two marine fish species, *Solea senegalensis* (Kaup) and *Sparus aurata* (L.). J. Exp. Biol. 204: 2175–2183.

Partch, C.L., C.B. Green and J.S. Takahashi. 2014. Molecular architecture of the mammalian circadian clock. Trends Cell Biol. 24: 90–99.

Pegoraro, M. and E. Tauber. 2011. Animal clocks: a multitude of molecular mechanisms for circadian timekeeping. Wiley Interdiscip. Rev. RNA. 2: 312–320.

Peirson, S.N., J.N. Butler, G.E. Duffield, S. Takher, P. Sharma and R.G. Foster. 2006. Comparison of clock gene expression in SCN, retina, heart, and liver of mice. Biochem. Biophys. Res. Commun. 351: 800–807.

Plautz, J.D., M. Kaneko, J.C. Hall and S.A. Kay. 1997. Independent photoreceptive circadian clocks throughout Drosophila. Science 278: 1632–1635.

Policansky, D. 1982. The asymmetry of flounders. Sci. Am. 246: 116–122.

Ponting, C.P. and L. Aravind. 1997. PAS: a multifunctional domain family comes to light. Curr. Biol. 7: 674–677.

Postlethwait, J.H., Y.L. Yan, M.A. Gates, S. Horne, A. Amores, A. Brownlie et al. 1998. Vertebrate genome evolution and the zebrafish gene map. Nat. Genet. 18: 345–349.

Power, D.M., L. Llewellyn, M. Faustino, M.A. Nowell, B.T. Bjornsson, I.E. Einarsdottir et al. 2001. Thyroid hormones in growth and development of fish. Comp. Biochem. Physiol. C Toxicol. Pharmacol. 130: 447–459.

Reebs, SG. 2002. Plasticity of diel and circadian activity rhythms in fishes. Rev. Fish Biol. Fish. 12: 349–371.

Rensing, L. and P. Ruoff. 2002. Temperature effect on entrainment, phase shifting, and amplitude of circadian clocks and its molecular bases. Chronobiol. Int. 19: 807–864.

Reppert, S.M. and D.R. Weaver. 2001. Molecular analysis of mammalian circadian rhythms. Annu. Rev. Physiol. 63: 647–676.

Rodríguez-Gómez, F.J, C. Sarasquete and J.A. Muñoz-Cueto JA. 2000b. A morphological study of the brain of *Solea senegalensis*. I. The telencephalon. Histol. Histopathol. 15: 355–364.

Rodríguez-Gómez, F.J., M.C. Rendón-Unceta, C. Sarasquete and J.A. Muñoz-Cueto. 2000a. Localization of tyrosine hydroxylase- immunoreactivity in the brain of the Senegalese sole, *Solea senegalensis*. J. Chem. Neuroanat. 19: 17–32.

Sánchez, J.A., J.A. Madrid and F.J. Sánchez-Vázquez. 2010. Molecular cloning, tissue distribution, and daily rhythms of expression of per1 gene in European sea bass (*Dicentrarchus labrax*). Chronobiol. Int. 27: 19–33.

Sánchez-Bretaño, A., M.M. Guequen, J. Cano-Nicolau, O. Kah, A.L. Alonso-Gómez, M.J. Delgado et al. 2015a. Anatomical distribution and daily profile of gper1b gene expression in brain and peripheral structures of goldfish (*Carassius auratus*). Chronobiol. Int. 32: 889–902.

Sánchez-Bretaño, A., A.L. Alonso-Gómez, M.J. Delgado and E. Isorna. 2015b. The liver of goldfish as a component of the circadian system: integrating a network of signals. Gen. Comp. Endocrinol. 221: 213–216.

Sánchez-Bretaño, A., M. Callejo, M. Montero, A.L. Alonso-Gómez, M.J. Delgado and E. Isorna. 2016. Performing a hepatic timing signal: glucocorticoids induce gper1a and gper1b expression and repress gclock1a and gbmal1a in the liver of goldfish. J. Comp. Physiol. B. 186: 73–82.

Sánchez-Vázquez, F.J., J.A. Madrid and S. Zamora. 1995. Circadian rhythms of feeding activity in sea bass, *Dicentrarchus labrax* L.: dual phasing capacity of diel demand-feeding pattern. J. Biol. Rhythms. 10: 256–266.

Sánchez-Vázquez, F.J., J.A. Madrid, S. Zamora, M. Iigo and M. Tabata. 1996. Demand feeding and locomotor circadian rhythms in the goldfish, *Carassius auratus*: dual and independent phasing. Physiol. Behav. 60: 665–674.

Schreiber, A.M. 2006. Asymmetric craniofacial remodeling and lateralized behavior in larval flatfish. J. Exp. Biol. 209: 610–621.

Shearman, L.P., M.J. Zylka, D.R. Weaver, L.F. Kolakowski, Jr. and S.M. Reppert. 1997. Two period homologs: circadian expression and photic regulation in the suprachiasmatic nuclei. Neuron. 19: 1261–1269.

Springer, A.D. and A.S. Mednick. 1984. Selective innervation of the goldfish suprachiasmatic nucleus by ventral retinal ganglion cell axons. Brain Res. 323: 293–296.

Stokkan, K.A., S. Yamazaki, H. Tei, Y. Sakaki and M. Menaker. 2001. Entrainment of the circadian clock in the liver by feeding. Science 291: 490–493.

Tamai, T.K., V. Vardhanabhuti, N.S. Foulkes and D. Whitmore. 2004. Early embryonic light detection improves survival. Curr. Biol. 14: R104–105.

Tamai, T.K., L.C. Young and D. Whitmore. 2007. Light signaling to the zebrafish circadian clock by Cryptochrome 1a. Proc. Natl. Acad. Sci. USA 104: 14712–14717.

Toh, K.L. 2008. Basic science review on circadian rhythm biology and circadian sleep disorders. Annu. Acad. Med. Singapore 37: 662–668.

Tosini, G., N. Pozdeyev, K. Sakamoto and P.M. Iuvone. 2008. The circadian clock system in the mammalian retina. Bioessays 30: 624–633.

Vallone, D., S.B. Gondi, D. Whitmore and N.S. Foulkes. 2004. E-box function in a period gene repressed by light. Proc. Natl. Acad. Sci. USA 101: 4106–4111.

Vallone, D., K. Lahiri, T. Dickmeis and N.S. Foulkes. 2007. Start the clock! Circadian rhythms and development. Dev. Dyn. 236: 142–155.

Vatine, G., D. Vallone, L. Appelbaum, P. Mracek, Z. Ben-Moshe, K. Lahiri et al. 2009. Light directs zebrafish period2 expression via conserved D and E boxes. PLoS Biol. 7: e1000223.

Vatine, G., D. Vallone, Y. Gothilf and N.S. Foulkes. 2011. It's time to swim! Zebrafish and the circadian clock. FEBS Lett. 585: 1485–1494.

Velarde, E., R. Haque, P.M. Iuvone, C. Azpeleta, A.L. Alonso-Gómez and M.J. Delgado. 2009. Circadian clock genes of goldfish, *Carassius auratus*: cDNA cloning and rhythmic expression of period and cryptochrome transcripts in retina, liver, and gut. J. Biol. Rhythms 24: 104–113.

Vera, L.M., C. Oliveira, J.F. López-Olmeda, J. Ramos, E. Mañanos, J.A. Madrid and F.J. Sánchez-Vázquez. 2007. Seasonal and daily plasma melatonin rhythms and reproduction in Senegal sole kept under natural photoperiod and natural or controlled water temperature. J. Pineal Res. 43: 50–55.

Vera, L.M., P. Negrini, C. Zagatti, E. Frigato, F.J. Sánchez-Vázquez and C. Bertolucci. 2013. Light and feeding entrainment of the molecular circadian clock in a marine teleost (*Sparus aurata*). Chronobiol. Int. 30: 649–661.

Vielhaber, E.L., D. Duricka, K.S. Ullman, D.M. Virshup. 2001. Nuclear export of mammalian PERIOD proteins. J. Biol. Chem. 276: 45921–45927.

Villamizar, N., A. García-Alcazar and F.J. Sánchez-Vázquez. 2009. Effect of light spectrum and photoperiod on the growth, development and survival of European sea bass (*Dicentrarchus labrax*). Aquaculture. 292: 80–86.

Villamizar, N., L.M. Vera, N.S. Foulkes and F.J. Sánchez-Vázquez. 2014. Effect of lighting conditions on zebrafish growth and development. Zebrafish 11: 173–181.

Vivanco, P., M.A. Rol and J.A. Madrid. 2010. Pacemaker phase control versus masking by light: setting the circadian chronotype in dual *Octodon degus*. Chronobiol. Int. 27: 1365–1379.

Wagemans, F., B. Focant and P. Vandewalle. 1998. Early development of the cephalic skeleton in the turbot. J. Fish Biol. 52: 166–204.

Wang, H. 2008a. Comparative analysis of period genes in teleost fish genomes. J. Mol. Evol. 67: 29–40.

Wang, H. 2008b. Comparative analysis of teleost fish genomes reveals preservation of different ancient clock duplicates in different fishes. Mar. Genomics. 1: 69–78.

Wang, H. 2009. Comparative genomic analysis of teleost fish bmal genes. Genetica. 136: 149–161.

Watanabe, N., K. Itoh, M. Mogi, Y. Fujinami, D. Shimizu, H. Hashimoto et al. 2012. Circadian pacemaker in the suprachiasmatic nuclei of teleost fish revealed by rhythmic period2 expression. Gen. Comp. Endocrinol. 178: 400–407.

Weger, M., B.D. Weger, N. Diotel, S. Rastegar, T. Hirota, S.A. Kay et al. 2013. Real-time in vivo monitoring of circadian E-box enhancer activity: a robust and sensitive zebrafish reporter line for developmental, chemical and neural biology of the circadian clock. Dev. Biol. 380: 259– 273.

Whitmore, D., N.S. Foulkes, U. Strahle and P. Sassone-Corsi. 1998a. Zebrafish Clock rhythmic expression reveals independent peripheral circadian oscillators. Nat. Neurosci. 1: 701–707.

Whitmore, D., P. Sassone-Corsi and N.S. Foulkes. 1998b. PASting together the mammalian clock. Curr. Opin. Neurobiol. 8: 635–641.

Whitmore, D., N.S. Foulkes and P. Sassone-Corsi. 2000. Light acts directly on organs and cells in culture to set the vertebrate circadian clock. Nature 404: 87–91.

Whitmore, D. 2010. Cellular clocks and the importance of light in zebrafish. pp. 125–153. *In*: Kulczykowska, E., W. Popek and B.G. Kapoor [eds.]. Biological Clock in Fish. Science Publishers, Enfield, New Hampshire, USA.

Wullimann, M.F. 1998. The central nervous system. pp. 245–282. *In*: Evans, D.H. [ed.]. The Physiology of Fishes. CRC Press, Boca Raton, Florida, USA.

Yagita, K., S. Yamaguchi, F. Tamanini, G.T.J van der Horst, J.H.J. Hoeijmakers, A. Yasui et al. 2000. Dimerization and nuclear entry of mPER proteins in mammalian cells. Genes Dev. 14: 1353–1363.

Yagita, K., F. Tamanini, M. Yasuda, J.H.J. Hoeijmakers, G.T.J. van der Horst and H. Okamura. 2002. Nucleocytoplasmic shuttling and mCRY-dependent inhibition of ubiquitylation of the mPER2 clock protein. EMBO J. 21: 1301–1314.

Yamamoto, T., Y. Nakahata, H. Soma, M. Akashi, T. Mamine and T. Takumi. 2004. Transcriptional oscillation of canonical clock genes in mouse peripheral tissues. BMC Mol. Biol. 5: 18.

Yamazaki, S., R. Numano, M. Abe, A. Hida, R. Takahashi, M. Ueda et al. 2000. Resetting central and peripheral circadian oscillators in transgenic rats. Science 288: 682–685.

Yáñez, J., J. Busch, R. Anadón and H. Meissl. 2009. Pineal projections in the zebrafish (*Danio rerio*): overlap with retinal and cerebellar projections. Neuroscience 164: 1712–1720.

Yoo, S.H., S. Yamazaki, P.L. Lowrey, K. Shimomura, C.H. Ko, E.D. Buhr et al. 2004. PERIOD2::LUCIFERASE real-time reporting of circadian dynamics reveals persistent circadian oscillations in mouse peripheral tissues. Proc. Natl. Acad. Sci. USA 101: 5339–5346.

Yoshimura, T., Y. Suzuki, E. Makino, T. Suzuki, A. Kuroiwa, Y. Matsuda et al. 2000. Molecular analysis of avian circadian clock genes. Brain Res. Mol. Brain Res. 78: 207–215.

Yúfera, M., G. Parra, R. Santiago and M. Carrascosa. 1999. Growth, carbon, nitrogen and caloric content of *Solea senegalensis* (Pisces: Soleidae) from egg fertilization to metamorphosis. Mar. Biol. 134: 43–49.

Zhdanova, I.V., L. Yu, M.A. López-Patiño, E. Shang, S. Kishi and E. Guelin. 2008. Aging of the circadian system in zebrafish and the effects of melatonin on sleep and cognitive performance. Brain Res. Bull. 75: 433–441.

Zhu, H., S. LaRue, A. Whiteley, T.D. Steeves, J.S. Takahashi and C.B. Green. 2000. The Xenopus clock gene is constitutively expressed in retinal photoreceptors. Brain Res. Mol. Brain Res. 75: 303–308.

Zhuang, M., Y. Wang, B.M. Steenhard and J.C. Besharse. 2000. Differential regulation of two period genes in the Xenopus eye. Brain Res. Mol. Brain Res. 82: 52–64.

Ziv, L., S. Levkovitz, R. Toyama, J. Falcón and Y. Gothilf. 2005. Functional development of the zebrafish pineal gland: light-induced expression of period2 is required for onset of the circadian clock. J. Neuroendocrinol. 17: 314–320.

Zylka, M.J., L.P. Shearman, D.R. Weaver and S.M. Reppert. 1998. Three period homologs in mammals: differential light responses in the suprachiasmatic circadian clock and oscillating transcripts outside of brain. Neuron. 20: 1103–1110.

B-2.2

Embryonic and Larval Ontogeny of the Senegalese sole, *Solea senegalensis*
Normal Patterns and Pathological Alterations

C. Sarasquete,[1,*] *E. Gisbert*[2] and *J.B. Ortiz-Delgado*[1]

1. Introduction

The Senegalese sole (*Solea senegalensis* Kaup 1858) is a valuable food fish and one of the best candidates for aquaculture development along Mediterranean Sea and South Atlantic Ocean coasts. This pleuronectiform species is exploited in both extensive and intensive rearing conditions (Rodríguez and Pascual 1982, Drake et al. 1984, Yúfera and Arias 2010) in different countries such as Tunisia (Bedoui 1995) and particularly, in Spain and Portugal (Dinis et al. 1999, Morais et al. 2016). Larval rearing systems for common sole, *Solea solea* and *S. senegalensis*, have been studied over the last for decades. Accordingly, the larval period has been described as a particularly sensitive and critical phase in which noticeable nutritional disorders, several pathologies and high mortalities have been frequently documented. Furthermore, the period of weaning of the juveniles onto an artificial diet has been considered as the most critical phase in terms of duration and mortality, and different studies have provided insights to the nutritional requirements of this species and provided

[1] Instituto de Ciencias Marinas de Andalucía-ICMAN-CSIC. Campus Universitario Rio San Pedro, Puerto Real, Cádiz, Spain.
[2] Institut de Recerca i Tecnologia Agroalimentàries -IRTA-, Sant Carles de la Rápita, Tarragona, Spain.
* Corresponding author: carmen.sarasquete@icman.csic.es

recommendations for improving rearing protocols over the last twenty years (Dinis et al. 1999, Imsland et al. 2003, Pittman et al. 2013, Morais et al. 2016).

A significant component of aquaculture is the production of good quality larvae and juveniles and, in the case of flatfish, this is linked with the change from a symmetric larva to an asymmetric juvenile. This complex metamorphic transformation is a thyroid hormone driven process, whereas the role of other hormones in the regulation of the process along with the interplay of abiotic factors are still relatively poorly characterized, as it is the extent of tissue and organ remodeling, which underlie the profound structural and functional modifications that accompany the transition from a larva to a juvenile phenotype. Metamorphosis drastically modifies the anatomy of Pleuronectiformes, since flatfish species have undergone settlement, behavioural and ecological changes associated with the transition from a pelagic to a benthic way of life. This process implies the migration of the eye towards the upper-pigmented side of the body and craniofacial remodeling, which determines its characteristic juvenile and adult asymmetry (Saele et al. 2006, Geffen et al. 2007, Power et al. 2001, 2008, Inui and Miwa 2012, Schreiber 2013, Darras et al. 2015).

Studying larval development at early life stages is of great importance, since it is a very sensitive and critical ontogenetic period in which the newly hatched specimen develops and forms most of its functional organ-systems; thus, potential deviations from the optimal biotic and abiotic rearing factors may impact the performance, survival and quality of the individual, and ultimately impact the efficiency and profitability of the rearing process. In this context, in early stages of life of Senegalese sole, different researches have focused at the morphological, histological and histochemical levels mainly, such as the yolk-resorption process, development of the brain, digestive system and associated glands, senso-visual, skeletal and thyroid systems, as well as the ontogenetic developmental patterns of many other organ-systems, i.e., respiratory, cardiovascular, excretory, and inmune tissues, among others (Sarasquete et al. 1996, Ribeiro et al. 1999a,b, Gavaia et al. 2000, 2002, Piñuela et al. 2004, Ortiz-Delgado et al. 2006, Gisbert et al. 2008, Zambonino-Infante et al. 2008, Padrós et al. 2011, among others). Flatfish metamorphosis also involves important and progressive modifications of the different organ-systems and tissues, as well as other noticeable ontogenetic changes in molecular, biochemical and functional tissue properties (Vazquez et al. 1984, Bayarri et al. 2004, Manchado et al. 2008, 2009, Navarro et al. 2009, Blanco-Vives et al. 2012, Darias et al. 2013a,b, Martín-Robles et al. 2013, Navarro-Guillén et al. 2016), which are typical of Pleuronectiformes. With the exception of metamorphosis, the process by which the symmetrical larva becomes an asymmetric juvenile, the challenges and problems associated with flatfish culture are common to the aquaculture of round fishes. However, abnormal development during metamorphosis is a frequent problem in flatfish farming-rearing systems, and Atlantic halibut, *Hippoglossus hippoglossus*, turbot, *Scophthalmus maximus* and *S. senegalensis* are no exception (Dinis et al. 1999, Power et al. 2001, 2008, Morais et al. 2016). Indeed, at the early life stages of flatfish and especially during the metamorphosis phase, many abnormalities, skeletal deformities, and pigmentation disorders, among other developmental disturbances, may occur, in both wild and cultured Senegalese sole specimens (Dinis et al. 1999, Gavaia et al. 2009, Fernández and Gisbert 2011, Boglione et al. 2013a,b). Comparable ontogenetic disorders and/or pathologies were induced when submitting the early stages of life to different experimental non-optimal nutritional conditions (Boglino et al. 2012, 2013, 2014a,b, Darias et al. 2013a,b, Fernández et al. 2017).

On the other hand, there are many cellular studies including some important histochemical and histophysiological findings of practically all the organ-systems and tissues (brain, neuroendocrine, senso-visual, digestive, musculature-skeleton, cardio-respiratory and vascular, and gonads, etc.) in Senegalese sole adult specimens (Gutiérrez et al. 1985, Rendón et al. 1997, Rodríguez-Gómez et al. 2000a,b, Arellano and Sarasquete 2005, Garcia-López et al. 2005, 2007, 2008, among others). However, from our knowledge there are few studies related with the embryonic development in soleidae fish species. In this sense, Ramos (1986) and more recently, Assem et al. (2012) have provided few morphological descriptions on the embryogenesis in common sole *S. solea*.

2. Eggs and Embryos

The general development patterns of marine fish eggs and larvae follows a similar pattern, but apparently there exists considerable differences between species with regard to egg size, yolk composition, developmental rates, egg incubation time, ontogenetic status and size at hatching, and the timing of development and functionality of various organ-systems and tissues. The differentiation of cells, tissues, organs and organ systems proceeds continuously from the early embryonic phase (egg stage) throughout metamorphosis in accordance with their gradual improvement of functionality (Blaxter 1986, Govoni et al. 1986, Falk-Petersen 2005).The morphogenesis of Senegalese sole from fertilization to embryo formation follows the same basic pattern described for other flatfishes and many other altricial marine species, with small temporal differences depending upon species and rearing temperature (Zambonino-Infante et al. 2008, Morais et al. 2016). According to different authors (Rodríguez and Pascual 1982, Bedoui 1995, Dinis and Reis 1995, Cañavate and Fernández-Díaz 1999, Yúfera et al. 1999, Navarro-Guillen et al. 2015), the incubation of *S. senegalensis* eggs is performed at water temperatures comprised between 17 and 21°C. The egg size ranges between 0.99 and 1.03 mm, and the meroblastic embryonic development is generally completed within 36–42 hours post fertilization (hpf). *Solea vulgaris* and *S. senegalensis* spawn pelagic eggs, which are fertilized externally and float individually near the water surface (Ramos 1986, Dinis et al. 1999, Imsland et al. 2003, Assem et al. 2012).The ripe and viable eggs of Senegalese sole are round and transparent, and the egg membrane is smooth and not separated from the yolk, whereas the cytoplasm is reduced to a thin layer covering the yolk. The eggs of sole are telolecithal with a large mass of yolk and a relatively small yolk-free blastodisc where cleavage proceeds. Fertilized eggs appear with active cytoplasm around the zygote nucleus as a disk shape cap at the animal pole. Around 30–35 minutes/min after fertilization, the egg membrane swells up and separate from perivitelline space, between chorion and yolk-mass. Then, the egg diameter increases (1.2–1.4 mm), and multicellular blastodisc is clearly visible. After 4–5 hpf, the peripheral periblast cells are evident and blastoderm expands, as a result of the peripheral periblast reduction to form the morula. After *ca.* 8–9 hpf, the early gastrulation stage starts, the blastoderm gets thinner and interface between the periblast and blastoderm curves. At this stage, the embryonic layers (ectoderm, endoderm and mesoderm) form and more than half of the yolk is covered by the blastoderm at the end of gastrulation. At 10–12 hpf, epiboly takes place and the embryonic shield is enlarged, whereas gastrulation is completed and the blastopore closes up. Organogenesis starts at *ca.* 18–28 hpf with the reorganization of the three germinal/embryonic layers (Falk-Petersen 2005). In altricial larvae, as Senegalese sole, the embryonic fold begins to differentiate into the head and trunk regions, and yellow pigmented cells, presumably xanthophores, are visible in the body regions. The first structures that are easily recognized in the embryo are the out-pocketings of the brain and the somites. The notochord, neural tube and gut first appear as three rods within the embryonic ridge. Eyes develop at both sides of the brain, and gill pouches form laterally, towards the anterior gut; the heart appears as a cone-shaped tube ventrally to the head.

From 30 hpf onwards, a fully formed embryo inside the egg is distinguishable; the yolk sac diameter is around 0.8 mm, the heart starts beating with weak cardiac contractions, and the blood is colorless due to the absence of hemoglobin as the exchange of gases is done passively through body surfaces. At the end of this embryonic stage, the chorion starts to weaken and begins to get loosen as a result of increased muscular and jerky trunk and tail movements. During the late embryonic stage, small chromathophores are distributed throughout the trunk and the yolk-sac and the optic vesicles are clearly visible. As the end of the embryonic period approaches, the fin folds, and somites, heart, optic and auditory vesicles with otoliths, and chromatophores become apparent. Similar to other altricial fish species described so far, some embryonic cells of a transitory nature (i.e., hatching gland cells) are found in a restricted belt over the front of the yolk sac, and these cells are responsible for secreting proteases that participate in the digestion and breakage of the chorion (Fuiman 2002, Falk-Petersen 2005).

Suboptimal rearing temperatures can critically affect the embryonic development in flatfish species, compromising their hatching and growth performance as it has been shown by Dionisio et al. (2012) in Senegalese sole. According to these and other authors (Blanco-Vives et al. 2010), altered water temperature has also been reported to affect the development of the skeleton by inducing metamorphosis-related morphological abnormalities, as well as affect the number of vertebrae composing the axial skeleton. In addition, other biotic and abiotic factors may affect the proper development of the embryos during this period and affect the hatching rate and larval quality; thus, egg incubation has to be conducted under the most optimal environmental conditions in order to not disturb normal embryogenesis and produce good quality animal for further stages of the rearing process.

Interestingly, researches have been performed on Senegalese sole fertilized-eggs and about effects of daily thermo-and photoperiod cycles of different light spectrum on embryogenesis, hatching rates, and larval development (present book).

3. The metamorphosis: a thyroid-driven process

Flatfish metamorphosis is not considered to be a common developmental strategy. Indeed, metamorphosis is considered a phase in the early developmental period of the larval life, and it is sub-divided into pre-metamorphosis, metamorphosis, climax, and post-metamorphosis. This distinction eliminates in fish some transitory forms that do not experience a pronounced climax event and excludes some juvenile transition (i.e., salmon parr-smolt), which involves a major change in habitat and inevitably a number of functional modifications. Moreover, this complex transformation is highly sensitive to intrinsic and extrinsic signals, leading to a tightly regulated climax event required for successful migration of one eye so that both eyes are on the same side of the head. In this sense, the most spectacular post-embryonic tissue remodelling process in vertebrates is the eye-migration in Pleuronectiformes occurring during larval metamorphosis (Saele et al. 2006). At hatching, pleuronectiform larvae are pelagic and display the principal external morphological features of other teleost larvae: one eye is present on either side of the head, the mouth is horizontal or sub-horizontal and the body is bilaterally symmetric. During larval ontogeny, one of the eyes migrate across the midline to the opposite side of the head and its displacement induces a major remodelling of the whole head, with the whole body structure also modified accordingly (Inui and Miwa 2012, Schreiber 2013). There is concomitant pigmentation of the ocular side of the fish and the transition from a pelagic larva to a benthic juvenile. Associated with the obvious changes in external morphology during metamorphosis, larvae also undergo other dramatic changes that drive the maturation of organs and also shape the morphology into that of the adult animal (Geffen et al. 2007, Power et al. 2001, 2008). Furthermore, flatfish metamorphosis is an energy-demanding process, and it generally interferes with growth in size if the developmental changes cause difficulties in larval feeding; thus, metamorphic specimens have to rely on stored reserves and reduce their growth performance during this process. In this sense, readers are encouraged to consult the revision to Geffen et al. (2007) about the cost of metamorphosis in flatfishes. The end of the metamorphic phase is then determined by the time that the fish is able to resume, begin or alter feeding, and this is clearly correlated to morphological, ontogenetic, physiological and behavioural changes. It is usually indicated that flatfish suffers during metamorphosis due to changes in their behavior or due to the inability to process visual information and thus, feed efficiently. Moreover, there are important inter-specific senso-visual differences and feeding behaviours between different flatfish species. Indeed, the visual acuity reaches its maximum after metamorphosis in some flatfish species such as in turbot, whereas in others like plaice *Pleuronectes platessa* (Geffen et al. 2007) and Senegalese sole, the eye is developed and functional before metamorphosis (Bejarano et al. 2010, Padrós et al. 2011).

As a typical flatfish species, Senegalese sole exhibits the most extreme features with respect to shape asymmetry and lateralized behavior of any vertebrate species. They begin life as a pelagic and bilaterally symmetrical larva but, as development proceeds, they undergo a spectacular ontogenetic metamorphosis, during which one eye migrates from one side of the head to the other, so that eventually both eyes are on the same side of the head, the ocular side. Senegalese sole is a typical dextral species. The larval period of the Senegalese sole ends with metamorphosis, during which the pelagic larva is remodelled to a juvenile form adapted to a benthic life (Dinis et al. 1999, Fernández-Díaz et al. 2001, Gavaia et al. 2002, 2006, Piñuela et al. 2004, Padrós et al. 2011, among others). The ontogenetic events involved in development of *S. senegalensis* indicates that the pelagic early life stages are characterized by a rapid development of the visual-sensory (early visual feeders) and digestive organ- systems (Ribeiro et al. 1999a,b, Bejarano et al. 2010), which occurred during the lecitotrophic or the so-called endotrophic phase and through pre- and pro-metamorphic stages. Vision plays the most important role in prey detection in early stages of fish life (Blaxter 1986). However, Senegalese sole larvae are capable of eating and growing independent of light or in situations with low light availability (Ronnestad et al. 2013). A recent study indicated that while pre-metamorphic larvae mostly consume food during the light period, post-metamophic specimens showed a continuous feeding pattern along the whole day, with a trend of higher gut content during the dark hours. Thus, it has been suggested that in this flatfish species, larvae change from visual consumers to chemical and mechanical predators after metamorphosis (Navarro-Guillén et al. 2015). The above-mentioned feeding pattern and behavior changes during the development of the Senegalese sole may be related with an early, progressive and gradual functionality of the retina and neural system, which is consequently associated with an early pigmentation of the eye and fast development and differentiation of cone and rod opsin photoreceptors, as well as with the early development of other sense organs and mechano-structures in larval phases and/or early metamorphic stages (Bejarano et al. 2010, Padrós et al. 2011), similar to what has been reported for many other altricial fish larvae (Fuiman 2002, Evans and Browman 2004, Falk-Petersen 2005).

S. *senegalensis* shows a fast and efficient larval development, and an early onset of exogenous feeding, thus a very short lecitotrophic (endogenous feeding) phase, as well as a short pre-metamorphic stage before it became asymmetric (Dinis et al. 1999, Fernández-

Díaz et al. 2001, Ribeiro et al. 1999a,b, Imsland et al. 2003, Padrós et al. 2011) compared to other flatfish species (Falk-Petersen 2005). Under intensive rearing conditions (i.e., at 19.5°C), newly hatched Senegalese sole larvae are able to attain the benthic phase and to complete metamorphosis in less than 3 weeks (Dinis et al. 1999, Imsland et al. 2003). In particular, Senegalese sole is known to start metamorphosis between 9 and 15 days post-hatch (dph) depending on the feeding regimes (Fernández-Díaz et al. 2001, Boglino et al. 2012). The onset of eye migration (stage S1, according to the staging method developed by Fernández-Díaz et al. (2001)), starts between 8 and 12 dph, at *ca.* 4.5–5.0 mm total length (TL). Two days later, > 50% of the larvae initiate metamorphosis and the left eye starts to displace. At 6.0–7.0 mm TL, the left eye reaches the middle of the dorsal surface and larvae change their swimming plane (S3). The stage S4 is when the eye-migration process is accomplished and generally occurs at body sizes longer than 8 mm TL. In fact, the size, when practically all specimens (95%) have completed metamorphosis, is *ca.* 9.4–9.8 mm TL, around 16 dph (Fernández-Díaz et al. 2001). In contrast, for the Atlantic halibut it has been shown that metamorphosis is related to the developmental stage rather than age (Saele et al. 2006). However, it has been shown that on the onset of metamorphosis Senegalese sole larvae is size dependant rather than age dependant, which means that those larvae growing faster will begin their metamorphosis earlier. In this context, and in order to characterize the process of development in Senegalese sole, different larval stages and metamorphic phases have been defined on the basis of morphological and ecological features and according to the position of the left eye. Thus, pre-metamorphic symmetric larvae (< 4.5 mm TL) show a vertical swimming plane, whereas in early pro-metamorphic larvae (4.5–5.5 mm TL), the left-eye starts to migrate towards the dorsal cephalic region until it touches the middle of the dorsal surface, and larvae lose their bilateral symmetry. During the middle of metamorphosis or the so-called pro-metamorphosis, larvae change their swimming plane and the eye continues to migrate within the ocular side of the animal. Finally, during late metamorphosis (*ca.* 9.5 mm, by 16 dph), eye translocation is completed and the orbital arch is clearly visible (Fernández-Diaz et al. 2001). As reported by Osse and van den Boogaardt (1997), flattening of the body with a strong positive allometry of the body depth, head size and mouth gape, is required to start the metamorphosis transformation. Moreover, these authors suggested that flatfish larvae should be inactive during metamorphosis to allow recalibration of binocular vision after eye migration, as it has been found in specimens measuring 4.5–5.5 mm TL (Fernández-Díaz et al. 2001).

On the other hand, early field studies indicated that S. *senegalensis* did not feed during metamorphosis (see review in Geffen et al. 2007), although more recently works have shown that feeding continues, but not as effectively (Yúfera et al. 1999), and prey ingestion and daily ration (food ingested by fish weight) as well as the C:N ratio decreases at early metamorphosis (Fernández-Díaz et al. 2001, Cañavate et al. 2006). The highest growth rates are detected before metamorphosis, from first-feeding to the onset of eye migration in Senegalese sole; afterwards, specific growth rates tend to decrease significantly when metamorphosis takes place (Fernández-Diaz et al. 2001). Interestingly, flatfish larvae may be particularly vulnerable to predation if vision and other senses are impaired during metamorphosis process (Geffen et al. 2007). As it has been previously mentioned, visual, neuronal and behavioural changes occur very early and are functional at early metamorphosis in Senegalese sole as (Piñuela et al. 2004, Bejarano-Escobar et al. 2010, Padrós et al. 2011); so the above-mention transition is a gradual process, rather than abrupt change, and larvae continue to feed on planktonic prey throughout metamorphosis, but at a lower rate (Fernández-Díaz et al. 2001, Fernández et al. 2009). During this period, food digestion and nutrient assimilation may be less efficient than at later stages because the gut, and mainly the gastric stomach, are not fully differentiated, although at this developmental period, Senegalese sole have enough digestive functionally, since intracellular digestion (pynocitosis and cathepsins) is detected at this larval developmental pre-metamorphic stage (Sarasquete et al. 1996, 2001, Ribeiro et al. 1999a,b, Zambonino-Infante et al. 2008), as it has been reported in many other altricial fish larvae (Govoni et al. 1986, Fuiman 2002, Falk-Petersen 2005). Furthermore, energy storage reserves increase in this species prior to metamorphosis and is followed by a decline in energy content (Yúfera et al. 1999). In particular, Senegalese sole accumulates large reserves like lipids and glycogen in the liver (Ribeiro et al. 1999a), which are used during metamorphosis (Geffen et al. 2007). Thus, larval development can be delayed if larvae have accumulated sufficient hepatic and other energetic reserves before metamorphosis (Osse and Van den Boogaart 1997). In summary, it is assumed that the developmental changes resulting in body asymmetry during metamorphosis take place at the cost of somatic growth, affecting especially the length of specimens.

Taking into account the above-mentioned and discussed points and our own results and considerations, the classification into different larval developmental phases and metamorphic stages in *S. senegalensis* may be considered a more accurate method to determine and standardize larval development than other criteria frequently used such as the age from hatching (dph) and/or larval size or total length (TL) because larval staging is independent of the rearing zootechnical conditions. Moreover, in different flatfish species, including Senegalese sole, the larval length (TL) is relatively constant and less variable than condition index (weight) or age at the onset of metamorphosis process (Amara and Lagardére 1995, Fernández-Díaz et al. 2001, Geffen et al. 2007). In addition, staging larval development using morphological characters is of special value when dealing with fish species with a great range of size, and *S. senegalensis* is not an exception (Padrós et al. 2011). Consequently, different developmental larval phases (P), such as the endotrophic or lecitotrophic and the exotrophic and pre-metamorphic phases, as well as several metamorphic stages (S), are described and/or indicated in this chapter to characterize the ontogeny and development of *S. senegalensis,* as it has been recently updated (Sarasquete et al. 2017), re-summarized in Table 1, and represented graphically in Figs. 1 to 5. The above-mentioned tabular overview also includes data of the median values of ranges of TL and dph that characterize each ontogenetic event (phases, P0–P9 and stages, S1–S4). These values were obtained from the synthesis of own observations and from the available literature in this flatfish species (Dinis et al. 1999, Ribeiro et al. 1999a,b, Fernández-Díaz et al. 2001, Piñuela et al. 2004, Villalta et al. 2005, Cañavate et al. 2006, Ortiz-Delgado et al. 2006, Bejarano et al. 2010, Padrós et al. 2011, Sarasquete et al. 2017, among others).

In fish, the embryogenesis, larval development and particularly the metamorphosis of flatfish is an endocrine-driven process in which the thyroid system plays a central role (Tanaka et al. 1995, Power et al. 2008, Inui and Miwa 2012, Schreiber 2013). Gross changes in external morphology occurring during metamorphosis in Senegalese sole, as in many other flatfish species, are regulated by molecular and biochemical processes that are directly or indirectly affected by endocrine factors, and mainly via thyroid hormones—THs—and thyroid receptors—TRs—(Power et al. 2001, 2008, Darras et al. 2015). Levels of THs peak at metamorphosis climax during the larvae-to-juvenile transition in different flatfish species including the Senegalese sole (Klaren et al. 2008, Manchado et al. 2008, 2009, Ribeiro et al. 2012). The presence of first thyroid follicles, which are detected by means of optical microscopy and inmunohistochemistry methods, appear as coinciding with the opening of the mouth (Ortiz-Delgado et al. 2006), whereas other authors such as Klaren et al. (2008) and Padrós et al. (2011) have reported their appearance at older ages

Table 1. Development of Senegalese sole, *S. senegalensis* from hatching to the completion of metamorphosis based on morphological, histological and ecological features and eye migration process described by many listed authors and re-summarized (Modified from Sarasquete et al. 2017).

Larval phases (P) and metamorphic stages (S)		TL (mm)	Dph
Phase P0: Yolk-sac larvae	Newly hatched larvae. Yolk sac present with several lipid globules. Bilateral symmetry, notochord straight. Digestive tube slightly curved, surrounding the upper zone of the yolk sac. Head bent downwards and attached to the yolk sac. Melanophores distributed along the longitudinal axis of the body and in the finfold that surrounds the larval body. Motionless larva, floating on its side or with the yolk sac upwards.	2.4–2.8	0
P1-P2: Endotrophic phases	Digestive tube enlarged in the central area and bent in the distal region Anus not completely open near the pelvic fin. Melanophores arranged in four groups in the dorsal finfold and one group in the pelvic, as well as along the trunk and the yolk sac. Yolk sac reduced. Eyes already pigmented. The dorso-cranial distension is noticeable.	2.9–3.0	1–2
P3-P4: Endo-Exotrophic phases	The size of the yolk sac substantially decreases. Absence of lipid globules. Digestive tube curved into a loop in the central area. Mouth and anus open. Eyes fully pigmented. The larva displays active hunting behavior.	3.1–3.8	3–4
P5-P9: Pre-metamorphic phases	Absence of yolk. Eyes fully pigmented. The dorso-cranial distension no longer visible. Preys visible inside the gut. Notochord slightly bent upwards. Hypurals appearing in the primordial caudal fin supporting the fin rays. Dorsal finfold covering the head at the beginning of this stage retracts back leaving the eyes uncovered. Myotome height augments. Melanophores occupying mostly the cephalic region and the trunk increase in number.	3.9–4.4	5–9
Stages of Metamorphosis (S)			
S1—Early metamorphic stage	Body asymmetry and eye migration starts. Notochord caudally bent upwards < 45°. The number of fin rays in the caudal fin increases and the caudal fin reaches its final form. The fin rays appear in all the fins. Pigmentation covers the body, but not the fin rays. Dorsal finfold behind the eyes. Olfactory cavities and nostrils present. Pelagic life. Asymmetrical larvae. Beginning of lateralized behavior. Left eye migration towards the dorsal margin of the larvae, but had not reached the dorsal margin. Pigmentation was still generally similar on both sides.	4.5–5.9	10–12
S2—Middle metamorphic stage	Left eye positioned upwards, partially visible from the right side. Swimbladder no longer observable. Notochord curves upwards > 45° and < 90°. Caudal fin ray fully develops with all the fin rays and final shape. The fin rays well developed. The larvae begin to change their swimming from vertical to benthic. Asymmetrical larvae. The left eye is on top of the head, just in the dorsal margin. Left eye visible from the right-ocular side. The body right side is more pigmented than the left side.	6.0–7.9	13–15

Table 1 contd. ...

...Table 1 contd.

Larval phases (P) and metamorphic stages (S)		TL (mm)	Dph
S3—Late metamorphic stage	Left eye on the right side. Notochord bent upwards 90°. Full pigmentation, including all the fins. Small melanophores (adult type) appear together with large melanophores (larval type). Dorsal fin anteriorly placed. Benthic life. Asymmetrical larvae. Left eye located entirely on the left side. The distance between the two eyes is greater than the diameter of a single eye. The orbital arch is clearly visible.	8.0–9.9	16–19
S4—Postmetamorphic stage	Post-metamorphic stage. Eye migration and metamorphosis completed. Left eye close to the right eye. The parafrontal bone extends and the gap left by the migrated eye is filled. Dorsal fin extends beyond the ocular zone. Barbels appear around the mouth. Pigmentation similar to that in juveniles showing the whole range of chromatophores. Benthic life. Asymmetrical juvenile. The distance between the two eyes is less than the diameter of a single eye. The left side of the body has almost totally lost its pigmentation.	10.0–12.00	20–40

(*ca.* 7–8 dph). Therefore, the thyroid system is essential for normal growth, differentiation and development, since THs and signalling pathways directly or indirectly stimulate processes of proliferation and apoptosis patterns, as well as resorption and remodelling of larval tissues (Power et al. 2001, 2008, Schreiber 2013, Darras et al. 2015, Alves et al. 2016, among others). In this context, some of the main steps for thyroid gland development, TH biosynthesis, metabolism and accumulation in the thyroid follicles through a negative feedback regulation of the HPT axis during development of the Senegalese sole are fairly well established (Campinho et al. 2015, Infante et al. 2007, 2008, Isorna et al. 2009, Izigar et al. 2010, Klaren et al. 2008, Manchado et al. 2008, 2009, Ortiz-Delgado et al. 2006, Ponce et al. 2010, Sarasquete et al. 2017). For example, among other neuroendocrine signalling patterns, the constitutive *TSH* transcript levels decrease significantly at the onset of metamorphosis. On the contrary, *THs*, *Tg*, *TRs*, and iodothyronine deiodinases (e.g., *Dio2*) increase significantly at the metamorphic climax, showing variable temporal pattern changes during the larval development of this flatfish species, as recently reviewed by Darras et al. (2015). Thyroid hormones have been firstly localized within the yolk matrix during the endogenous feeding stage, by means of immunohistochemical methods, which indicates that THs at this stage are of maternal origin since no functional thyroid tissue is developed at embryogenesis (Tanaka et al. 1995). In particular, an immunohistochemical staining against T4 and T3 antisera was still strong at 1 dph (yolk-sac matrix) and decreased its staining intensity at 4–5 dph, suggesting a gradual utilization and/or mobilisation of these hormones as organogenesis progresses (Ortiz-Delgado et al. 2006). The thyroid hormone T3 has a direct action on the production of retinoic derivates and therefore on the visual system function for successful switching from endogenous to exogenous energy sources and foraging behaviour (Power et al. 2008). The thyroid follicles are apparently functional at first feeding and the thyroid hormone T4 can be immunolocalized inside the colloid content by 6 dph, while T3 appears later, suggesting that the active form of the thyroid hormone (T3) appears late in the development while its inactive form (T4) appears at earlier ages (Klaren et al. 2008). These results are in agreement to those described in

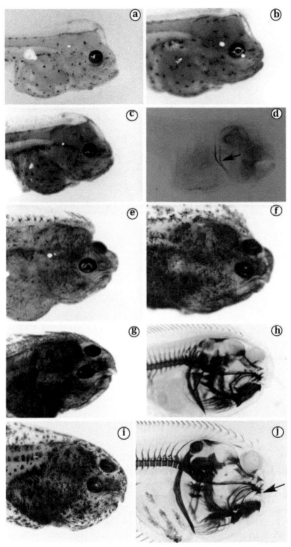

Fig. 1. Morphological features and changes during larval phases (P) and metamorphic stages (S) of Senegalese sole, *Solea senegalensis* larvae, following the classification previously re-summarized, in this review.

Some morphological characteristics of different larval phases: P3 (a), P4 (b), P5 (c and d) and metamorphic stages: S1 (e), S2 (f), S3 (g and h) and S4 (i and j) are shown. A detailed description of each morphological event during larval development and metamorphosis is summarized in the Table 1. Note mouth opened and eyes fully pigmented in larvae from opening of mouth at P3until P5 phases (a to c). *Cleithrum* (arrow) is the first-calcified structure in larvae at phase P5 (Alcian Blue/Alizarin Red staining) (d). During metamorphic stages, at S1–S2, the body assymetry and eye migration started (e and f). At the end of metamorphosis, at S3 (g and h) and in postmetamorphic S4 (i and j) specimens, the body was fully pigmented and the eye migration completed. Premaxillary from skull was completely calcified (j; arrow) in postmetamorphic specimens, at S4 stage (30 dph).

other fish species where levels of T4 increase at first feeding, but T3 concentrations remain low. Evidence supporting a critical role for thyroid hormones during the transition from larvae to juveniles has been provided from different studies, in which metamorphosis was accelerated by the administration of exogenous thyroid hormone and inhibited by the blockage of thyroglobulin iodination with thionamide and thiourea. In addition, exogenous

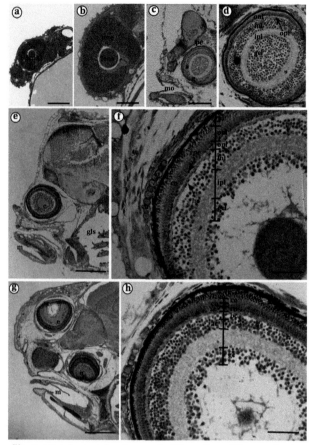

Fig. 2. Histological development of the retina of *S. senegalensis*.

(a and b). Larvae at P1 phase (1dph) showing a developing retina of undiferentiated cells. Lens is clearly distinguished at this stage. (c and d) Larvae after first exogenous feeding, at phase P4, showing the mouth opening and eyes fully pigmented. Differentiated retinal layers were detected at this early larval phase: a pigmented epithelium followed by an outer nuclear layer, an outer plexiform layer and the inner nuclear layer differentiated into two substrates, amacrine and bipolar cell layer, and an inner plexiform layer and finally a ganglion cell layer were observed. (e and f) In premetamorphic larvae (P9), rod precursor cells were clearly distinguished in the photoreceptor cell layer. Between the outer plexiform and the inner nuclear layers, a layer of horizontal cells is clearly distinguished. (g and h) In postmetamorphic Senegalese sole specimens at S4 stage, the retina showed the histological characteristics and features of a fully developed and functional eye. Scale bars: a, b, c, d, e, g represents 50 μm; e and g represents 150 μm. ac: amacrine cells; bc: bipolar cells; ce: cone ellipsoids; ch: choroid; cn: cone nucleus; e: eye; gcl: ganglion cell layer; gls: gills; hc: horizontal cells; inl: internal nuclear layer; ipl: internal plexiform layer; m: mouth; mo: mouth opening; onl: outher nuclear layer; opl: outher plexiform layer; pe: pigmentary epithelium; rn: rod nucleus; rpc: rod precursor cells; une: undiferentiated epithelium.

thyroid hormone induces the premature differentiation of pectoral fins, whereas treatment with chemical inhibitors (goitrogens) of the thyroid gland function prevent the transition of larvae to juveniles (Tanaka et al. 1995, Manchado et al. 2008, 2009, Ribeiro et al. 2012). Around the first month of life (30 dph), the thyroid follicles present a colloid-filled lumen surrounded by a cuboidal epithelium, suggesting a high hormonal activity (Ortiz-Delgado et al. 2006). Inside the colloid matrix, many vacuoles adjacent to the epithelium are also detected; this colloid comprises the reserve of potential hormones for the thyroid, which is the only endocrine organ in vertebrates with such an extracellular storage system. The

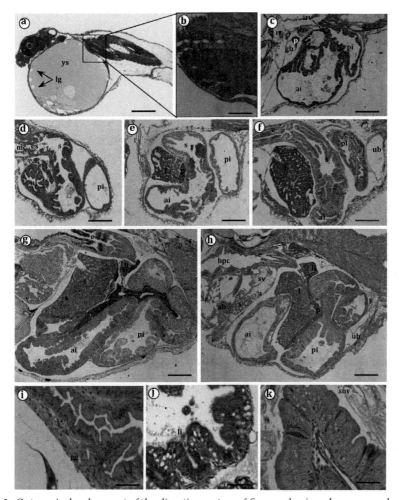

Fig. 3. Ontogenic development of the digestive system of *S. senegalensis* and accessory glands.
Larvae at phase P1 (a and b), showing a large yolk sac with several lipid droplets and a gut consisted on a straight tube dorsal to the yolk-sac are clearly detected. An incipient liver and pancreas between the digestive tract and the yolk sac are also evidenced at this early developmental phase. With larval growth, a gut elongation and the progressive development of accessory glands are also observed, during larval phases and metamorphic stages: P4 (c), P7 (d), P8 (e), S1 (f), S3 (g) and S4 (h) stages. Gastric glands are developed in the stomach of postmetamorphic specimens, as evidenced at stage S4 (i). Mucosal epithelium lining the anterior (j) and posterior (k) portions of the intestine in premetamorphic larvae (P9). Scale bars represents: a to h, 100 µm; I to k, 50 µm. a: auricle; ab: aortic bulb; ai: anterior intestine; bpc: bucopharyngeal cavity; e: eye; es: esophagus; gb: gall bladder; gg: gastric gland; irv: ileorectal valve; k: kidney; l: liver; lg: lipid globules; li: lipid inclusions; lpc: liver primordial cells; mc: mucous cells; p: pancreas; pi: posterior intestine; ppc: pancreas primordial cells; rt: renal tubules; s: stomach; sb: swim bladder; snv: supranuclear vacuoles; sv: senus venosus; ub: urinary bladder; v: venricle; ys: yolk sac.

presence of neutral glycoproteins within the colloid content of the thyroid follicles has been demonstrated since it was strongly stained due to the Periodic Acid-Schiff (PAS) reaction without any changes in the coloration when using the combined Alcian Blue at pH 2.5/PAS technique (absence of acid rich glycoproteins). These histochemical results suggest the presence of thyroglobulin, a large glycoprotein formed by the combination of iodinated molecules of tyrosine and the main component of this colloid, which is the precursor of thyroid hormones. Moreover, the colloid of thyroid gland is rich in proteins,

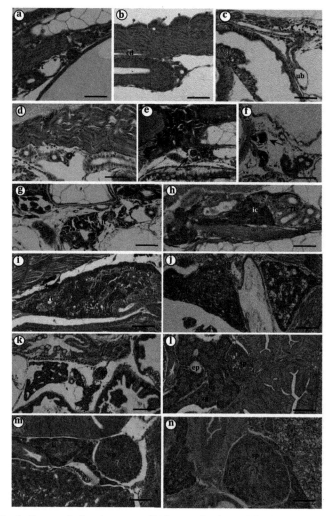

Fig. 4. Ontogenic development of kidney and spleen of the *S. senegalensis.*
Two primordial renal tubules and the collecting duct were detected in larvae at early larval phase P1 (a) and (b). The ureter and urinary bladder are evident in larvae at phase P4 (c) in endo-exotrophic larvae. The haematopoietic tissue is still scarce and the first glomerulus is detected in larvae at thistime P4 (arrow) (d) and (e). Convoluted renal tubules, Stannius corpuscles (arrow) and haematopoietic tissue are present in the kidney in premetamorphic larvae at phase P8(f). The haematopoietic tissue increased in pronephros at P9 and interrenal cells were detected during metamorphic stage S1 (g and h respectively), while the mesonephros (i) and pronephros (j) were easily differentiated at the end of metamorphosis, in larvae at S3 stage. In early larvae, at phase P4 (k), the spleen anlage appeared as a small cluster of mesenchymal cells adjacent to the wall of the midgut. This organ increased in size and volume from metamorphosis S1(l), S3(m) and in postmetamorphic stage S4 in which the typical structure of adult specimens, with numerous splenic sinusoids and ellipsoids was detected (n) particularly in posmetamorphic specimens. Scale bars represents 100 μm. cd: collecting duct; ep: exocrine pancreas; ic: interrenal cells; it: interstitial tissue, li: Langerhans islet; rt: renal tubules, sp: spleen, ub: urinary bladder; zg: zymogen granules.

and specially in proteins rich in tyrosine, arginine and tryptophan. However, the presence of proteins within thyrocytes, which formed the follicular wall is scarce (Ortiz-Delgado et al. 2006, Sarasquete et al. 2017). Recent interesting reviews about flatfish metamorphosis and hormonal regulation have been published (Power et al. 2001, 2008, Geffen et al. 2007,

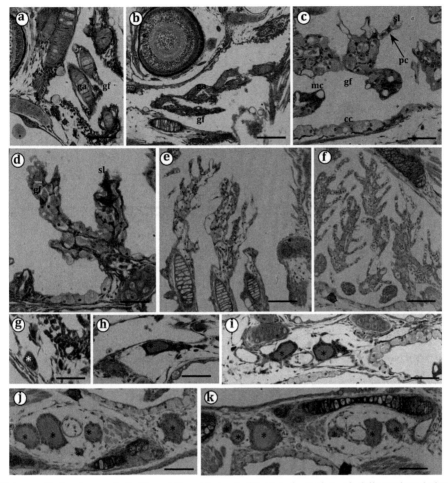

Fig. 5. Ontogenetic development of gills and thyroid gland of S. *senegalensis* through different larval phases and metamorphic stages.

Gill development of larvae at phases P4 (a) and P7 (b) showing gill arches and primary filaments. In premetamorphic larvae, at P8 (c), gill secondary filaments emerging from the base of primary filaments are clearly recognized. At this larval phase, blood capillaries and pillar cells were present in the inner part of filaments. An increase in the number of chloride cells is clearly visible at this development larval phase. During metamorphic and postmetamorphic stages: S1 (d), S3 (e) to S4 (f), a proliferation of primary and secondary filaments is clearly recognized. Thyroid development in early larvae at phase P4 (g) showing a detail of the first thyroid follicle adjacent to ventral aorta and bulbus arteriosus (asterisk). A progressive increase in number of follicles (asterisks) was detected in premetamorphic larvae, at P9 (h), during metamorphosis stages: S2 (i), S3 (j) and in postmetamorphic specimens, at S4 (k). Scale bars: a, b, e, f, g, h, i, j and k represents 50 µm; c and d represents 100 µm. cc: chloride cells; ga: gill arches; gf: gill filament; mc: mucous cells; pc: pillar cells; sl: secondary lamellae.

Inui and Miwa 2012, Schreiber 2013, Darras et al. 2015), where readers may find a deeper insight on this topic.

Interestingly, it has been described that altricial fish larvae only develop a functional stomach (gastric glands), such as evidenced in S *senegalensis* (Ribeiro et al. 1999a) during the metamorphosis period (Table 1) since T4 strongly stimulates the differentiation of gastric glands and the production of pepsinogen (precursor of pepsin, the main gastric enzyme), whereas thiourea (a potent inhibitor of thyroid hormone synthesis) has the converse effect

(Miwa et al. 1992, Tanaka et al. 1995, Manchado et al. 2008, Ribeiro et al. 2012). Indeed, during metamorphosis, important molecular and morphological modifications of the gastrointestinal tract (GI) occur, in parallel with the highest thyroid hormonal levels (Power et al. 2008). An elongation of the digestive tract and an increase in the absorption surface of the intestinal mucosa are detected during metamorphosis in flatfish species. In this sense, fully developed and functional gastric glands are only evidenced in metamorphosed *S. senegalensis* specimens (Ribeiro et al. 1999a, Sarasquete et al. 2001), usually after the first month of life in different flatfish as well as in other marine altricial fish species (Govoni et al. 1986, Fuiman 2002, Zambonino-Infante et al. 2008, Gomes et al. 2014). In different flatfish species, pepsinogen can be only detected after metamorphosis phase. Indeed, Murray et al. (2006) indicated that this enzyme precursor appears when gastric glands are completely developed. Nonetheless, it remains to be demonstrated that the abundance of pepsinogen transcripts correlates with the acid proteolytic activity of pepsin. The synchronous expression of *pepsinogen* and H+/K+ *ATPase* genes has been proposed to be a physiological strategy to promote the quick conversion of pepsinogen into pepsin (Gomes et al. 2014). Acid digestion and proteolysis in the stomach seems to be residual in the Senegalese sole; the digestion occurs primarily in its long intestine in a slightly alkaline environment. Indeed, in larvae, juveniles and adults of this flatfish, the gastric pH never decreased below 6.0 irrespective of the age (Yúfera and Darias 2007, Morais et al. 2016). In other flatfish species, such as the Atlantic halibut, *Hippoglossus hippoglossus*, the pH in the intestinal regions remains alkaline in all developmental stages. During the climax of metamorphosis, a rapid colour change from acidic to alkaline is observed in the midgut when the pH indicator solution passes through the pyloric sphincter. However, a gradual stomach acidification is detected in the stages corresponding to the climax of metamorphosis and the stomach lumen is clearly in the acidic range (pH < 3.5) at the end of the metamorphosis phase. Furthermore, there is a co-ordination between morphological changes and the key elements essential for the stomach proteolytic activity such as the H^+/K^+-ATPase and HCl production, and the thyroid hormonal system most likely orchestrate this change (Gomes et al. 2014, Alves et al. 2016).

4. Brain development

The development of the Senegalese sole forebrain is interesting due to the coincidence of several features. As for other Actinopterygian, the part of the neural tube rising from the forebrain goes through an eversion process, with later migration and neural differentiation. On the other hand, the forebrain goes through a dramatic metamorphosis, changing from a symmetrical to an asymmetrical shape, which implies brain transformations; this process mainly relies on the forebrain that undergoes significant changes in shape and is responsible for regulating such big modifications (Piñuela et al. 2004). During larval development, most of the neural proliferation is restricted to the ventral part of the central nervous system (CNS). Thus, from hatching until the onset of exogenous feeding phase in *S. senegalensis* larvae, the CNS is formed by a developing encephalon connected to the spinal cord. Although the encephalon is rudimentary, four different structures are identified in the cranial area: (1) the forebrain or telencephalon; (2) a prominent and rounded mesencephalic lobe in the dorsal area; (3) the midbrain or diencephalon in the ventral area separated from the telencephalon by a ventricle and, posterior to these structures; (4) the hindbrain or myelencephalon, which is elongated and connected with the neural tube that runs the larval length, dorsal to the notochord (Padrós et al. 2011). From 1 dph onwards, the brain is constituted by vesicles; the forebrain appears as a single vesicle organized as a three-laminar structure, and at this time the germinal layer is continuous. These germinal cells are small, with round or elongated bodies, and are strongly stained and densely packed. The sheath-like germinal zone inserts into the next medial layer, in the form of finger-like branches. The medial layer is composed of bigger, more round-shaped cells, less densely packed, and more weakly stained. The most internal sheath is characterized by a lack of cells except those that already have ventrally migrated. The cell bodies of these migrated neurons are large, round or polyhedric shaped, weakly stained in the cytoplasm and stronger in the nucleus. Although there is an increment of the nuclear complexity coinciding with first exogenous feeding, the three laminar structures are still present, but there is a higher differentiation of cells. The prosencephalon is seen as a unique vesicle, without division between telencephalon and diencephalon. The inferior lobes of the hypothalamus are already present as a rudimentary primordium, round shaped with large cells distributed around the adjacent to the periphery. Progressively, in premetamorphic Senegalese sole specimens, around 6–7 dph, there is a clear separation between the telencephalon and diencephalon. The hypothalamus, particularly the inferior lobes, shows higher complexity in which the cells start grouping as discrete nuclei. It may be expected that *S. senegalensis* has a longer embryological time since goes it through more events, but in reality there is a first phase, just before they become asymmetric, which is shorting-lasting than in other species. Indeed, the olfactory bulb and the differentiation between telencephalon and diencephalon are present, coinciding with the first exogenous feeding. The pituitary is already visible at 1 dph, and the inferior lobes of the hypothalamus around 3 dph. From this time, the olfactory bulbs differentiate from the telencephalon. The CNS is covered by an elongated layer of cells, the so-called meninges, which are clearly visible due to the presence of melanin pigmented cells. The pineal gland is also visible, coinciding with the onset of exogenous feeding, as a frontal extension of the diencephalon. It is formed by a stalk and an end-vesicle, and together with the habenular ganglions compose the epithalamus. It has been suggested that even the biggest transformations regarding the telencephalon take place in a quite short period of time, its complete metamorphosis ends much later. Interestingly, the proliferation zones described (preoptic region, dorsal thalamus, posterior tubercle, hypothalamus, etc.) in pre-metamorphic specimens are conserved until later developmental stages, long after metamorphosis. The eye migration implies changes in the brain shape and its rearrangement that mainly applies to the rostral part of the brain. It is assumed that most ethological and ecological changes seen during these transitional stages occur in parallel with appropriate neural innervation of the brain. According to Piñuela et al. (2004), although the biggest transformation regarding the telencephalon of Senegalese sole take place in quite a short time, the complete metamorphosis ends much later, possibly after three months of life, coinciding with the sexual determination and gonad development in this flatfish species (Pacchiarini et al. 2013, Viñas et al. 2013).

5. Eye development and retinal functionality

The range of activity in fish larvae increases with growth, as the function of sensory organs becomes more refined and locomotory structures develop. In this context, the eye is one of the major sensory organs in fish. It detects light stimuli, forms images and shows a wide range of structural adaptations to the visual environment. In most fish species, vision is considered the dominant sense during early life stages of development as it is required for feeding, orientation, schooling and avoiding potential predators (Blaxter 1986, Osse and Van den Boogaart 1997, Evans and Browman 2004).Vision is the major sense used by fish larvae in feeding, and thus, eyes become well pigmented and functional before the onset of first exogenous feeding (Blaxter 1986, Govoni et al. 1986, Falk-Petersen 2005). Indeed, at hatching of Senegalese sole, the retina is occupied by a proliferating neuroepithelium with no apparent morphological signs of differentiated neurons. Mitotic figures are located in the sclera surface of the presumptive neural retina. The eyes are un-pigmented, but an extremely flat presumptive epithelium is observed and becomes more evident during the lecitotrophic stage (Table 1). From this time onwards, the lens structure has two differentiated layers: an external layer of epithelial cells surrounded by a non-nucleated lens fiber layer formed by highly modified epithelial cells. The inner and outer plexiform layers become visible in the central and mid-peripheral regions of the retina. The retinal layering is completed in central and peripheral regions, coinciding with the first exogenous feeding in Senegalese sole. Thus, four presumptive retinal layers are observed, from the exterior to interior of the retina: the outer nuclear, formed by a row of nuclear segments of the presumptive single cones composed by columnar nuclear bodies; the inner nuclear layer consists by presumptive round bipolar cells; the inner plexiform layer, the first that appears and consists of the neural dendrites located in both the inner nuclear and the ganglion cell layer, which is composed of several rows of amacrine cells (Bejarano-Escobar et al. 2010, Padrós et al. 2011). Coinciding with mouth opening (Table 1, Fig. 2), most retinal cells are distinguished by morphological and topographic features; the ganglion cell layer presents four or five rows of ganglion cells. At this time, a continuous row of highly opsin immunoreactive photoreceptor outer segments is present in the central and mid-peripheral region of the retina, which probably corresponds to a new subset of cones (Bejarano-Escobar et al. 2010). During early metamorphosis at stage 1, around 10–12 dph, both types of photoreceptors, cones and rods are detected in the outer nuclear layer, and the cornea appears surrounded by an external epithelium, the same that covered the larval body. In the cornea, a fibrobastic layer (choroid) is present in the inner zone next to the lens and surrounds the entire eye. At the middle and late metamorphosis of Senegalese sole (Table 1), the extracellular matrix of chondroblasts and eye muscles are observed (Padrós et al. 2011). The proliferation of rods and emergence of scotopic vision plays an important role in the retina development because they make the larvae more capable of feeding under low illumination conditions and increase the motion sensitivity, contributing to predator avoidance (Evans and Browman 2004).

As in other altricial fish species (Blaxter 1986, Evans and Browman 2004), all retinal layers are developed during the first month of larval life. At this time, the postlarvae acquire a better vision in deeper waters allowing extending the hunting area. According to Bejarano-Escobar et al. (2010) and Padrós et al. (2011), and our own observations in Senegalese sole (Table 1, Fig. 1), at stage S4 (*ca.* 30 dph), such as it is showed in the Fig. 2, the eye consists basically of the typical retinal layers and cell types: (1) Pigment epithelium; (2) photoreceptor layer (cone and rods with inner and outer segments); (3) outer nuclear layer; (4) outer plexiform layer; (5) inner nuclear layer; (6) inner plexiform layer; (7) ganglion cell layer and (8) nerve fiber layer. The nuclear layers and the ganglion cell layer are separated by synaptic interactions of the two plexiform layers. The outer nuclear layer, holding all the somata of photoreceptors arranged into several rows is prominent, and cone and rod nuclei are visualized. The inner nuclear layer is composed of three cell types: horizontal cells, contacting with photorreceptors, round diffuse bipolar cells, synapsing with several photoreceptors and amacrine cells, which are in contact with ganglion cells from the ganglion cell layer, and displace the monolayer of amacrine cells. The glial or somata Müller cells appear to send processes to both inner and outer plexiform layers. An internal limiting membrane, the basal lamina of the Müller cells, separates the vitreous body from the retina, and an external limiting membrane separates the photorreceptor layer (cone and rod inner segments) from the outer nuclear layer.

6. Digestive system development and functionality

Similarly to other altricial fish larvae species, nutrient acquisition in Senegalese sole from hatching (0 dph) until the end of the endogenous feeding phase relies exclusively on the yolk-oil reserves assimilated through the yolk syncytial layer. Yolk-sac and lipid absorption is the result of the intracellular activity of the periblast in this flatfish (Sarasquete et al. 1996, Dinis et al. 1999, Ribeiro et al. 1999a, Zambonino-Infante et al. 2008). Because newly hatched larvae have relatively small yolk reserves and energetic sources, they must start feeding as soon as possible, and this requires the development of systems, organs and tissues involved in food capture and digestion. Thus, the opening of the buccal cavity and anus is one of the first major developmental events of the digestive system development, coupled with an early visual functionality (Blaxter 1986, Govoni et al. 1986, Pittman et al. 2013). Moreover, the early appearance of pharyngeal and mandibular teeth and the formation of longitudinal oesophageal folds, as well as the presence of goblet or mucous cells in the oesophagus are directly linked to the ability of fish larvae for the early onset of exogenous feeding (Govoni et al. 1986, Fuiman 2002, Falk-Petersen 2005, Zambonino-Infante et al. 2008). In newly hatched Senegalese sole larvae, the digestive tract consists of a tubular segment, histologically undifferentiated, laying dorsally to the yolk-sac with mouth and anus still closed. The epithelium consists of a monostratified layer of cubic or columnar cells, and the lumen of the digestive tract is narrow with a tendency to widen at both extremities. The yolk sac contains several peripheral oil globules and exhibits a homogenous acidophilic yolk surrounded by a squamous epithelium. Between the digestive tract and the yolk sac, two groups of round cells with spherical nucleus, prominent nucleolus and basophilic cytoplasm are observed, which may correspond to the incipient accessory digestive glands, the liver and pancreas. During the endogenous feeding phase (Table 1, Fig. 3), the posterior portion of digestive tube bents slightly approximately 45°, and the yolk-sac volume decreases, being completely absorbed by the end of this larval phase. At this time, the digestive tract is differentiated in two portions. The anterior portion is lined with a squamous epithelium, and groups of cells with circular disposition, which later develop into the gill arches, are observed in the pavement of the last third of this portion. The following portion of the digestive tract is lined with a columnar epithelium with a basal nucleus and cytoplasmic projections to the lumen, which disappear at first feeding. An elongation of the digestive tract is observed during this period, but larvae still do not possess a functional anus. The esophagus is lined by a cubic epithelium and a muscular layer is observed surrounding this structure, which becomes thicker with larval growth. An incipient undifferentiated stomach, which will differentiate at the end of the esophagus exhibiting a pouched shape, is lined by a cubic epithelium, with an ellipsoid, and centrally located nucleus, a prominent nucleolus and heterogeneous cytoplasm. At this larval stage, a slight constriction at the end of the region when the future stomach will differentiate indicates the beginning of the intestine. The intestinal valve is visible, dividing the anterior and the posterior intestinal regions. These two portions are lined by a columnar epithelium, with basal nuclei and prominent nucleolus. Nevertheless, a brush border, a more basophilic cytoplasm with several apical vesicles and dense granules are observed in the anterior intestinal portion. With larval growth, the gut elongates, and an increase in dense granules, vesicles and vacuoles and also in the thickness of the brush border in the anterior intestine is observed. The intestinal lumen appears more corrugated due to development of folds. During the lecitotrophic phase of Senegalese sole, the basophilic exocrine pancreas shows an increase in the number of acidophilic zymogen granules (precursors of pancreatic digestive enzymes) and several internal divisions. At this phase, the endocrine pancreas, or islet of Langerhans, is also observed as one islet inside the exocrine pancreas. The liver is located in opposition to the yolk sac border on the larval left side and the pancreas faces the posterior border of the liver, occupying both sides of the abdominal cavity. Hepatocytes remain spherical, but acinar cells become polyedric, with spherical and basal nuclei and several acidophilic non-dense vesicles and dense granules in apical cytoplasm (Fig. 3).

At the onset of exogenous feeding, both the mouth and the anus open (Table 1) and a loop is observed in the third portion of the digestive tract; residues of the yolk sac and oil globules remain visible in the frontal part of the abdominal cavity. By this time, the abdominal cavity has a prominent round shape with a ventral position. Until the start of a benthic life style, a similar disposition of all digestive structures is maintained apart from their increase in size. However, the liver is the only exception, since its growth takes place in an area left by the endogenous reserves that become occupied by the hepatic tissue. The first oesophageal mucous cells are observed coinciding with first feeding, along with prominent longitudinal folds (Ribeiro et al. 1999a, Sarasquete et al. 2001). As larval development proceeds and larvae grow during Senegalese sole pre-metamorphosis, the hepatocytes' cell morphology appears better defined, with a big spherical and central nucleus, a prominent nucleolus and a granular cytoplasm. Hepatocytes are arranged along sinusoids forming the hepatic parenchyma. The liver presents a granular and less basophilic cytoplasm than the exocrine pancreas. Acidophilic zymogen granules are observed in pancreatic acini. Both liver and pancreas increase their size and complexity with larval development. The presence of vacuoles in the anterior portion of the intestine occurs mainly at 2–3 days after exogenous feeding, whereas lipid vacuoles are more abundant in the posterior rather than the anterior intestinal segment. Biliary and pancreatic ducts open in the anterior intestine just after the stomachal region. At early metamorphosis or Stage 1 (Table 1, Fig. 1), the left eye migration and also the migration of the anus toward the pelvic fin is observed, which is accompanied by the reorganisation and widening of the buccal and gill cavities. The stomach increases in size in conjunction with the development of deep mucosa rugae during the middle of metamorphosis (Stage 2). From this period onwards, the epithelium changes to columnar and the submucosa becomes thicker. The portions of the digestive tract located in the abdominal cavity are the stomach, still poorly developed, the anterior intestine, which delineates the lower part of the abdominal cavity bending over itself to reach the posterior part of the cavity, and the posterior intestine, which delineates the posterior part of the abdomen. The lining of the buccal and the gill cavities progressively changes to stratified squamous epithelium and the mucosa gets thicker. Once metamorphosed, the abdominal cavity is positioned on the right side of the dorsal side, beginning just after the gill arch; it exhibits a triangular shape, wider near the head region and becoming thinner in its posterior part. The stomach has a sac-shaped form and limited in its concave side by the liver, occupying the wider part of the abdominal cavity. At the level of the pyloric valve, the intestine bents slightly over the stomach and delineates the periphery of the abdominal cavity, when it reaches the frontal right side it bends over again following in a diagonal direction until it reaches the end of the cavity where it turns to the end near the pelvic fin. At this stage, the epithelial cells from the stomach mucosa have a wider apical portion than basal. In post-metamorphic specimens at Stage 4 (Table 1, Fig. 1), the submucosa is thicker than at the previous stages of development, whereas between 24 and 27 dph several transverse sections of canals, lined with cubic epithelium progressively appear beneath the mucosa, indicating the development of tubular gastric glands (Ribeiro et al. 1999a, Zambonino-Infante et al. 2008).

During the first month of larval life, the appearance of functional gastric glands and the increase in size and complexity of the digestive system and associated structures and glands (i.e., liver, pancreas) are the more important ontogenetic modifications detected in Senegalese sole, similar to most altricial fish larvae described so far (Govoni et al. 1986, Falk-Petersen 2005). In early juveniles like adults, specimens of Senegalese sole have a small stomach, a relatively long intestine and lack pyloric caeca (Arellano and Sarasquete 2005, Yúfera and Darias 2007). Regarding the activity of digestive enzymes, during the endogenous feeding phase, alkaline phosphase activity is detected in the periphery of the yolk-sac and also in the intestinal epithelium (Ribeiro et al. 1999a,b, Zambonino-Infante et al. 2008) indicating the presence of nutrient absorption and transport processes through membranes (Govoni et al. 1986). The yolk-sac matrix is constituted mainly by glycoproteins and lipids, especially neutral lipids and waxes within oil globules (Sarasquete et al. 1996, Ribeiro et al. 1999a). During the endogenous feeding phase, larvae accumulate in the exocrine pancreas, the precursors of the main pancreatic digestive enzymes in the form of zymogens, which guarantees that larvae are able to digest food at the onset of exogenous feeding. At histochemical level, zymogen granules are strongly acidophilic, and contain proteins rich in tyrosine, lysine and arginine (Sarasquete et al. 2001). Moreover, the early presence of peptidase activity (Ribeiro et al. 1999b) confirmed that Senegalese sole larvae possess the required enzymes to perform protein digestion, and for the transport of aminoacids through membranes (Govoni et al. 1986). Although an incipient stomach is distinguished during the lecitotrophic phase, it is only at the end of the first month of larval life when the presence of the first noticeable developing gastric glands is evidenced (Ribeiro et al. 1999a,b, Sarasquete et al. 2001) that suggests the beginning of the juvenile period (Govoni et al. 1986, Zambonino-Infante et al. 2008). Interestingly, the increased secretion of mucous digestive cells, and also the strong muscular layer, during larval development, as in *S. solea* (Boulhic and Gabaudan 1992), would facilitate food ingestion and transport of food particles, since these secretions can participate in the protection of the digestive mucosa from proteolytic degradation, as well as in the prevention of mechanical injuries and bacterial infections (Sarasquete et al. 2001, Zambonino-Infante et al. 2008). The presence of neutral lipids in the anterior intestinal mucosa after the beginning of exogenous feeding in Senegalese sole has been described as an indicator of luminal lipid absorption and storage due to poor lipid metabolism, which improves with larval growth (Govoni et al. 1986). Pinocytotic invaginations and acidophilic supranuclear vesicles that contain a mixture of cathepsin-like enzymes and acid phosphatases are prominent in the enterocytes of the posterior intestine after 3 days of exogenous feeding (Ribeiro et al. 1999a,b, Sarasquete et al. 2001, Zambonino-Infante et al. 2008), and are involved in the intracellular digestion of proteins. In many species, these structures are present only until formation of the gastric stomach, suggesting a transient mechanism for protein digestion, related to the immaturity of the digestive tract, as well as for compensation of a weak or absent extracellular protein digestion (Govoni et al. 1986, Pittman et al. 2013).

Such as above indicated, the ability of the fish larvae, as Senegalese sole, to digest proteins increases with larval development and age, as has been confirmed by several enzymatic and histochemical analyses (Govoni et al. 1986, Ribeiro et al. 1999b, Zambonino-Infante et al. 2008). According to several authors (Yúfera and Darias 2007, Morais et al. 2016), acid digestion and proteolysis in the stomach seems to be residual in the Senegalese sole, since the gastric pH was always above 6.0 even in the adult specimens. However, in other flatfish species, like the Atlantic halibut, a transition from an alkaline to an acidic pH in the lumen stomach was evident at the end of the metamorphosis phase, suggesting the stomach proteolytic potential (Gomes et al. 2014). Additionally, the link between ghrelin and the acquisition of gastric-stomach proteolytic function in flatfishes has been recently confirmed (Gomes et al. 2014, Navarro-Guillen et al. 2016). Ghrelin is a hormone that acts via its receptors named growth hormone secretagogue receptors, and it is mainly produced in the stomach. The first discovered action of ghrelin was connected with regulating growth hormone (GH) secretion, has the putative role of a fully functional stomach and appetite regulation (Xu and Volkoff 2009), and is typically known as the "hunger hormone". In this sense, it has been suggested that during pre-metamorphosis and at the post-climax in flatfish species, there may be multiple correlations between *pepsinogen,* gastric proton pump transcripts, HCl production, pH, pepsin activity, and thyroid and ghrelin hormonal levels. Furthermore, the presence of a "physiological" pyloric sphincter, specific region with a strong muscular loop-area in the midgut of the GI-tract, allowed it to assume a reservoir function or at least to delay the chyme transit so that sufficient mixing with bile and digestive enzymes from the pancreas can occur. The lack of a fully developed stomach, during determined larval stages, to mix the ingested food may be functionally compensated by the strong peristaltic activity (anterograde/retrograde contractions) observed in the midgut, which contributes to the mechanical degradation of the ingested food (Gomes et al. 2014). This supports earlier notions acquired in zebrafish, *Danio rerio,* a stomachless species, where the retrograde contractions observed in the anterior part of the midgut generate a similar mechanical mixing as the stomach (Holmberg et al. 2003). Considered in the context of a chemical reactor (Horn and Messer 1992), for instance, the Atlantic halibut GI-tract changed from a plug-flow reactor (PFR) operating system, in which ingested food flowed continuously through the intestine to a continuous-flow stirred-tank reactor (CTSR), where food entered and exited continuously through the reaction vessel-acid stomach. As it has been pointed out by Gomez et al. (2014), the generally conserved nature of the post-embryonic modifications of the GI-tract in altricial gastric species are likely a general characteristic of teleost fish and potentially other vertebrates.

7. The respiratory system

Compared with other organ systems, the gills form relatively late in development and gradually displace the skin as the site of most exchange activities (Rombough 2004). In Senegalese sole, four gill arches without primary filaments, but already provided with chloride cells, are visible (Padrós et al. 2011). Thus, from hatching until the onset of exogenous feeding phase, the first and fourth gill arches comprised a mass of undifferentiated cells surrounded by epithelial cells, and gill filaments were absent. In the second and third gill arches, chondrogenesis starts and a cartilaginous central axis (chondroblasts and extracellular matrix) and an epithelial layer are distinguished. Only during the early metamorphic stage, at the beginning of the eye migration process, secondary lamellae appear and a higher number of chloride cells are evident in the gill filaments, indicating a major shift of the respiratory function from the skin to the gills. These histomorphological changes are coupled with the notochord flexion and the onset of development of the caudal fin complex, indicating a change in larval swimming performance, and consequently, a greater oxygen demand for supporting swimming activity. During metamorphosis, the skin thickens and chloride cells, as well as dermal spaces disappear, and the gills become the main site of osmoregulation due to the vicinity to the circulatory system (Rombough 2004). During the middle metamorphic stage, gill arches show their characteristic well-developed structure with a central cartilaginous axis, and functional blood vessels, as well as groups of small muscular fibres; in the axis of the primary filaments, some early developed cartilaginous bars are observed. Chloride cells are only present in the interlamellar spaces, whereas some dispersed mucous cells are observed within the epithelial layer of the filaments (Sarasquete et al. 2001, Padrós et al. 2011). The pseudobranch is not observed until late metamorphosis and this structure is formed by piles of epithelial cells with small developing blood vessels and some chloride cells on top of the piles. After metamorphosis, gill filaments are already formed and the formation of lamellae is visible. Blood vessels and chloride cells are also present on the filament and on and around the lamellae. Most of the lamellae show the typical pillar cell network with clearly visible erythrocytes between the pillar cells, whereas the pseudobranch is already developed. As it has been described in many other fish larvae (Falk-Petersen 2005), the process of chondrogenesis in post-metamorphic Senegalese sole specimens is observed to occur in the filaments and the pseudobranch that comprised five filaments and an agglutination of their own lamellae and vascular structure.

In Table 1 and Figs. 4 and 5, some ontogenetic cell-tissue characteristics of gill development of Senegalese sole, among other ontogenetic events (heart, kidney, spleen, thyroid, pancreas, etc.), are summarized and shown.

8. The cardiovascular system

As in other altricial fish species, the development of the cardiovascular system is regulated by regional and systemic needs for transporting oxygen, carbon dioxide and nutrients, which are driven by morphological, physiological, behavioural and ecological changes associated with the transition from a pelagic to a benthic life (Falk-Petersen 2005). In *S. senegalensis*, the heart appears very poorly developed during the endogenous feeding stage. It is observed just below the cephalic area, and it is formed by a straight tube open to the yolk-sac sinus. The formation of four chambers and valves is delayed until metamorphosis; thus, at earlier stages of development, only a single inner endocardial layer and an external myocardial layer are observed and no segmentation is visible. Coinciding with mouth opening at *ca.* 3 dph, the heart is located inside the cardiac cavity and the septum transversum is already formed. At this age, four well-differentiated segments could be observed: the sinus venosus, atrium, ventricle and bulbus arteriosus, which are externally identified by a constriction between them. No sinus-atrial valve is developed, although a thin and poorly developed atrial-ventricular valve is present. The ventricle is the most developed heart chamber at this stage, whereas the *bulbus arteriosus* is formed by a spherical group of undifferentiated cells surrounding a small lumen. In pre-metamorphic larvae, the atrium and ventricle are clearly separated by the atrial-ventricular valve and a sinus-atrial valve begins to develop. The ventricle has developed extensive trabeculation and appears covered by endocardic tissue. Elastic connective tissue appears in the exterior zone of the bulbus. As the heart develops, the number of mature erythrocytes increases. At early metamorphosis, the connection between the hepatic vein and the sinus venosus, now clearly separated from the atrium by the sinus-atrial and the bulbus-ventricular valves, is evident. In the bulbus, the lumen is formed by a layer each of endocardic, myocardic and elastic connective tissues, and is transformed into the ventral aorta in the distal zone. During the middle metamorphosis or stage S2 (Table 1), the sinus venosus and atrium are clearly separated by a completely developed sinus-atrial valve, and trabeculae begin to appear in the atrium, whereas they have already extensively proliferated in the ventricle. The atrial-ventricular valve resembles post-metamorphic specimens and all the segments of the heart are now occupied by mature erythrocytes. In post-metamorphic specimens, the structure of the heart is the same as observed in juveniles and the subsequent changes observed are only due to changes in body growth. Elastic fibres, similar to those observed in the bulbus, are evidenced in the caudal zone. The caudal vein, cardinal veins, and other venous and arterial structures are formed by a single endothelium layer and do not present other structures such as connective tissue or smooth muscle layers through the entire larval development (Padrós et al. 2011).

9. Immune system, and haematopoietic and excretory organs

The interrenal gland of fish species is not well differentiated as a compact organ as in higher vertebrates (Fuiman 2002, Falk-Petersen 2005, Yamano 2005). Few polygonal-shaped cells with round and prominent nuclei are observed in the anterior part of the kidney after the complete absorption of the yolk sac (Tanaka et al. 1995). In *S. senegalensis*, the development of the interrenal tissue during the metamorphosis may indicate a major change in the corticosteroidogenic activity at pre-metamorphosis, which is in parallel to the early development of the hypophysis at 1 dph (Piñuela et al. 2004) and associated with pituitary hormonal levels and developmental events such as eye-pigmentation (Tanaka et al. 1995). According to the above-mentioned authors, cortisol and thyroid hormonal levels exhibit similar patterns during larval fish development, and the thyroideal system participates in metamorphosis, eye-migration, muscular and stomach development, and maturation of blood cells in flatfishes (Power et al. 2008). From hatching and during early life stages of *S. senegalensis*, the excretory kidney is comprised of a pair of poorly differentiated straight primary pronephric tubules developing from the caudal and mid-part of the larval body, and running below the notochord. In the caudal region, these tubules join in a primary urinary bladder that opens to the exterior next to the anal opening. The urinary bladder has a large lumen with a flattened epithelium in its posterior region. In the cranial region, each tubule is slightly convoluted and developed in a rudimentary Bowman's capsule with a reduced space that encloses an undeveloped renal corpuscle.

At early metamorphosis of Senegalese sole or stage S1 (Table 1, Fig. 4), convolution of the pronephric tubules both in the cranial and the caudal regions of pronephric ducts substantially increases. At the end of metamorphosis, mesonephric tubules and rudimentary renal corpuscles become evident. Furthermore, the development of Stannius corpuscles is not evident until metamorphosis when a small group of undifferentiated cells is observed in the dorso-posterior area of the kidney. During larval development, the spleen appears, *ca.* 4–5 dph, as a weakly differentiated spherical mass of cells with erythropoietic properties. Later in development, both the kidneys and spleen become lymphoid, with lymphocytes observed first at middle-metamorphosis phase within the haematopoietic cells of the spleen. The thymus is not observed until metamorphosis phase with thymocytes allocated in the outer part and lymphoblasts in the inner. Progressively, granular and lymphoid cells are observed in the thymus (Padrós et al. 2011). The timing of appearance of the lymphohaematopoietic tissue in *S. senegalensis* has been previously reported by Cunha et al. (2003). Immature haematopoietic cells that can be observed at hatching in the kidney region are probably derived from the intermediate cell mass observed during the embryonic period. From hatching and during the lecitotrophic phase, some undifferentiated cells are noted around the mesodermal mass and the early pronephric tubules. In pre-metamorphic larvae, the number of haematopoietic cells increases and small groups of these cells are present between the sinusoids. The number of erythrocytes in the renal blood vessels and the number of cellular haematopoietic aggregates increases from early metamorphosis onwards, especially in the cranial region during metamorphosis phase (Padrós et al. 2011). Finally, according to all previous results (Cunha et al. 2003, Padrós et al. 2011) and our observations, it may be hypothesized that the high survival rates of *S. senegalensis* after metamorphosis could be also attributed to the fast development of the immune cells and lympho-haematopoietic functional systems.

10. Gonad differentiation

In the last two decades, the crucial role played by the *Vasa* gene and the expression of protein-RNA helicases in the germ-cell line has made this gene a useful marker for the sexual determination and gonad development in fish species (Devlin and Nagahama 2002). In Senegalese sole, first signs of sex determination-differentiation, at histological, biochemical and molecular levels, are discernible at around 98–100 dph (Guzman et al. 2009, Pacchiarini et al. 2013). Furthermore, sex differentiation in females started before 98 dpf when fish had an average TL of 33 mm. In contrast, males began sex differentiation by 127 dpf with a mean TL of 44 mm. At this time, an incipient growth advantage in favour of females was already observed. All fishes were sexually differentiated by 48 mm (Viñas et al. 2013). Interestingly, several *vasa* transcripts (*Ss1-2* and *Ssvasa* 3–4) are differentially expressed during embryogenesis, larval development and metamorphosis in Senegalese sole. Thus, during the embryonic and larval development, a switch between the longest and the shortest transcripts is detected; while *Ss1-2* are maternally supplied, *Ss3-4* depend on the *novo* expression program of the growing juveniles. In larvae undergoing metamorphosis, the expression of *Ssvasa* 3–4 homologues decreases, probably due to the metabolic changes in which the developing larvae are involved. The expression profile of *Ssvasa* 3–4 increases again in post-metamorphic larvae and juveniles (60 and 124 dph). In juveniles aged 150 dph, the expression of *Vasa* is found in the germinal region of early developing gonads (Pacchiarini et al. 2013). All these results confirm the usefulness of *vasa* gene products as biomarkers of the germinal lineage in Senegalese sole, as it has been pointed out for many other fish species (Devlin and Nagahama 2002).

On the other hand, it is largely known that in fish species, as in higher vertebrates, gonadotropins (GTHs), the follicle stimulating hormone (FSH) and luteinizing hormone (LH), determine the reproductive competence of adult breeders, but they also participate in the establishment of the reproductive axis at early stages of life. In Senegalese sole, there is very recent information on the FSH protein levels at different development stages (Cerda et al. 2008, Guzman et al. 2009). Furthermore, anatomical and histological studies have shown an early development of the pituitary, which is visible in yolk-sac Senegalese sole larvae (Piñuela et al. 2004). This information, together with the results obtained by Guzman et al. (2009), show an early parallel expression in larvae aged 1–3 dph of both FSHβ and GPα subunit genes, which suggest a relevant function of the FSH proteins during the first stages of larval development in this flatfish species. The LHβ transcripts are first detected in larvae at 5 dph and although maintained at very low levels, they are detectable until the end of the metamorphic period. The results observed in Senegalese sole, showing higher FSHβ and LHβ mRNA levels throughout larval development, may suggest that FSH rather than LH would participate in the early development of the reproductive axis of this flatfish species. The transcript levels of all three GTH subunits peak simultaneously at 15 dph, coinciding with the period of middle-metamorphosis. At the end of metamorphosis, GTH mRNA levels drop sharply and are maintained low (FHSβ) or undetectable (LHβ) in post-metamorphic specimens. Interestingly, there is a second rise in FSHβ gene expression, and also GPα at around 90–100 dph (Guzman et al. 2009). This increment on GTH expression is also detected in specimens of Senegalese sole aged 1 year old, mainly on FSHβ and GPα, but also on LHβ transcripts. The preliminary data obtained by these last authors might suggest that FSH, rather than LH, would play a major role in this early establishment of the reproductive axis with a possible participation in the process of metamorphosis of Senegalese sole larvae.

11. The development of the skeleton and the problem of deformities

The skeleton is now largely recognized for a wide set of functions such as maintaining body shape, movement and muscle attachment, feeding habits, reproduction, protecting vital organs and, in some cases, acting as a calcium and phosphorus reservoir for mineral homeostasis. A major particularity of fish skeleton compared with that of mammals is that it has a continuous growth throughout the life and considering that fish hatch at a much earlier stage of development than other vertebrates (Haga et al. 2009), its development is much more sensitive to known disruptors. Skeletal deformities mainly originate in fish at the larval and juvenile stages, but they are only clearly identified during the juvenile stage (on-growing period) at the earliest; thus, an accurate study of skeletogenesis during larval development is of utmost importance for the recognition and identification of abnormalities in skeletal structures (Boglione et al. 2013a,b).

The vertebral column in Senegalese sole is composed of 45 vertebrae, separated in 8 abdominal (prehaemal) and 37 caudal (haemal) vertebrae, including the urostyle. Each abdominal vertebra is equipped dorsally with a neural arch and neural spine and ventrally with a pair of parapophysis from the fourth to the eighth vertebrae. The first five neural spines are generally thicker than the others. Caudal vertebrae are equipped dorsally with a neural arch and neural spine, and ventrally with a haemal arch and haemal spine; the neural spines of preural vertebrae 1 and 2 and the haemal spine of the first preural vertebra are elongated and modified to help support the caudal fin ray. Both abdominal and caudal vertebrae exhibited in the neural arch an anterior neural prezygapophysis and a posterior neural poszygapophysis (Gavaia et al. 2002). All vertebrae are formed by intramembranous ossification except for the arches and spines of the four preural centra, which appeared initially as cartilaginous structures and calcified by endochondral ossification (Gavaia et al. 2002, Fernández and Gisbert 2010). The development of the vertebral column in Senegalese sole reared under standard protocols at 16–18°C may be summarized as follows (Gavaia et al. 2002, Padrós et al. 2011): At hatching and during the first early stages of development, no vertebral elements were observed in Senegalese sole larvae; structurally, the notochord is the sole skeletal support tissue in the embryo and in early life stages, a stiffened rod against which muscular contraction can drive motility (Boglione et al. 2013b). At 4.3 mm in standard length (LS), the first elements of the vertebral column start to differentiate and correspond to the neural arches 2–6 from the first abdominal vertebrae. After the arch is formed, its corresponding spine also appears by intramembranous ossification and elongates dorsally. Vertebral development continues in caudal direction for all the remaining neural arches. Haemal arches also develop from the anterior to the posterior region of the vertebral column, even though they appear at larger sizes (4.55 mm). At 4.7 mm, 28 neural arches and 17 haemal arches are visible along the notochord and the first neural arch is completely formed. The cartilaginous buds of the arches in preural vertebra are already present near the tip of the notochord (future urostyle), which has not yet begun its bending upwards. The mineralization of the notochord sheath establishes the identity of vertebral bodies (Boglione et al. 2013a); mineralization of the vertebral column extended from the base of the vertebral arches and formed the vertebral centra that finally surrounded the notochord. As Boglione et al. (2013a) reviewed, the formation of schelerotome-derived bone is only the second step of teleost vertebral body formation, and anomalies/mutations affecting specification of the early notochord cells provoke profound defects on vertebral body patterning in teleost fish. At 5.65 mm, the anterior vertebrae and arches are almost completely formed and the notochord appears completely surrounded by calcified tissue, while in the posterior caudal vertebrae, the calcified tissue starts to spread around the notochord from the site of insertion of the vertebral arches. The urostyle starts to bend upwards (notochord flexion), and also appears partially calcified. In juvenile specimens (at 8.3 mm TL), all vertebral elements are formed and calcified. In detail, the vertebral body completely surrounds the notochord and the adjacent arches are all closed around the ventral aorta in the haemal side and around the nervous chord in the neural side. After the vertebral bodies have developed fully, notochord tissue in the intervertebral spaces can transform into cartilage under pathological conditions, as it has been recently described in this species by Cardeira et al. (2015). The arches exhibit a spine that formed in the median plan of the body, and elongated from the proximal to the distal region, whereas the parazygapophyses of vertebral bodies are still forming and the characteristic holes in the neural arches are not yet visible in the posterior most caudal vertebrae. In larger animals, the vertebral elements and processes only increase in complexity and size, and the spines elongate distally.

The tail is the second region of the Senegalese sole's body showing the major concentration of skeletal deformities. The skeletal elements forming the tail or the so-called caudal fin complex are formed either by endochondral or by intramembranous ossification (Gavaia et al. 2002, Fernández and Gisbert 2010). As Gavaia et al. (2002) described, the first group of skeletal structures included the five hypurals and epural, the preural arches and modified haemal and neural spines; the second group included the urostyle, the preural centra and the caudal fin rays. Although five hypurals form the caudal fin complex, their development is asynchronic, the hypural 1 being the first visible element, appearing as cartilaginous tissue in specimens measuring between 4.3 and 4.65 mm, whereas hypurals 2 and 3 appear at 4.4 and 4.45 mm, respectively. Hypural 4 appears in larvae larger than 4.5 mm, coinciding with the appearance of the parhypural and the cartilaginous buds of the arches. From the bases of these arches, the preural centra 1, 2 and 3, on each latero-dorsal face of the notochord in specimens measuring 4.7 mm, begin to form. At this stage, the upward flexion of the urostyle is initiated and the caudal rays started to form by intramembranous ossification within the primordial fin, in the area adjacent to the forming cartilaginous hypurals. The epural appears during the flexion of the urostyle, between 5.2 and 5.65 mm. The last skeletal structure to differentiate in the caudal fin complex is the hypural 5 appearing next to the tip of the notochord in larvae longer than 5.5 mm, when flexion of the urostyle is almost concluded. At 6.6 mm, all the caudal plates had increased in size and calcification is visible in the urostyle, extending to the proximal bases of hypurals 1–4, which began to fuse with the urostyle. The neural and haemal arches of the preural vertebrae start to calcify, as well as the corresponding centrum, which gradually surrounds the notochord. At this stage, all fin rays are already present, with alizarin red staining decreasing in intensity from the longer central to the shorter lateral rays. In Senegalese sole juveniles, at 7.8 mm, all the plates are largely ossified, with greater intensity of alizarin red staining observed in the proximal parts, and progressively decreasing in intensity towards the distal parts. Their posterior end is still cartilaginous and thus stained in blue, whereas calcification also extended to the spines in the preural vertebrae. At this stage, the caudal rays are all ossified and articulated with hypurals, and the caudal fin is totally separated from the anal and dorsal fins. At larger sizes, most of the changes in the skeletal elements forming the caudal fin complex consist of the appearance of vertical fissures in the posterior region of the parahypural and hypural 4, as well as their branching. The proper mineralization of the skeleton along larval development is of importance, since a less mineralized skeleton has been shown to contribute to the development of skeletal disorders in fish, as less mineralized bones tend to be more fragile and prone to abnormal development or get more easily deformed (Boglione et al. 2013b).

In contrast to the vertebral column and appendicular skeleton, there is scarce information regarding the description of the development of the cranium in Senegalese sole. According to Padrós et al. (2011), no cartilaginous or bony structures are noticed in the cephalic region of this species at hatching (0 dph). At 2.7–2.9 mm, Meckel's cartilages are detected in the mandibular arch, and ceratobranchial and trabecular bars are also identified. After the onset of exogenous feeding at 3.1 mm (3 dph), other cartilaginous structures, such as the palatoquadrate and the otic capsule, are observed in the cranium. Until 3.4 mm (4 dph), very few bone structures are detected, most of them in the maxillar and opercular areas forming the splanchnocranium. From age and until the post-metamorphosis, a substantial increase in the development of bone structures is observed, and otic capsules are completely surrounded by cartilaginous structures, although the formation of a thin bone layer was observed over the cartilage. At this stage, both the neurocranium and splanchnocranium are fully developed. Authors interested in the skeletogenesis of the cranial structures in Soleidae fish are recommended to consult work from Wagemans and Vandewalle (2001) on the common sole (*S. solea*). Considering that the sequence in bone formation in both species is similar due to their same response to functional demands and their close phylogenetic position, the skeletal development of the cranium from common sole may be considered as a reference for Senegalese sole, even though readers need to take into account possible differences due to different rearing protocols.

Similarly to many other marine fish species cultured so far, skeletal deformities are one of the main factors affecting the intensive hatchery production of Senegalese sole larvae (Koumoundouros 2010, Fernández and Gisbert 2011, Boglione et al. 2013a,b, among others). The incidence of fish with medium to severe skeletal anomalies may vary greatly, not only among the different farms using different rearing protocols, but also among different production lots within the same hatchery or even within the same batch of eggs (Boglione et al. 2013a). Hatcheries grade out malformed fingerlings in order to save production costs, while those farms that grow-out abnormal fish to market size have to either downgrade the product to a lower value or discard them before their commercialization. The extent of losses at either point is different, and dependent on the hatchery and husbandry practices. In whatever circumstance, these losses may be substantial, in terms of both productivity and profitability. However, the real economic losses are difficult to estimate due to the reluctance of farmers to provide data that could compromise the farm's reputation (Boglione et al. 2013b). In particular, different studies have shown a high incidence of skeletal deformities in hatchery-reared early juveniles of Senegalese sole, with values ranging from 44% (Gavaia et al. 2002) to 80% (Engrola et al. 2009, Fernández et al. 2009, Boglino et al. 2012, Lobo et al. 2014). A comparison of rearing methodologies revealed a relatively high incidence of skeletal deformities in larvae captured in nature (20%), which indicates that deformities have other causative factors than just the ones reported under captivity (Gavaia et al. 2009). In those studies, most of the skeletal anomalies were found along the vertebral column and caudal fin complex of pro-metamorphic larvae and early juveniles. The vertebral column in Senegalese sole is generally composed by 45 vertebrae, divided in 8 prehaemal and 37 haemal vertebrae (including the urostyle). The haemal region of the vertebral column, especially the pleural vertebrae, is mostly affected by the fusion and/or compression of vertebral bodies, and abnormalities of the vertebral arches (Gavaia et al. 2002, Engrola et al. 2009, Fernández et al. 2009). Among different body regions in Senegalese sole, the head is the area less affected by skeletal disorders in comparison to the vertebral column and caudal fin complex, which differs from round finfish species where jaw abnormalities are very common (Boglione et al. 2013a). Such difference in the incidence of skeletal abnormalities affecting the splanchnocranium might be due to the different temporal ossification pattern existing between Senegalese sole and gilthead sea bream or European sea bass, since both jaws ossify soon after the onset of exogenous feeding, *ca.* 2 dph at 17–18°C (Wagemans and Vandewalle 2001, Fernández et al. 2009), whereas in the other species this process takes place at later stages of development (Koumoundouros 2010, Gisbert et al. 2014). In Senegalese sole, lower incidences of jaw anomalies were observed when fish were reared at a thermal cycle of 22.1°C day/19.0°C night, instead of 19.2°C day/22.0°C night (Blanco-Vives et al. 2010).

Most of the commonest problems regarding skeletal deformities appear during the larval period, although unsuitable rearing conditions during the on-growing phase may also affect bone homeostasis and result into skeletal anomalies. Skeletal deformities may impair the ecophysiological performance of fish larvae (Pimentel et al. 2014), and consequently their performance under the aquaculture condition. Thus, vertebral curvatures (kyphosis, lordosis and scoliosis) and fin deformities may affect larval swimming behaviour, feeding efficiency and the capacity to maintain their position in a current (Powell et al. 2009). Cranial deformities, such as ocular migration anomalies, probably will have affect fish capability to feed, attack prey and avoid predators, whereas larvae with opercular deformities may increase gill's susceptibility to pathogens (Powell et al. 2008) and, as a result, their swimming and cardiovascular performance might be compromised (Powell et al. 2009), as well as their welfare (Noble et al. 2012, Boglione et al. 2013a). Additionally, fish with deformities in the jaw apparatus affecting the dental, premaxillar or maxillar bone may have problems in adducting their mandible and, besides having potential feeding restrictions, the buccal-opercular pumping of water across gills is also likely to be impaired and compromised (Lijalad and Powell 2009). Regarding their etiology, skeletal anomalies appearing at early life stages of development are generally the result of genetic factors and/ or the incapacity of homeorhetic mechanisms to compensate for stressful environmental and nutritional conditions affecting larval morphogenesis (Boglione et al. 2013b). In this sense, readers are invited to consult recent contributions on this topic in Senegalese sole (Morais et al. 2016), in addition to other studies where different rearing conditions (Gavaia et al. 2009, Dionisio et al. 2012, Losada et al. 2014) and nutritional factors like liposoluble vitamins (Fernández and Gisbert 2011, Richard et al. 2014, Fernández et al. 2017), essential polyunsaturated fatty acids (Boglino et al. 2012, 2013, 2014a,b) and probiotics (Lobo et al. 2014) are assessed with regard to the proper development of the skeleton.

In the Fig. 1 are represented graphically noticeable, morphological features and crucial ontogenetic changes during larval phases (P) and metamorphic stages (S) of the Senegalese sole, following the classification previously updated (Sarasquete et al. 2017) and it is also enclosed in this review (Table 1). As it was previously reported (Gavaia et al. 2002, Padrós et al. 2011), during pre-metamorphic larval phases, the first-calcified structure is the *cleithrum*, and pre-maxillary from skull was completely calcified in those post-metamorphosed specimens. Besides, in these post-larvae (S4), the eye migration is completed, and the body appeared fully pigmented.

12. The development of skin and pigmentation disorders

The skin of teleost larvae includes protective, osmoregulatory, respiratory and sensory structures. Thus, after hatching, gas exchange is performed passively over the whole body surface, and chloride cells that appear dispersed over the epidermis are responsible for ion balance (Rombough 2004). In newly hatched Senegalese sole larvae, the epidermis consists of a simple epithelium with two layers of squamous epithelial cells with scattered chloride and sacciform or saccular cells (Sarasquete et al. 1998, 2001, Padrós et al. 2011) that are homogeneously distributed throughout the whole epidermis. In addition, a large fluid-filled subdermal space is observed between the epidermis and the mesoderm. The dermis is absent and only embryonic pigment cells are observed between the epidermis and the underlying tissues. In pre-metamorphic larvae, a high number of melanophores proliferate at the base of the subdermal space, and some fibroblasts in loose connective layers are evidenced. In metamorphic Senegalese sole specimens, most subdermal spaces disappear and some fibroblast in loose connective layers occupies the space between the epidermis and subjacent structures. In post-metamorphic specimens, mucous cells are visible at the base of the epidermis. Subdermal spaces are normally associated with buoyancy, reducing the specific gravity of larvae; thus, they tend to disappear when larval fins are developed and the swimming and floating abilities of the larva are completed during metamorphosis, coinciding with their settlement (Rombough 2004). Sacciform cells, observed in pre-metamorphic Senegalese sole larvae (Sarasquete et al. 1998, 2001, Padrós et al. 2011), are a characteristic feature of early larval fish stages when the skin epithelial barrier is weak, similarly to what has also been described in the epidermis of different fish larvae (Falk-Petersen 2005).

Interestingly, colouration in fish has diverse functions and is the result of the spatial combination and changes in numbers of different pigment cells: brown-black melanophores, yellow-orange xanthophores, red erythrophores, iridescent iridophores, white leucophores and blue cyanophores (Sugimoto 2002). Senegalese sole, as a pleuronectiform fish species, is known for its ability to match their body colour to the background by adjusting their ocular-side pigmentation, which is achieved by distribution changes of pigment granules in chromatophores (Inui and Miwa 2012). The development of the asymmetric pigmentation in flatfishes and their ability to adapt to background colour changes occurs during metamorphosis and is regulated by endocrine and nutritional factors (Bolker and Hill 2000, Villalta et al. 2005, Inui and Miwa 2012, Darias et al. 2013a). The description of the development of the skin pigmentation in Senegalese sole has been recently reported by Darias et al. (2013a) and it may be summarized as follows: at 2 dph, two lines of dendritic black melanophores, white leucophores and orange-yellowish xanthophores overlay the dorsal and ventral flanks of the body skin in the bilateral symmetric larva, with the exception of the future region of the caudal fin. These chromatophores also cover the skin of the head and abdominal area. At this age, Senegalese sole larvae present around 6% of the maximum amount of melanophores counted until the early juvenile stage, which remain invariable until 22 dph. This distribution of skin pigment cells remained similar until the pre-metamorphosis stage. In metamorphosing Senegalese sole larvae, at 16 dph, round-shaped xanthophores are also visible, whereas patches of larval chromatophores located in the fins begin to disappear. At 19 dph, the linear pattern of allocation of the body skin chromatophores begins to disorganize, and at 22 dph, some iridophores are already visible in the head of some individuals. At 27 dph, the amount of iridophores increases and they begin to organize to conform to the distribution pattern of pigment cells in adult specimens. The distribution of chromatophores is restricted to two bands on either side of the vertebral column and in the distal parts of the trunk, close to the beginning of the dorsal and anal fins. Some melanophores group to form a patch in the middle of the trunk, whereas two patches of chromatophores are distinguishable in the dorsal fin and another one in the ventral fin. In postmetamorphic specimens, at 33 dph, coinciding with the complete migration of the left eye, three lines of melanophores and xanthophores are found in the dorsal and ventral regions of the trunk, from both sides of the vertebral column to the end of the trunk. Patches of chromatophores, composed of a mixture of melanophores, xanthophores and leucophores, alternating with patches of iridophores, are visible in the skin of the post-metamorphic specimens. In addition, five patches of chromatophores on the trunk appear at the level of the vertebral column, as well as in the margin of the dorsal and ventral trunk, whereas a higher number of patches of chromatophores, surrounded by iridophores are visible in the dorsal and ventral fins. This pattern of chromatophore distribution is preserved until 47 dph when the pattern of skin pigmentation starts to resemble that of adults. At 33 dph, leucophores cover most of the trunk, where there were no iridophores present, and in the dorsal and anal fins, leucophores are found in the patches and also in the border of the dorsal fin. Xanthophores are the most abundant pigmentary cell in the skin of Senegalese sole, followed by melanophores and finally by iridophores. The shape of xanthophores is no longer dendritic, but round. At 35 dph, the number of iridophores increases, while the amount of melanophores and xanthophores becomes equal and still represents the 20% of the total pigment cells counted at post-metamorphosis already observed at younger ages. The blind side of the larvae is composed of melanophores and few xanthophores, and some iridophores are observed at the level of the head. The above-mentioned phenotypic changes that occur during morphogenesis and skin pigmentation in Senegalese sole correspond well with the main transitions/changes in gene expression that occur; thus, alternating actions of both melanophore differentiation- and melanogenesis related genes coordinate melanophore ontogenesis in this pleuronectids species. Further information about the regulation of key genes involved in skin pigmentation in Senegalese sole early development may be found in Darias et al. (2013b).

In flatfish species, pigmentation abnormalities are characterized by either a deficiency of pigment cells (albinism, pseudoalbinism or hypomelanism) on portions of the ocular side, or excess pigmentation (staining, spotting, or ambicolouration) on the blind side (Bolker and Hill 2000). Pigmentary disorders, especially albinism and ambicolouration, can affect up to 61% of the reared fish of different flatfish species (Estévez and Kanazawa 1995, Bolker and Hill 2000, Villalta et al. 2005, Hamre et al. 2007, Darias et al. 2013a, Boglino et al. 2014a,b). However, pigmentation abnormalities are not as critical in Senegalese sole as it has been reported in other flatfish species like summer flounder (*Paralichthys dentatus*), Atlantic halibut (*Hippoglossus hippoglossus*) and Japanese flounder (Bolker and Hill 2000, Hamre et al. 2007 among others). In Senegalese sole, most of the data regarding pigmentation disorders is from experimental studies in which pseudoalbinism was dietary induced by an excess of arachidonic acid—ARA—(Villalta et al. 2005, 2008, Boglino et al. 2013, Darias et al. 2013a), although environmental rearing conditions and other nutrients like DHA, EPA and VA have been investigated as factors possibly related to the occurrence of albinism in hatchery production in other flatfish species (Bolker and Hill 2000, Hamre et al. 2008). Morphological studies in Senegalese sole revealed that ARA do not affect larval pigmentation at the pre-metamorphic stage, but prevent chromatophore terminal differentiation at metamorphosis, leading to the appearance of pseudo-albinism (Darias et al. 2013a). According to these last authors, those larvae later becoming pseudo-albino and pigmented individuals develop pigmentation in the same way, but once metamorphosed, the future pseudo-albinos began to show different relative proportions, allocation patterns, shapes and sizes of skin chromatophores.

Finally, recent researches reported that several doses of vitamin A, enclosed in diets, affected development of the Senegalese sole in a thyroid hormone-signalling mode, but also dependant on the stage of metamorphosis (pre-, pro- and postmetamorphosis), inducing several skeletal disorders, and some other noticeable disturbances, at molecular, biochemical and cellular levels. These results provide new and key knowledge to better understand how retinoid derivatives and thyroid hormones pathways interact at tissue, cellular and nuclear level at different developmental periods in Senegalese sole, unveiling how dietary modulation might determine juvenile phenotype and physiology (Fernández et al. 2017). On the other hand, several xenoestrogens, which act as hormonal disruptors, could induce a generalized failure in the normal progress of metamorphosis in flatfishes. Indeed, at early larval stages of different fish species, including Senegalese sole, noticeable alterations were also detected, which were induced by different lipophilic xenobiotics that provoked severe ontogenetic disturbances, such as lack of eye migration, notochord disorders, skeletal deformities and/or pigmentation abnormalities (Sarasquete and Ortiz-Delgado 2015). Moreover, it was pointed out that sublethal concentrations of some pesticides (i.e., malathion) induced noticeable disturbances which altered the normal development of the notochord, trunk-musculature, and collagen-fibres, and it also provoked pericardial and yolk-sac oedemas, among other digestive and metabolic alterations (Ortiz-Delgado et al. 2017). Interestingly, a recent study performed during the first month of larval life of the Senegalese sole showed that phytochemical isoflavonoids (i.e., genistein), induced weak and temporal thyroid and estrogen receptor signalling disruptions, although these transcriptional hormonal imbalances were quickly restored to the baseline levels (Sarasquete et al. 2017). Accordingly, the isoflavone genistein, currently considered as a selective estrogen receptor modulator, did not affect the normal development, metamorphosis and growth (i.e., size, length). In all experimental designs, just as in controls, mortality rates and pigmentation anomalies were about 20–30%. Besides, most Senegalese sole specimens, at around 70–80%, finished the metamorphosis process, and as was expected, they temporarily were adapted to benthic life.

Conclusions

Marine fish larvae hatch earlier in their development than other vertebrates, and flatfish species as *S. senegalensis* are no exception, suggesting that the spatial and temporal sequences of larval development in teleoteans are quite different from higher vertebrates. Consequently, during the first weeks of larval life, marine fish and particularly altricial flatfish larvae like *S. senegalensis* undergo significant morphological, structural, and physiological modifications to acquire all the juvenile/adult features by the end of the larval period. Thus, the early life stages represent a transitional ontogenetic period of simultaneous differentiation and growth, during which fish undergoes the transition from endogenous (lecitotrophic phase) to exogenous feeding, i.e., from yolk consumption to ingestion of external food. In Senegalese sole, this transitional endo-exotrophic feeding phase is very short, in comparison with other fish species. This transition implies profound anatomical, physiological changes in different organ-systems, organs and tissues, including a very fast development and functionality of the visual-sensory structures (visual feeders), neuroendocrine (i.e., pituitary, thyroid), locomotory and digestive systems. Whereas these processes are almost completed at the onset of exogenous feeding, additional morphological, physiological and behavioural changes occur during metamorphosis—a thyroid-driven process, which particularly includes eye-migration, differentiation of the skin and gills (a shift from cutaneous to branchial respiration), functional gastric stomach and acidic digestion, skeletal development, functional retinal vision, among other ontogenetic events (cardiovascular, excretory, immune organ-systems, sex-differentiation, etc.), as well as the adaptation to benthic life style. Furthermore, the early life stages may be considered as the most critical events of fish development, from fertilization and hatching and specially during the transitional feeding phase, at the weaning, particularly during the complex thyroid-driven metamorphosis process. As a consequence, several non-optimal zootechnical rearing conditions (i.e., temperature, salinity, oxygen), feeding and nutritional disorders (i.e., starvation, unbalanced diets, nutrient deficiencies), as well as environmental stress factors (i.e., chemical xenobiotics, phytochemicals, etc.), and the presence of infectious agents are of great relevance during these ontogenetic phases of the larval development and metamorphosis of the flatfish Senegalese sole.

References

Alves, R.N., A. Gomes, K. Stueber, M. Tine, M.A.S. Thorne, H. Smáradóttir, R. Reinhard, M.S. Clark, I. Ronnestad and D.M. Power. 2016. The transcriptome of metamorphosing flatfish. BMC Genomics 17: 413.

Amara, R. and F. Lagardere. 1995. Size and age at onset of metamorphosis in sole (*Solea solea*) of the gulf of Gascogne, Ices. J. Nar. Sci. 52: 247–256.

Arellano, J.M. and C. Sarasquete. 2005. Atlas histomorfológico del lenguado senegalés, *Solea senegalensis*, Biblioteca de Ciencias. Dpto de Publicaciones CSIC (Madrid, España), 206 pp.

Assem, S.S., A.A. El-Dahhr, H.S. El-Syed, M El.Salama and M.M. Mourad.2012, Induced Spawning, Embryonic and larval developmental stages of Solea vulgaris in the Mediterranean water. J. Arabian Aquaculture Soc. 7: 51–74.

Bayarri, M.J., J.A. Muñoz Cueto, J.F. López-Olmeda, L. Vera, M.A. Rol de Lama, J.A. Madrid and F.J. Sánchez Vázquez. 2004. Daily locomotor activity and melatonin rhythms in Senegal Sole (*Solea senegalensis*). Physiology and Behaviour 81: 577–583.

Bedoui, R. 1995. Rearing of *Solea senegalensis* (Kaup, 1958) in Tunisia. Cahier Options Mediterranee 16: 31–39.

Bejarano-Escobar, R., M. Blasco, W.J. Degrip, J.A. Oyola-Velasco, G. Martín-Partido and J.F. Morcillo. 2010. Eye development and retinal differentiation in an altricial fish species, the Senegalese Sole (*Solea senegalensis*). J. Exp. Zool. 314B: 580–605.

Blanco-Vives, B., N. Villamizar, J. Ramos, M.J. Bayarri, O. Chereguini and F.J. Sánchez-Vázquez. 2010. Effect of daily thermo- and photo-cycles of different light spectrum on the development of Senegal sole (*Solea senegalensis*) larvae. Aquaculture 306: 137–145.

Blanco-Vives, B., M. Aliaga-Guerrero and J.P. Cañavate, G. García-Mateos, A.J. Martín-Robles, P. Herrera, J.A. Muñoz-Cueto and F.J. Sánchez-Vezquez. 2012. Metamorphosis induces a light-dependence switch in Senegalese sole (*Solea senegalensis*) from diurnal to nocturnal behavior. J. Biol. Rhythms 27: 135–144.

Blaxter, J.H.S. 1986. Development of sense organs and behaviour of teleost larvae with special reference to feeding and predator avoidance. Trans. Am. Fish. Soc. 115: 98–114.

Boglino, A., M.J. Darias, J.B. Ortiz-Delgado, F. Ozcan, A. Estévez, K.B. Andree, F. Hontoria and E. Gisbert. 2012. Commercial products for Artemia enrichment affect growth performance, digestive system maturation, ossification and incidence of skeletal deformities in Senegalese sole (*Solea senegalensis*) larvae. Aquaculture 324-325: 290–302.

Boglino, A., A. Wishkerman, M.J. Darias, K.B. Andree, P. De la Iglesia, A. Estévez and E. Gisbert. 2013. High dietary arachidonic acid levels affect the process of eye migration and head shape in pseudoalbino Senegalese sole *Solea senegalensis* early juveniles. J. Fish Biol. 83: 1302–1320.

Boglino, A., A. Wishkerman, M.J. Darias, P. De la Iglesia, K.B. Andree, E. Gisbert and A. Estévez. 2014a. Senegalese sole (*Soleasenegalensis*) metamorphic larvae are more sensitive to pseudo-albinism induced by high dietary arachidonic acid levels than postmetamorphic larvae. Aquaculture 433: 276–287.

Boglino, A., M.J. Darias, A. Estevez, K.B. Andree, C. Sarasquete, J.B. Ortiz-Delgado, J.B. , M. Solé and E. Gisbert. 2014b. The effect of dietary oxidized lipid levels on growth performance, antioxidant exzyme activities, intestinal lipid deposition and skeletogenesis in Senegalese sole (*Soleasenegalensis*) larvae. Aquac. Nutr. 20: 692–711.

Boglione, C., P.J. Gavaia, G. Koumoundouros, E. Gisbert, M. Moren, S. Fontagné and P.E. Eckard. 2013a. A review on skeletal anomalies in reared European fish larvae and juveniles. 1: normal and anomalous skeletogenic processes. Rev. Aquaculture 5: s99–s120.

Boglione, C., E. Gisbert, P.J. Gavaia, P.E. Witten, M. Moren, S. Fontagné, S. and G. Koumoundouros. 2013b. Skeletal anomalies in reared European fish larvae and juveniles. Part 2: main typologies, occurrences and causative factors. Rev. Aquaculture 5: S121–S167.

Bolker, J.A. and C.R. Hill. 2000. Pigmentation development in hatchery-reared flatfishes. J. Fish Biol. 56: 1029–1052.

Boulhic, M. and J. Gabaudan. 1992. Histological study of the organogenesis of the digestive system and swimbladder of the Dover sole, *Solea solea* (Linnaeus 1758). Aquaculture 102: 373–396.

Burggren, W.W. and B. Bagatto. 2008. Cardiovascular anatomy and physiology. pp. 119–162. *In*: Finn, R.N. and B.G. Kapoor (eds.). Fish Larval Physiology.

Campinho, M.A., N. Silva, J. Roman-Padilla, M. Ponce, M. Manchado and D.M. Power. 2015. Flatfish metamorphosis: A hypothalamic independent process?. Mol. Cell. Endocrinol. 404: 16–25.

Cañavate, J.P. and C. Fernández-Díaz. 1999. Influence of co-feeding larvae with live and inert diets on weaning the Soleasenegalensis onto commercial dry feeds. Aquaculture 174: 255–263.

Cañavate, J.P., R. Zerolo and C. Fernández-Díaz. 2006. Feedings and development of Senegal sole (*Solea senegalensis*) larvae reared in different photoperiods. Aquaculture 258: 368–377.

Cardeira, J., A.C. Mendes, P. Pousão-Ferreira, M.L. Cancela and P.J. Gavaia. 2015. Micro-anatomical characterization of vertebral curvatures in Senegalese sole *Solea senegalensis*. J. Fish Biol. 86: 1796–1810.

Chambers, R.C. and W.C. Leggett. 1987. Size and age at metamorphosis in marine fishes: an analysis of laboratory-reared winter flounder (*Pseudopleuronectes americanus*) with a review of variation in other species. Can. J. Fish. Aquat. Sci. 44: 1936–1947.

Cerdá, J., F. Chauvigne, M.J. Agulleiro, S. Halm, G. Martínez-Rodríguez and F. Prat. 2008. Molecular cloning of Senegalese sole (*Solea senegalensis*) follicle-stimulating hormone and luteining hormone subunits and expression pattern during spermatogenesis. Gen. Comp. Endocrinol. 156: 470–481.

Conceiçao, L.E., L. Ribeiro, S. Engrola, C. Aragao, S. Morais and M.T. Dinis. 2007. Nutritional physiology during development of Senegalese sole (*Solea senegalensis*). Aquaculture 268: 64–81.

Cunha, M., F. Soares, P. Makridis, J. Skjermo and M.T. Dinis. 2003. Development of the immune system and use of immunostimulants in Senegalese sole (*Solea senegalensis*). The big Fish Bang. Proceedings of the 26 th Annual Larval Fish Conference, pp. 189–192.

Darias, M.J., K.B. Andree, A. Boglino, I. Fernández, A. Estévez and E. Gisbert. 2013a Coordinated regulation of chromatophore differentiation and melanogenesis during the ontogeny of skin pigmentation of *Soleasenegalensis* (Kaup, 1858). PLoS ONE 8: e63005.

Darias, M.J., K.B. Andree, A. Boglino, J. Rotllant, J.M. Cerdá-Reverter, A. Estévez and E. Gisbert. 2013b. Morphological and molecular characterization of dietary-induced pseudo-albinism during post-embryonic development of *Soleasenegalensis* (Kaup, 1858). PLoS ONE 8: e68844.

Darras, V.M., A.M. Houbrechts and S.L.J. Van Herck. 2015. Intracellular thytoid hormone metabolism as a local regulator of nuclear thyroid hormone receptor-mediated impact on vertebrate development. Biochim. Biophys. Acta 1849: 130–141.

Devlin , R. and Y. Nagahama. 2002. Sex determination and sex differentiation in fish: an overview of genetic, physiological and environmental influences. Aquaculture 208: 191–364.

Dinis, M.T. and J. Reis. 1995, Culture of *Solea* sp., Workshop on Diversification in Aquaculture. Cyprus June. Cahiers Option Mediterraneés 16, 16pp.

Dinis, M.T., L. Ribeiro, F. Soares and C. Sarasquete. 1999. A review on the cultivation potential of *Solea senegalensis* in Spain and in Portugal. Aquaculture 176: 27–38.

Dionísio, G., C. Campos, L.M.P. Valente, L.E.C. Conceição, M.L. Cancela and P.J. Gavaia. 2012. Effect of egg incubation temperature on the occurrence of skeletal deformities in *Solea senegalensis*. J. Appl. Ichthyol. 28: 471–476.

Drake, P., A.M. Arias and R.B. Rodríguez. 1984. Cultivo extensivo de peces marinos en los esteros de las salinas de San Fernando (Cádiz) II, Informes Técnicos de Investigación Pesquera, 116, 23 pp.

Engrola, S., L. Figueira, L.E.C. Conceicão, P.J. Gavaia and M.T. Dinis. 2009. Co-feeding in Senegalese sole larvae with inert diet from mouth opening promotes growth at weaning. Aquaculture 288: 264–272.

Estévez, A. and A. Kanazawa. 1995. Effect of (n-3) PUFA and vitamin A Artemia enrichment on pigmentation success of turbot, *Scophthalmus maximus* (L.) Aquacult. Nutr. 1: 159–168.

Evans, B. and H. Browman. 2004. Variation in the development of the fish retina. pp. 145–166. *In*: Development of form and function in fishes and the question of larval adaptation. Norway; Bergen.

Falk Petersen, I.B. 2005. Comparative organ differentiation during early life stages of marine fish. Fish Shellfish Immunol. 19: 397–412.

Fernández, I., M.S. Pimentel, J.B. Ortiz-Delgado, F. Hontoria, C. Sarasquete, A. Estévez and E. Gisbert. 2009. Effect of dietary vitamin A on Senegalese sole (*Solea senegalensis*) skeletogenesis and larval quality. Aquaculture 295: 250–265.

Fernández, I. and E. Gisbert. 2010. Senegalese sole bone tissue originated from chondral ossification is more sensitive than dermal bone to high vitamin A content in enriched *Artemia*. J. Appl. Ichthyol. 26: 344–349.

Fernández, I. and E. Gisbert. 2011. The effect of vitamin A on flatfish development and skeletogenesis: a review. Aquaculture 315: 34–48.

Fernández, I., J.B. Ortiz-Delgado, M.J. Darias, F. Hontoria, K.B. Andree, M. Manchado, C. Sarasquete and E. Gisbert. 2017. Vitamin A affects flatfish development in a thyroid hormone signaling and metamorphic stage dependent manner. Front Physiol. 8: 458.

Fernández-Díaz, C., M. Yúfera, J.P. Cañavate, F.J. Moyano and F. Alarcón. 2001. Growth and physiological changes during metamorphosis of Senegal sole reared in the laboratory. J. Fish Biol. 58: 1086–1097.

Fuiman, L.A. 2002. Special considerations of fish eggs and larvae. *In*: Fuiman, L.A. and R.G. Werner [eds.]. Fishery Science The Unique Contributions of Early Life Stages. Wiley-Blackwell, 340 pp.

Gallo, V.P. and A.P. Civinini. 2005. The development of adrenal homolog of rainbow trouutOncorhunchus mykiss: an immunohistochemical and ultrastructural study. Anat. Embriol. 209: 233–242.

García-López, A., G. Martínez-Rodríguez and C. Sarasquete. 2005. Male reproductive system in Senegalese sole *Solea senegalensis* (Kaup): anatomy, histology and histochemistry. Histol. Histopathol. 2: 1179–1189.

García-López, A., E. Couto, A.V.M. Canario, C. Sarasquete and G. Martínez-Rodríguez. 2007. Ovarian development and plasma sex steroid levels in cultured female Senegalese sole *Solea senegalensis*. Comparative Biochemistry and Physiology Part A: Molecular & Integrative Physiology 146: 342–354.

García-López, A., G. Martínez-Rodríguez and C. Sarasquete. 2008. Temperature manipulation stimulates gonadal maturation and sex steroid production in Senegalese sole *Solea senegalensis* Kaup kept under two different light regimes. Aquac. Res. 40: 103–111.

Gavaia, P.J., C. Sarasquete, C. and M.L. Cancela. 2000. Detection of mineralized structures in early stages of development of marine Teleostei using a modified alcian blue-alizarin red double staining technique. B iotechnic. Histochem. 75: 79–84.

Gavaia, P.J., M.T. Dinis and M.L. Cancela. 2002. Osteological development and abnormalities of the vertebral column and caudal skeleton in larval and juvenile stages of hatchery-reared Senegal sole *(Solea senegalensis)*. Aquaculture 211: 305–323.

Gavaia, P.J., S. Simes, J.B. Ortiz-Delgado, C. Viegas, J.P. Pinto, R. Kelsh, C. Sarasquete and M.L. Cancela. 2006. Osteocalcin and matrix Gla protein in zebrafish (Danio rerio) and Senegal sole (*Solea senegalensis*): Comparative gene and protein expression during larval development through adulthood. Gene Expression Patterns 6: 637–652

Gavaia, P.J., S. Domingues, S. Engrola, P. Drake, C. Sarasquete and M.T. Dinis. 2009. Comparing skeletal development of wild and hatchery-reared Senegalese sole (*Solea senegalensis*, Kaup 1858): evaluation in larval and postlarval stages. Aquac. Res. 40: 1585–1593.

Geffen, A.J., H.W. van der Veer and R.D.M. Nash. 2007. The cost of metamorphosis in flatfishes. J. Sea Res. 58: 25–45.

Gisbert, E., J.B. Ortiz-Delgado and C. Sarasquete. 2008. Nutritional cellular biomarkers in early life stages of fish. Histol. Histopathol. 23: 1525–1539.

Gisbert, E., I. Fernández, N. Villamizar, M.J. Darias, J.L. Zambonino-Infante and A. Estévez. 2014. European sea bass larval culture. pp. 162–206. *In*: Sánchez Vázquez, F.J. and J.A. Muñoz-Cueto [eds.]. Biology of European sea bass.CRC Press, Boca Raton, Florida, USA.

Gomes, A.S., Y. Kamisaka, T. Harboe, D.M. Power and I. Ronnestad. 2014. Functional modifications associated with gastrointestinal tract organogenesis during metamorphosis in Atlantic Halibut (*Hippoglossus hippoglossus*), BMC Dev. Biol. 14: 11.

Govoni, J.J., G.W. Boehlert and Y. Watanabe. 1986. The physiology of digestion in fish larvae, Env. Biol. Fish. 16: 59–77.

Gutiérrez, M., C. Sarasquete and R.B. Rodríguez. 1985. Caracteres citohistoquímicos de carbohidratos y proteínas durante la ovogénesis del lenguado, *Solea senegalensis* Kaup, 1858. Inv. Pesq. 49(3): 353–363.

Guzman, J.M., M.J. Bayarri, J. Ramos, Y. Zohar, C. Sarasquete and E. Mañanos. 2009. Follicle stimulating hormone (FSH) and luteinizing hormone (LH) gene expression during larval development in Senegalese sole (*Solea senegalensis*). Comp. Biochem. Physiol. A 164: 37–43.

Haga, Y., V.J. Dominique and S.J. Du. 2009. Analyzing notochord segmentation and intervertebral disc formation using the twhh: gfp transgenic zebrafish model. Transgenic Res. 18: 669–683.

Hamre, K., E. Holen and M. Moren. 2007. Pigmentation and eye-migration in Atlantic halibut (*Hippoglossus hippoglossus* L.) larvae: new findings and hypotheses. Aquac. Nutr. 13: 65–80.

Hamre, K. and T. Harboe. 2008. Critical levels of essential fatty acids for normal pigmentation in Atlantic halibut (*Hippoglossus hippoglossus* L.) larvae. Aquaculture 277: 101–108.

Holmberg, A., T. Schwerte, R. Fritsche, B. Pelster and S. Holmgren. 2003. Ontogeny of intestinal motility in correlation to neuronal development in zebrafish embryos and larvae. J. Fish Biol. 63(2): 318–331.

Horn, M.H. and K.S. Messer. 1992. Fish guts as chemical reactors: a model of the alimentary canals of marine herbivorous fishes. Mar. Biol. 113: 527–535.

Howell, B.R. 1997. A re-appaisal of the potential of the sole, *Solea solea* (L.) for commercial cultivation. Aquaculture 155: 355–365.

Imsland, A.K., A. Foss, L.E.C. Conceiçao, M.T. Dinnis, D. Delbare, E. Schram and A. Kamstra. 2003. A review of the culture potential of *Solea solea* and *Solea senegalensis*, Rev. Fish Biol. Fish. 13: 379–407.

Infante, C., M. Manchado, E. Asensio and J.P. Cañavate. 2007. Molecular characterization gene expression and dependence on thyroid hormones of two type I keratin genes (sse Ker 1 and sse Ker2) in the flatfish, Senegalese sole (*Solea senegalensis*, Kaup). BMC Dev. Biol. 7: 118.

Infante, C., E. Asensio, J.P. Cañavate and M. Manchado. 2008. Molecular characterization and expression analysis of five different elongation factor 1 alpha genes in the flatfish Senegalese sole (*Solea senegalensis* Kaup): differential gene expression and thyroid hormones dependence during metamorphosis. BMC Mol. Biol. 9: 19.

Inui, Y. and S. Miwa. 2012. Metamorphosis of Flatfish (Pleuronectiformes). pp. 105–152. *In*: Dufour, S., K. Rousseau and B.G. Kapoor [eds.]. Metamorphosis in Fish. CRC pubications, Boca Raton, Florida, USA.

Isorna, E., M.J. Obregón, R.M. Calvo, R. Vázquez, C. Pendón, J. Falcó and J.A. Muñoz-Cueto. 2009. Iodothyronine deiodinases and thyroid hormone receptors regulation during flatfish (*Solea senegalensis*) metamorphosis. J. exp. Zool. 312B: 231–246.

Izigar, R., M. Ponce, C. Infante, L. Rebordinos, J.P. Cañavate and M. Manchado. 2010. Molecular characterization and gene expression of thyrotropin-releasing hormone in Senegalese sole (*Solea senegalensis*). Comp. Biochem. Physiol. B. Biochem. Mol. Biol. 157: 167–174.

Klaren, P.H., Y.S. Wunderink, M. Yúfera, J.M. Mancera and G. Flik. 2008. The thyroid gland and thyroid hormones in Senegalese sole (*Solea senegalensis*) during early development and metamorphosis. Gen. Comp. Endocr. 155: 686–94.

Koumoundouros, G. 2010. Morpho-anatomical abnormalities in Mediterranean marine aquaculture. pp. 125–148. *In*: Koumoundouros, G. [ed.]. Recent Advances in Aquaculture Research, Transworld Research Network, Kerala, India.

Lijalad, M. and M.D. Powell. 2009. Effects of lower jaw deformity on swimming performance and recovery from exhaustive exercise in triploid and diploid Atlantic salmon *Salmo salar* L. Aquaculture 290: 145–154.

Lobo, C., P.J. Gavaia, I. García-de-la-Banda, J.R. Gutiérrez, M. Oria, X. Moreno-Ventas, M.A. Moriñigo and S.T. Tapia-Paniagua. 2014. Effects of Shewanellaputrefaciens Pdp11 on Senegalese sole skeletogenesis. Proceedings of the Aquaculture Europe 2014 - Donostia–San Sebastián, Spain. https://www.was.org/easOnline/AbstractDetail.aspx?i=3468. Last accessed: 18 January 2016.

Losada, A.P., A.M. de Azevedo, A. Barreiro, J.D. Barreiro, I. Ferreiro, A. Riaza, M.I. Quiroga and S. Vázquez. 2014. Skeletal malformations in Senegalese sole (*Solea senegalensis* Kaup, 1858): gross morphology and radiographic correlation. J. Appl. Ichthyol. 30: 804–808.

Manchado, M., C. Infante, E. Asensio, J. Planas and J.P. Cañavate. 2008. Thyroid hormones down regulate thyrotropin subunit and thyroglobulin during metamorphosis in the flatfish Senegalese sole (*Solea senegalensis*). Gen. Comp. Endocr. 155: 447–455.

Manchado, M., C. Infante, L. Rebordinos and J.P. Cañavate. 2009. Molecular characterization, gene expression and transcriptional regulation of thyroid hormone receptors in Senegalese sole. Gen. Comp. Endocr. 160: 139–147.

Martínez, I., F.J. Moyano, C. Fernández-Díaz and M. Yúfera. 1999. Digestive enzyme activity during larval development of the Senegal sole (*Solea senegalensis*). Fish Physiol. Biochem. 21: 317–323.

Martín-Robles, A. J.D. Whitmore, C. Pendón and J.A. Muñoz-Cueto. 2013. Differential effects of transient constant light-dark conditions on daily rhythms of Period and clock transcripts during Senegalese sole metamorphosis, Chronobiol. Int. 30: 699–710.

Miwa, S., K. Yamano and Y. Inui. 1992. Thyroid hormone stimulates gastric development in flounder larvae during metamorphosis. J. Exp. Zool. 261: 424–430.

Morais, S., C. Aragão, E. Cabrita, L.E.C. Conceição, M. Constenla, B. Costas, J. Dias, N. Duncan, S. Engrola, A. Estévez, E. Gisbert, E. Mañanós, L.M.P. Valente, M. Yúfera and M.T. Dinis. 2016. New developments and biological insights into the farming of *Solea senegalensis* reinforcing its aquaculture potential. Rev. Aquaculture (in press).

Murray, H.M., J.W. Gallant, S.C. Johnson and S.E. Douglas. 2006. Cloning and expression analysis of three digestive enzymes from Atlantic halibut (*Hippoglossus hippoglossus*) during early development: predicting gastrointestinal functionality. Aquaculture 252: 394–408.

Navarro, D. B., V.C. Rubio, R.K. Luz, J.A. Madrid and F.J. Sánchez-Vázquez. 2009. Daily feeding rhythms of Senegalese sole under laboratory and farming conditions using self-feeding systems. Aquaculture 291: 130-135.

Navarro-Guillen, C., F.J. Moyano and M. Yúfera. 2015. Diel food intake and digestive enzyme production patterns in Soleasenegalensis larvae. Aquaculture 435: 33–42.

Navarro-Guillén, C., M. Yúfera and S. Engrola. 2016. Ghrelin in Senegalese sole (*Solea senegalensis*) post-larvae: Paracrine effects on food intake. Como. Biochem. Physiol. A. Mol. Integr. Physiol. 204: 85–92.

Noble, C., H. Cañon Jones, B. Damsgård, M. Flood, K. Midling, A. Roque, B. Sæther and S. Cottee. 2012. Injuries and deformities in fish: their potential impacts upon aquacultural production and welfare. Fish Physiol. Biochem. 38: 61–83.

Ortiz-Delgado, J.B., N.M. Ruane, P. Pousao Ferreira, M.T. Dinis and C. sarasquete. 2006. Thyroid gland development in Senegal sole (*Solea senegalensis* Kaup 1858) during early life stages: a histochemical and immunohistochemical approach. Aquaculture 260: 346–356.

Ortiz-Delgado, J.B., E. Scala, J.M. Arellano, M. Úbeda-Manzanaro and C. Sarasquete. 2017. Toxicity of malathion at early life stages of the Senegalese sole, *Solea senegalensis* (Kaup, 1858): notochord and somatic disruptions. Histol Histopathol. doi:10.14670/HH-11-899.

Osse, J.W.M. and J.G.M. van den Boogaart. 1997. Size of flatfish larvae at transformation, functional demands and historical constraints. J. Sea Res. 37: 229–239.

Pacchiarini, T., I. Cross, R.B. Leite, P.J. Gavaia, J.B. Ortiz-Delgado, P. Pousão-Ferreira, L. Rebordinos, C. Sarasquete and E. Cabrita. 2013. *Solea senegalensis* vasa transcripts: molecular characterisation, tissue distribution and developmental expression profiles. Reprod. Fertil. Dev. 25: 646–660.

Padrós, F., M. Villalta, E. Gisbert and A. Estévez. 2011. Morphological and histological study of larval development of the Senegal sole *Solea senegalensis*: an integrative study J. Fish Biol. 79: 3–32.

Pimentel, M.S., F. Faleiro, G. Dionísio, T. Repolho, P. Pousão-Ferreira, J. Machado and R. Sosa. 2014. Defective skeletogenesis and oversized otoliths in fish early stages in a changing ocean. J. Exp. Biol. 217: 2062–2070.

Piñuela, C., C. Rendón, M.L. González de Canales and C. Sarasquete. 2014. Development of the Senegal sole, *Solea senegalensis* forebrain, Eur. J. Histochem. 48: 377–384.

Pittman, K., M. Yúfera, M. Pavlidis, A.J. Geffen, W. Koven, L. Ribeiro, J.L. Zambonino-Infante and T. Tandler. 2013. Fantastically plastic: fish larvae equipped for a new world. Rev. Aquaculture 5: S224–S267.

Ponce, M., C. Infante and M. Manchado. 2010. Molecular characterization and gene expression of thyrotropin receptor (TSHR) and a truncated TSHR-like in Senegalese sole. Gen. Comp. Endocr. 168: 431–439.

Powell, M.D., M.J. Leef, S.D. Roberts and M.A. Jonesk. 2008. Neoparamoebic gill infections: host response and physiology in salmonids. J. Fish Biol. 73: 2161–2183.

Powell, M.D., M.A. Jones and M. Lijalad. 2009. Effects of skeletal deformities on swimming performance and recovery from exhaustive exercise in triploid Atlantic salmon. Dis. Aquat. Organ. 85: 59–66.

Power, D.M., L. Llewellyn and M. Faustino. 2001. Thyroid hormones in growth and development of fish. Comp. BiochemP hysiol C. Toxicol Pharmacol. 130: 447–459.

Power, D., I.E. Einarsdottir, K. Pittman, G.E. Sweeney, J. Hildahl, M.A. Campinho and N. Silva. 2008. The molecular and endocrine basis of flatfish metamorphosis. Rev. Fish. Sci. 16: 95–111.

Ramos, J. 1986. Desarrollo embrionario en el lenguado, *Solea vulgaris* (Quensel, 1806). Misc. Zool. 10: 395–400.

Rendón, C., F.J. Rodriguez-Gomez, J.A. Muñoz-Cueto, C. Piñuela and C. Sarasquete. 1997. An immunocytochemical study of pituitary cells of the Senegalese sole, *Solea senegalensis* (Kaup 1858). Histochem. J. 29: 813–822.

Ribeiro, L., C. Sarasquete and M.T. Dinis. 1999a. Histological and histochemical development of the digestive system of *Solea senegalensis* (Kaup, 1858) larvae. Aquaculture 171: 293–298.

Ribeiro, L., J.L. Zambonino-Infante, C. Cahu and M.T. Dinis. 1999b. Development of digestive enzymes in larvae of *Solea senegalensis*, Kaup, 1858, Aquaculture 179: 465–473.

Ribeiro, A.R.A., L. Ribeiro, Ø. Saele, K. Hamre, M.T. Dinis and M. Moren. 2012. Seleniun supplementation changes gluthatione peroxidase activity and thyroid hormone production in Senegalese sole (*Solea senegalensis*) larvae. Aquac. Nutr. 18: 559–567.

Richard, N., I. Fernández, T. Wulff, K. Hamre, M.L. Cancela, L.E.C. Conceição and P.J. Gavaia. 2014. Dietary supplementation with vitamin k affects transcriptome and proteome of Senegalese sole, improving larval performance and quality. Mar. Biotechnol. 16: 522–537.

Rodríguez, R.B. and E. Pascual. 1982. Primeros ensayos sobre utilización de la hipófisis del atún (*Thunnus thynnus*) en la maduración y puesta de *Solea senegalensis* y *Sparus aurata*. Inv. Pesq. 97: 1–11.

Rodríguez-Gómez, F.J., C. Sarasquete and J.A. Muñoz-Cueto. 2000a. A morphological study of the brain of *Solea senegalensis*. I. The telencephalon. Histol. Histopathol. 15: 355–364.

Rodríguez-Gómez, F.J., M.C. Rendón, C. Sarasquete and J.A. Muñoz-Cueto. 2000b. Localization of tyrosine hydroxylase-immunoreactivity in the brain of the Senegalese sole, *Solea senegalensis*. J. Chem. Neuroanat. 19: 17–32.

Rombough, P.J. 2004. Gas exchange, ionoregulation and the functional development of the teleost gill. Am. Fish. Soc. Symp. 40: 47–83.

Rønnestad, I., M. Yúfera, B. Ueberschär, L. Ribeiro, Ø. Sæle and C. Boglione. 2013. Feeding behaviour and digestive physiology in larval fish: current knowledge, and gaps and bottlenecks in research. Rev. Aquaculture 5: 59–98.

Saele, Ø., N. Silva and K. Pittman. 2006. Post-embryonic remodelling of neurocranial elements: A comparative study of normal versus abnormal eye migration in a flatfish, the Atlantic halibut. J. Anat. 209: 31–41.

Sarasquete, C., M.L. González de Canales, J.M. Arellano, J.A. Muñoz-Cueto, L. Ribeiro and M.T. Dinis. 1996. Histochemical aspects of the yolk-sac and digestive tract of larvae of the Senegal sole, *Solea senegalensis*. Histol. Histopathol. 11: 881–888.

Sarasquete, C., M.L. González de Canales, J.M. Arellano, J.A. Muñoz-Cueto, L. Ribeiro and M.T. Dinis. 1998. Histochemical study of skin and gills of Senegalese sole, *Solea senegalensis* larvae and adults. Histol. Histopathol. 13: 727–735.

Sarasquete, C., E. Gisbert, L. Ribeiro, L. Vieira and M.T. Dinis. 2001. Glyconjugates in epidermal, branchial and digestive mucous cells and gastric glands of gilthead sea bream, *Sparusaurata*, Senegal sole, *Solea senegalensis* and Siberian sturgeon, *Arcipenserbaeri* development, Eur. J. Histochem. 45: 267–278.

Sarasquete, C. and M. Gutiérrez. 2005. New tetrachromic VOF Stain (Type -III G.S.) for normal and pathological fish tissues. Eur. J. Histochem. 49: 105–114.

Sarasquete, C. and J.B. Ortiz-Delgado. 2015. Disfuncion ontogenética tiroidea y efectos en la metamorfosis de lenguado senegales, *Solea senegalensis*. SEA, Congreso Nacional Acuicultura, pp: 280–281, Huelva, España.

Sarasquete, C., Úbeda-Manzanaro, M. and J.B. Ortiz-Delgado. 2017. Effects of the soya isoflavone genistein in early life stages of the Senegalese sole, *Solea senegalensis*: Thyroid, estrogenic and metabolic biomarkers. Gen. Comp Endocrinol. 250: 136–151.

Schreiber, M. 2013. Flatfish: an asymmetric perspective on metamorphosis. Curr. Top. Dev. Biol. 103: 167–194.

Sugimoto, M. 2002. Morphological color changes in fish: regulation of pigment cell density and morphology. Microsc. Res. Tech. 58: 496–503.

Tanaka, M., J.B. Tanangonan, M. Tagawa, E.G. De Jesús, H. Nishida, M. Isaka and T. Hirano. 1995. Development of the pituitary, thyroid and interrenal glands and applications of endocrinology to the improved rearing of marine fish larvae. Aquaculture 135: 111–126.

Vázquez, R., S. González, A.R. Rodríguez and G. Mourente. 1994. Biochemical composition and fatty acid content of fertilized eggs, yolk sac stage larvae and first feeding larvae of the Senegal sole (*Solea senegalensis*), Aquaculture 119: 273–285.

Villalta, M., A. Estévez and M.P. Bransden. 2005. Arachidonic acid enriched live prey induces albinism in Senegalese sole (*Solea senegalensis*) larvae. Aquaculture 245: 193–209.

Villalta, M., A. Estévez, M.P. Bransden and J.G. Bell. 2008. Arachidonic acid, arachidonic/eicosapentaenoic acid ratio, stearidonic acid and eicosanoids are involved in dietary-induced albinism in Senegal sole (*Solea senegalensis*). Aquac. Nutr. 14: 120–128.

Viñas, J., E. Asensio, J.P. Cañavate and F. Piferrer. 2013. Gonadal sex differentiation in the Senegalese sole (*Solea senegalensis*) and first data on the experimental manipulation of its sex ratios. Aquaculture 384-387: 74–81.

Wagemans, F. and P. Vandewalle. 2001. Development of the bony skull in common sole: brief survey of morphofunctional aspects of ossification sequence. J. Fish Biol. 59: 1350–1369.

Xu, M. and H. Volkoff. 2009. Molecular characterization of ghrelin and gastrin-releasing peptide in atlantic cod (Gadus morhua): cloning, localization, devbelopmental profile and role in food intake regulation. Gen. Comp. Endocrinol. 160: 250–258.

Yamano, K. 2005. The role of Thyroid Hormone in fish development with reference to aquaculture. Review JARQ 39: 161–168.

Youson, J.H. 1988. First metamorphosis. pp. 135–198. *In*: Hoar, W.S and D.J. Randall [eds.]. Fish Physiology, vol. XI B. Academic Press, Toronto.

Yúfera, M., G. Parra, R. Santiago and M. Carrascosa. 1999. Growth, carbon, nitrogen and caloric content of *Solea senegalensis* Kaup (Pisces, Soleidae) from egg fertilization to metamorphosis.Mar.Biol. 134: 43–49.

Yúfera, M. and M.J. Darias. 2007. Changes in the gastrointestinal pH from larvae to adult in Senegal sole (*Solea senegulensis*). Aquaculture 267: 94–99.

Yúfera, M. and A.M. Arias. 2010. Traditional polyculture in "esteros" in the Bay of Cádiz (Spain). Hopes and expectancias for the prevalence of a unique activity in Europe. Aquac. Eur. 35: 22-25.

Zambonino-Infante, J., E. Gisbert, C. Sarasquete, I. Navarro, J. Gutierrez and C. Cahu. 2008. Ontogeny and physiology of the digestive system of marine fish larvae. pp. 277–344. *In*: Cyrino, J.E.O., D. Bureau and B.G. Kapoor [eds.]. Feeding and Digestive Functions of Fish. SciencePublishers. Inc, Enfield, USA.

B-2.3

Effects of Light and Temperature Cycles during Early Development

Juan Fernando Paredes, José Fernando López-Olmeda
and *Francisco Javier Sánchez-Vázquez**

1. Introduction

Geophysical cycles such as tides, day/night alternation, moon phases and seasons, have fostered the evolution of biological clocks that keep track of time and let living organisms anticipate these dependable events. In the aquatic environment, the axial rotation of the Earth causes linked environmental cycles of light and water temperature, generating daily photo- and thermo-cycles: during the day the temperature raises—thermophase, while during the night temperature drops—cryophase. Actually, in the wild water temperature peaks by the end of the day and it reaches minimum values in the early hours (Fig. 1), which can serve as powerful time cues to synchronize biological clocks.

Circadian rhythmicity is explained by the existence of self-sustained pacemakers. These clocks require daily adjustment by means of exogenous synchronizing factors called zeitgebers which include light-dark (Panda et al. 2002), temperature (Rensing and Ruoff 2002), feeding (Mistlberger 2009) and food (Sherman et al. 2012, Mistlberger 2009). The light-dark and temperature cycles are the most powerful abiotic factors that entrain biological rhythms in animals. In vertebrates, the most important is the central pacemaker, also called the light-entrainable oscillator (LEO), which is synchronized by light-dark cycles. In mammals, reptiles and birds a master circadian LEO has been found in the suprachiasmatic nucleus of the hypothalamus (SCN) (Bertolucci et al. 2008, Shibata and Tominaga 1991, Welsh et al. 2010).

Department of Physiology. Faculty of Biology. University of Murcia. Campus de Espinardo, E30100, Murcia. Spain.
* Corresponding author: javisan@um.es

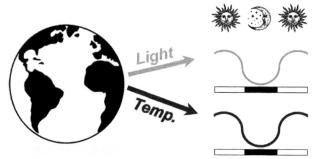

Fig. 1. Environmental cycles of light and water temperature. Daily photo- and thermo-cycles.

Although most biological processes are temperature-dependent, biological clocks are temperature-compensated to avoid running faster at higher temperatures and slower at lower temperatures, which would make them unreliable to keep time (Pittendrigh 1954). In fish, circadian locomotor activity rhythms entrain to daily thermocycles, although when conflicting light and temperatures zeitgebers are applied (light T = 25 h, and temperature T = 23 h), relative coordination appears, as locomotor activity increases when light and thermophase coincide, suggesting the existence of weakly coupled light- and temperature-entrainable oscillators (López-Olmeda and Sánchez-Vázquez 2009).

Circadian regulation controls most physiological activities during the course of a day (Panda et al. 2002). Comparative analysis of transcriptome, metabolome, proteome and epigenome has revealed a clock-dependent control in a number of metabolic pathways (Storch et al. 2002, Panda et al. 2002, Eckel-Mahan et al. 2012, Robles et al. 2014, Mauvoisin et al. 2014). Such clock-regulation covers multiple activities as lipogenesis, xenobiotic detoxification, cholesterol synthesis, ribosome biogenesis, mitochondrial respiration, sleep-wake rhythms, hormone secretion or reproduction (Dibner et al. 2010, Gerstner et al. 2009, Jouffe et al. 2013, Mauvoisin et al. 2014, Menet and Rosbash 2011, Peek et al. 2013, Sahar and Sassone-Corsi 2012, Mañanós et al. 2008).

Light and temperature cycles are the main synchronizers of seasonal (migration, reproduction) and daily (feeding, spawning) rhythms in fish (Bromage et al. 2001, Sánchez-Vázquez et al. 1997, 2001). Moreover, during early development light characteristics and thermocycle have been described as determining factors in relation with larval growth, yolk sac development, metamorphosis, survival, fin development and jaw malformations (Villamizar et al. 2009).

2. Effects of light spectrum and photoperiod

In the underwater photo-environment, light spectrum is modified as the radiant energy from the sun is selectively absorbed and scattered by particles present in the water column. The light filtering effect depends greatly on the water characteristics (oceanic, coastal or continental waters). In this dynamic landscape, fishes have efficiently adapted to live and develop, adapting their photosensitivity to the available light spectrum (Kusmic and Gualtieri 2000). Recent research in Senegalese sole has revealed that spawning, hatching rhythms, larval development, and growth performance are strongly influenced by light conditions. For instance, Blanco-Vives et al. (2010) tested the effect of photoperiod and light spectrum on growth, yolk sac, jaw malformations, eye migration and complete metamorphosis in Senegalese sole. In this study, fertilized eggs were collected and submitted to five different light regimes from day one to day 30 post hatching (DPH): 12

Light:12 Dark cycle with red light (LDr, λ = 592–668 nm); 12 L:12 D with blue light (LDb, λ = 435–500 nm); and with white light (λ = 367–757 nm) at 12 L:12 D (LDw), continuous light 24 L:0 D (LL) and continuous darkness 0 L: D 24 (DD).

2.1 Effects on growth

These experiments showed that larvae exposed to different light conditions presented differences in size before the beginning of metamorphosis. At 9 DPH, larvae submitted to LDb were longer than those under LDw or LDr (Fig. 2A). At 30DPH larvae under LDb proved to be the longest of all groups and those exposed to constant photoperiod conditions (LL and DD) were the smallest ones. These results suggest that the blue light is the most efficient spectrum for the development of Senegalese sole. Accordingly, another report on Senegalese sole also described different effects of light spectrum (red, violet and white) on plasma melatonin, the shortest wavelengths being the ones with impact (Oliveira et al. 2007).

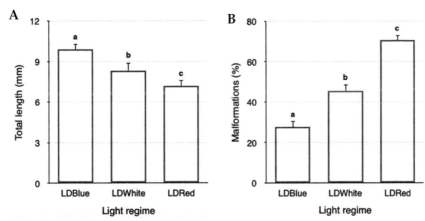

Fig. 2. Effect of light spectrum on Senegalese sole larvae. (A) Total length of larvae submitted under different light regime. (B) Jaw malformations depending on light spectrum. Lower case letters indicate statistical differences between sampling points. Data are expressed as the mean ± S.D (modified from Blanco-Vives et al. 2010).

2.2 Effects on jaw formation

The relationship between light conditions and jaw formation has also been studied. For instance, in previous reports in several fish species, Cobcroft and Battaglene (2009) described a link between the red color of the rearing tank and a major appearance of jaw malformations in striped trumpeter (*Latris lineatea*); Morrison and McDonald (1995) reported a relation between light conditions and abrasion of larval jaw tissues in atlantic halibut larvae (*Hippoglossus hippoglossus*); Villamizar et al. (2009) also described the detrimental effects of certain light spectra on jaw formation and swim bladder hypertrophy appearance in sea bass larvae (*Dicentrarchus labrax*). Similarly, in the case of Senegalese sole, Blanco-Vives et al. (2010) also observed a negative relationship between certain light spectrum and jaw formations in sole larvae, with those animals exposed to the LDr, LL and DD treatments exhibiting the highest proportion of jaw malformations. On the other hand, larvae exposed to LDb and LDw treatment showed the lowest jaw malformations (Fig. 2B). The results in Senegalese sole suggest that malformations may be due to the impaired photoperiod and light spectrum as previously reported on sea bass (Villamizar et al. 2009). However, little is known about the exact mechanism connecting light with jaw malformations, though possible explanations have been proposed regarding excess of retinoic acid (Geay et al. 2009), high levels of vitamin A in feeding (Villeneuve et al. 2006) or abnormal feeding activity (Blaxter et al. 1986). Furthermore, whether the malformations come from one single parameter or from multifactorial interactions still require further investigation.

2.3 Effects on yolk sac absorption

Light spectrum also affects yolk absorption in Senegalese sole: larvae exposed to DD and LL absorbed the yolk sac much slower than LDb, LDw and LDr, respectively. Villamizar et al. (2009) reported similar results in experiments with sea bass larvae in which the yolk sac in DD and LDr was still present 48 hours after it had been completely depleted in the LDb, LDw and LL treatments. This consideration is of high importance for larvicultures that systematically have a continuous darkness or dim light conditions protocol for early larval development, thus diminishing the use of endogenous reserves and delaying the onset of exogenous feeding (Downing and Litavak 1999).

2.4 Effects on eye migration and completed metamorphosis

Another important characteristic in the development of Senegalese Sole is metamorphosis. According to the results presented by Blanco-Vives et al. (2010), larvae exposed to LDb treatment presented an earlier eye migration (9 DPH) with respect to other light treatments: at 25 DPH all of sole larvae exposed to LDb had completed metamorphosis whereas the other groups at LD (LDw and LDr) finished at day 27 DPH. For DD and LL photoperiods, larvae died at the beginning of metamorphosis: 15 and 17 DPH, respectively.

In conclusion, several studies have reported the considerable effects of light's conditions and photoperiod in fish development. Many other examples, such as the oscillation of melatonin due to different light spectrum (Bayarri et al. 2002), differences in growth depending on the photoperiod (Tandler et al. 1958, Puvanendran et al. 2002, Trotter et al. 2003), and malformations because of the color of the rearing tank (Cobcroft et al. 2009), contribute to consider light conditions and photoperiod as crucial factors when trying to achieve optimal fish development during larviculture.

3. Effects of temperature cycles

3.1 Effects of constant temperature and thermocycles

The rotation of the Earth generates a cyclic environmental change of day and night that is associated with changes in temperature. Thus, water cools during the night (cryophase) and warms during the day (thermophase). The thermocycle has proved to be by itself strong enough to entrain rhythms such as hatching and gene expression, even in the presence of light oscillations (Boothroyd et al. 2007, Villamizar et al. 2012). Regarding fish, there is data reporting that temperature affects most aspects of behavior, physiology, foraging ability, growth and sex differentiation (Bennett et al. 1997, Bergman et al. 1987, Goolish et al. 1984, Ospina-Álvarez and Piferrer 2008). In Senegalese sole, temperature and thermocycles have been reported to have a relevant influence on growth, jaw malformations, yolk sac and metamorphosis (Blanco-Vives et al. 2010). In this study, the authors tested the effect of constant temperature vs. two types of thermocycles: a thermocycle in phase with the LD cycle (TC) (12 h thermophase-light:12 h cryophase-dark) and an inverse thermocycle (CT) (12 h cryophase-light: 12 h thermophase-dark).

3.1.1 Effects on growth

At 13 DPH, larvae under CT (19°C day/22°C night) treatment presented the lowest growth with respect to the other treatments. Larvae under TC treatment at 30 DPH reached a size (8.5 ± 0.5 mm) significantly higher than those under constant temperature (7.7 ± 0.2 mm), which in turn were significantly higher than those under CT treatment (7.2 ± 0.2 mm) (Fig. 3A). A similar experiment playing with temperature oscillations (Villamizar et al. 2012) also revealed that the highest temperature favored embryonic development in zebrafish.

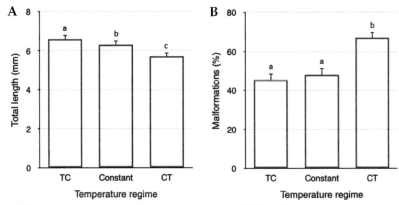

Fig. 3. Effect of different thermocycle on Senegalese sole larvae. (A) Total length of larvae reared under different thermocycles. (B) Jaw malformations depending on different temperature regimes. Lower case letters indicate statistical differences between sampling points. Data are expressed as the mean ± S.D (modified from Blanco-Vives et al. 2010).

3.1.2 Effects on jaw formation

Temperature oscillations proved to be relevant for the correct jaw formation in Senegalese sole. In the experiment performed by Blanco-Vives et al. (2010), jaw malformations at 9 DPH were significantly greater in CT treatment than in the others. AT 30 DPH, larvae exposed to TC treatment presented fewer malformations than constant temperature treatment which in turn exhibited less malformations than the CT treatment (Fig. 3B).

3.1.3 Effects on yolk sac absorption

The effects of temperature on yolk sac absorption were clearly noticeable (Blanco-Vives et al. 2010). Larvae kept under TC and constant temperature exhibited faster yolk sac absorption with respect to CT larvae treatment. Going further in comparing absorption times, at 5 DPH larvae under TC and constant temperature had mostly absorbed the yolk sac, and at 9 DPH these larvae had already completed the sac absorption whereas CT larvae finished at 11 DPH.

3.1.4 Effects on eye migration and completed metamorphosis

The effects of temperature marked significant differences in the onset and finish of metamorphosis depending on the treatment (Blanco-Vives et al. 2010). The TC treatment presented the earliest process of metamorphosis beginning and finishing eye migration at 9–17 DPH, respectively. Larvae under constant temperature begun eye migration at 11 DPH and finished at 17 DPH, and larvae under CT treatment started at 11DPH and finished at 19 DPH.

Most investigations concerning Senegalese sole development have dealt with constant temperature (Martínez et al. 1999, Parra and Yúfera 1999, Ribeiro et al. 1999, Yúfera et al. 1999, Cañavate et al. 2006); however, little interest has been paid to the natural temperature fluctuations. According to Yamashita et al. (2001), in flatfish larvae the environmental factors are crucial during the metamorphosis settlement period to such extent of determining growth and mortality. In the same line, other studies have described the existence of diel rhythms in temperature selection, that is, daily fish migration to preferred temperatures for physiological and developmental activities (Gibson et al. 1998, Sims et al. 2006). Therefore, Blanco-Vives et al. (2010), paying much attention to natural temperature oscillations, revealed that Senegalese sole exhibited higher growth, lower jaw malformations and faster yolk sac absorption under TC treatment and worse development under CT or constant temperature. In this way, temperature variations in phase with the LD cycle (matching what happens in the natural environment) reveal to be a crucial factor when dealing with developmental stages in Senegalese sole farming.

3.1.5 *Effects of temperature on sex ratio*

Sex determination in fish has been a subject of much investigation. Environmental factors can determine gonadal development and lead to skewed sex ratios in both wild and farmed fish (Siegfried et al. 2010). In addition, undifferentiated gonads are highly susceptible to these external factors to the point of overriding the genetic sex determination, shifting the destiny of the gonad towards the opposite sex (Baroiller et al. 2009). However, the precise mechanism and the thermosensitive window (time window of susceptibility) are yet not fully understood.

The most important external factors thought to determine sex ratio are photoperiod and temperature (Bromage et al. 1987, Aida and Amano 1995, Taranger et al. 1995, Blázquez et al. 1998, Colombo et al. 1998, Pavlidis et al. 2000, Blázquez et al. 2009). In the case of temperature, this factor has further relevance because of the daily thermocycles generated by the cyclic radiations from the sun due to the rotation of the Earth. Taking in consideration these temperature variations, Blanco-Vives et al. (2011) investigated the effects of temperature and thermocycles on sex determination, gonad development and sex steroid levels in juvenile Senegalese sole. For this purpose, sole larvae and juveniles were submitted to different temperature cycles (natural thermocycle (TC) or reverse thermocycle (CT) vs. constant temperature).

Fish under TC treatment presented a higher proportion of females (79.8%) than males (21.2%), whereas those exposed to CT showed a greater proportion of males (82.5%) than females (17.5%) (Fig. 4) (Blanco-Vives et al. 2011). Fish exposed to constant temperature showed a greater proportion of males (61.6%) than females (38.3%). The results of sex determination correlated with those of sex steroid production: fish exposed to TC during development showed a greater estradiol (E_2) and lower testosterone (T) and 11-keto testosterone (11-KT) production than fish exposed to either CT or constant temperature (Fig. 4). These results clearly describe the importance of temperature oscillations in the culturing of Senegalese sole, species extensively exploited in aquaculture, and the need for optimizing raring protocols.

4. Hatching rhythms

Fish embryogenesis has long been considered a progression of developmental stages primarily controlled by temperature and that successfully concludes with hatching. Recent

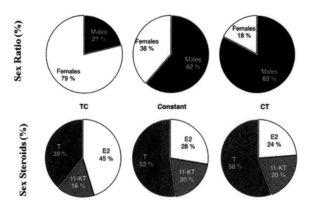

Fig. 4. Sex and steroids ratios of the Senegalese sole population in percentage submitted to three different experimental temperature regimes. Data are expressed as mean ± SEM (Blanco-Vives et al. 2011).

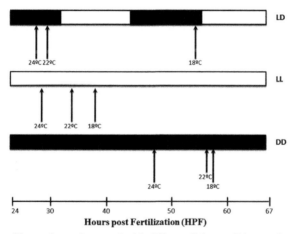

Fig. 5. Hatching rhythms of Senegalese sole treated with different light conditions and temperatures. Light: Dark cycle (LD), continuous light (LL), continuous darkness (DD). Black and white bars indicate the dark or light phase of the cycle, respectively. Arrows indicate the hatching times depending on the temperature. The button line indicates the hours after fertilization.

investigations have revealed the existence of constant temporal control during embryonic development (Gorodilov et al. 2010), suggesting a connection with the biological clock. However, little is known about the rhythms during embryonic development and hatching process, especially, about its synchronization to the light-dark cycle.

In a comparative study between zebrafish, Senegalese sole and Somalian cavefish, Villamizar et al. (2013) described the existence of daily hatching rhythms synchronized to the LD cycle with different acrophases depending on the species. Zebrafish displayed a diurnal acrophase, Senegalese sole a nocturnal one and Somalian cavefish a nocturnal acrophase during the first night and a diurnal (but very close to the beginning of the dark phase) on the second day (Fig. 5). These results suggest that hatching rhythms are apparently controlled by a clock mechanism restricting or "gating" hatching to a specific time of day/night (window). Hence, embryos that have reached a certain point of development hatch while those that have not wait to hatch until the next available window (Villamizar et al. 2012). In conclusion, the hatching time is determined by interplay between the developmental stage and the circadian endogenous clock with an acrophase depending on the natural behavior of the species.

Acknowledgements

This research was funded by the project "SOLEMBRYO" (AGL2013-49027-C3-1-R) granted by the Spanish Ministry of Economic Affairs and Competitiveness (MINECO) to FJSV. J.F.LO was funded through a research fellowship granted by MINECO (Juan de la Cierva Program).

References

Aida, K. and M. Amano. 1995. Salmon GnRH gene expression following photoperiod manipulation in precocious male masu salmon. pp. 161–163. *In*: Goetz, F.W. and P. Thomas [eds.]. Proceedings of the Fifth International Symposium on the Reproductive Physiology of Fish. TX: University of Texas, Port Aransas, Austin.

Baroiller, J.F., H. D'Cotta, E. Bezault et al. 2009. Tilapia sex determination: where temperature and genetics meet. Comp. Biochem. Physiol. A. Mol. Integr. Physiol. 153: 30–38.

Bayarri, M.J., J.A. Madrid and F.J. Sánchez-Vázquez. 2002. Influence of light intensity, spectrum and orientation on sea bass plasma and ocular melatonin. J. Pineal Res. 32: 34–40.

Bennett, W.A. and T.L. Beitinger. 1997. Temperature tolerance of the sheeps head minnow, *Cyprinodon variegatus*. Copeia. 1997: 77–87.

Bergman, E. 1987. Temperature-dependent differences in foraging ability of two percids, *Perca fluviatilis* and *Gymnocephalus cernuus*. Environ. Biol. Fish 19: 45–53.

Bertolucci, C., F. Fazio and G. Piccione. 2008. Daily Rhythms of Serum Lipids in Dogs: Influences of Lighting and Fasting Cycles. Comparative Med. 58: 485–489.

Blanco-Vives, B., N. Villamizar, J. Ramos et al. 2010. Effect of daily thermo- and photo-cycles of different light spectrum on the development of Senegal sole (*Solea senegalensis*) larvae. Aquaculture 306: 137–145.

Blanco-Vives, B., L.M. Vera, J. Ramos, M.J. Bayarri et al. 2011. Exposure of larvae to daily thermocycle affects gonad development, sex ratio and sexual steroids in *Solea senegalensis*. J. Exp. Zool. 315: 162–169.

Blaxter, J.H.S. 1986. Development of sense organs and behaviour of teleost larvae with special reference to feeding and predator avoidance. T. Am. Fish Soc. 115: 98–114.

Blazquez, M., S. Zanuy, M. Carrillo et al. 1998. Effects of rearing temperature on sex differentiation in the European sea bass *Dicentrarchus labrax* L. J. Exp. Zool. (Mol. Dev. Evol.) 282: 207–216.

Blázquez, M., L. Navarro-Martín and F. Piferrer. 2009. Expression profiles of sex differentiation-related genes during ontogenesis in the European sea bass acclimated to two different temperatures. J. Exp. Zool. (Mol. Dev. Evol.) 312: 686–700.

Bolla, S. and I. Holmefjord. 1988. Effect of temperature and light on development of Atlantic halibut larvae. Aquaculture 74: 355–358.

Boothroyd, C.E., H. Wijnen, F. Naef et al. 2007. Integration of Light and Temperature in the Regulation of Circadian Gene Expression in Drosophila. PLoS Genet 3: e54.

Bromage, N.R. 1987. The advancement of puberty or time of first spawning in female rainbow trout *Salmo gairdneri* maintained on altered seasonal-light cycles. *In*: Idler, D.L., L.W. Crim and J.M. Walsh [eds.]. Proceedings of the Third International Symposium on the Reproductive Physiology of Fish. Canada: Memorial University of Newfoundland, St. John's, Newfoundland. p. 303.

Bromage, N.R., M.J.R. Porter and C.F. Randall. 2001. The environmental regulation of maturation in farmed finfish with special reference to the role of photoperiod and melatonin. Aquaculture 197: 63–68.

Cañavate, J.P., P. Zerolo and C. Fernandez-Diaz. 2006. Feeding and development of Senegal sole (*Solea senegalensis*) larvae reared in different photoperiods. Aquaculture 258: 368–377.

Cobcroft, J.M. and S.C. Battaglene. 2009. Jaw malformation in striped trumpeter *Latris lineata* larvae linked to walling behavior and tank colour. Aquaculture 289: 274–282.

Colombo, L., A. Barbaro, A. Francescon et al. 1998. Towards an integration between chromosome set manipulations, intergeneric hybridization and gene transfer in marine fish culture. Cah. Opt. Mediterran 34: 77–122.

Dibner, C., U. Schibler and U. Albrecht. 2010. The mammalian circadian timing system: organization and coordination of central and peripheral clocks. Annu. Rev. Physiol. 72: 517–549.

Downing, G. and M.K. Litvak. 1999. The influence of light intensity on growth of larval haddock. N. Am. J. Aquacult. 61: 135–140.

Eckel-Mahan, K., V. Patel, R. Mohney, K.S. Vignola et al. 2012. Coordination of the transcriptome and metabolome by the circadian clock. Proc. Natl. Acad. Sci. USA 109: 5541–5546.

Geay, F., M.J. Darias, E. Santigosa et al. 2009. Cloning of endothelin-1 (ET-1) from European sea bass (*Dicentrarchus labrax*) and its gene expression analysis in larvae with retinoic acidinduced malformations. Aquaculture 287: 169–173.

Gerstner, J.R., L.C. Lyons, K.P. Wright et al. 2009. Cycling behavior and memory formation. J. Neurosci. 29: 12824–12830.

Gibson, R.N., L. Pihl, M.T. Burrows, J. Modin et al. 1998. Diel movements of juvenile plaice, *Pleuronectes platessa*, in relation to predators, competitors, food availability and abiotic factors on a microtidal nursery ground. Mar. Ecol. Prog. Ser. 165: 145–159.

Goolish, E.M. and I.R. Adelman. 1984. Effects of ration size and temperature on the growth of juvenile common carp (*Cyprinus carpio* L.). Aquaculture 36: 27–35.

Gorodilov, Y.N. 2010. The biological clock in vertebrate embryogenesis as a mechanism of general control over the developmental organism. Russ. J. Dev. Biol. 41: 201–216.

Jouffe, C., G. Cretenet, L. Symul, E. Martin et al. 2013. The circadian clock coordinates ribosome biogenesis. Plos Biol. 11, e1001455.

Kusmic, C. and P. Gualtieri. 2000. Morphology and spectral sensitivities of retinal and extraretinal photoreceptors in freshwater teleosts. Micron. 31: 183–200.

López-Olmeda, J.F. and F.J. Sánchez-Vázquez. 2009. Zebrafish temperature selection and synchronization of locomotor activity circadian rhythm to ahemeral cycles of light and temperature. Chronobiol. Int. 26: 200–218.

Mañanós, E., N. Duncan and C.C. Mylonas. 2008. Methods in Reproductive Aquaculture: Marine and Freshwater Species. Taylor & Francis, eds. Reproduction and control of ovulation, spermiation and spawning in cultured fish. FL, USA. pp. 3–80.

Martínez, I., F.J. Moyano, C. Fernández-Díaz et al. 1999. Digestive enzyme activity during larval development of the Senegal sole (*Solea senegalensis*). Fish Physiol. Biochem. 21: 317–323.

Mauvoisin, D., J. Wang, C. Jouffe et al. 2014. Circadian clock-dependent and independent rhythmic proteomes implement distinct diurnal functions in mouse liver. Proc. Natl. Acad. Sci. USA 111: 167–172.

Menet, J.S. and M. Rosbash. 2011. When brain clocks lose track of time: cause or consequence of neuropsychiatric disorders. Curr. Opin. Neurobiol. 21: 849–857.

Mistlberger, R.E. 2009. Food-anticipatory circadian rhythms: concepts and methods. Eur. J. Neurosci. 30: 1718–1729.

Morrison, C.M. and C.A. MacDonald. 1995. Normal and abnormal jaw development of the yolksac larva of Atlantic halibut (*Hippoglossus hippoglossus*). Dis. Aquat. Org. 22: 173–184.

Oliveira, C., A. Ortega, J.F. López-Olmeda, Vera et al. 2007. Influence of constant light and darkness, light intensity, and light spectrum on plasma melatonin rhythms in Senegal sole. Chronobiol. Int. 24: 615–627.

Ospina-Álvarez, N. and F. Piferrer. 2008. Temperature-dependent sex determination in fish revisited: Prevalence, a single sex ratio response pattern, and possible effects of climate change. PLoS ONE 3: e2837.

Panda, S., J.B. Hogenesch and S.A. Kay. 2002. Circadian rhythms from flies to human. Nature 417: 329–335.

Parra, G. and M. Yúfera. 1999. Tolerance response to ammonia and nitrite exposure in larvae of two marine fish species (gilthead seabream *Sparus aurata* L. and Senegal sole *Solea senegalensis* Kaup). Aquac. Res. 30: 857–863.

Pavlidis, M., G. Komoundouros, A. Sterioti et al. 2000. Evidence of temperature dependent sex determination in the European sea bass *Dicentrarchus labrax* L. J. Exp. Zool. (Mol. Dev. Evol.) 287: 225–232.

Peek, C., A.H. Affinati, K.M. Ramsey et al. 2013. Circadian clock NAD+ cycle drives mitochondrial oxidative metabolism in mice. Science 342 (6158).

Pittendrigh, C.S. 1954. On temperature independence in the clock system controlling emergency time in Drosophila. Proc. Natl. Acad. Sci. USA 40: 1018–1029.

Puvanendran, V. and J.A. Brown. 2002. Foraging, growth and survival of Atlantic cod, *Gadus morhua*, larvae reared in different light intensities and photoperiods. Aquaculture 214: 131–151.

Rensing, L. and P. Ruoff. 2002. Temperature effect on entrainment, phase shifting, and amplitude of circadian clocks and its molecular bases. Chronobiol. Int. 19: 807–864.

Ribeiro, L., J.L. Zambonino-Infante, C. Cahu et al. 1999. Development of digestive enzymes in larvae of *Solea senegalensis*, Kaup 1858. Aquaculture 179: 465–473.

Robles, M., J. Cox and M. Mann. 2014. *In-vivo* quantitative proteomics reveals a key contribution of post-transcriptional mechanisms to the circadian regulation of liver metabolism. Plos Genet. 10, e1004047.

Sahar, S. and P. Sassone-Corsi. 2012. Regulation of metabolism: the circadian clock dictates the time. Trends Endocrinol. Metab. 23: 1–8.

Sánchez-Vázquez, F.J., J.A. Madrid, S. Zamora et al. 1997. Feeding entrainment of locomotor activity rhythms in the goldfish is mediated by feeding-entrainable circadian oscillator. J. Comp. Physiol. 181: 121–132.

Sánchez-Vázquez, F.J., A. Aranda and J.A. Madrid. 2001. Differential effects of meal size and food energy density on feeding entrainment in goldfish. J. Biol. Rhythms 16: 58–65.

Sherman, H., Y. Genzer, R. Cohen et al. 2012. Timed high-fat diet resets circadian metabolism and prevents obesity. Faseb. J. 26: 3493–3502.

Shibata, S. and K. Tominaga. 1991. Brain neuronal mechanisms of circadian rhythms in mammalians. Yakuga Zasshi. 111: 270–283.

Siegfried, K.R. 2010. In search of determinants: gene expression during gonadal sex differentiation. J. Fish Biol. 76: 1879–1902.

Sims, D.W., V.J. Wearmouth, E.J. Southall et al. 2006. Hunt warm, rest cool: bioenergetic strategy underlying diel vertical migration of a benthic shark. J. Anim. Ecol. 75: 176–190.

Storch, K.F., O. Lipan, I. Leykin et al. 2002. Extensive and divergent circadian gene expression in liver and heart. Nature 417: 78–83.

Tandler, A. and S. Helps. 1985. The effects of photoperiod and water exchange on growth and survival of gilthead sea bream (*Sparus aurata*, Linnaeus; Sparidae) from hatching to metamorphosis in mass rearing systems. Aquaculture 48: 71–82.

Taranger, G.L., H. Daae, K.O. Jørgensen et al. 1995. Effects of continuous light on growth and sexual maturation in sea water reared Atlantic salmon. *In*: Goetz, F.W. and P. Thomas [eds.]. Proceedings of the Fifth International Symposium on the Reproductive Physiology of Fish. TX: University of Texas, Port Aransas, Austin. p. 200.

Villamizar, N., A. Garcia-Alcazar and F.J. Sánchez-Vázquez. 2009. Effect of light spectrum and photoperiod on the growth, development and survival of European sea bass. Aquaculture 292: 80–86.

Villamizar, N., L. Ribas, F. Piferrer et al. 2012. Larval performance and sex differentiation of zebrafish. PLoS ONE 7(12): e52153.

Villamizar, N., B. Blanco-Vives, C. Oliveira et al. 2013. Circadian rhythms of embryonic development and Hatching in fish: A comparative study of zebrafish (Dirunal), Senegalese Sole (Nocturnal), and Somalian Cavefish (Blind). Chronobiol Int. 1–12.

Villeneuve, L.A.N., E. Gisbert, J. Moriceau et al. 2006. Intake of high levels of vitamin A and polyunsaturated fatty acids during different developmental periods modifies the expression of morphogenesis genes in European sea bass (*Dicentrarchus labrax*). Br. J. Nutr. 95: 677–687.

Welsh, D., J. Takahashi and S. Kay. 2010. Suprachiasmatic nucleus: cell autonomy and network properties. Annu. Rev. Physiol. 72: 551–577.

Yamashita, Y., M. Tanaka and J.M. Millar. 2001. Ecophysiology of juvenile flatfish in nursery grounds. J. Sea Res. 45: 205–218.

Yúfera, M., G. Parra, R. Santiago et al. 1999. Growth, carbon, nitrogen and caloric content of *Solea senegalensis* (Pisces: Soleidae) from egg fertilization to metamorphosis. Mar. Biol. 134: 43–49.

B-2.4

Larval Production Techniques

Sofia Engrola, * *Cláudia Aragão* and *Maria Teresa Dinis*

1. Introduction

Senegalese sole is a species of high commercial value and during the last few years, the high interest in industrial production has led to major achievements in larval nutrition and rearing techniques (Campos et al. 2013a,b, Canada et al. 2016a, Engrola et al. 2009, 2010, Morais et al. 2014, Pinto et al. 2010b). However, some aspects of the biology make Senegalese sole culture particularly challenging such as the change from pelagic symmetrical larva into an asymmetrical benthic flatfish. Consequently, this shift from pelagic to benthic life will condition the rearing conditions, feeding regimes and nutritional requirements of Senegalese sole larvae and postlarvae.

2. Larval rearing systems

The current zootechnical parameters that are being used or should be considered for sole culture are as follows:

2.1 Pelagic vs. benthonic systems

Due to the larva metamorphosis, the rearing systems, in particular the type of tanks, change within the larvae development. Basically, the larval rearing tanks for the pelagic phase are from glass reinforced plastic and circular, with a diameter from 2–4 m, with flat or sloping bottom. During pre-metamorphosis, when the benthic phase starts, tanks need to have a flat bottom and might be circular or rectangular shallow raceways type. In both

Centre of Marine Sciences of Algarve, University of Algarve, Campus de Gambelas, 8005-139 Faro, Portugal.
* Corresponding author: sengrola@ualg.pt

cases, recirculating is the most common water circulation as it provides a stable physical environment, with 10–20% water exchange per day.

2.2 Pelagic rearing: Green algae technique

The green water technique is nowadays a usual procedure in marine larviculture, and consists in adding microalgae to the rearing tanks. The improvement in the larval performance when using microalgae has been described by several authors, who pointed out that microalga might provide nutrients to larvae (Moffatt 1981), contribute to preserve live prey nutritional quality (Makridis and Olsen 1999), promote changes of light conditions in water and changing the colour of the medium and its chemical composition (Naas and Huse 1996, Naas et al. 1992), contribute for the microflora diversification either in the tank or in the larval gut (Nicolas et al. 1989, Reitan et al. 1997, Skjermo and Vadstein 1993, 1999) and promote feeding ability in sole (Rocha et al. 2008).

At hatching, the gut of larvae contains only small numbers of bacteria. Yolk-sac larvae of marine fish species drink seawater to maintain homeostasis and thereby ingest microalgae cells together with bacteria present in the water (Reitan et al. 1998). These microalgae accumulate in the gut, and may trigger the digestion process, or modify the newly established microbiota of the larval gut (Hjelmeland et al. 1988, Olafsen 2001). Microalgae cells may as well produce a number of bioactive compounds which, when excreted in the medium, may stimulate the immune system of the larvae and improve the digestive capacity of the larvae (Makridis et al. 2010). Study has shown that addition of microalgae in the rearing of sole larvae does not play such an important role as in the rearing of larvae of other marine fish species. These findings indicated that the addition of either microalgae supernatant or bacteria isolated from microalgae cultures did not influence the survival and growth of Senegalese sole larvae.

2.3 Salinity

Senegalese sole is a marine teleost that uses estuarine areas as nursery grounds (Dinis et al. 1999, Ramos et al. 2010, Vinagre et al. 2009). Coastal areas of Portugal and Spain (earth ponds and wet areas), where salinity shows quite important variation, are the main environments for the species (Drake et al. 1984). The possibility of successfully rearing sole larvae under reduced salinity conditions would lead to new options for hatchery facilities where only brackish water is available (Leiton et al. 2012). In sole high hatching rates (above 80%) were registered when incubating eggs for 48h at salinities of 10, 18, 27 and 33 g L^{-1}, but salinities of 5 g L^{-1} contributed to a delay of 24 h on the hatching time and at 0 g L^{-1} no hatching was obtained (Leiton et al. 2012). In standard protocols salinity of 35–38 g L^{-1} is commonly used (Pousão-Ferreira 2007).

Concerning the larval rearing, sole larvae might be successfully cultivated until complete metamorphosis at 10 g L^{-1}, provided that this salinity is carried out at first feeding when the mouth opening process has been completed (2–3 days after hatching, DAH). Feeding activity is also not affected if rearing is between salinities of 10 and 33 g L^{-1}. Both salinities led to similar individual dry weights till the benthic stage. With the exception of 14DAH, salinity did not induce differences in the development of metamorphosis—a process completed at 21DAH under both conditions. A higher final survival was, in contrast, obtained in larvae cultivated at 33 g L^{-1} salinity (Leiton et al. 2012). Anatomical abnormalities during early development, mainly mouth deformities, occurred when larvae were hatched and subsequently kept at 10 g L^{-1} salinity. The results reveal a key time window (0–2DAH) where salinity higher than 10 g L^{-1} is required for adequate mouth development and further functionality of larvae.

2.4 Temperature

The current temperature for sole rearing is 18–20°C (Dinis et al. 1999) but it is known that in Teleost fish, temperature conditions during early stages of development can produce different phenotypes regarding muscle fibre composition and growth patterns in the adults (Johnston 2006, Valente et al. 2013). Recent works from Campos et al. (2013a,b,c,d) demonstrate that differences in embryonic temperature were sufficient to produce differences in growth, muscle phenotype and gene expression in Senegalese sole larvae up to 30DAH, even if the larvae were reared at the same temperature after hatching. Muscle formation being a consequence of a series of complex, concerted events (Steinbacher et al. 2006, Valente et al. 2013), the changes in temperature during critical developmental windows have the potential to irreversibly alter muscle cellularity (Campos et al. 2013b). For this reason, larva rearing protocols consider the hatching temperature as the most adequate temperature.

Senegalese sole eggs are normally obtained from natural spawning of wild broodstock kept in captivity, and spawning takes place at a wide range of temperatures, from 13 to 23°C but with higher fecundities between 15 and 21°C (Anguis and Cañavate 2005). Normally, egg incubation is performed at hatching temperature. The incubation of embryos at 15°C produced smaller larvae with less and smaller fibres throughout growth compared to those incubated at 18 and 21°C, but a great increase in fast fibre number was seen particularly after mouth opening, and the highest percentage of hypertrophic growth was found by 30DAH (Campos et al. 2013b). These results pointed out that incubation of *S. senegalensis* eggs at 15°C produced a delay in embryonic development and in smaller larvae with less and smaller fibres throughout on-growth compared to those incubated at 18 and 21°C. Larvae reared at 15°C took more than twice the time to acquire a benthic lifestyle than larvae from 21°C (35DAH and 16DAH, respectively). During metamorphosis, larvae from 21°C had a similar total length (6.9 ± 0.8 mm) to larvae from 18°C (6.7 ± 0.8 mm); however, both had a larger body length than larvae from 15°C (5.5 ± 0.8 mm). These results show that temperature during specific time frames of ontogeny has both short- and long-term effects on growth and muscle cellularity of Senegalese sole. Nevertheless, Senegalese sole also seems to rapidly adapt to environmental temperature through a set of molecular mechanisms and physiological responses such as regulation of feed intake, even at early developmental stages (Campos et al. 2014b).

3. Feeding regimes

The onset of exogenous feeding is a critical moment for all fish larvae (Hamre et al. 2013, Yúfera and Darias 2007). Marine fish larvae have small body size and high metabolic demands due to high growth rates (Conceição et al. 1998, Rønnestad et al. 2013). Senegalese sole larva may present, during the pelagic phase, growth rates around 21–29% a day (Campos et al. 2013a, Canada et al. 2016a, Cañavate and Fernández-Díaz 1999, Engrola et al. 2009). Fast growth is of vital importance for larval fish as predation susceptibility decreases with increasing body size (Blaxter 1988). In order to grow, larvae should eat and be able to digest the feed, meaning that at mouth opening mouth gape is a vital factor to account when creating a feeding protocol. At first-feeding, 2DAH larvae presents 3.0–3.3 mm of total length (Dinis et al. 1999) and a mouth gape of 350 µm

(Parra and Yúfera 2001). Two similar feeding protocols are commonly adopted for sole rearing: ST 1—Shorter period with rotifers, and ST 2—Longer period with rotifers (Table 1). Both feeding protocols offer rotifers (*Brachionus plicatilis*) as first food and after some days will shift to larger preys like *Artemia* nauplii and metanauplii. In a feeding protocol, it is always assumed that gradually the larvae will be fed with larger and more energy-rich prey (Rønnestad et al. 2013).

As Senegalese sole is an altricial species, with limited swimming capacity at early stage, feeding protocols need to include slow movement preys like rotifers or *Artemia* nauplli. However, the inclusion of *Artemia* nauplii should be considered very carefully because depending on the *Artemia* strain the nauplii size will change, as well as the nutritional content.

The values presented in Table 1 are only reference values to help establishing a feeding protocol. Several factors like temperature, water renewal, feeding frequency, larvae density, among others, also need to be considered when feeding fish. Larvae performance is quiet similar for both feeding protocols, around 1.1–1.3 mg of dry weight at 20DAH (Cañavate and Fernández-Díaz 1999, Engrola et al. 2009, Navarro-Guillen et al. 2015, Villalta and Estévez 2005). Live preys, such as rotifers and Artemia, are normally offered to larvae in marine hatcheries at first-feeding. Similar growth between the feeding protocols may indicate that despite a longer period with a low protein content prey like the rotifers (ST 2)

Table 1. Feeding protocol with live feed for Senegalese sole larvae from 2 to 25 days after hatching, ST1—Standard feeding and ST2—Standard feeding with a longer period with rotifers.

Treatments						
	ST 1			ST 2		
DAH	Rot	Na	Meta	Rot	Na	Meta
2	5000			5000		
3	5000			5000		
4	2000	3000		7000		
5	2000	4000		9000		
6		5000		11000		
7		7000		13000		
8		8000		15000	1500	
9		3000	3000	15000	3000	
10			5000		3000	2000
11			7000			7000
12			9000			9000
13			10000			10000
14			11000			11000
15			12000			12000
16			13000			13000
17			15000			15000
18–25			16000			16000

DAH: days after hatching; Rot: Rotifers; Na: *Artemia* nauplii; Meta: *Artemia* metanauplii. Rotifers and *Artemia* are expressed as "number of prey/L tank volume/day".

(around 28%, Conceição et al. 2010), sole larvae is probably ingesting more preys in order to grow at similar rates.

Since the nutrient composition of *Artemia* metanauplii is inadequate to sustain growth of sole postlarvae at later stages (Engrola et al. 2009), weaning is pivotal to sustain larva growth and to produce quality juveniles. The transition from live feeds to inert diets is a critical period that must be tightly controlled and will depend from the handler sensitivity. Senegalese sole weaning was considered a major bottleneck for large scale production for several years (Conceição et al. 2007, Dinis 1992). Nevertheless, with the vast research efforts made during the last decade, successful weaning protocols have appeared that made possible large scale production. Sole age, around 40DAH, was considered during many years as an indicator to start weaning (Cañavate and Fernández-Díaz 1999, Dinis 1992, Dinis et al. 1999, Engrola et al. 2005, Engrola et al. 2009, Ribeiro et al. 2002). At that time survival rate of weaned fish were up to 70%, but growth rate was very low. The feeding regime for weaning could include a co-feeding regime (1 week with live feed and inert diet) or a feeding regime with only inert diet (sudden weaning). Both feeding regimes were able to wean sole; however, the co-feeding with live prey period should be brief because otherwise growth could be impaired. In fact (Engrola et al. 2007) observed that sole larvae fed live feed alone were three- to sevenfold smaller than sole larvae fed with inert diet, at 60DAH. In the same paper, the authors suggested that the feeding strategy for weaning should be based on postlarvae weight rather larval age. So, as an indicator 5–10 mg sole weight could be used to start weaning. In the last years, the improvements made in feed technology and feeding protocols made possible to continuously anticipate sole weaning and to increase weaning survival rates to around 90% in sole hatcheries.

At the moment, it is not possible to sustain growth and development when sole larvae are fed inert diet at mouth opening. Thus, early feeding regimes that combine live preys and inert diets at mouth opening are critical to fully exploit fish larvae growth potential. In several marine fish larvae, co-feeding regimes were able to promote survival and growth. In Senegalese sole, considerable experiments were performed with inert diets at mouth opening or at early stages of development using a co-feeding regime (Canada et al. 2016a, Cañavate and Fernández-Díaz 1999, Engrola et al. 2009, 2010, Fernández-Díaz et al. 2006, Gamboa-Delgado et al. 2008, Ribeiro et al. 2005, Yúfera et al. 1999, 2005). Nevertheless, inert diets have experienced major improvements regarding physical and biochemical properties during the past years. In more recent years, the inclusion of protein hydrolysates increased diet acceptability by the larvae and promoted growth at early stages (Canada et al. 2014, Pinto et al. 2015). Therefore, in order to improve larval performance co-feeding regimes, based on live feeds and inert diet, may be proposed for *Solea senegalensis* larvae (Table 2).

Co-feeding with live feeds and inert diet from mouth opening has been shown to depress growth at sole earlier stages, but promotes better postlarval quality after full weaning (Engrola et al. 2009). Currently, it is possible to start sole weaning at mouth opening with the available commercial diets. Besides, all the attention that offering inert diet to fish larvae implies, handlers should choose the size of the pellet carefully. Accordingly, at mouth opening sole larvae should be fed with pellets around 300 µm and pellet size should increase with larval weight (Fig. 1).

Feeding regimes are one of the most important factors during larval rearing. Offering larger preys that the larvae are not able to eat or inert diet that larvae are not attracted to may influence, quiet rapidly, the rearing outcome. Even when offering the proper diet to larvae, according to the feeding protocol, handlers should always inspect visually the larvae to assure intake and fish wellbeing.

Table 2. Weaning strategies for Senegalese sole larvae from 2 to 25 days after hatching based on a Standard feeding regime (ST 1); CoFL—Co-feeding regime with low *Artemia* replacement with inert diet (dry matter basis) from mouth opening; CoFH—Co-feeding regime with high *Artemia* replacement with inert diet (dry matter basis) from mouth opening.

Treatments								
	ST 1			CoFL		CoFH		
DAH	Rot	Na	Meta	Live prey (%)	Inert diet (%)	Live prey (%)	Inert diet (%)	Inert diet (µm)
2 to 5	3500	3500		80	20	52	48	200–400
6 to 9		5750	3000	67	33	42	58	200–400
10 to 13			7750	45	55	27	73	400–600
14 to 17			12750	45	55	23	77	400–600
18 to 25			16000	30	70	14	86	400–600

DAH: days after hatching; Rot: Rotifers; Na: *Artemia* nauplii; Meta: *Artemia* metanauplii; Rotifers and *Artemia* are expressed as the "average number of prey/L tank volume/day between the age intervals" based on the feeding protocol ST1 described in Table 1.

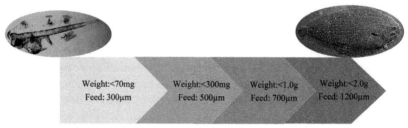

Fig. 1. Senegalese sole feeding schedule for inert diet, since mouth opening until 2g.

4. Early nutrition of sole larvae

Fish larvae have tremendous growth potential, with relative growth rates much higher than juvenile and adult fish (Rønnestad and Conceição 2005). Feeding and nutrition are the keys to unlock this growth potential. Thus, several studies have addressed the nutritional requirements of Senegalese sole in order to maximize growth potential, produce high quality larvae and postlarvae and achieve a successful transition from live feeds to inert microdiets.

The amino acid profile of Senegalese sole changes during ontogenesis (Aragão et al. 2004b, Cara et al. 2007), suggesting changes in the amino acid requirements along the development. These changes are more pronounced in Senegalese sole than in other fish species, which seems to be linked to the marked metamorphosis that sole undergoes, involving drastic variations in body shape, habitat, and feeding regime. In this sense, Senegalese sole is highly sensitive to dietary imbalances in amino acid profiles (Cara et al. 2007). Balancing the dietary amino acid profile increases protein retention (Aragão et al. 2004c, Canada et al. 2016b), but translation of this effect to higher growth performance is yet to be demonstrated (Canada et al. 2016a,b).

Dietary supplementation with specific amino acids showed positive effects. Tyrosine supplementation has been proposed to be beneficial for Senegalese sole during metamorphosis, as a way to increase tyrosine availability for coping with metamorphosis-related processes, such as the production of thyroid hormones (Pinto et al. 2010c).

Table 3. Summary of dietary supplements provided to *Solea senegalensis* larvae with positive impacts on growth or physiological parameters.

Dietary supplement	Diet provided	Feeding period	Reference
Taurine	Microencapsulated diet (9.3 mg taurine/g diet)	3 to 16DAH	Pinto et al. (2010a)
Arachidonic acid	Enriched *Artemia* metanauplii (ARA at 4.5% of total fatty acids in experimental emulsions)	8 to 50DAH	Boglino et al. (2012a)
Vitamin C	Microencapsulated diets (3500–5700 µg ascorbyl palmitate/g diet)	2 to 7DAH	Jimenez-Fernandez et al. (2015)
Vitamin K	Enriched rotifers and *Artemia* metanauplii (250 mg phylloquinone/kg commercial emulsion)	2 to 39DAH	Richard et al. (2014)
Selenium	Enriched rotifers and *Artemia* metanauplii (3 mg Sel-Plex[®1]/million individuals)	2 to 31DAH	Ribeiro et al. (2011)
Iodine	Enriched rotifers and *Artemia* metanauplii (780 mg NaI/g commercial emulsion)	2 to 34DAH	Ribeiro et al. (2012b)

DAH = days after hatching; [1]selenized yeast.

Apparently, this species seems to be unable to biosynthesize tyrosine at a sufficient rate to supply its physiological requirements until after metamorphosis (Pinto et al. 2010c). Dietary tyrosine supplementation seems also to help postlarvae cope with stressful conditions (Aragão et al. 2010).

Inclusion of taurine in microdiets (Table 3) during the pelagic phase has been suggested in order to increase larval growth potential and guarantee metamorphosis success (Pinto et al. 2010a). Based on the results from Pinto et al. (2010a), taurine supplementation may also be advisable when adopting the feeding regime Standard 2 described in the previous section (9 days of feeding with rotifers), since taurine content in rotifers is seven-fold lower than in *Artemia* metanauplii (when both are enriched with commercial products; Aragão et al. 2004a) and taurine transport is enhanced by metamorphosis (Pinto et al. 2012).

The molecular form of dietary nitrogen has been long recognized as an important factor to enhance larval diet attractability and maturation of the digestive tract (Cahu and Zambonino Infante 2001). In a recent work with Senegalese sole larvae, in which co-feeding of live-feeds and experimental microdiets containing different levels of protein hydrolysates were tested, the results suggested that high dietary protein hydrolysate levels may be more adequate for sole only at first-feeding and during the early developmental stages (Pinto et al. 2015). These results reinforce the point that due to the peculiar development of sole, diets for pelagic larvae may not be the more adequate to post-metamorphic stages, since nutritional requirements also change (Aragão et al. 2004b, Pinto et al. 2010c).

In terms of lipid nutrition, experiments suggest that an increase in dietary neutral lipid content increases lipid accumulation within the gut epithelium and results in lower fatty acid absorption (Morais et al. 2005a,b, 2006). In line with this, oil sources rich in long-chain polyunsaturated fatty acids (PUFA) and n-3 PUFA were superior to those having high n-6 PUFA or monounsaturated fatty acid levels in the larval nutrition of Senegalese sole. Navarro-Guillén et al. (2014) claimed that it is advantageous to use enrichment emulsions based on vegetable oils in postlarval Senegalese sole nutrition, since these present high levels of monounsaturated fatty acids and supply higher levels of energy substrates.

However, supplementation with DHA-rich oil is advisable, in order to achieve a correct balance between dietary energy and essential fatty acids (Navarro-Guillén et al. 2014).

Several studies have focused on the requirements of essential fatty acids of Senegalese sole larvae. Docosahexaenoic acid (DHA) requirement has been suggested to be low or negligible in sole during the early rearing stages (Morais et al. 2004, Villalta et al. 2005b), which was reinforced by recent studies indicating an ability of Senegalese sole to biosynthesise DHA from its precursors during the larval period (Morais et al. 2012, Navarro-Guillén et al. 2014). However, the incorporation of low DHA and lipid levels in weaning diets for Senegalese sole has been shown to compromise postlarval growth performance (Pinto et al. 2016).

Senegalese sole larvae have been considered to present a very low eicosapentaenoic acid (EPA) requirement during the live feeding period (Villalta et al. 2008b), although the ratios of arachidonic acid-to-eicosapentaenoic acid (ARA/EPA) are of importance. Acute stress coping response is more efficient in Senegalese sole postlarvae fed low ARA/EPA ratios (Alves Martins et al. 2011). Furthermore, studies on dietary ARA supplementation showed that the effects on growth performance were quite variable and dependent on the ARA/DHA and especially ARA/EPA ratios (Boglino et al. 2012a, Villalta et al. 2005a, 2008a). Dietary ARA levels also affect larval quality. Increasing dietary ARA levels slowed metamorphic events, such as eye migration, increasing the incidence of cranial deformities, and also resulted in pigmentation disorders (Boglino et al. 2014a,b, Villalta et al. 2005a). The latter is also associated with an increase in dietary ARA/EPA ratios (Boglino et al. 2014a, Villalta et al. 2005a). Larvae sensitivity to dietary ARA levels seems to be higher during pre- and pro-metamorphosis than post-metamorphosis (Boglino et al. 2014a). Nevertheless, dietary supplementation up to a certain level can be beneficial (Table 3), as Boglino et al. (2012a) found that it promoted the best larval growth and proper skeletogenesis.

Although, in general, several studies pointed out that Senegalese sole seem to have an unusually very low requirement for n-3 highly unsaturated fatty acids during the live feeding period compared with other fish species, care should be taken when developing microdiets for sole, as deleterious effects on juvenile performance may arise (Dâmaso-Rodrigues et al. 2010). Furthermore, slight variation in the essential fatty acid levels (e.g., EPA, DHA or ARA) or in their ratios can modify the metabolism of lipids, leading to intestinal and hepatic steatosis (Boglino et al. 2012b). Moreover, the importance of dietary fatty acids goes beyond effects on growth and larval quality. As a recent study showed, dietary fatty acid composition modulates the proactive-reactive behavioural dimension of stress coping style of sole larvae, which opens the possibility of producing organisms with behavioural styles that could ultimately result in improved aquaculture productivity (Ibarra-Zatarain et al. 2015).

Some studies have also analysed the importance of micronutrients on Senegalese sole nutrition. Vitamins have been shown to play a critical role not only in growth of Senegalese sole larvae, but also in important developmental aspects as skeletogenesis or metamorphosis success. From the water-soluble vitamins, only the effects of dietary vitamin C supplementation, under the form of ascorbyl palimitate, have been studied. Microcapsules containing vitamin C supplements (Table 3) provided during early developmental stages enabled a faster metamorphosis progress and reduced the number of some malformations. These vitamin C supplements enhanced larval growth, but this result was only seen after 20DAH, so a later effect of dietary supplementation was observed (Jimenez-Fernandez et al. 2015).

The effects of some liposoluble vitamins, such as vitamin A and vitamin K, have also been addressed in Senegalese sole larvae. Dietary vitamin K supplementation has a positive effect on larval growth and skeletogenesis. Live prey grown in a commercial enriching emulsion supplemented with a source of vitamin K (Table 3) significantly improved larval growth performances and postlarval skeletal quality (Richard et al. 2014). However, experiments addressing the effect of dietary vitamin A supplementation revealed a remarkable negative impact in skeleton morphogenesis with increasing dietary retinyl palmitate content, although growth and survival were not affected (Fernández et al. 2009).

Studies on the impact of minerals on Senegalese sole larvae are very scarce with reports published only for the effects of iodine and selenium supplements. Iodine-enriched live prey (Table 3) enhanced growth and prevented larvae from developing goitre (Ribeiro et al. 2011, 2012a). The importance of enriching live feed with iodine when rearing fish in a recirculation system using ozone injection has been emphasized as crucial to sustain normal larval development. Ribeiro et al. (2012b) detected effects of selenium supplementation (Table 3) on oxidative status and thyroid hormone production, suggesting that live prey, enriched in commercial emulsions, present deficient or suboptimal levels of selenium for Senegalese sole larvae. However, selenium supplementation seems to be more important before metamorphosis climax (Ribeiro et al. 2012a).

Although, as described above, several studies analysed the effects of macro- and micronutrients on sole larval performance and quality (summarized in Table 3), few studies on quantitative requirements have been done up to now, due to methodological difficulties of performing these studies in larvae (e.g., Conceição et al. 2011, Izquierdo and Koven 2011). Thus, the data summarized in Table 3 is only indicative of the dietary needs, although it represents a fruitful and long research path. Furthermore, the drastic metamorphosis that this species undergoes has definitely a significant impact on the nutritional requirements and has to be carefully considered when developing inert microdiets.

Acknowledgements

Sofia Engrola and Cláudia Aragão acknowledge a FCT investigator grant (IF/00482/2014/CP1217/CT0005) and post-doc grant (SFRH/BDP/65578/2009), respectively, funded by the European Social Fund, the Operational Programme Human Potential and the Foundation for Science and Technology of Portugal (FCT – Portugal). This study received Portuguese national funds from FCT - Foundation for Science and Technology through project UID/Multi/04326/2019.

References

Alves Martins, D., S. Engrola, S. Morais, N. Bandarra, J. Coutinho, M. Yúfera et al. 2011. Cortisol response to air exposure in *Solea senegalensis* post-larvae is affected by dietary arachidonic acid-to-eicosapentaenoic acid ratio. Fish Physiol. Biochem. 37: 733–743.

Anguis, V. and J.P. Cañavate. 2005. Spawning of captive Senegal sole (*Solea senegalensis*) under a naturally fluctuating temperature regime. Aquaculture 243: 133–145.

Aragão, C., L.E.C. Conceição, M.T. Dinis and H.J. Fyhn. 2004a. Amino acid pools of rotifers and *Artemia* under different conditions: nutritional implications for fish larvae. Aquaculture 234: 429–445.

Aragão, C., L.E.C. Conceição, H.J. Fyhn and M.T. Dinis. 2004b. Estimated amino acid requirements during early ontogeny in fish with different life styles: gilthead seabream (*Sparus aurata*) and Senegalese sole (*Solea senegalensis*). Aquaculture 242: 589–605.

Aragão, C., L.E.C. Conceição, D. Martins, I. Rønnestad, E. Gomes and M.T. Dinis. 2004c. A balanced dietary amino acid profile improves amino acid retention in post-larval Senegalese sole (*Solea senegalensis*). Aquaculture 233: 293–304.

Aragão, C., B. Costas, L. Vargas-Chacoff, I. Ruiz-Jarabo, M.T. Dinis, J.M. Mancera et al. 2010. Changes in plasma amino acid levels in a euryhaline fish exposed to different environmental salinities. Amino Acids 38: 311–317.

Blaxter, J.H.S. 1988. Pattern and variety in development. pp. 1–58. *In*: Hoar, W.S. and D.J. Randall [eds.]. Fish Physiology Vol XI, The Physiology of Developing Fish Part A: Eggs and Larvae. Academic Press, San Diego, CA.

Boglino, A., M.J. Darias, A. Estevez, K.B. Andree and E. Gisbert. 2012a. The effect of dietary arachidonic acid during the *Artemia* feeding period on larval growth and skeletogenesis in Senegalese sole, *Solea senegalensis*. Journal of Applied Ichthyology 28: 411–418.

Boglino, A., E. Gisbert, M.J. Darias, A. Estévez, K.B. Andree, C. Sarasquete et al. 2012b. Isolipidic diets differing in their essential fatty acid profiles affect the deposition of unsaturated neutral lipids in the intestine, liver and vascular system of Senegalese sole larvae and early juveniles. Comp. Biochem. Physiol. A 162: 59–70.

Boglino, A., A. Wishkerman, M.J. Darias, P. de la Iglesia, K.B. Andree, E. Gisbert et al. 2014a. Senegalese sole (*Solea senegalensis*) metamorphic larvae are more sensitive to pseudo-albinism induced by high dietary arachidonic acid levels than post-metamorphic larvae. Aquaculture 433: 276–287.

Boglino, A., A. Wishkerman, M.J. Darias, P. de la Iglesia, A. Estévez, K.B. Andree et al. 2014b. The effects of dietary arachidonic acid on Senegalese sole morphogenesis: A synthesis of recent findings. Aquaculture 432: 443–452.

Cahu, C.L. and J.L. Zambonino Infante. 2001. Substitution of live food by formulated diets in marine fish larvae. Aquaculture 200: 161–180.

Campos, C., M.F. Castanheira, S. Engrola, L.M.P. Valente, J.M.O. Fernandes and L.E.C. Conceicao. 2013a. Rearing temperature affects Senegalese sole (*Solea senegalensis*) larvae protein metabolic capacity. Fish Physiol. Biochem. 39: 1485–1496.

Campos, C., J.M.O. Fernandes, L.E.C. Conceicao, S. Engrola, V. Sousa and L.M.P. Valente. 2013b. Thermal conditions during larval pelagic phase influence subsequent somatic growth of Senegalese sole by modulating gene expression and muscle growth dynamics. Aquaculture 414: 46–55.

Campos, C., L.M.P. Valente, L.E.C. Conceicao, S. Engrola and J.M.O. Fernandes. 2013c. Temperature affects methylation of the myogenin putative promoter, its expression and muscle cellularity in Senegalese sole larvae. Epigenetics 8: 389–397.

Campos, C., L.M.P. Valente, L.E.C. Conceicao, S. Engrola, V. Sousa, E. Rocha et al. 2013d. Incubation temperature induces changes in muscle cellularity and gene expression in Senegalese sole (*Solea senegalensis*). Gene 516: 209–217.

Campos, C., A.Y.M. Sundaram, L.M.P. Valente, L.E.C. Conceicao, S. Engrola and J.M.O. Fernandes. 2014a. Thermal plasticity of the miRNA transcriptome during Senegalese sole development. BMC Genomics 15.

Campos, C., L.M.P. Valente, L.E.C. Conceicao, S. Engrola and J.M.O. Fernandes. 2014b. Molecular regulation of muscle development and growth in Senegalese sole larvae exposed to temperature fluctuations. Aquaculture 432: 418–425.

Canada, P., S. Engrola, L.E.C. Conceição, R. Teodósio, S. Mira, V. Sousa et al. 2014. A high inclusion of fish protein hydrolysate on Senegalese sole larval diet affects growth and is associated with altered expression of DNA methyltransferases. Proceedings of the Epiconcept Conference 2014—Epigenetic and Periconception Environment Vilamoura, Portugal.

Canada, P., S. Engrola, S. Mira, R. Teodósio, J.M.O. Fernandes, V. Sousa et al. 2016a. The supplementation of a microdiet with crystalline indispensable amino-acids affects muscle growth and the expression pattern of related genes in Senegalese sole (*Solea senegalensis*) larvae. Aquaculture 458: 158–169.

Canada, P., S. Engrola, N. Richard, A.F. Lopes, W. Pinto, L.M.P. Valente et al. 2016b. Dietary indispensable amino acids profile affects protein utilization and growth of Senegalese sole larvae. Fish Physiol. Biochem. 1–16.

Cañavate, J.P. and C. Fernández-Díaz. 1999. Influence of co-feeding larvae with live and inert diets on weaning the sole *Solea senegalensis* onto commercial dry feeds. Aquaculture 174: 255–263.

Cara, J.B., F.J. Moyano, J.L. Zambonino Infante and F.J. Alarcón. 2007. The whole amino acid profile as indicator of the nutritional condition in cultured marine fish larvae. Aquacult. Nutr. 13: 94–103.

Conceição, L., C. Aragão and I. Rønnestad. 2011. Proteins. pp. 83–116. *In*: Holt, G.J. [ed.]. Larval Fish Nutrition. John Wiley & Sons, Inc., Oxford, UK.

Conceição, L.E.C., Y. Dersjant-Li and J.A.J. Verreth. 1998. Cost of growth in larval and juvenile African catfish (*Clarias gariepinus*) in relation to growth rate, food intake and oxygen consumption. Aquaculture 161: 95–106.

Conceição, L.E.C., L. Ribeiro, S. Engrola, C. Aragão, S. Morais, M. Lacuisse et al. 2007. Nutritional physiology during development of Senegalese sole (*Solea senegalensis*). Aquaculture 268: 64–81.

Conceição, L.E.C., M. Yúfera, P. Makridis, S. Morais and M.T. Dinis. 2010. Live feeds for early stages of fish rearing. Aquacult. Res. 36: 1–16.

Dâmaso-Rodrigues, M.L., P. Pousão-Ferreira, L. Ribeiro, J. Coutinho, N.M. Bandarra, P.J. Gavaia et al. 2010. Lack of essential fatty acids in live feed during larval and post-larval rearing: effect on the performance of juvenile *Solea senegalensis*. Aquacult. Int. 18: 741–757.

Dinis, M.T. 1992. Aspects of the potential of *Solea senegalensis* Kaup for aquaculture: larval rearing and weaning to an artificial diet. Aquacult. Fish. Manage. 23: 515–520.

Dinis, M.T., L. Ribeiro, F. Soares and C. Sarasquete. 1999. A review on the cultivation potential of *Solea senegalensis* in Spain and in Portugal. Aquaculture 176: 27–38.

Drake, P., M.P. Arias and A. Rodriguez. 1984. Cultivo extensivo de peces mariños en los esteros de San Fernando (Cádiz): II. Características de la producción de peces. Informe Técnico del Instituto de Investigaciones Pesqueras 116: 1–23.

Engrola, S., L.E.C. Conceição, P.J. Gavaia, M.L. Cancela and M.T. Dinis. 2005. Effects of pre-weaning feeding frequency on growth, survival, and deformation of Senegalese sole, *Solea senegalensis* (Kaup, 1858). Isr. J. Aquacult.-BAMID. 57: 10–18.

Engrola, S., L.E.C. Conceição, L. Dias, R. Pereira, L. Ribeiro and M.T. Dinis. 2007. Improving weaning strategies for Senegalese sole: effects of body weight and digestive capacity. Aquacult. Res. 38: 696–707.

Engrola, S., L. Figueira, L.E.C. Conceição, P.J. Gavaia, L. Ribeiro and M.T. Dinis. 2009. Co-feeding in Senegalese sole larvae with inert diet from mouth opening promotes growth at weaning. Aquaculture 288: 264–272.

Engrola, S., M.T. Dinis and L.E.C. Conceição. 2010. Senegalese sole larvae growth and protein utilization is depressed when co-fed high levels of inert diet and Artemia since first feeding. Aquacult. Nutr. 16: 457–465.

Fernández-Díaz, C., J. Kopecka, J.P. Cañavate, C. Sarasquete and M. Solé. 2006. Variations on development and stress defences in *Solea senegalensis* larvae fed on live and microencapsulated diets. Aquaculture 251: 573–584.

Fernández, I., M.S. Pimentel, J.B. Ortiz-Delgado, F. Hontoria, C. Sarasquete, A. Estévez et al. 2009. Effect of dietary vitamin A on Senegalese sole (*Solea senegalensis*) skeletogenesis and larval quality. Aquaculture 295: 250–265.

Gamboa-Delgado, J., J.P. Cañavate, R. Zerolo and L. Le Vay. 2008. Natural carbon stable isotope ratios as indicators of the relative contribution of live and inert diets to growth in larval Senegalese sole (*Solea senegalensis*). Aquaculture 280: 190–197.

Hamre, K., M. Yúfera, I. Rønnestad, C. Boglione, L.E.C. Conceição and M. Izquierdo. 2013. Fish larval nutrition and feed formulation: knowledge gaps and bottlenecks for advances in larval rearing. Reviews in Aquaculture 5: S26–S58.

Hjelmeland, K., B.H. Pedersen and E.M. Nilssen. 1988. Trypsin content in intestines of herring larvae, Clupea harengus, ingesting inert polystyrene spheres or live crustacea prey. Mar. Biol. 98: 331–335.

Ibarra-Zatarain, Z., S. Morais, K. Bonacic, C. Campoverde and N. Duncan. 2015. Dietary fatty acid composition significantly influenced the proactive–reactive behaviour of Senegalese sole (*Solea senegalensis*) post-larvae. Applied Animal Behaviour Science 171: 233–240.

Izquierdo, M. and W. Koven. 2011. Lipids. pp. 47–81. *In*: Holt, G.J. [ed.]. Larval Fish Nutrition. John Wiley & Sons, Inc., Oxford, UK.

Jimenez-Fernandez, E., M. Ponce, A. Rodriguez-Rua, E. Zuasti, M. Manchado and C. Fernandez-Diaz. 2015. Effect of dietary vitamin C level during early larval stages in Senegalese sole (*Solea senegalensis*). Aquaculture 443: 65–76.

Johnston, I.A. 2006. Environment and plasticity of myogenesis in teleost fish. J. Exp. Biol. 209: 2249–2264.

Leiton, E.A.S., A. Rodriguez-Rua, E. Asensio, C. Infante, M. Manchado, C. Fernandez-Diaz et al. 2012. Effect of salinity on egg hatching, yolk sac absorption and larval rearing of Senegalese sole (*Solea senegalensis* Kaup 1858). Reviews in Aquaculture 4: 49–58.

Makridis, P. and Y. Olsen. 1999. Protein depletion of the rotifer *Brachionus plicatilis* during starvation. Aquaculture 174: 343–353.

Makridis, P., L. Libeiro, R. Rocha and M.T. Dinis. 2010. Influence of microalgae supernatant, and bacteria Isolated from microalgae cultures, on microbiology, and digestive capacity of larval gilthead seabream, *Sparus aurata*, and Senegalese sole, *Solea senegalensis*. J. World Aqua. Soc. 41: 780–790.

Moffatt, N.M. 1981. Survival and growth of Northern anchovy larvae on low zooplankton densities affected by the presence of Chlorella sp. bloom. Rapp. p.-v. réun. - Cons. int. explor. mer 178: 475–486.

Morais, S., L. Narciso, E. Dores, P. Pousão-Ferreira. 2004. Lipid enrichment for Senegalese sole (*Solea senegalensis*) larvae: effect on larval growth, survival and fatty acid profile. Aquacult. Int. 12: 281–298.

Morais, S., W. Koven, I. Rønnestad, M.T. Dinis and L.E.C. Conceição. 2005a. Dietary protein/lipid ratio affects growth and amino acid and fatty acid absorption and metabolism in Senegalese sole (*Solea senegalensis* Kaup 1858) larvae. Aquaculture 246: 347–357.

Morais, S., W. Koven, I. Rønnestad, M.T. Dinis and L.E.C. Conceição. 2005b. Dietary protein:lipid ratio and lipid nature affects fatty acid absorption and metabolism in a teleost larva. Br. J. Nutr. 93: 813–820.

Morais, S., M.J. Caballero, L.E.C. Conceição, M.S. Izquierdo and M.T. Dinis. 2006. Dietary neutral lipid level and source in Senegalese sole (*Solea senegalensis*) larvae: Effect on growth, lipid metabolism and digestive capacity. Comp. Biochem. Physiol. B 144: 57–69.

Morais, S., F. Castanheira, L. Martinez-Rubio, L.E.C. Conceição and D.R. Tocher. 2012. Long chain polyunsaturated fatty acid synthesis in a marine vertebrate: Ontogenetic and nutritional regulation of a fatty acyl desaturase with Δ4 activity. Biochimica et Biophysica Acta (BBA)—Molecular and Cell Biology of Lipids 1821: 660-671.

Morais, S., C. Aragão, E. Cabrita, L.E.C. Conceição, M. Constenla, B. Costas et al. 2014. New developments and biological insights into the farming of *Solea senegalensis* reinforcing its aquaculture potential. Reviews in Aquaculture 6: 1–37.

Naas, K. and I. Huse. 1996. Illumination in first feeding tanks for marine fish larvae. Aquacult. Eng. 15: 291–300.

Naas, K.E., T. Næss and T. Harboe. 1992. Enhanced 1st feeding of halibut larvae (*Hippoglossus hippoglossus*) in green water. Aquaculture 105: 143–156.

Navarro-Guillén, C., S. Engrola, F. Castanheira, N. Bandarra, I. Hachero-Cruzado, D.R. Tocher et al. 2014. Effect of varying dietary levels of LC-PUFA and vegetable oil sources on performance and fatty acids of Senegalese sole post larvae: Puzzling results suggest complete biosynthesis pathway from C18 PUFA to DHA. Comparative Biochemistry and Physiology Part B: Biochemistry and Molecular Biology 167: 51–58.

Navarro-Guillen, C., F.J. Moyano and M. Yufera. 2015. Diel food intake and digestive enzyme production patterns in *Solea senegalensis* larvae. Aquaculture 435: 33–42.

Nicolas, J.L., E. Robic and D. Ansquer. 1989. Bacterial-flora associated with a trophic chain consisting of microalgae, rotifers and turbot larvae - influence of bacteria on larval survival. Aquaculture 83: 237–248.

Olafsen, J.A. 2001. Interactions between fish larvae and bacteria in marine aquaculture. Aquaculture 200: 223–247.

Parra, G. and M. Yúfera. 2001. Comparative energetics during early development of two marine fish species, *Solea senegalensis* (Kaup) and *Sparus aurata* (L.). J. Exp. Biol. 204: 2175–2183.

Pinto, W., L. Figueira, L. Ribeiro, M. Yúfera, M.T. Dinis and C. Aragão. 2010a. Dietary taurine supplementation enhances metamorphosis and growth potential of *Solea senegalensis* larvae. Aquaculture 309: 159–164.

Pinto, W., L. Ribeiro, M. Yúfera, M.T. Dinis and C. Aragão. 2010b. Dietary taurine supplementation enhances metamorphosis and growth potential of *Solea senegalensis* larvae. Aquaculture 309: 159–164.

Pinto, W., V. Rodrigues, M.T. Dinis and C. Aragão. 2010c. Can dietary aromatic amino acid supplementation be beneficial during fish metamorphosis? Aquaculture 310: 200–205.

Pinto, W., I. Rønnestad, A.-E. Jordal, A. Gomes, M. Dinis and C. Aragão. 2012. Cloning, tissue and ontogenetic expression of the taurine transporter in the flatfish Senegalese sole (*Solea senegalensis*). Amino Acids 42: 1317–1327.

Pinto, W., S. Engrola, H. Teixeira, A. Santos, J. Dias and L.E.C. Conceição. 2015. Towards *Artemia* replacement in first feeding Senegalese sole larvae, Aquaculture Europe 2015. European Aquaculture society, Roterdam, The Netherlands.

Pinto, W., S. Engrola, A. Santos, N.M. Bandarra, J. Dias and L.E.C. Conceição. 2016. Can Senegalese sole post-larvae effectively grow on low dietary DHA and lipid levels during weaning? Aquaculture 463: 234–240.

Pousão-Ferreira, P. 2007. Cultivo larvario de lenguado. pp. 41–52. *In:* C.y.E. Instituto de Investigación y Formación Agraria y Pesquera. Consejería de Innovación [eds.]. Manual de cultivo de lenguado y otros peces planos. Junta de Andalucía, Consejería de Agricultura y Pesca.

Ramos, S., P. Re and A.A. Bordalo. 2010. Recruitment of flatfish species to an estuarine nursery habitat (Lima estuary, NW Iberian Peninsula). J. Sea Res. 64: 473–486.

Reitan, K.I., J.R. Rainuzzo, G. Oie and Y. Olsen. 1997. A review of the nutritional effects of algae in marine fish larvae. Aquaculture 155: 207–221.

Reitan, K.I., C.M. Natvik and O. Vadstein. 1998. Drinking rate, uptake of bacteria and microalgae in turbot larvae. J. Fish Biol. 53: 1145–1154.

Ribeiro, A.R.A., L. Ribeiro, Ø. SÆLe, K. Hamre, M.T. Dinis and M. Moren. 2011. Iodine-enriched rotifers and *Artemia* prevent goitre in Senegalese sole (*Solea senegalensis*) larvae reared in a recirculation system. Aquacult. Nutr. 17: 248–257.

Ribeiro, A.R.A., L. Ribeiro, Ø. Sæle, M.T. Dinis and M. Moren. 2012a. Iodine and selenium supplementation increased survival and changed thyroid hormone status in Senegalese sole (*Solea senegalensis*) larvae reared in a recirculation system. Fish Physiol. Biochem. 38: 725–734.

Ribeiro, A.R.A., L. Ribeiro, Ø. SÆLe, K. Hamre, M.T. Dinis and M. Moren. 2012b. Selenium supplementation changes glutathione peroxidase activity and thyroid hormone production in Senegalese sole (*Solea senegalensis*) larvae. Aquacult. Nutr. 18: 559–567.

Ribeiro, L., J.L. Zambonino-Infante, C. Cahu and M.T. Dinis. 2002. Digestive enzymes profile of *Solea senegalensis* post larvae fed Artemia and a compound diet. Fish Physiol. Biochem. 27: 61–69.

Ribeiro, L., S. Engrola and M.T. Dinis. 2005. Weaning of Senegalese sole (*Solea senegalensis*) postlarvae to an inert diet with a co-feeding regime. Cienc. Mar. 31: 327–337.

Richard, N., I. Fernández, T. Wulff, K. Hamre, L. Cancela, L.E.C. Conceição et al. 2014. Dietary supplementation with vitamin K affects transcriptome and proteome of Senegalese sole, improving larval performance and quality. Mar Biotechnol 16: 522–537.

Rocha, R.J., L. Ribeiro, R. Costa and M.T. Dinis. 2008. Does the presence of microalgae influence fish larvae prey capture? Aquacult. Res. 39: 362–369.

Rønnestad, I. and L.E.C. Conceição. 2005. Aspects of protein and amino acids digestion and utilization by marine fish larvae. pp. 389–416. *In*: Starck,J.M. and T. Wang [eds.]. Physiological and Ecological Adaptations to Feeding in Vertebrates. Science Publishers, Enfield, NH, USA.

Rønnestad, I., M. Yúfera, B. Ueberschaer, L. Ribeiro, O. Sæle and C. Boglione. 2013. Feeding behaviour and digestive physiology in larval fish: current knowledge, and gaps and bottlenecks in research. Reviews in Aquaculture 5: S59–S98.

Skjermo, J. and O. Vadstein. 1993. Characterization of the bacterial-flora of mass cultivated Brachionus plicatilis. Hydrobiologia 255: 185–191.

Skjermo, J. and O. Vadstein. 1999. Techniques for microbial control in the intensive rearing of marine larvae. Aquaculture 177: 333–343.

Steinbacher, P., J.R. Haslett, M. Six, H.P. Gollmann, A.M. Sanger and W. Stoiber. 2006. Phases of myogenic cell activation and possible role of dermomyotome cells in teleost muscle formation. Developmental Dynamics 235: 3132–3143.

Valente, L.M.P., K.A. Moutou, L.E.C. Conceição, S. Engrola, J.M.O. Fernandes and I.A. Johnston. 2013. What determines growth potential and juvenile quality of farmed fish species? Reviews in Aquaculture 5: S168–S193.

Villalta, M. and A. Estévez. 2005. Culture of Senegal sole larvae without the need for rotifers. Aquacult. Int. 13: 469–478.

Villalta, M., A. Estévez and M.P. Bransden. 2005a. Arachidonic acid enriched live prey induces albinism in Senegal sole (*Solea senegalensis*) larvae. Aquaculture 245: 193–209.

Villalta, M., A. Estévez, M.P. Bransden and J.G. Bell. 2005b. The effect of graded concentrations of dietary DHA on growth, survival and tissue fatty acid profile of Senegal sole (*Solea senegalensis*) larvae during the *Artemia* feeding period. Aquaculture 249: 353–365.

Villalta, M., A. Estévez, M.P. Bransden and J.G. Bell. 2008a. Arachidonic acid, arachidonic/eicosapentaenoic acid ratio, stearidonic acid and eicosanoids are involved in dietary-induced albinism in Senegal sole (*Solea senegalensis*). Aquacult. Nutr. 14: 120–128.

Villalta, M., A. Estévez, M.P. Bransden and J.G. Bell. 2008b. Effects of dietary eicosapentaenoic acid on growth, survival, pigmentation and fatty acid composition in Senegal sole (*Solea senegalensis*) larvae during the Artemia feeding period. Aquacult. Nutr. 14: 232–241.

Vinagre, C., A. Maia, P. Reis-Santos, M.J. Costa and H.N. Cabral. 2009. Small-scale distribution of *Solea solea* and *Solea senegalensis* juveniles in the Tagus estuary (Portugal). Estuar. Coast. Shelf Sci. 81: 296–300.

Yúfera, M., G. Parra, R. Santiago and M. Carrascosa. 1999. Growth, carbon, nitrogen and caloric content of *Solea senegalensis* (Pisces: Soleidae) from egg fertilization to metamorphosis. Mar. Biol. 134: 43–49.

Yúfera, M., C. Fernández-Díaz and E. Pascual. 2005. Food microparticles for larval fish prepared by internal gelation. Aquaculture 248: 253–262.

Yúfera, M. and M.J. Darias. 2007. The onset of exogenous feeding in marine fish larvae. Aquaculture 268: 53–63.

B-3.1

Macronutrient Nutrition and Diet Formulation

Luisa M.P. Valente,[1,2,*] *Luis Conceição,*[3]
Francisco Javier Sánchez-Vázquez[4] *and Jorge Dias*[3]

1. Macronutrient requirements

As most flatfish species, *Solea* spp. have a high protein requirement. According to Rema et al. (2008), at a fixed dietary lipid level of 12%, Senegalese sole, *Solea senegalensis,* diets should include a high crude protein level (53% DM) to maintain good overall growth performance. But the protein requirement for maximum protein accretion (N gain) in Senegalese sole juveniles was met by a diet containing 60% crude protein. In most marine fish, a significant protein sparing can be achieved by increasing digestible energy levels through an increase of fats and/or carbohydrates (Helland and Grisdale-Helland 1998, Kaushik 1998). But the ability of Senegalese sole juveniles to efficiently use high dietary lipid levels seems limited, in both juvenile (Dias et al. 2004, Borges et al. 2009, Guerreiro et al. 2012) and market-sized fish (Valente et al. 2011). Borges et al. (2009) clearly demonstrated a low lipid tolerance of this species, and recommended a dietary lipid inclusion up to 8% for optimal growth and feed utilisation efficiency at a protein level of 57% (DM basis). Dietary lipids do not seem to be a good energy source for promoting growth in *S. senegalensis* as there is no clear evidence of a protein-sparing effect by increasing dietary lipid levels,

---
[1] CIMAR/CIIMAR, Centro Interdisciplinar de Investigação Marinha e Ambiental, Universidade do Porto, Av. General Norton de Matos s/n, 4050-208 Matosinhos, Portugal.
[2] ICBAS, Instituto de Ciências Biomédicas de Abel Salazar, Universidade do Porto, Rua de Jorge Viterbo Ferreira 228, 4050-313 Porto, Portugal.
[3] SPAROS Lda, Área Empresarial de Marim, Lote C, 8700-221 Olhão, Portugal.
[4] Department of Physiology, Faculty of Biology, Regional Campus of International Excellence "Campus Mare Nostrum", University of Murcia, 30100 Murcia, Spain.
* Corresponding author: lvalente@icbas.up.pt

even when the dietary protein level is below this species' requirement (Mandrioli et al. 2012, Borges et al. 2013a). Irrespective of the rearing temperature (16 vs. 22°C), Guerreiro et al. (2012) also found that feed efficiency, N retention and energy retention were the highest in sole juveniles which were fed a diet containing 55% protein and 8% lipids. The activity of enzymes involved in key metabolic pathways points towards a lack of metabolic adaptation to high lipid levels, reflecting the high protein requirement of the species. The lipogenic pathway was depressed by elevated levels of dietary lipids (Dias et al. 2004, Borges et al. 2013a), whereas data on the activity of enzymes involved in amino acid catabolism reinforce the assumption that high-lipid diets do not promote protein sparing in Senegalese sole resulting in reduced protein accretion regardless of the dietary protein levels (Borges et al. 2013a). Moreover, Campos et al. (2010) observed a decrease in the expression of myogenic regulatory factors and myosins in the muscle of Senegalese sole fed with increasing dietary lipid levels, supporting the hypothesis that high lipid levels depress growth by reducing protein accretion. The activity of alanine aminotransferase (ALAT) and glutamate dehydrogenase (GDH) in the liver of sole was not affected by dietary protein or lipids levels (Guerreiro et al. 2012, Borges et al. 2013a). Mandrioli et al. (2012) showed that a concomitant decrease of dietary protein and an increase in dietary lipids was associated with a massive storage of unused lipid within sole hepatocytes. Similarly, Valente et al. (2011) reported moderate steatosis and some cellular necrosis in large-sized sole fed with high lipid levels. On the other hand, the incorporation of carbohydrates (CHO) at the expense of lipids in isonitrogenous diets improved Senegalese sole's growth performance (Borges et al. 2009, 2013b). This species was shown to develop an efficient metabolic response under induced short-term hyperglycaemia, restoring glucose homeostasis (Conde-Sieira et al. 2015b). Nevertheless, Salas-Leiton et al. (2017) have recently evidenced a high dependency from energy contained in dietary protein, avoiding an effective protein-sparing effect from CHO. This study also demonstrated that when Senegalese sole is fed low fat diets, a dietary protein content of 52% is able to induce maximal growth rate.

2. Lipid metabolism

Lipids are not efficiently used as non-protein energy sources in Senegalese sole, but its digestion and absorption seem identical to other marine species. It was firstly hypothesised that impaired nutrient digestibility or lipid absorption and uptake, induced by high fat diets, would limit the availability of dietary lipids for metabolic pathways. Several studies characterizing intestine morphology and physiology (Arellano et al. 1999, 2002, Conceição et al. 2007, Yúfera and Darías 2007, Darias et al. 2012) suggest this is not the case. Moreover, Senegalese sole's lipid digestion capacity assessed in juveniles showed that lipase activity was high and not affected by dietary lipid level (Dias et al. 2010, Borges et al. 2013b). Distinct dietary lipid levels (4 vs. 17%) were further shown to equally induce high lipid digestibility (83%–87%) and intestinal lipase activity (19–27 mU/mg protein) (Borges et al. 2013b). Dietary lipids were mainly absorbed 5h after feeding and fish fed with the high fat diet presented significantly higher plasma triglyceride concentrations compared to those that were fed the low fat diet, also demonstrating effective lipid absorption (Borges et al. 2013b). Still, high (18%) dietary lipid level has also been shown to cause large lipid accumulations in the enterocytes of Senegalese sole, together with a down-regulation of fatty acid synthase expression in liver and intestine, and increased carnitine palmitoyltransferase 1 (that imports FA-CoA into the mitochondria for oxidation) mRNA in liver (Bonacic et al. 2016). These authors suggest a poor capacity of sole to adapt to high dietary lipid level, as most genes involved in intestinal absorption were not regulated in response to the diet. Bonvini et al. (2015) have also demonstrated that feeding common sole (*Solea solea*) juveniles with increasing dietary lipid levels affects growth, feed utilization and gut health. Common sole grow better on diets with 8 and 12% lipid, and fish with 16 and 20% dietary lipid had large lipid accumulations in the enterocytes (Bonvini et al. 2015).

S. senegalensis is a lean fish (1–3 g of fat/100 g of flesh) with a scarce capacity to accumulate fat, even when fed with lipid levels above 16% DM basis (Dias et al. 2004, Borges et al. 2009). The low whole body fat content (< 8% ww) and quite passive behavior suggests a low daily metabolic budget, that might justify the low fat depots. Previous studies (Rueda-Jasso et al. 2004, Borges et al. 2009, Valente et al. 2011, Fernandes et al. 2012) reported that liver is the preferential local site for fat deposition (5.5–37% of fat). *S. senegalensis* liver seems to have an important role in clearing high plasma triglycerides that high fat diets lead to, increasing expression of VLDLr, proteins related to lipid transport (microsomal triglyceride transfer protein, MTP), trafficking (Fatty acid binding protein 11, FABP11) and fatty acid uptake (VLDL-r) in juvenile fish fed with high fat diets (Borges et al. 2013b). Moreover, the most abundant protein in high-density lipoprotein (ApoC1) was not nutritionally regulated in sole by dietary fat levels, at least at the lipid levels tested. Borges et al. (2013b) results also suggest that FABP11 facilitates fatty acid intracellular transport and metabolism in the liver of *S. senegalensis*, although no effects were encountered at muscular level, implying low fatty acid utilization. This study also reported no differences in liver PPARβ and RXR expression between low and high fat treatments suggesting that, in this species, these genes are not directly controlled by dietary lipid intake. Likewise in rainbow trout, fatty acid (FA) sensing systems were recently demonstrated to be present in Senegalese sole hypothalamus (Conde-Sieira et al. 2015a). Moreover, these FA sensing systems were shown to be activated both by a MUFA, oleate (C18:1 n9), and an n-3 PUFA (ALA), but not by the saturated stereate (C18:0) or EPA (C20:5 n-3). Thus, ALA may be sensed in the hypothalamus and possibly enhance sole anorexigenic capacity. But further studies are needed to clearly evaluate PUFA's role in key regulatory pathways of lipid metabolism.

In conclusion, the failure in growth improvement by increasing dietary lipids in Senegalese sole does not seem to be caused by impairment of lipid digestibility or transport as sole can digest equally well high fat and low fat diets. Overall, the use of dietary lipids seems independent of protein content and high dietary lipid level does not impair lipid digestion even if absorption might be somewhat affected. The metabolic fate of dietary lipids remains to be fully understood but could be linked to the competition with carbohydrates for energy supply. The recommended diets for Senegalese sole have a low lipid level (< 12%), and therefore, most energy is derived from other substrates.

3. Carbohydrates metabolism

Sole has omnivorous digestive profile, and a glucose clearance rate comparable to omnivorous species. Conde-Sieira et al. (2015b) have initially suggested that carbohydrates might eventually bring protein sparing in this species, but Salas-Leiton et al. (2017) have recently demonstrated a low ability of Senegalese sole to digest starch compared to other carnivorous fish species. Dietary carbohydrates (CHO) are generally poorly used as an energy source in marine carnivorous fish. But improved growth performance was reported in Senegalese sole juveniles with the progressive incorporation of CHO at the expense of dietary lipids (Borges et al. 2009, 2013a). Prolonged hyperpglycaemia was reported in sole fed with high lipids/low CHO diets, probably due to increasing endogenous glucose production (Borges et al. 2014b). On the other hand, the same study showed that sole fed with a high fat/low CHO diet had a lower protein accretion apparently associated to a down regulation of the AKT-mTor nutrient signalling pathway, and a possible insulin resistance state. Moreover, increased activity of the phosphofructokinase 1, related to glucose catabolism, was observed in the muscle of fish fed with high CHO as dietary lipid substitutes (Borges et al. 2013a), suggesting a role for glucose as an energy source in this species muscle. These studies evidenced that dietary lipids strongly affect glucose metabolism in Senegalese sole.

The progressive incorporation of CHO at the expense of dietary lipids seems to promote Senegalese sole growth, but the same is not observed when dietary protein is replaced by CHO. Protein/CHO ratios did not affect Senegalese sole growth performance (Guerreiro et al. 2014, Salas-Leiton et al. 2017), although it increased voluntary feed intake and feed conversion ratio (Salas-Leiton et al. 2017). This species was shown to have adequate capacity to deal with high dietary CHO contents and restore glucose homeostasis under different glycaemia conditions, indicating a functional metabolic machinery to utilize CHO (Conde-Sieira et al. 2015b, 2016). This tolerance to glucose is reflected by an increased use of glucose through glycolysis in liver, but without relevant changes in its lipogenic potential. Moreover, no significant correlation between dietary CHO content and glycolysis activation could be depicted in muscle (indicated by hexokinase activity) although an incremented glycogen synthase activity was noticed in fish fed with the high CHO diet (Conde-Sieira et al. 2016).

Under low dietary protein/CHO ratios, Senegalese sole increases feed intake and therefore higher levels of energy are expended due to this elevated feeding activity (Conde-Sieira et al. 2016, Salas-Leiton et al. 2017). Such compensatory mechanism seems to be triggered to satisfy a specific protein metabolic requirement for energy purposes as tissue accretion remained unchanged. In short, according to the present knowledge, a high dependency from energy contained in dietary protein may be inferred from overall results, avoiding an effective protein-sparing effect from increasing CHO levels, even when the dietary protein level is below the requirement level.

4. Alternative protein sources

High quality fishmeal (FM) still remains the major dietary amino acid source currently used in sole farming. But supplies of fish meal are finite, and the choice of ingredients from available plant sources in aquafeed formulations is a major trend in aquaculture (FAO 2017). Senegalese sole capacity to cope with practical diets in which the marine-derived protein are replaced by plant-protein ingredients was demonstrated by Silva et al. (2010). It was further suggested that fish meal could be totally replaced by a mixture of plant-protein (PP) sources, without any adverse effects on growth, feed or protein utilization provided that the dietary amino acids are balanced by the addition of small amounts of crystalline amino acids (AA) (Silva et al. 2009). Still, the only available direct estimate for indispensable amino acids (IAA) requirement in sole refers to lysine (Lys). Silva (2009) estimated an optimum dietary supply of 4.7 Lys 16 g^{-1} N for maximum protein accretion. Moreover, Costas (2011) further refined the ideal protein profile in diets for juvenile sole by estimating bioavailability of the 10 IAA. The utilization of the 10 individual IAA was evaluated by tube-feeding Senegalese sole juveniles with compound feed containing 14C-labelled IAA as tracers. Differences in digestibility, retention and catabolism between individual IAA were demonstrated, and individual IAA bioavailabilities relative to lysine could then be calculated. High relative bioavailabilities were found for histidine, leucine,

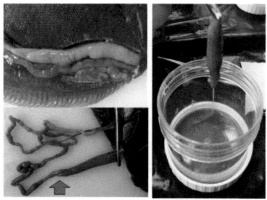

Fig. 1. Left: S-shaped and posterior end of intestine of Senegalese sole and identification of the ileocaecal valve (red arrow); Right: Liquid faeces being collected from the posterior intestine.

isoleucine, valine, methionine, threonine, phenylalanine and arginine, meaning that these IAA are retained more efficiently than lysine, while tryptophan had the lowest relative bioavailability among IAA (Costas 2011).

The feasibility of using new feedstuffs to replace the unsustainable marine sources in aqua feeds largely depends on a species' ability to digest them. Nevertheless, the *in vivo* determination of nutrient apparent digestibility coefficients (ADC) in Senegalese sole is a huge challenge compared to other marine fish species and explains the scarce number of ADC studies on this species. The intestine of sole is multiple S-shaped, very long and produces faeces with extremely low cohesion (Fig. 1), impossible to collect either by netting methods or using the traditional decanter systems (e.g., Guelph system). Attempts to collect faeces by classic stripping also failed due to the coiled and very fragile morphology of the intestine, although Peres et al. (2013) employed a simultaneous stripping and anal sphincter stimulation in large-sized fish (294 g). Most studies had to rely on intestine dissection method for faeces collection (Dias et al. 2010, Cabral et al. 2011, Borges et al. 2013a,b). The single morphological feature that can be visually identified along the intestine is the posterior ileocaecal valve (Fig. 1, red arrow). Preliminary studies showed that dry matter ADC value calculated using faeces collected from the posterior end (from the ileocaecal valve to anus) was similar to that observed in faeces from the second third of the intestine (44%), but higher than that from the anterior intestine (27%) (L.M.P. Valente, unpublished data). Nevertheless, due to the small amount of fish used in this evaluation, most published studies from this group just relied on faeces collected after the ileocaecal valve only, to make sure that reported ADC values reflect the full digestion capacity of the species.

Dias et al. (2010) conducted a study to determine the apparent digestibility coefficients (ADCs) of protein, phosphorus and energy in practical feed ingredients, as well as their digestible energy (DE) content (Table 1); protein digestibility was high (above 91%) for fishmeal and corn gluten, intermediate for soybean meal (87%) and moderate for wheat meal (59%); energy digestibility varied between 88 and 93% for soybean meal, corn gluten and anchovy fishmeal and was reduced in wheat meal (73%); phosphorus digestibility was the highest in fishmeal (58%) and greatly reduced in vegetable ingredients (28 to 33%).

Senegalese sole, despite its high dietary protein requirement, digests vegetable ingredients relatively well, which paved the way for the development of practical feeds with high levels of plant-protein sources. Following an intense research effort throughout the

Table 1. Apparent digestibility coefficients of feed ingredients by Senegalese sole.

ADC Ingredients (%)	Fish meal	Soybean meal	Corn gluten meal	Wheat meal
Dry matter	98.6	86.9	98.7	96.4
Organic matter	99.8	97.6	99.3	99.8
Protein	94.4	86.7	90.6	58.6
Phosphorus	58.1	27.6	32.8	31.2
Energy	92.9	88.3	89.9	72.6
DE (MJ·kg^{-1})	16.2	13.5	15.1	10.1

last decade, it is now clear that replacement of marine-derived protein sources by practical plant protein (PP) ingredients in Senegalese sole feeds with minimal AA supplementation is feasible in both juvenile (Cabral et al. 2011) and large-sized fish (Valente et al. 2011, Cabral et al. 2013). Salas-Leiton et al. (2015) reported a high tolerance of Senegalese sole to low P content diets, supporting the utilization of practical plant protein-rich diets. Data indicate that Senegalese sole can effectively use diets with high levels of PP sources, up to 75% of FM replacement (Cabral et al. 2013), but growth rate and nutrient gain in juveniles seems to mainly depend on the selection of adequate plant protein blends, rather than on the plant protein incorporation level (Cabral et al. 2011). The type of dietary protein source was shown to modify both the amount and composition of the pancreatic proteases secreted into the intestinal lumen of juveniles, without involving growth reduction (Rodiles et al. 2012). This result evidences Senegalese sole capability to modulate digestive protease secretion when dietary protein sources are modified. This is further supported by data on large-sized sole, evidencing similar nutrient intake and utilization (including similar ADCs of nutrients) in fish fed with either PP or FM base-diets (Cabral et al. 2013).

Further studies on ingredients' digestibility are warranted to help selecting the most adequate plant protein combination for this species. Moreover, increasing FM replacement level promotes good growth rates and have a positive environmental impact as it reduces nitrogen (N) losses and faecal phosphorus (P) waste, and decreases the fishmeal used per kg of sole produced (Fi:Fo ratio) (Cabral et al. 2011, 2013). However, sole seems to have a low tolerance for processed animal proteins (J. Dias, unpublished data).

Replacement of FM by alternative protein sources may also have impact on flesh quality, namely on its nutritional value and sensorial properties for humans. The whole body lipid content of sole fed PP diets decreases with the increasing replacement of FM (Silva et al. 2009, Cabral et al. 2013), but the same does not apply to liver. Muscle lipid content is generally low (1.3–4% WW) and not affected by the inclusion of PP. The replacement of FM by PP sources, even at extremely high levels (75 and 100%), is still effective in producing an n-3 PUFA rich product. Muscle fatty acid profile of fish fed with PP sources evidenced increasing levels of linoleic acid (C18:2 n-6) while docosahexaenoic acid (DHA) is selectively retained in muscle. DHA/EPA (eicosapentaenoic acid) and EPA/ARA (arachidonic acid) ratios were not affected by the dietary inclusion of PP sources (Cabral et al. 2013). Substitution of fishmeal by plant protein hence seems possible without major differences on the lipid content and fatty acid profile of the main edible portion of the fish—the muscle. Nevertheless, the long term impact of high PP incorporation levels is needed to determine its practical viability.

Senegalese sole is a highly-appreciated fish in Mediterranean countries due to the sensorial characteristics of its flesh. The sensory evaluation of cooked slices showed that the replacement of marine protein sources (FM) by PP blends did not have a significant impact on the majority of volatile compounds (Silva et al. 2012, Moreira et al. 2014) or sensory descriptors (Cabral et al. 2013). But total FM substitution resulted in decreased stiffness and small sized muscle fibres cells associated with reduced expression of key genes involved in myogenesis and muscle growth (Valente et al. 2016). This study suggests a modulation of such genes by increasing levels of PP sources that alter textural properties of Senegalese sole when the level of substitution is total.

Recent studies suggested that high plant protein inclusion may stimulate innate immune defence of Senegalese sole in short-term, but after 73 days of feeding a reduced number of goblet cells and low hepatic glycogen content suggests a lower capacity to overcome stress conditions (Batista et al. 2016). In spite of all these promising results in using PP to substitute FM in sole feeds, the long term impact of high PP incorporation levels on gut integrity, liver function and immune status needs to be evaluated to validate its practical application.

5. Alternative lipid sources

Fish is the main dietary source of DHA and EPA for humans and these fatty acids are best known for preventing cardiovascular and inflammatory diseases (Williams 2000, Ruxton et al. 2004, von Schacky 2006, De Caterina 2011). Consequently, there is a legitimate concern about a possible loss of healthy beneficial effects for humans, when replacing fish oil (FO), rich in EPA and DHA, by vegetable oils (VO) which lack these fatty acids. Additionally, this species has a relatively low dietary requirement for DHA and EPA as suggested by the negligible amounts of DHA required for optimal larvae growth (Villalta et al. 2005, 2008). Nonetheless, a Δ4 desaturase (with Δ5 activity for the n-3 fatty acids) and an Elov5 with the potential to synthetize DHA from EPA have recently been cloned and functionally described in Senegalese sole (Morais et al. 2012).

In *S. senegalensis* juveniles, it seems possible to substitute up to 100% of supplemental fish oil by either linseed oil (Benítez-Dorta et al. 2013) or blends of rapeseed, soybean and linseed oil (Borges et al. 2014a), without compromising growth performance and feed utilization. Although such high substitutions alter muscle fatty acid profile reflecting the dietary composition, a selective deposition and retention of some LC-PUFA is observed. The muscle level of both ARA and DHA in fish fed with either linseed oil or VO blends remains similar to that observed in FO fed fish (Benítez-Dorta et al. 2013, Borges et al. 2014a). But, total substitution of FO by soybean oil was shown to depress growth and significantly decrease ARA level in Senegalese sole juvenile's muscle, in spite of keeping ARA/EPA and EPA/DHA ratios (Benítez-Dorta et al. 2013). In another study comparing diets containing either 100% FO or 25% FO and 75% VO blend (rapeseed, linseed and soybean oils), at 8 and 18% lipid levels, Bonacic et al. (2016) have shown that hepatic lipid deposits were higher in fish fed VO, associated with increased hepatic ATP citrate lyase activity and up-regulated carnitine palmitoyltransferase 1 (cpt1) mRNA levels post-prandially. Moreover, in that study lipid level had a larger effect on gene expression of metabolic (lipogenesis and β-oxidation) genes than lipid source.

Recommended daily intake (RDI) of EPA+DHA is estimated to be at least 0.25 g per day for healthy human individuals (EFSA 2010) and even fish fed with VO based diets can provide consumers between 1.5 and 2 times the RDI, confirming their good nutritional value (Borges et al. 2014a, Reis et al. 2014). In a long term study (5 months), it was shown that it seems possible to substitute up to 100% of FO by a VO blend, as well as concomitantly substituting 50% FO and FM by vegetable sources, in on-growing Senegalese sole diets, without compromising growth performance and feed utilization (Reis et al. 2014). Moreover, in this same study fish fed with 50% VO diet showed a muscle n-3 HUFA profile similar to those that were fed the FO-based diet. But, total FO substitution resulted in a strong reduction of muscle EPA content that was totally recovered after 26 days of re-feeding with a FO based diet. Therefore, it seems possible to re-establish the fatty acid profile to a FO type completely with a finishing diet (Reis et al. 2014). Senegalese sole shows a high effectiveness to accumulate highly unsaturated fatty acids (ARA and DHA) and seems to adapt well to a low dietary supply of these FAs. Moreover, sole fillets from fish fed 100% VO blends were very well accepted by fish consumers and still are good nutritional value end-products for human consumption, providing 1.5 times the RDI level (0.4 g per 100 g of muscle) of EPA + DHA (Reis et al. 2014). Further studies are still required to better evaluate the long term impact of the substitution of FO in PP-based diets.

6. Feeding behaviour and dietary selection

The early development of self-feeding systems, in which fish can be trained to obtain food by triggering a food-demand sensor, allowed research on feeding behaviour being centred on fish rather than humans (Rozin and Mayer 1961). Using these devices, fish can decide when and how much feed they want to ingest, so that many relevant questions can be answered within a short time and considering simultaneously many variables (Madrid et al. 2001). Senegalese sole is a nocturnal species and so displays most (83%) of their locomotor activity rhythms during darkness (Bayarri et al. 2004). When given access to self-feeders, sole revealed their ability to use self-feeders efficiently under both indoors conditions and under farming outdoor conditions (Navarro et al. 2009). Since this is a benthic species, the most suitable food-demand sensor (rod, string or optical switch) was tested, the most efficient being the string sensor, since this led to the lowest amount of food waste and the highest food demand levels, which occurred mostly (81%) at night. Interestingly, when the food reward level was increased or reduced, fish displayed a compensatory feeding behaviour, increasing two-fold their feeding activity when the reward level was reduced to half or decreasing their demands by 50% when the reward level was doubled (Navarro et al. 2009).

If fish can choose between two or three different diets through self-feeders, they can also design their own diet in accordance with their requirements (Aranda et al. 2000). Dietary selection has become a powerful tool to study nutritional needs as well as signals involved in regulating macronutrient intake and responses to nutritional challenges (da Silva et al. 2016). In the case of *S. senegalensis*, macronutrient self-selection and responses to dietary protein dilution have been investigated. When fish are given free choice to feed on three different diets (PC-75% crude protein (CP) and 25% carbohydrate (CHO), PF-75% CP and 25% crude fat (CF), and PFC-10% CP, 45% CF and 45% CHO), they selected a diet containing 68% CP, 16% CF and 16% CHO (Fig. 2). The composition of this diet approaches to that self-selected by other carnivorous species, such as sharpsnout seabream (64% CP, 17% CHO, 19% CF), rainbow trout (64% CP, 18% CHO, 18% CF) and European sea bass (59% CP, 22% CHO, 19% CF) (Sanchez-Vazquez et al. 1999, Aranda et al. 2000, Vivas et al. 2006).

The methodology of the demand feeder provides a means for assessing the feeding preferences of fish and defining the target of consumption of each nutrient following a nutritional challenge. If one nutrient is diluted (e.g., with cellulose), fish responds by increasing the nutrient intake to regulate its consumption and "defend" a given nutritional target (da Silva et al. 2016). Senegalese sole accurately regulated energy intake despite

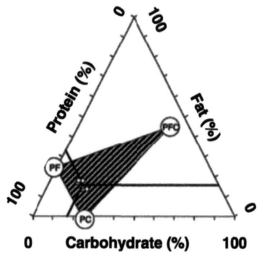

Fig. 2. Dietary selection made by Senegalese sole groups given free access to three experimental diets: PC-75% crude protein and 25% carbohydrate, PF-75% crude protein and 25% crude fat, and PFC-10% crude protein, 45% crude fat and 45% carbohydrate (modified from Rubio et al. 2009).

such nutritional challenges. For instance, when the PF diet was diluted 50% with cellulose, fish increased their feed intake, modifying the macronutrient selection pattern and mainly increasing protein intake (52% increase) to sustain the previous energy intake (5.0 kJ/kg BW/day). When both PC and PF were diluted 50% with cellulose, the sole increased their feed intake again to sustain their energy intake (5.2 kJ/kg BW/day), with protein being the most demanded nutrient (58%), followed by carbohydrate (24%) and fat (18%) (Rubio et al. 2009). This suggests that sole seems to defend the percentage of energy obtained from protein rather than their total energy intake, provided that a minimum fat intake is guaranteed. Similarly, in a nutritional trial Salas-Leiton et al. (2017) have recently shown that dietary crude protein seems to be the main factor regulating feed intake in Senegalese sole.

7. Diet formulation

Senegalese sole seems to tolerate and adapt to a wide range of formulation, both in what concerns macronutrient composition and ingredient formulation (Table 2). Most commercial feeds used for sole to date are still high in fish meal and fish oil, and their crude protein and crude lipid contents range from 48–60% and 12–18% feed basis, respectively. This wide variation in protein and lipid contents derives from different results at laboratorial-scale and commercial operations, which are likely to be mostly caused by differences in feed intake, associated with difficulties in monitoring sole feed consumption due to its relatively passive feeding behaviour. Laboratorial scale controlled trials clearly suggest that a diet with > 52% protein and < 12% lipid would maximize growth (Table 3). Senegalese sole tolerates well vegetable protein sources but seems to have a low tolerance for processed animal proteins (Dias, personal communication). Moreover, this species shows high tolerance to vegetable oils, with very good results obtained with 50% replacement of both fish meal and fish oil during on-growing stages (Reis et al. 2014). Senegalese sole shows high effectiveness to accumulate DHA in the muscle, probably associated with this species' high capacity to n-3 HUFA biosynthesis. A high dependency from energy contained in

Table 2. Ingredients and proximate composition of experimental diets from several studies giving good overall performance with Senegalese sole.

Trial/objective	Fish meal replacement				Lipid Level				Fish oil replacement				
Initial weight (g)	10	10	106	106	6	6	29	29	12	12	152	152	152
Diet name	F45	F5+IAA	FM	PP75	55P8L	55P16L	54P4L	54P17L	CTR	VO100	CTRL	VO100	VO50PP50
Ingredients (%)													
Fish meal	37	-	51.5	5.5	40	40	30	30	51.5	51.5	51.5	51.5	21
Fish soluble protein conc.	3	2	5	5	5	5	8.5	8.5	5	5	5	5	5
Squid meal	5	3	5	5	-	-	-	-	5	5	5	5	5
Soy Protein Concentrate	-	12.5	-	6	-	-	-	-	-	-	-	-	4
Soybean meal 48	16	22.5	12.5	9	-	-	12.5	10.6	12.5	12.5	12.5	12.5	9.8
Whole peas	-	-	-	-	-	-	-	-	-	-	-	-	11.5
Corn gluten	13	13.5	-	9	19	20.7	9	9	-	-	-	-	7.5
Wheat meal	18	14.6	11	9.6	25.6	15.8	27.2	14.1	10	10	10	10	8.8
Wheat gluten	3.5	16	-	7	5	5	11.8	14.8	-	-	-	-	4.3
Potato protein Concentrate	-	-	-	6	-	-	-	-	-	-	-	-	2.5
Dehulled pea meal	-	-	11	11.5	-	-	-	-	11	11	11	11	8.9
Fish oil	4	6.6	2	5.7	1.9	10	-	12	2.5	2.5	2.5	-	3
Rapeseed oil	-	-	-	-	-	-	-	-	-	0.75	-	0.75	0.9
Soybean oil	-	-	-	-	-	-	-	-	-	0.5	-	0.5	0.6
Linseed oil	-	-	-	-	-	-	-	-	-	1.25	-	1.25	1.5
Soy lecithin	-	-	-	-	-	-	-	-	-	-	0.5	0.5	0.5
L-Lysine HCl	-	1.7	-	0.5	-	-	-	-	-	-	-	-	0.5
DL-Methionine	-	-	-	0.2	-	-	-	-	-	-	-	-	0.2

Table 2 contd. ...

...*Table 2 contd.*

Trial/objective	Fish meal replacement				Lipid Level				Fish oil replacement				
Initial weight (g)	10	10	106	106	6	6	29	29	12	12	152	152	152
Diet name	F45	F5+IAA	FM	PP75	55P8L	55P16L	54P4L	54P17L	CTR	VO100	CTRL	VO100	VO50PP50
Amino acid mixture	-	3.41	-	-	-	-	-	-	-	-	-	-	-
Taurine	-	0.3	-	-	-	-	-	-	-	-	-	-	-
Di/Mono Ca Phosphate	-	3.4	-	4	-	-	0.5	0.5	-	-	-	-	2.5
Min & Vit sources	0.44	0.44	1	1	2.5	2.5	0.43	0.43	1	1	1	1	1
Betaine	0.07	0.07	-	-	-	-	0.07	0.07	-	-	-	-	-
Binder	-	-	1	1	1	1	-	-	1	1	1	1	1
Proximate composition													
Dry matter (% DM)	93.1	96	91.4	93.4	86.6	90.6	93.1	96	91.1	92.8	94.9	94.9	93.5
Ash (% DM)	9	6.2	13.5	8.8	8	8.1	7.8	7.4	13.5	13.5	14.4	14.4	11.1
Crude protein (% DM)	57.2	55.8	54.2	54.7	55.3	55.7	55.5	54.1	56.8	57.4	55.6	55.6	55.7
Crude fat (% DM)	6.6	5.5	8.8	9.4	8.2	17	4.7	17.3	8.7	9.4	8.4	8.7	10.2
Gross Energy (kJ g-1 DM)	21.2	23	20.2	20.8	18.9	21.5	20.4	23	20.6	20.3	19.9	20.1	21
NFE (% DM)	20.3	28.5	nd	nd	nd	nd	20.1	13.5	nd	nd	nd	nd	Nd
Reference	Silva et al. (2009)		Cabral et al.(2013)		Guerreiro et al. (2012)		Borges et al.(2013a)		Borges et al. (2014a)		Reis et al.(2014)		
Final Weight (g)	40	38	106	106	26	23	60	52	39	41	152	152	152
FCR	0.85	0.94	1.48	1.46	0.91	1.23	1.23	1.14	1.13	1.08	1.42	1.35	1.33
Duration (days)	84	84	140	140	74	74	88	88	89	89	140	140	140
Temp (°C)	20	20	19	19	22	22	20	20	20	20	19	19	19

Table 3. Recommend ingredient formulations and proximate compositions for experimental diets to be used in Senegalese sole.

Fish size	5–50 g		50–300 g	
Feed type	Marine	Plant-based	Marine	Plant-based
Ingredients (%)				
Fish meal	37.0	-	51.5	5.5
Fish soluble protein conc.	3.0	2.0	5.0	5.0
Squid meal	5.0	3.0	5.0	5.0
Soy Protein Concentrate	-	12.5	-	6.0
Soybean meal 48	16.0	22.5	12.5	9.0
Whole peas	-	-	-	-
Corn gluten	13.0	13.5	-	9.0
Wheat meal	18.0	14.6	11.0	9.6
Wheat gluten	3.5	16.0	-	7.0
Potato protein Concentrate	-	-	-	6.0
Dehulled pea meal	-	-	11.0	11.5
Fish oil	4.0	6.6	2.0	5.7
Rapeseed oil	-	-	-	-
Soybean oil	-	-	-	-
Linseed oil	-	-	-	-
Soy lecithin	-	-	-	-
L-Lysine HCl	-	1.7	-	0.5
DL-Methionine	-	0.5	-	0.2
Amino acid mixture	-	2.91	-	-
Taurine	-	0.3	-	-
Di/Mono Ca Phosphate	-	3.4	-	4.0
Min & Vit sources	0.44	0.44	1.0	1.0
Betaine	0.07	0.07	-	-
Binder	-	-	1.0	1.0
Proximate composition				
Dry matter (% DM)	93.1	96.0	91.4	93.4
Ash (% DM)	9.0	6.2	13.5	8.8
Crude protein (% DM)	57.2	55.8	54.2	54.7
Crude fat (% DM)	6.6	5.5	8.8	9.4
Gross Energy (kJ.g^{-1} DM)	21.2	23.0	20.2	20.8

[1] Amino acids in % of mixture: Arg 0.60, His 0.20, Ile 0.35, Leu 0.50, Thr 0.50, Trp 0.06, Val 0.70.

dietary protein may be inferred from overall results, avoiding an effective protein-sparing effect either from lipid or CHO sources.

Acknowledgments

The authors acknowledge the R&D&I Project INNOVMAR—"Innovation and Sustainability in the Management and Exploitation of Marine Resources" (ref. NORTE-01-0145-FEDER-000035) within the research line "INSEAFOOD—Innovation and valorisation of seafood products: meeting local challenges and opportunities", founded by the Northern Regional Operational Programme (NORTE2020) through the European Regional Development Fund (ERDF).

References

Aranda, A., F.J. Sanchez-Vazquez, S. Zamora and J.A. Madrid 2000. Self-design of fish diets by means of self-feeders: validation of procedures. J. Physiol. Biochem. 56: 155–66.

Arellano, J., M.T. Dinis and C. Sarasquete. 1999. Histomorphological and histochemical characteristics of the intestine of the Senegal sole, *Solea senegalensis*. Eur. J. Histochem. 43: 121–33.

Arellano, J.M., V. Storch and C. Sarasquete. 2002. Ultrastructural study on the intestine of Senegal sole, *Solea senegalensis*. J. Appl. Ichthyol. 18: 154–158.

Batista, S., A. Medina, M.A. Pires, M.A. Moriñigo, K. Sansuwan, J.M.O. Fernandes et al. 2016. Innate immune response, intestinal morphology and microbiota changes in Senegalese sole fed plant protein diets with probiotics or autolysed yeast. Appl. Microbiol. Biotechnol. 100: 7223–7238.

Bayarri, M.J., J.A. Munoz-Cueto, J.F. Lopez-Olmeda, L.M. Vera, M.A. Rol de Lama, J.A. Madrid et al. 2004. Daily locomotor activity and melatonin rhythms in Senegal sole (*Solea senegalensis*). Physiology and Behavior 81: 577–83.

Benítez-Dorta, V., M. Caballero, M. Izquierdo, M. Manchado, C. Infante, M. Zamorano et al. 2013. Total substitution of fish oil by vegetable oils in Senegalese sole (*Solea senegalensis*) diets: effects on fish performance, biochemical composition, and expression of some glucocorticoid receptor-related genes. Fish Physiol Biochem 39: 335–349.

Bonacic, K., A. Estévez, O. Bellot, M. Conde-Sieira, E. Gisbert and S. Morais. 2016. Dietary fatty acid metabolism is affected more by lipid level than source in Senegalese sole juveniles: interactions for optimal dietary formulation. Lipids 51: 105–122.

Bonvini, E., L. Parma, L. Mandrioli, R. Sirri, C. Brachelente, F. Mongile et al. 2015. Feeding common sole (*Solea solea*) juveniles with increasing dietary lipid levels affects growth, feed utilization and gut health. Aquaculture 449: 87–93.

Borges, P., B. Oliveira, S. Casal, J. Dias, L. Conceição and L.M.P. Valente. 2009. Dietary lipid level affects growth performance and nutrient utilisation of Senegalese sole (*Solea senegalensis*) juveniles. Br. J. Nutr. 102: 1007–1014.

Borges, P., F. Medale, J. Dias and L.M.P. Valente. 2013a. Protein utilisation and intermediary metabolism of Senegalese sole (*Solea senegalensis*) as a function of protein: lipid ratio. Br. J. Nutr. 109: 1373–1381.

Borges, P., F. Medale, V. Veron, M.d.A. Pires, J. Dias and L.M.P. Valente. 2013b. Lipid digestion, absorption and uptake in *Solea senegalensis*. Comp. Biochem. Physiol. A-Mol. Integr. Physiol. 166: 26–35.

Borges, P., B. Reis, T.J.R. Fernandes, Â. Palmas, M. Castro-Cunha, F. Médale et al. 2014a. Senegalese sole juveniles can cope with diets devoid of supplemental fish oil while preserving flesh nutritional value. Aquaculture 418–419: 116–125.

Borges, P., L.M.P. Valente, V. Véron, K. Dias, S. Panserat and F. Médale. 2014b. High Dietary Lipid Level Is Associated with Persistent Hyperglycaemia and Downregulation of Muscle Akt-mTOR Pathway in Senegalese Sole (*Solea senegalensis*). PLoS ONE 9: e102196.

Cabral, E.M., M. Bacelar, S. Batista, M. Castro-Cunha, R.O.A. Ozório and L.M.P. Valente 2011. Replacement of fishmeal by increasing levels of plant protein blends in diets for Senegalese sole (*Solea senegalensis*) juveniles. Aquaculture 322–323: 74–81.

Cabral, E.M., T.J.R. Fernandes, S.D. Campos, M. Castro-Cunha, M.B.P.P. Oliveira, L.M. Cunha et al. 2013. Replacement of fish meal by plant protein sources up to 75% induces good growth performance without affecting flesh quality in ongrowing Senegalese sole. Aquaculture 380–383: 130–138.

Campos, C., L.M. Valente, P. Borges, T. Bizuayehu and J.M.O. Fernandes. 2010. Dietary lipid levels have a remarkable impact on the expression of growth-related genes in Senegalese sole (*Solea senegalensis* Kaup). J. Exp. Biol. 213: 200–9.

Conceição, L.E.C., L. Ribeiro, S. Engrola, C. Aragão, S. Morais, M. Lacuisse et al. 2007. Nutritional physiology during development of Senegalese sole (*Solea senegalensis*). Aquaculture 268: 64-81.

Conde-Sieira, M., K. Bonacic, C. Velasco, L.M.P. Valente, S. Morais and J.L. Soengas. 2015a. Hypothalamic fatty acid sensing in Senegalese sole (*Solea senegalensis*): response to long-chain saturated, monounsaturated, and polyunsaturated (n-3) fatty acids. Am. J. Physiol. Regul. Integr. Comp. Physiol. 309: R1521–R1531.

Conde-Sieira, M., J.L. Soengas and L.M.P. Valente. 2015b. Potential capacity of Senegalese sole (*Solea senegalensis*) to use carbohydrates: Metabolic responses to hypo- and hyper-glycaemia. Aquaculture 438: 59–67.

Conde-Sieira, M., E. Salas-Leiton, M.M. Duarte, N.F. Pelusio, J.L. Soengas and L.M. Valente. 2016. Short- and long-term metabolic responses to diets with different protein:carbohydrate ratios in Senegalese sole (*Solea senegalensis*, Kaup 1858). Br. J. Nutr. 115: 1896–910.

Costas, B. 2011. Stress mitigation in sole (*Solea senegalensis*) through improved nitrogen nutrition: amino acid utilization, disease resistance and immune status. Ph.D. Thesis. University of Porto, Portugal.

da Silva, R.F., A. Kitagawa and F.J. Sánchez Vázquez. 2016. Dietary self-selection in fish: a new approach to studying fish nutrition and feeding behavior. Reviews in Fish Biology and Fisheries 26: 39–51.

Darias, M.J., A. Boglino, M. Manchado, J.B. Ortiz-Delgado, A. Estevez, K.B. Andree et al. 2012. Molecular regulation of both dietary vitamin A and fatty acid absorption and metabolism associated with larval morphogenesis of Senegalese sole (*Solea senegalensis*). Comp. Biochem. Physiol. A. Mol. Integr. Physiol. 161: 130–9.

De Caterina, R. 2011. n–3 Fatty Acids in Cardiovascular Disease. New England Journal of Medicine 364: 2439–2450.

Dias, J., R. Rueda-Jasso, S. Panserat, L.E.C. da Conceicao, E.F. Gomes and M.T. Dinis. 2004. Effect of dietary carbohydrate-to-lipid ratios on growth, lipid deposition and metabolic hepatic enzymes in juvenile Senegalese sole (*Solea senegalensis*, Kaup). Aquac. Res. 35: 1122–1130.

Dias, J., M. Yúfera, L.M.P. Valente and P. Rema. 2010. Feed transit and apparent protein, phosphorus and energy digestibility of practical feed ingredients by Senegalese sole (*Solea senegalensis*). Aquaculture 302: 94–99.

FAO. 2017. Regional Review on Status and Trends in Aquaculture Development in Europe – 2015. FAO Fisheries and Aquaculture Circular No. 1135/1, 43.

Fernandes, T.J.R., R.C. Alves, T. Souza, J.M.G. Silva, M. Castro-Cunha, L.M.P. Valente et al. 2012. Lipid content and fatty acid profile of Senegalese sole (*Solea senegalensis* Kaup, 1858) juveniles as affected by feed containing different amounts of plant protein sources. Food Chem. 134: 1337–1342.

Guerreiro, I., H. Peres, M. Castro-Cunha and A. Oliva-Teles. 2012. Effect of temperature and dietary protein/lipid ratio on growth performance and nutrient utilization of juvenile Senegalese sole (*Solea senegalensis*). Aqua. Nutr. 18: 98–106.

Guerreiro, I., H. Peres, C. Castro, A. Pérez-Jiménez, M. Castro-Cunha and A. Oliva-Teles. 2014. Water temperature does not affect protein sparing by dietary carbohydrate in Senegalese sole (*Solea senegalensis*) juveniles. Aquac. Res. 45: 289–298.

Helland, S.J. and B. Grisdale-Helland. 1998. Growth, feed utilization and body composition of juvenile Atlantic halibut (*Hippoglossus hippoglossus*) fed diets differing in the ratio between the macronutrients. Aquaculture 166: 49–56.

Kaushik, S.J. 1998. Nutritional bioenergetics and estimation of waste production in non-salmonids. Aquatic Living Resources 11: 211–217.

Madrid, J.A., T. Boujard and F.J. Sánchez-Vázquez. 2001. Feeding Rhythms. pp. 189–215. *In*: Houlihan, D., T. Boujard and M. Jobling [eds.]. Food Intake in Fish. Blackwell Science Ltd, Oxford, UK.

Mandrioli, L., R. Sirri, P.P. Gatta, F. Morandi, G. Sarli, L. Parma et al. 2012. Histomorphologic hepatic features and growth performances of juvenile Senegalese sole (*Solea senegalensis*) fed isogenertic practical diets with variable protein/lipid levels. J. Appl. Ichthyol. 28: 628–632.

Morais, S., F. Castanheira, L. Martinez-Rubio, L.E. Conceicao and D.R. Tocher. 2012. Long chain polyunsaturated fatty acid synthesis in a marine vertebrate: ontogenetic and nutritional regulation of a fatty acyl desaturase with Delta4 activity. Biochim Biophys Acta 4: 660–71.

Moreira, N., S. Soares, L.M.P. Valente, M. Castro-Cunha, L.M. Cunha and P. Guedes de Pinho. 2014. Effect of two experimental diets (protein and lipid vegetable oil blends) on the volatile profile of Senegalese sole (*Solea senegalensis* Kaup, 1858) muscle. Food Chem. 153: 327–333.

Navarro, D.B., V.C. Rubio, R.K. Luz, J.A. Madrid and F.J. Sánchez-Vázquez. 2009. Daily feeding rhythms of Senegalese sole under laboratory and farming conditions using self-feeding systems. Aquaculture 291: 130–135.

Peres, H., I. Guerreiro, A. Pérez-Jiménez and A. Oliva-Teles. 2013. A non-lethal faeces collection method for Senegalese sole (*Solea senegalensis*) juveniles. Aquaculture 414–415: 100–102.

Reis, B., E.M. Cabral, T.J.R. Fernandes, M. Castro-Cunha, M.B.P.P. Oliveira, L.M. Cunha et al. 2014. Long-term feeding of vegetable oils to Senegalese sole until market size: Effects on growth and flesh quality. Recovery of fatty acid profiles by a fish oil finishing diet. Aquaculture 434: 425–433.

Rema, P., L.E.C. Conceicao, F. Evers, M. Castro-Cunha, M.T. Dinis and J. Dias. 2008. Optimal dietary protein levels in juvenile Senegalese sole (*Solea senegalensis*). Aqua. Nutr. 14: 263–269.

Rodiles, A., E. Santigosa, M. Herrera, I. Hachero-Cruzado, M.L. Cordero, S. Martínez-Llorens et al. 2012. Effect of dietary protein level and source on digestive proteolytic enzyme activity in juvenile Senegalese sole, *Solea senegalensis* Kaup 1850. Aqua. Int. 20: 1053–1070.

Rozin, P. and J. Mayer. 1961. Regulation of food intake in the goldfish. American Journal of Physiology - Legacy Content 201: 968–974.

Rubio, V.C., D. Boluda Navarro, J.A. Madrid and F.J. Sánchez-Vázquez. 2009. Macronutrient self-selection in *Solea senegalensis* fed macronutrient diets and challenged with dietary protein dilutions. Aquaculture 291: 95–100.

Rueda-Jasso, R., L.E.C. Conceição, J. Dias, W. De Coen, E. Gomes, J.F. Rees et al. 2004. Effect of dietary non-protein energy levels on condition and oxidative status of Senegalese sole (*Solea senegalensis*) juveniles. Aquaculture 231: 417–433.

Ruxton, C.H.S., S.C. Reed, M.J.A. Simpson and K.J. Millington. 2004. The health benefits of omega-3 polyunsaturated fatty acids: a review of the evidence. Journal of Human Nutrition and Dietetics 17: 449–459.

Salas-Leiton, E., J. Dias, P. Gavaia, A. Amoedo and L.M.P. Valente. 2015. Optimization of phosphorus content in high plant protein practical diets for Senegalese sole (*Solea senegalensis*, Kaup 1858) juveniles: influence on growth performance and composition of whole body and vertebrae. Aqua. Nutr., n/a-n/a.

Salas-Leiton, E., M. Conde-Sieira, N. Pelusio, A. Marques, M.R.G. Maia, J.L. Soengas et al. 2017. Dietary protein/carbohydrate ratio in low-lipid diets for Senegalese sole (*Solea senegalensis*, Kaup 1858) juveniles. Influence on growth performance, nutrient utilization and flesh quality. Aqua. Nutr., n/a-n/a.

Sanchez-Vazquez, F.J., T. Yamamoto, T. Akiyama, J.A. Madrid and M. Tabata. 1999. Macronutrient self-selection through demand-feeders in rainbow trout. Physiology and Behavior 66: 45–51.

Silva, J.M.G. 2009. Use of alternative protein sources in diets for Senegalese sole (*Solea senegalensis* kaup, 1858) juveniles [Documento electrónico]. Porto: [Edição do Autor]. 2009.

Silva, J.M.G., M. Espe, L.E.C. Conceição, J. Dias and L.M.P. Valente. 2009. Senegalese sole juveniles (*Solea senegalensis* Kaup, 1858) grow equally well on diets devoid of fish meal provided the dietary amino acids are balanced. Aquaculture 296: 309–317.

Silva, J.M.G., M. Espe, L.E.C. Conceição, J. Dias, B. Costas and L.M.P. Valente. 2010. Feed intake and growth performance of Senegalese sole (*Solea senegalensis* Kaup, 1858) fed diets with partial replacement of fish meal with plant proteins. Aquac. Res. 41: e20–e30.

Silva, J.M.G., L.M.P. Valente, M. Castro-Cunha, M. Bacelar and P. Guedes de Pinho. 2012. Impact of dietary plant protein levels on the volatile composition of Senegalese sole (*Solea senegalensis* Kaup, 1858) muscle. Food Chem. 131: 596–602.

Valente, L.M.P., F. Linares, J.L.R. Villanueva, J.M.G. Silva, M. Espe, C. Escórcio et al. 2011. Dietary protein source or energy levels have no major impact on growth performance, nutrient utilisation or flesh fatty acids composition of market-sized Senegalese sole. Aquaculture 318: 128–137.

Valente, L.M.P., E.M. Cabral, V. Sousa, L.M. Cunha and J.M.O. Fernandes. 2016. Plant protein blends in diets for Senegalese sole affect skeletal muscle growth, flesh texture and the expression of related genes. Aquaculture 453: 77–85.

Villalta, M., A. Estévez, M.P. Bransden and J.G. Bell. 2005. The effect of graded concentrations of dietary DHA on growth, survival and tissue fatty acid profile of Senegal sole (*Solea senegalensis*) larvae during the Artemia feeding period. Aquaculture 249: 353–365.

Villalta, M., A. Estévez, M.P. Bransden and J.G. Bell. 2008. Arachidonic acid, arachidonic/eicosapentaenoic acid ratio, stearidonic acid and eicosanoids are involved in dietary-induced albinism in Senegal sole (*Solea senegalensis*). Aqua. Nutr. 14: 120–128.

Vivas, M., V.C. Rubio, F.J. Sánchez-Vázquez, C. Mena, B. García García and J.A. Madrid. 2006. Dietary self-selection in sharpsnout seabream (*Diplodus puntazzo*) fed paired macronutrient feeds and challenged with protein dilution. Aquaculture 251: 430–437.

von Schacky, C. 2006. A review of omega-3 ethyl esters for cardiovascular prevention and treatment of increased blood triglyceride levels. Vascular Health and Risk Management 2: 251–262.

Williams, C.M. 2000. Dietary fatty acids and human health. Ann. Zootech. 49: 165–180.

Yúfera, M. and M.J. Darías. 2007. Changes in the gastrointestinal pH from larvae to adult in Senegal sole (*Solea senegalensis*). Aquaculture 267: 94–99.

B-4.1

Welfare, Stress and Immune System

José Fernando López-Olmeda,[1,*] *Yvette S. Wunderink,*[2]
Benjamín Costas,[3] *Juan Miguel Mancera*[2] and
Francisco Javier Sánchez-Vázquez[1]

1. Welfare issues

Animal welfare is an issue of growing concern for both scientists and society. There are over 30,000 entries dealing with this topic in the web of science. Actually, there are books and guidance documents on standard procedures, assessment tools and management to prevent suffering and maximize the wellbeing of animals (e.g., Grandin 2015).

Terrestrial livestock welfare has become a priority area and the focus of many research programs for many years. Recently, the World Organization for Animal Health (OIE) has overviewed and developed international standards, adopted by consensus by all 180 OIE Member Countries, publishing guidelines for the transport and slaughter in dairy, pork and poultry farms (http://www.oie.int/).

1.1 Welfare definitions

The concept of animal welfare is quite complex because this term is used in many different ways, raising both scientific and philosophical issues. Basically, a good state of welfare is achieved if (1) the animal is free from pain, fear and hunger, (2) is in good health and

[1] Department of Physiology, Faculty of Biology, Regional Campus of International Excellence "Campus Mare Nostrum", University of Murcia, 30100 Murcia, Spain.
[2] Department of Biology, Faculty of Marine and Environmental Sciences, Campus de Excelencia Internacional del Mar (CEI-MAR), University of Cádiz, 11510 Puerto Real, Spain.
[3] Interdisciplinary Centre of Marine and Environmental Research of the University of Porto (CIIMAR), Novo Edifício do Terminal de Cruzeiros do Porto de Leixões, Avenida General Norton de Matos, S/N, 4450-208 Matosinhos, Portugal.
* Corresponding author: jflopez@um.es

can adapt to its environment, and (3) can express natural behavior as it would be in the wild (Segner et al. 2012). Many concerns arise from the fact that some stressors may not be removed in intensive fish farming, as fish are confined in captivity at high densities with restricted feeding. Furthermore, fear and distress in fish are poorly understood and identified. Since there is no single stress indicator to define fish welfare, a combination of physiological, physical and behavioral signs should be considered (Crousos and Gold 1992, Huntingford et al. 2006). As most fish farmers realize, they should monitor not only how much fish grow and feed, but also how fish look like and swim.

The coexistence within the same fish population of individuals with different "coping strategies" further complicates matters. Some fish strains may show higher/lower tendency to take risks in stressful environments. This seems to be the case of Senegalese sole, as proactive individuals exhibited shorter feeding latency, higher duration of escape attempts and lower cortisol levels than passive individuals (Mota-Silva et al. 2010).

The stress response can be considered as an adaptive strategy to cope with changes in the environment and safeguard animal's homeostasis. In this chapter, we will review the functioning of the stress control axis as well as its connections with the immune system of Senegalese sole.

2. The stress response in Senegalese sole

2.1 The hypothalamus-pituitary-interrenal (HPI) axis

In teleosts, the regulation of the neuroendocrine stress response is generally conceptualized by the hypothalamus-pituitary-interrenal (HPI) axis (Wendelaar Bonga 1997, Flik et al. 2006, Bernier et al. 2009). In Senegalese sole, four key central peptides of the endocrine stress axis (i.e., corticotropin-releasing hormone—CRH, CRH binding protein—CRH-BP, and two forms of proopiomelanocortin:– POMC-A and POMC-B) have been characterized (Wunderink et al. 2011, 2012a). The deduced amino acid sequences of both sseCRH and sseCRH-BP appear to be well-conserved and highly homologous to those in other vertebrates (Wunderink et al. 2011). The mature peptide of sseCRH, for instance, shows a high homology of up to 72% identity with human, whereas the deduced sseCRH-BP amino acid sequence contains the typical well-conserved 10 cystein residues, which are involved in the formation of 5 Cys-Cys bridges that are important for ligand binding activity (Fischer et al. 1994).

The roles of sseCRH and sseCRH-BP in stress responses have been respectively investigated in Senegalese sole juveniles (Wunderink et al. 2011) and larvae (Wunderink et al. 2012b). Juveniles kept at high stocking density elevated plasma cortisol levels and *crh* expression mildly, and showed a tendency to decrease *crh-bp* expression (Wunderink et al. 2011). Food-deprived larvae also increased total cortisol concentrations and *crh* expression, but significantly reduced *crh-bp* expression (Wunderink et al. 2012b). These conditions seem typical for Senegalese sole in chronic stress situations and it is plausible that the enhanced plasma cortisol levels are maintained by an upregulation of *crh* and down-regulation of *crh-bp* expression.

A big surprise came from two transcripts' coding for ssePOMC (Wunderink et al. 2012a). As a precursor peptide, POMC can be cleaved into a number of peptides with roles in several physiological processes such as stress response regulation, pigmentation, and rewarding system. In Senegalese sole, the capacity of functional POMC-derived peptides is distributed over the two *pomc* genes, two paralogues which show a trend towards subfunctionalization (Wunderink et al. 2012a). Differential expression of ssePOMC-A and ssePOMC-B following a chronic stress response also indicates different physiological roles for both paralogues, which is important when investigating POMC's function in endocrine responses. The benefit of a simpler regulation, e.g., a non-functional β endorphin (β-END), eliminates β-END's side effects when only the production of melanocyte stimulating hormone (MSH) is needed and may have driven the subfunctionalization between the two copies. Specializations of *pomc* are also reported for other vertebrates, suggesting an evolutionary trend towards a different form of POMC peptide regulation (de Souza et al. 2005). In the teleost lineage, however, these specializations are the result of the fish specific 3R whole genome duplication, whereas the paralogues of Senegalese sole *pomc* are most likely the result of a far more recent duplication (Wunderink et al. 2012a).

2.2 The stress response in Senegalese sole

Stress is defined as a condition in which the animal's equilibrium, homeostasis, is threatened or disturbed as a result of the actions of intrinsic or extrinsic stimuli (Wendelaar Bonga 1997, McEwen and Wingfield 2010). Therefore, the adaptive response to stressors is fundamental to cope with them and increase survival rate, which of course is a major benefit for the organism.

When an individual experiences stress, its normal homeostasis levels are altered to adapt to the stressful situation: allostasis. Due to the energy spent on allostasis, stress can negatively affect growth, reproduction and immunological responses (Wendelaar Bonga 1997). The degree of stress, or allostatic load, depends on the chronicity of stress conditions, as well as a combination of the nature and extent of the stressor (McEwen and Wingfield 2010). Acute stress is defined by short exposures to a stressor such as transport, weighing, sorting/grading and sudden environmental changes (e.g., in water temperature or salinity) (Pankhurst 2011). Hence, acute stressors evoke pronounced, though, short peaks of increased plasma cortisol levels in teleosts (Flik et al. 2006). In teleosts, acute stress enhanced hypothalamic-sympathetic-chromaffin (HSC) axis, increasing plasma catecholamine levels and stimulating oxygen uptake and transfer, as well as mobilization of energy substrates (Reid et al. 1996, Wendelaar Bonga 1997). However, to our knowledge, no data exist on HSC in Senegalese sole and further studies are necessary on this topic. In addition, acute stress also increased plasma cortisol levels inducing metabolic as well as osmoregulatory alterations (Arjona et al. 2007, Costas et al. 2011a, Herrera et al. 2012).

Long term exposure to stressors, on the other hand, leads to chronic stress which affects the HPI-axis less pronounced, but with longer duration. Consequently, exposure to stressors can lead to allostatic overload, which is when the individual is no longer able to cope with stress. In that situation, stress negatively affects immune functions leading to diseases, reduced animal welfare and eventually mortality (Wendelaar Bonga 1997). In Senegalese sole submitted to chronic stress situation (i.e., high stocking densities, inadequate environmental salinity or temperature), an activation of HPI system is observed with high plasma cortisol levels, and altered metabolism due to the higher demand for energy production in order to cope with stressful rearing conditions (Arjona et al. 2007, 2008, 2009, Costas et al. 2008, 2013c, Salas-Leiton et al. 2008, 2010). Molecular analysis of HPI in Senegalese sole showed that high stocking density or food-deprivation enhanced *crh* expression while decreased *crh-bp* expression (Wunderink et al. 2009, 2012b), suggesting a concomitant *crh* expression upregulation and *crh-bp* expression down-regulation in order to maintain the high plasma cortisol levels observed in chronic stress situations.

2.3 Stress in early development

Fishes are extremely vulnerable during early life stages. Since larval stages have two functions, viz. (i) to fulfill the actual needs of the organisms and (ii) to build the final set of organs, larval development is mostly characterized by a huge increase of biomass and changes in morphology, physiology and behavior (Osse and Van den Boogaart 1997). Therefore, larvae are much more sensitive to subsequent exposure to stressors than adults are, and suffer more rapidly of allostatic overload. As a result, stressors, but also stressful events like metamorphosis, during the larval phase can lead to lower survival rates and disturbed development (see Chapter B-2.2). Successful larval development is species-specific and depends on the vigor of the individual, but also on important limiting abiotic factors such as light, temperature, salinity and feeding conditions. Those who work with *Solea* sp. know that when their fish turn black in the tank, something is going wrong, which makes it also obvious that the endocrine system is strongly involved (Ruane et al. 2005). At least, experimenting with Senegalese sole larvae and chronic or subsequent ambient stressors resulted to be very complicated, since most larvae indeed die when they are exposed to different stressors, like low-salinity water, or are transferred from tank to tank directly after transport (Wunderink et al. 2012b, Morais et al. 2014).

The importance of the stress system is not only reflected by their gene conservation (Wunderink et al. 2011, 2012a), but also by its development in very early life stages of teleosts. CRH was quantified in larvae of Mozambique tilapia (*Oreochromis mossambicus*) as young as 5 dph, and a more sensitive immunohistochemistry extrapolated these results to even younger stages, 2 dph (Peppels and Balm 2004, Flik et al. 2006). In zebrafish (*Danio rerio*), expression of *crh* was shown by quantitative PCR and *in situ* hybridization in the telencephalon and hypothalamus of embryos at 1 day post fertilization (dpf) and in the NPO at 2 dpf, around hatching point (Chandrasekar et al. 2007). In the same species, other components of the HPI-axis were found in early development, like POMC, urotensin I (UI), CRH-R1 and 2 (Alderman and Bernier 2009, Alsop and Vijayan 2009). Moreover, an increase of the two cortisol receptors (glucocorticoid and mineralocorticoid, GR and MR, respectively) was detected already around hatching (Alsop and Vijayan 2008). Based on immunohistochemistry studies, there are indications that very young Senegalese sole larvae are equipped with an endocrine stress axis already upon or right after the hatching event, showing positive anti-ACTH and anti-CRH staining in Senegalese sole larvae of only 1 and 3 dph, respectively (Klaren et al. 2008). However, it will be necessary to combine gene expression analyses with physiological studies in order to determinate the ontogeny of the HPI-axis more closely, as well as whether the axis is already functional. Although the components of the stress system are already present before hatching, in zebrafish an adequate cortisol response following a stressor was first seen at the point of first feeding (Alsop and Vijayan 2009). Common carp (*Cyprinus carpio*) embryos, however, exhibit a cortisol response and increase in ACTH to handling stress, just prior to hatching (Stouthart et al. 1998). This means that the HPI-axis could already be operational at such early stages of life in Senegalese sole as well.

2.4 Saturation of stress

Surprisingly, when chronically stressed juveniles were submitted to a subsequent acute stressor, viz. transfer to high salinity water, no additional cortisol or *crh* response was seen (Wunderink et al. 2009). The osmoregulatory capacity was disturbed, indicating that the animals were not able to cope with the subsequent stressor. Unfortunately, not much is known about the influence of consecutive stressors in fish. The lack of additional cortisol or CRH response could indicate saturation of the involved glands, with corticotropic or interrenal cells producing no more hormones (Rotllant and Tort 1997, Rotllant el al. 2000) and/or a putative increased relevance of the receptors involved herein. The stress system seems to have its limits and at some point, the individual loses its capacity to adapt to stress situations which is when stress changes from being *"eustress"* to *"distress"* (Wendelaar Bonga 1997, McEwen and Wingfield 2010). Therefore, chronic stress in aquaculture settings could have serious effects on the animal when exposed to a second stressor such as handling, and it is recommended to prevent subsequent stressors and introduce periods of acclimation to new conditions.

2.5 Interactions between thyroid and stress systems

In teleosts, depending on the species, the stress and the thyroid systems seem to be interwoven. In this sense, the pituitary thyrotropin releasing hormone (TSH), perhaps surprisingly, is stimulated (production and release) by hypothalamic factors such as the CRH (de Groef et al. 2006, Bernier et al. 2009). The dual function of CRH as a corticotropin- and thyrotropin-releasing hormone suggests a functional relationship between the HPI and hypothalamus-pituitary-thyroid (HPT) axes. Indeed, in *C. carpio* T4 inhibits the HPI axis via upregulation of *crh-bp* expression in the hypothalamus (Geven et al. 2006). Reciprocally, it was shown that cortisol affects the thyroid system, i.e., by stimulating the conversion of T4 to T3 in brook charr (*Salvelinus fontinalis*) liver *in vitro* (Vijayan et al. 1988). Conversely, in Nile tilapia (*Oreochromis niloticus*), administration of the synthetic glucocorticoid dexamethasone decreased the activity of hepatic D1 and D2 (Walpita et al. 2007). During larval ontogeny, stimulating effects of cortisol on the thyroid system have been described, as cortisol enhances *in vitro* the effect of thyroid hormones on metamorphosis of Japanese flounder (*Paralichthys olivaceus*) (de Jesus et al. 1990). However, the interaction of stress and thyroid system during larval development is poorly understood, and contradictory results makes it difficult to establish a model that is valid for all teleost species (Walpita et al. 2007, Bernier et al. 2009).

For fish farmers, the improvement of the knowledge of the interactions between stress and thyroid systems could be interesting, since thyroid hormones are strongly involved in growth, reproduction, development and metabolism in fish, factors which could negatively be affected under stress conditions (Power et al. 2001). Understanding of the involved mechanisms of the influence of stress on thyroid function, but also the effect of thyroid hormone on stress, could help improve these factors.

Interactions between HPT- and HPI-axes are indeed present in Senegalese sole post-larvae and juveniles (Arjona et al. 2008, 2010, 2011; summarized in Fig. 1). But overall, it seems that cortisol has a stronger impact on the HPT-axis than thyroid hormones have on the HPI-axis. The relationship between these two axes could be merely unidirectional for cortisol, or stressors, towards thyroid function. The suggestion that cortisol has a more pronounced effect on the HPT-axis than the other way around, indicates the value of the endocrine stress response. The value of a well-functioning stress response was already reflected in the strong conservation of components of the HPI-axis during evolution (Wunderink et al. 2011, 2012b). In case of emergency, the fight or flight mechanism is crucial; to live or let die depends strongly on the first reaction of the organism. Therefore, every other physiological process, like growth or reproduction, becomes second when an individual is stressed, explaining the powerful inhibiting impact of cortisol on other endocrine axes.

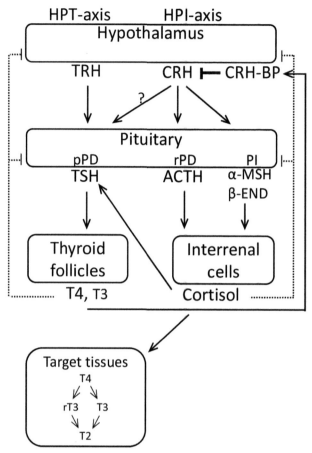

Fig. 1. Proposed diagram of interactions between the hypothalamus-pituitary-thyroid (HPT) axis and hypothalamus-pituitary-interrenal (HPI) axis in Senegalese sole. Arrows and dotted lines indicate stimulatory or inhibitory function, respectively (modified from Wunderink 2011).

Remarkably, cortisol initially stimulates *tshβ* subunit expression and hepatic T4-to-T3 conversion, reducing plasma fT4 levels although plasma fT3 concentrations remain stable (Arjona et al. 2011). For this reason, it is known that stress does have negative effects on thyroid function, and could even induce or aggravate thyroid disorders. The HPI- and HPT-axes are both intimately involved in energy metabolism and expenditure. The overall function of a stress response is to prepare the organism for fight or flight and to this end, redistributes energy by mobilizing glucose, fatty acids and catabolism of proteins and lipids to increase gluconeogenesis (Wendelaar Bonga 1997, Kühn et al. 1998). The thyroid system on the other hand is more involved in anabolic processes like growth and reacts strongly on differences in nutritional status (van der Geyten 1998, Eales 2006). Indeed, both T4 and T3 treatment decreased plasma glucose and triglyceride levels in Senegalese sole, indicating the stimulatory function of both hormones on energy metabolism (Arjona et al. 2008, 2011). Reciprocally, plasma glucose levels are elevated in stressed animals and the catabolizing processes are also illustrated by the decrease of carbon and nitrogen contents of fasted Senegalese sole post-larvae (Wunderink et al. 2012b).

In aquaculture, one of the biggest expenses goes to the costs of food, which makes it interesting to maximize the relationship between growth and feed intake as much as possible. However, the only way in which heterotrophic organisms can take in energy is through food intake. Food intake, therefore, is extensively regulated, and limited food availability, fasting or starvation induce a number of physiological processes that have evolved to preserve energy homeostasis in animals (McCue 2010). Hence, feeding conditions demonstrate well the relationship between stress and thyroid hormones and their roles in energy expenditure (Wunderink et al. 2012b). The thyroid system responds to a broad range of environmental variables linked to activation in anabolic states, explaining the inhibition on stressful or catabolic conditions caused by thyroid hormones, but also the inhibiting effects of cortisol on the thyroid system (Eales 2006). It is plausible that an adequate response of an animal to starvation must, of necessity, isolate actions of cortisol and thyroid hormones, these being two endocrine systems basically involved in energy metabolism. Therefore, it is likely to suggest that the stress system inhibits anabolic processes by inhibiting the thyroid system, and could explain the responsiveness of thyroid hormones to stress caused by starvation as seen in Senegalese sole post-larvae (Wunderink et al. 2012b). The contradictory functions of thyroid and stress axis seem not to be compatible, explaining the inhibiting effect of one towards the other.

3. Rhythms in the HPI axis

3.1 Synchronization of the HPI axis

Cyclic variations are present in a variety of environmental factors such as photoperiod or temperature. To cope with such environmental cycles, the biological rhythms have evolved in living organisms, driven by endogenous oscillators that affect physiological variables (behavioral, endocrine, enzymatic, and metabolic). Among these variables, daily rhythms in the mammalian hypothalamus-pituitary-adrenal axis controlling cortisol secretion have been reported (Haus 2007). Plasma cortisol shows daily variations, displaying the maximum values before the active phase of the animal. Hence, diurnal species present the peak of cortisol at the beginning of the light phase, whereas nocturnal species have the greatest values at the beginning of the dark phase (Mohawk and Lee 2005, Dickmeis 2009). This seems to be related with the role of cortisol as an arousal signal, which prepares the organism for the active phase (Haus 2007). Interestingly, the adrenal gland presents a daily rhythm in sensitivity to ACTH, which means that cortisol response varies depending on the time of day (Haus 2007). In mammals, the greatest sensitivity to adrenocorticotropic hormone (ACTH) coincides with the highest glucocorticoid production and release (Sage et al. 2002, Engeland and Arnhold 2005).

In fish, daily rhythms in cortisol have been described in many species, most of them of importance in aquaculture (Ellis et al. 2012). The amplitude of the variations and the time of day when the highest values occur are species-dependent (Ellis et al. 2012). Cortisol values in fish are correlated with rhythms in the hormones involved in the HPI axis such as ACTH, which would in turn generate the rhythms of cortisol in the blood (Singley and Chavin 1976). Although cortisol rhythms have been studied mostly in unstressed animals, when fish are submitted to a stressor the rhythm is maintained with the same acrophase but the average values and the amplitude increase (Pickering and Pottinger 1983). Moreover, as reported in mammals, the response to a stressor is different depending on the time of the day, as was observed in the green sturgeon (*Acipenser medirostris*) in which the acute stress of air exposure for a short period of time elicited higher cortisol values when applied at night than during the day (Lankford et al. 2003). In addition to the daily rhythmicity, fish cortisol can display seasonal variations in fish. These variations can involve changes in both the amplitude and the acrophase of the cortisol rhythm (Ellis et al. 2012) and they seem to be controlled by the annual variations in both photoperiod and water temperature (Kühn et al. 1986, Planas et al. 1990, Pavlidis et al. 1999).

In Senegalese sole, the existence of daily rhythms has been recently reported (López-Olmeda et al. 2013). In that study, daily rhythms were found in variables from all the components of the HPI axis of Senegalese sole (Fig. 2A–D): hypothalamic *crh* and *crhbp* expression, pituitary *pomca* and *pomcb* expression, as well as plasmatic cortisol, glucose and lactate. Both hypothalamic *crh* and plasma cortisol present a similar profile with the highest values displayed around the end of the day and beginning of the night (Fig. 2A, 2C). This correlation suggests that cortisol daily rhythm can be directly induced by the rhythmic production of hypothalamic CRH—a correlation that has been observed in other studies on the stress response of Senegalese sole (Wunderink et al. 2011). In addition, Senegalese sole is a nocturnal species and thus the acrophases of *crh* and cortisol were located at the beginning of active phase of this species (Bayarri et al. 2004, López-Olmeda et al. 2013), which is in agreement with the cortisol rhythms observed in other vertebrate species, including teleosts (Mohawk and Lee 2005, Ellis et al. 2012). The acrophase of *crhbp* was located towards the end of the night (ZT 22.3 h), 8 h after the acrophases of *crh* and cortisol (which were located at 14.5 and 14.4 h, respectively). CRH-BP in fish is part of a negative feedback process, which is stimulated by high cortisol levels in order to avoid an over-stimulation of HPI axis (Huising et al. 2004). The high *crhbp* expression levels coincided with the decline in *crh* expression and cortisol values at 22 h, thus pointing to an involvement of this protein in the regulation of cortisol daily rhythms in the Senegalese sole (López-Olmeda et al. 2013), as it has been suggested in juvenile specimens under chronic stress (Wunderink et al. 2011).

Since Senegalese sole HPI axis presents daily rhythms at all its levels, the response to stress could vary depending on the time of the day in which an acute stressor is applied. Following that hypothesis, in the same study by López-Olmeda et al. (2013), the influence of the time of day on the stress response was explored. For that purpose, sole were subjected to an acute stress consisting of 30 s of air exposure and sampled 1 h after the stress, as described previously (Costas et al. 2011a). The stress challenge was applied at two different times: ZT 1 and ZT 13 h. These times were selected because they coincided with the highest (ZT 13 h) and the lowest (ZT 1 h) *crh* and cortisol levels observed in the daily rhythms. In this second experiment, a clear effect of the time of day on the stress response was observed (Fig. 2E–H). This response was stronger when the stressor was applied during the light phase, eliciting a greater response of pituitary *pomca* expression and plasma cortisol, glucose and lactate levels at ZT 1 h compared with ZT 13 h (Fig. 2F–H).

3.2 Stress responses depend on the time of day

To date, little is known on the influence of the time of day in the neuroendocrine stress response in fish. Besides Senegalese sole, this time-dependent effect has been described in two fish species: green sturgeon and gilthead sea bream (*Sparus aurata*) (Lankford et al. 2003, Vera et al. 2014). In these two reports, the stressor applied was air exposure as in Senegalese sole, and both species presented a greater stress response during the dark phase of the LD cycle, which suggests that the time of day of higher stress sensitivity may be species dependent. Interestingly, *S. aurata* presented a diurnal behavior and hence the greatest stress response was observed during the resting phase (dark phase) (Vera et al. 2014). This observation coincides with the results obtained in Senegalese sole and could point that in fish the sensitivity to stress could depend on the daily activity patterns of

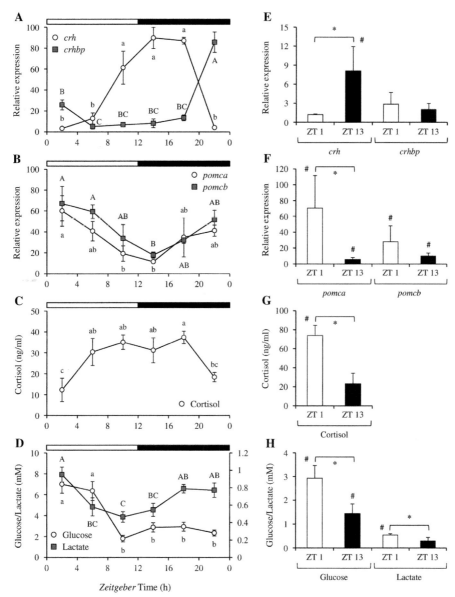

Fig. 2. Daily rhythms of several factors from the HPI axis of Senegalese sole (A–D) and effects of the time of day on the response of these factors to acute stress (E–H). Variables from the HPI axis analyzed were hypothalamic *crh* and *chrbp* expression (A, E), pituitary *pomca* and *pomcb* expression (B, F), and plasmatic cortisol (C, G), glucose and lactate (D, H). Data from daily rhythms were subjected to a one-way ANOVA ($p < 0.05$), followed by a Tukey's *post hoc* test. Different letters indicate significant differences between time points within the same variable. When two variables are represented in the same graph, upper and lower case letter are used to differentiate Tukey's *post hoc*'s results. White and black bars above the daily rhythms graphs represent the light and dark periods, respectively, of the LD cycle. Regarding the effects of the time of day on the stress response (E–H), Senegalese sole were subjected to a stress challenge (air exposure), either at the beginning of the light phase (ZT 1 h, white bars) or at the beginning of the dark phase (ZT 13 h, black bars). Samples were collected 1 h after stress. Data (mean ± S.E.M., n = 6) are represented as the increase with respect to the mean value from a control group. Data from each variable were subjected to a one-way ANOVA, followed by a Tukey's *post hoc* test. Hash symbols (#) indicate significant differences between the stressed and the control group; asterisks indicate significant differences between the stressed group at ZT 1 h and the stressed group at ZT 13 h. Figure modified from López-Olmeda et al. (2013).

the species, although further research is required to elucidate this hypothesis. In addition, melatonin could also play a role in the lower stress response of Senegalese sole observed at night. A treatment with melatonin attenuates the response induced by stressors such as high stocking density, low water replacement and chasing in Senegalese sole, reducing the plasmatic levels of stress indicators such as cortisol and lactate (López-Patiño et al. 2013, Gesto et al. 2016).

The existence of rhythms at all levels of the HPI axis of Senegalese sole should be considered in both research and sole aquaculture. For instance, during the evaluation of the stress state of animals under normal rearing conditions, samples may be collected to analyze some parameters from the HPI axis such as cortisol. In those studies, it is recommended to pay special attention to the time of the day in which samples for the evaluation are collected, especially if data are going to be compared with samples collected at different dates, from different tanks, stressors, etc. If the time of day in which samples are collected is not considered, then the error introduced to the protocol due to the normal daily variations can lead to over- or under- estimating the stress state of the animals.

Similarly, the different responses of the HPI axis of Senegalese sole to a stressor depending on the time of the day is also of great importance for the aquaculture of this species. Since the time of day affects the stress response, the moment in which stressors such as handling, sorting, grading or vaccination are performed should be specially considered. Some of these protocols are routinely used in aquaculture and are a cause of stress for the animals, compromising their welfare (Ellis et al. 2012); hence, reducing their impact as much as possible since many of them cannot be avoided is important. In addition, in the case of Senegalese sole, these time-dependent effects are not exclusive from the stress response and have been shown in other treatments such as the response of the somatotropic axis to exogenous GH administration (López-Olmeda et al. 2016) and the egg production and fertilization after GnRH injection to females (Rasines et al. 2013), highlighting the importance of the time of day and daily rhythms for aquaculture.

4. The immune system of Senegalese sole

The fish immune system not only gives some sort of protection through preventing the attachment of microorganisms to the external body surface, but also avoids microbes' invasion and multiplication inside the host tissues. Therefore, there is a need to understand the immunology behind the disease or infection, which is essential to develop important diagnostic and therapeutic tools. However, and keeping in mind the high incidence of diseases observed during Senegalese sole industrial production (Morais et al. 2014), few studies have focused on host's immune mechanisms against invading pathogens. The recently identified transcriptomes and molecular markers for both common and Senegalese sole will certainly be of assistance in future gene expression studies (Benzekri et al. 2014).

Most experiments aiming to study immune mechanisms in sole mainly focused on innate immune responses. A number of humoral substances and cell secretions are known to contribute to the natural resistance of fish to infectious agents, including complement, transferrins, anti-proteases, various lytic enzymes (e.g., lysozyme), lectins, interferon and enzyme inhibitors (Ellis 1999). Moreover, some of these factors, such as lysozyme and complement, appear to be more potent in fish than in mammals (Ellis 2001). In vertebrates, two main lysozyme types have been identified, commonly designated as the c-type (chicken-type) and the g-type (goose-type), and both molecules are known to be functional in the Senegalese sole. For instance, Fernández-Trujillo and colleagues (2008a) reported that c-type lysozyme transcripts barely increased in the head-kidney and liver from individuals stimulated by intraperitoneal injection of lipopolysaccharides (30 mg/kg body weight), an endotoxin found on the bacterial cell membrane of a bacterium, or *Photobacterium damselae* subsp. *damselae* (2×10^3 cells/fish), while g-type lysozyme mRNA levels increased up to 6 and 20 fold in the spleen and head-kidney, respectively, from Senegalese sole exposed to several pathogen-associated molecular patterns (PAMPs) (Ponce et al. 2011). Lipopolysaccharides also increased hepcidin mRNA levels in intraperitoneally injected Senegalese sole (Osuna-Jiménez et al. 2009) as well as in primary head-kidney leucocytes culture (Costas et al. 2013a). The level of expression of g-type lysozyme also increased in Senegalese sole blood leucocytes following incubation with UV killed *Tenacibaculum maritimum* for 24 and 48 h (Costas et al. 2013a). Senegalese sole phagocytes presented diverse innate immune responses following challenge with *Ph. damselae* subsp. *piscicida* and *T. maritimum* strains, suggesting that pattern recognition receptors from Senegalese sole macrophages may have detected PAMPs associated with unique structures for each bacterium isolate (Costas et al. 2013b, 2014). Moreover, Senegalese sole head-kidney leucocytes incubated with L15 medium alone showed a typical round monocyte morphology (Fig. 3A), whereas incubation with *T. maritimum* extra-cellular products (100 μg mL^{-1}) for 24 h appears to increase cellular vacuolization and elongation (Fig. 3B) (Costas et al. unpublished results).

Studies of cell migration to an inflammatory focus have also been conducted in the Senegalese sole. Costas et al. (2013b) reported leucocyte responses to inflammation in both peripheral blood and peritoneal cavity of Senegalese sole following challenge with *Ph. damselae* subsp. *piscicida*. Moreover, infected fish increased plasma lysozyme and peroxidase activities as well as circulating monocytes and neutrophils (Costas et al. 2013b). This study also suggested an important role of the alternative complement pathway against that particular bacterium. In contrast, Mabrok and co-workers (2016) showed that heat-inactivated plasma samples from Senegalese sole presented bactericidal activities against *T. maritimum* similar to untreated plasma, and speculated that the complement system may have no major role against this pathogen, at least through the alternative pathway.

The Senegalese sole's innate immune machinery also increased following a mixed leucocyte reaction. The level of expression of pro-inflammatory cytokines (i.e., interleukin-1β and interleukin-8), hepcidin antimicrobial peptide and g-type lysozyme increased after incubating peripheral blood leucocytes from three different individuals at 24 and 48 h (Costas et al. 2013c). Mx protein, an interferon-induced protein that protects against viral infections, showed in this species different expression profiles following both poly I:C injection and solevirus inoculation (Fernández-Trujillo et al. 2008b,c).

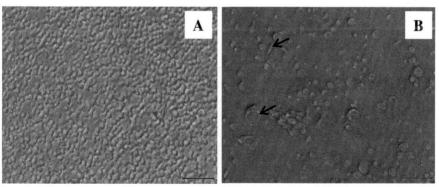

Fig. 3. Characteristic phenotypes of Senegalese sole head-kidney leukocytes following incubation with medium alone (A) or *Tenacibaculum maritimum* extra-cellular products (100 μg mL^{-1}) for 24 h (B). Arrows show increased vacuolization and cellular elongation.

Indeed, an effective disease control management is tightly linked to understanding the etiology of major diseases, being certainly of utmost importance within the intensive aquaculture sector. *De novo* transcriptomes from common and Senegalese sole recently available through the SoleaDB will assist this endeavor with new powerful tools to assess the complexity of these flatfish species' immune responses. Moreover, disease prevention by vaccination is, on an economic and environmental basis, the most appropriate method for pathogen control currently available to the aquaculture sector, and efforts should be also directed towards the understanding and characterization of acquired immunity in sole, as well as to the development of effective vaccines against main pathogens causing major disease outbreaks (e.g., *T. maritimum*).

4.1 Relation between stress and immune system

The vertebrate immune system is also linked to the endocrine system and it is currently known that an elaborate bi-directional communication exists. It has also been reported that stressful husbandry conditions not always induce suppressive effects on the immune system, and thus it does not necessarily translate in decreased resistance to infection in both mammals and fish (Dhabhar 2009, Tort 2011). In fact, immune responses can increase due to an augmentation in host glucocorticoid levels. However, glucocorticoids could be considered as a double-edged sword since similar hormone levels may also suppress immune function depending on the duration and severity of the stressor. For instance, acute and chronic stressful situations induced different responses in Senegalese sole innate immune parameters. Plasma lysozyme activity decreased in stressed specimens at 4 h following handling stress (Costas et al. 2011a) and after 18 days reared at high stocking density (Costas et al. 2013c), whereas those levels increased after 14 and 28 days in fish submitted to air exposure once a week (Costas et al. 2012) or after 14 days following daily handling (Costas et al. 2011b). Similarly, the alternative complement pathway decreased in an inverse linear relationship compared to plasma cortisol levels in specimens exposed to acute stress or reared at high density (Costas et al. 2011a, 2013c), whereas daily handled fish augmented complement levels after 14 days (Costas et al. 2011b). Interestingly, daily handling stress for 14 days from the latter study increased host resistance to *Ph. damselae* subsp. *piscicida* compared to undisturbed fish, while Senegalese sole treated with dexamethasone (a potent glucocorticoid) for 14 days appeared to be more susceptible to the same pathogen (Salas-Leiton et al. 2012). Likewise, several *in vitro* studies suggested that cortisol alone inhibits lipopolysaccharides-induced expression of several immune-related genes (Saeij et al. 2003, Fast et al. 2008, Castillo et al. 2009). In contrast, pre-metamorphosed 3-day old larvae incubated at two salinities (10 and 36 ppt) showed a total of 2,816 probes differentially expressed, of which 2,641 were up-regulated and 175 down-regulated at 10 ppt compared to 36 ppt. Among the differentially expressed genes, transcripts were involved in inflammatory mechanisms and innate immune system, including, for instance, cytokines and their receptors, genes of the complement and interferon pathways, g-type lysozyme and the prostaglandin biosynthesis pathway (Benzekri et al. 2014).

The modulation of the innate immune machinery after stressful situations was also observed by other authors. For instance, cortisol can stimulate apoptosis of B-cells (Weyts et al. 1998a), and inhibit apoptosis of neutrophils (Weyts et al. 1998b), an adaptive response to prolong the life span of neutrophils that forms the first-line of defense against pathogens. Moreover, short-term, but not long-term, stress resulted in increased plasma cortisol levels in Atlantic salmon (*Salmo salar*), which on a short-term basis translated in stimulatory effects on immune cells, such as augmented expression levels of the pro-inflammatory gene interleukin-1β, followed by inhibitory effects on a longer-term basis, with decreased stimulation of leucocytes by lipopolysaccharides and decreased leucocyte survival in the presence of *Aeromonas salmonicida* (Fast et al. 2008). *In vivo* neuroendocrine-immune interactions are thus dependent on the actions of various hormones (i.e., catecholamines, cortisol, adrenocorticotropic hormone, β-endorphin) and cytokines (i.e., interleukin-1β, interleukin-6 and tumor necrosis factor-α), as well as on their interactions (Tort 2011). This hypothesis could explain the increased immune responses observed in Senegalese sole chronically stressed by repeated air exposure (Costas et al. 2011b, 2012). Other hormones released during HSC and HPI axes activation may have influenced innate immune mechanisms in a higher degree, decreasing the suppressive effects of cortisol. In contrast, a different situation probably occurs during chronic situations such as high stocking density, where an allostatic load is probably imposed, thus compromising the immune function and cortisol playing its attributed immunosuppressive role. This hypothesis is supported by the increased susceptibility to opportunistic pathogens observed in Senegalese sole reared at high stocking density (Costas et al. 2008), in line with a general decrease in immune function (Salas-Leiton et al. 2010, Costas et al. 2013c). Although several trials have studied the effects of either glucocorticoids or stress on immune competence in sole through *in vivo* or *in vitro* models, evidence of its consequent effects on disease resistance are scarce. For instance, the exact mechanisms by which stressed Senegalese sole resisted better to photobacteriosis infection remain to be elucidated (Costas et al. 2011b). Future studies addressing mechanisms of action and regulation of cytokines and other immune-related genes following bacterial challenge in previously stressed fish deserve further attention.

References

Alderman, S.L. and N.J. Bernier. 2009. Ontogeny of the corticotropin-releasing factor system in zebrafish. Gen. Comp. Endocrinol. 164: 61–69.

Alsop, D. and M.M. Vijayan. 2008. Development of the corticosteroid stress axis and receptor expression in zebrafish. Am. J. Physiol. Regul. Integr. Comp. Physiol. 294: R711–719.

Alsop, D. and M.M. Vijayan. 2009. Molecular programming of the corticosteroid stress axis during zebrafish development. Comp. Biochem. Physiol. A 153: 49–54.

Arjona, F.J., L. Vargas-Chacoff, I. Ruiz-Jarabo, M.P. Martín del Río and J.M. Mancera. 2007. Osmoregulatory response of Senegalese sole (*Solea senegalensis*) to changes in environmental salinity. Comp. Biochem. Physiol. A 148: 413–421.

Arjona, F.J., L. Vargas-Chacoff, M.P. Martín del Río, G. Flik, J.M. Mancera and P.H.M. Klaren. 2008. The involvement of thyroid hormones and cortisol in the osmotic acclimation of *Solea senegalensis*. Gen. Comp. Endocrinol. 155: 796–803.

Arjona, F.J., L. Vargas-Chacoff, I. Ruiz-Jarabo, O. Gonçalves, I. Páscoa, M.P. Martín del Río et al. 2009. Tertiary stress responses in Senegalese sole (*Solea senegalensis* Kaup, 1858) to osmotic challenge: Implications for osmoregulation, energy metabolism and growth. Aquaculture 287: 419–426.

Arjona, F.J., I. Ruiz-Jarabo, L. Vargas-Chacoff, M.P. Martín del Río, G. Flik, J.M. Mancera et al. 2010. Acclimation of *Solea senegalensis* to different ambient temperaturas: implications for thyroidal status and osmoregulation. Mar. Biol. 157: 1325–1335.

Arjona, F.J., L. Vargas-Chacoff, M.P. Martín Del Río, G. Flik, J.M. Mancera and P.H.M. Klaren. 2011. Effects of cortisol and thyroid hormone on peripheral outer ring deiodination and osmoregulatory parameters in the Senegalese sole (*Solea senegalensis*). J. Endocrinol. 208: 323–330.

Bayarri, M.J., J.A. Muñoz-Cueto, J.F. López-Olmeda, L.M. Vera, M.A. Rol de Lama, J.A. Madrid et al. 2004. Daily locomotor activity and melatonin rhythms in Senegal sole (*Solea senegalensis*). Physiol. Behav. 81: 577–583.

Benzekri, H., P. Armesto, X. Cousin, M. Rovira, D. Crespo, M.A. Merlo et al. 2014. *De novo* assembly, characterization and functional annotation of Senegalese sole (*Solea senegalensis*) and common sole (*Solea solea*) transcriptomes: integration in a database and design of a microarray. BMC Genomics 15: 952.

Bernier, N.J., G. Flik and P.H.M. Klaren. 2009. Regulation and contribution of the corticotropic, melanotropic and thyrotropic axes to the stress response in fishes. pp. 235–311. *In*: Bernier, N.J., G. van der Kraak, A.P. Farrell and C.J. Brauner [eds.]. Fish Physiology. Academic Press, New York, NY, USA.

Castillo, J., M. Teles, S. Mackenzie and L. Tort. 2009. Stress-related hormones modulate cytokine expression in the head kidney of gilthead seabream (*Sparus aurata*). Fish Shellfish Immun. 27: 493–499.

Chandrasekar, G., G. Lauter and G. Hauptmann. 2007. Distribution of corticotropin-releasing hormone in the developing zebrafish brain. J. Comp. Neurol. 505: 337–351.

Chrousos, G.P. and P.W. Gold. 1992. The concepts of stress and stress system disorders: Overview of physical and behavioral homeostasis. J. Am. Med. Ass. 267: 1244–1252.

Costas, B., C. Aragão, J.M. Mancera, M.T. Dinis and L.E.C. Conceição. 2008. High stocking density induces crowding stress and affects amino acid metabolism in Senegalese sole *Solea senegalensis* (Kaup 1858) juveniles. Aquac. Res. 39: 1–9.

Costas, B., L.E.C. Conceição, C. Aragão, J.A. Martos, I. Ruiz-Jarabo, J.M. Mancera et al. 2011a. Physiological responses of Senegalese sole (*Solea senegalensis* Kaup 1858) alter stress challenge: Effects of non-specific immune parameters, plasma free amino acids and energy metabolism. Aquaculture 316: 68–76.

Costas, B., L.E.C. Conceição, J. Dias, B. Novoa, A. Figueras and A. Afonso. 2011b. Dietary arginine and repeated handling increase disease resistance and modulate innate immune mechanisms of Senegalese sole (*Solea senegalensis* Kaup, 1858). Fish Shellfish Immun. 31: 838–847.

Costas, B., C. Aragão, J.L. Soengas, J.M. Míguez, P. Rema, J. Dias et al. 2012. Effects of dietary amino acids and repeated handling on stress response and brain monoaminergic neurotransmitters in Senegalese sole (*Solea senegalensis* Kaup, 1858) juveniles. Comp. Biochem. Physiol. A 161: 18–26.

Costas, B., I. Simões, M. Castro-Cunha and A. Afonso. 2013a. Antimicrobial responses of Senegalese sole (*Solea senegalensis*) primary head-kidney leucocytes against *Tenacibaculum maritimum*. Fish Shellfish Immun. 34: 1702–1703.

Costas, B., P.C.N.P. Rêgo, I. Simões, J.F. Marques, M. Castro-Cunha and A. Afonso. 2013b. Cellular and humoral immune responses of Senegalese sole (*Solea senegalensis* Kaup, 1858) following challenge with two *Photobacterium damselae* subsp. *piscicida* strains from different geographical origins. J. Fish Dis. 36: 543–553.

Costas, B., C. Aragão, J. Dias, A. Afonso and L.E.C. Conceição. 2013c. Interactive effects of a high quality protein diet and high stocking density on the stress response and some innate immune parameters of Senegalese sole *Solea senegalensis*. Fish Physiol. Biochem. 39: 1141–1151.

Costas, B., I. Simões, M. Castro-Cunha and A. Afonso. 2014. Non-specific immune responses of Senegalese sole, *Solea senegalensis* (Kaup), head-kidney leucocytes against *Tenacibaculum maritimum*. J. Fish Dis. 37: 765–769.

de Groef, B., S. van der Geyten, V.M. Darras and E.R. Kühn. 2006. Role of corticotropin-releasing hormone as a thyrotropin-releasing factor in non-mammalian vertebrates. Gen. Comp. Endocrinol. 146: 62–68.

de Jesus, E.G., T. Hirano and Y. Inui. 1990. Cortisol enhances the stimulating action of thyroid hormones on dorsal fin-ray resorption of flounder larvae *in vitro*. Gen. Comp. Endocrinol. 79: 167–173.

de Souza, F.S.J., V.F. Bumaschny, M.J. Low and M. Rubinstein. 2005. Subfunctionalization of expression and peptide domains following the ancient duplication of the proopiomelanocortin gene in teleost fishes. Mol. Biol. Evol. 22: 2417–2427.

Dhabhar, F.S. 2009. A hassle a day may keep the pathogens away: The fight-or-flight stress response and the augmentation of immune function. Integr. Comp. Biol. 49: 215–236.

Dickmeis, T. 2009. Glucocorticoids and the circadian clock. J. Endocrinol. 200: 3–22.

Eales, J.G. 2006. Modes of action and physiological effects of thyroid hormones in fish. pp. 767–808. *In*: Reinecke, M., G. Zaccone and B.G. Kapoor [eds.]. Fish Endocrinology. Science Publishers, Enfield, NH, USA.

Ellis, A.E. 1999. Immunity to bacteria in fish. Fish Shellfish Immun. 9: 291–308.

Ellis, A.E. 2001. Innate host defense mechanisms of fish against viruses and bacteria. Dev. Comp. Immunol. 25: 827–839.

Ellis, T., H.Y. Yildiz, J. López-Olmeda, M.T. Spedicato, L. Tort, Ø. Øverli et al. 2012. Cortisol and finfish welfare. Fish Physiol. Biochem. 38: 163–188.

Engeland, W.C. and M.M. Arnhold. 2005. Neural circuitry in the regulation of adrenal corticosterone rhythmicity. Endocrine 28: 325–332.

Fast, M.D., S. Hosoya, S.C. Johnson and L.O.B. Afonso. 2008. Cortisol response and immune-related effects of Atlantic salmon (*Salmo salar* Linnaeus) subjected to short- and long-term stress. Fish Shellfish Immun. 24: 194–204.

Fernández-Trujillo, M.A., J. Porta, M. Manchado, J.J. Borrego, M.C. Alvarez and J. Bejar. 2008a. c-Lysozyme from Senegalese sole (*Solea senegalensis*): cDNA cloning and expression pattern. Fish Shellfish Immun. 25: 697–700.

Fernández-Trujillo, M.A., E. Garcia-Rosado, M.C. Alonso, J.J. Borrego, M.C. Álvarez, C. Infante et al. 2008b. *In vitro* inhibition of sole aquabirnavirus by Senegalese sole Mx. Fish Shellfish Immun. 24: 187–193.

Fernández-Trujillo, M.A., P. Ferro, E. Garcia-Rosado, C. Infante, M.C. Alonso, J. Bejar et al. 2008c. Poly I:C induces Mx transcription and promotes an antiviral state against sole aquabirnavirus in the flatfish Senegalese sole (*Solea senegalensis* Kaup). Fish Shellfish Immun. 24: 279–285.

Fischer, W.H., D.P. Behan, M. Park, E. Potter, P.J. Lowry and W. Vale. 1994. Assignment of disulfide bonds in corticotropin-releasing factor-binding protein, J. Biol. Chem. 269: 4313–4316.

Flik, G., P.H.M. Klaren, E.H. van den Burg, J.R. Metz and M.O. Huising. 2006. CRF and stress in fish. Gen. Comp. Endocrinol. 146: 36–44.

Gesto, M., R. Álvarez-Otero, M. Conde-Sieira, C. Otero-Rodiño, S. Usandizaga, J.L. Soengas et al. 2016. A simple melatonin treatment protocol attenuates the response to acute stress in the sole *Solea senegalensis*. Aquaculture 452: 272–282.

Geven, E.J.W., F. Verkaar, G. Flik and P.H.M. Klaren. 2006. Experimental hyperthyroidism and central mediators of stress axis and thyroid axis activity in common carp (*Cyprinus carpio* L.). J. Mol. Endocrinol. 37: 443–452.

Grandin, T. 2015. An introduction to implementing an effective animal welfare program. *In*: Improving Animal Welfare: a practical approach. 2nd edition (T. Grandin). Colorado State University, USA.

Haus, E. 2007. Chronobiology in the endocrine system. Adv. Drug Deliv. Rev. 59: 985–1014.

Herrera, M., C. Aragão, I. Hachero, I. Ruiz-Jarabo, L. Vargas-Chacoff, J.M. Mancera et al. 2012. Physiological short-term response to sudden salinity change in the Senegalese sole (*Solea senegalensis*). Fish Physiol. Biochem. 38: 1741–1751.

Huntingford, F., C. Adams, V.A. Braithwaite, S. Kadri, T.G. Pottinger, P. Sandoe and J.F. Turnbull. 2006. Current issues in fish welfare. J. Fish Biol. 70: 1311–1316.

Klaren, P.H.M., Y.S. Wunderink, M. Yúfera, J.M. Mancera and G. Flik. 2008. The thyroid gland and thyroid hormones in Senegalese sole (*Solea senegalensis*) during early development and metamorphosis. Gen. Comp. Endocrinol. 155: 686–694.

Kühn, E.R., S. Corneillie and F. Ollevier. 1986. Circadian variations in plasma osmolality, electrolytes, glucose, and cortisol in carp (*Cyprinus carpio*). Gen. Comp. Endocrinol. 61: 459–468.

Kühn, E.R., K.L. Geris, S. Van Der Geyten, K.A. Mol and V.M. Darras. 1998. Inhibition and activation of the thyroidal axis by the adrenal axis in vertebrates. Comp. Biochem. Physiol. A 120: 169–174.

Lankford, S.E., T.E. Adams and J.J. Cech. 2003. Time of day and water temperature modify the physiological stress response in green sturgeon, *Acipenser medirostris*. Comp. Biochem. Physiol. A 135: 291–302.

López-Olmeda, J.F., B. Blanco-Vives, I.M. Pujante, Y.S. Wunderink, J.M. Mancera and F.J. Sánchez-Vázquez. 2013. Daily rhythms in the hypothalamus-pituitary-interrenal axis and acute stress responses in a teleost flatfish, *Solea senegalensis*. Chronobiol Int. 30: 530–539.

López-Olmeda, J.F., I.M. Pujante, B. Blanco-Vives, L.S. Costa, A. Galal-Khallaf, J.M. Mancera and F.J. Sánchez-Vázquez. 2016. Daily rhythms in the somatotropic axis of Senegalese sole (*Solea senegalensis*): The time of day influences the response to GH administration. Chronobiol. Int. 33: 257–267.

López-Patiño, M.A., M. Conde-Sieira, M. Gesto, M. Librán-Pérez, J.L. Soengas and J.M. Míguez. 2013. Melatonin partially minimizes the adverse stress effects in Senegalese sole (*Solea senegalensis*). Aquaculture 388–391: 165–172.

Mabrok, M., M. Machado, C.R. Serra, A. Afonso, L.M.P. Valente and B. Costas. 2016. Tenacibaculosis induction in the Senegalese sole (*Solea senegalensis*) and studies of *Tenacibaculum maritimum* survival against host mucus and plasma. J. Fish Dis. in press. DOI:10.1111/jfd.12483.

McEwen, B.S. and J.C. Wingfield. 2010. What is in a name? Integrating homeostasis, allostasis and stress. Horm. Behav. 7: 105–111.

Mohawk, J.A. and T.M. Lee. 2005. Restraint stress delays reentrainment in male and female diurnal and nocturnal rodents. J. Biol. Rhythms 20: 245–256.

Morais, S., C. Aragão, E. Cabrita, L.E.C. Conceição, M. Constenla, B. Costas et al. 2014. New developments and biological insights into the farming of *Solea senegalensis* reinforcing its aquaculture potential. Rev. Aquacult. 6: 1–37.

Mota-Silva, P.I., C.I.M. Martins, S. Engrola, G. Marino, O. Overli and L. Conceicao. 2010. Individual differences in cortisol levels and behaviour of Senegalese sole (*Solea senegalensis*) juveniles: Evidence for coping styles. App. Anim. Behav. 124: 75–81.

Osse, J.W.M. and J.G.M. Van den Boogaart. 1997. Size of flatfish larvae at transformation, functional demands and historical constraints. J. Sea Res. 37: 229–239.

Osuna-Jiménez, I., T.D. Williams, M.J. Prieto-Álamo, N. Abril, J.K. Chipman and C. Pueyo. 2009. Immune- and stress-related transcriptomic responses of *Solea senegalensis* stimulated with lipopolysaccharide and copper sulphate using heterologous cDNA microarrays. Fish Shellfish Immun. 26: 699–706.

Pankhurst, N.W. 2011. The endocrinology of stress in fish: An environmental perspective. Gen. Comp. Endocrinol. 170: 265–275.

Pavlidis, M., L. Greenwood, M. Paalavuo, H. Mölsa and J.T. Laitinen. 1999. The effect of photoperiod on diel rhythms in serum melatonin, cortisol, glucose, and electrolytes in the common dentex, *Dentex dentex*. Gen. Comp. Endocrinol. 113: 240–250.

Pepels, P.P.L.M. and P.H.M. Balm. 2004. Ontogeny of corticotropin-releasing factor and of hypothalamic-pituitary-interrenal axis responsiveness to stress in tilapia (*Oreochromis mossambicus*; Teleostei). Gen. Comp. Endocrinol. 139: 251–265.

Pickering, A.D. and T.G. Pottinger. 1983. Seasonal and diel changes in plasma cortisol levels of the brown trout, *Salmo truta* L. Gen. Comp. Endocrinol. 49: 232–239.

Planas, J., J. Gutiérrez, J. Fernández, M. Carrillo and P. Canals. 1990. Annual and daily variations of plasma cortisol in sea bass, *Dicentrarchus labrax* L. Aquaculture 91: 171–178.

Ponce, M., E. Salas-Leiton, A. Garcia-Cegarra, A. Boglino, O. Coste, C. Infante et al. 2011. Genomic characterization, phylogeny and gene regulation of g-type lysozyme in sole (*Solea senegalensis*). Fish Shellfish Immun. 31: 925–937.

Power, D.M., L. Llewellyn, M. Faustino, M.A. Nowell, B.Th Björnsson, I.E. Einarsdottir, A.V.M. Canario and G.E. Sweeney. 2001. Thyroid hormones in growth and development of fish. Comp. Biochem. Physiol. C 130: 447–459.

Rasines, I., M. Gómez, I. Martín, C. Rodríguez, E. Mañanós and O. Chereguini. 2013. Artificial fertilisation of cultured Senegalese sole (*Solea senegalensis*): Effects of the time of day of hormonal treatment on inducing ovulation. Aquaculture 392–395: 94–97.

Reid, S.G., M.M. Vijayan and S.F. Perry. 1996. Modulation of catecholamine storage and release by the pituitary–interrenal axis in the rainbow trout (*Oncorhynchus mykiss*). J. Comp. Physiol. B 165: 665–676.

Rotllant, J. and L. Tort. 1997. Cortisol and glucose responses after acute stress by net handling in the sparid red porgy previously subjected to crowding stress. J. Fish Biol. 51: 21–28.

Rotllant, J., P.H.M. Balm, N.M. Ruane, J. Pérez-Sánchez, S.E. Wendelaar Bonga and L. Tort. 2000. Pituitary proopiomelanocortin-derived peptides and hypothalamus-pituitary-interrenal axis activity in gilthead sea bream (*Sparus aurata*) during prolonged crowding stress: Differential regulation of adrenocorticotropin hormone and alpha-melanocyte-stimulating hormone release by corticotropin-releasing hormone and thyrotropin-releasing hormone. Gen. Comp. Endocrinol. 119: 152–163.

Ruane, N.M., P. Makridis, P.H.M. Balm and M.T. Dinis. 2005. Skin darkness is related to cortisol, but not MSH, content in post-larval *Solea senegalensis*. J. Fish Biol. 67: 577–581.

Saeij, J.P.J., L.B.M. Verburg-van Kemenade, W.B. van Muiswinkel and G.F. Wiegertjes. 2003. Daily handling stress reduces resistance of carp to *Trypanoplasma borreli*: *in vitro* modulatory effects of cortisol on leukocyte function and apoptosis. Dev. Comp. Immunol. 27: 233–245.

Sage, D., D. Maurel and O. Bosler. 2002. Corticosterone-dependent driving influence of the suprachiasmatic nucleus on adrenal sensitivity to ACTH. Am. J. Physiol Endocrinol. Metab. 282: E458–E465.

Salas-Leiton, E., V. Anguís, M. Manchado and J.P. Cañavate. 2008. Growth, feeding and oxygen consumption of Senegalese sole (*Solea senegalensis*) juveniles stocked at different densities. Aquaculture 285: 84–89.

Salas-Leiton, E., V. Anguis, B. Martín-Antonio, D. Crespo, J.V. Planas, C. Infante et al. 2010. Effects of stocking density and feed ration on growth and gene expression in the Senegalese sole (*Solea senegalensis*): Potential effects on the immune response. Fish Shellfish Immun. 28: 296–302.

Salas-Leiton, E., O. Coste, E. Asensio, C. Infante, J.P. Cañavate and M. Manchado. 2012. Dexamethasone modulates expression of genes involved in the innate immune system, growth and stress and increases susceptibility to bacterial disease in Senegalese sole (*Solea senegalensis* Kaup, 1858). Fish Shellfish Immun. 32: 769–778.

Singley, J.A. and W. Chavin. 1975. Serum cortisol in normal goldfish (*Carassius auratus* L.). Comp. Biochem. Physiol. 50A: 77–82.

Segner, H., H. Sundh, K. Buchmann, J. Douxfils, K. Snuttan, C. Mathieu, N. Ruane, F. Jufelt, H. Toften and L. Vaughan. 2012. Health of farm fish: its relation to fish welfare and its utility as welfare indicators. Fish. Physiol. Biochem. 38: 85–105.

Stouthart, A.J., E.C. Lucassen, F.J. van Strien, P.H. Balm, R.A. Lock and S.E. Wendelaar Bonga. 1998. Stress responsiveness of the pituitary-interrenal axis during early life stages of common carp (*Cyprinus carpio*). J. Endocrinol. 157: 127–137.

Tort, L. 2011. Stress and immune modulation in fish. Dev. Comp. Immunol. 35: 1366–1375.

van der Geyten, S., K.A. Mol, W. Pluymers, E.R. Kühn and V.M. Darras. 1998. Changes in plasma T3 during fasting/refeeding in tilapia (*Oreochromis niloticus*) are mainly regulated through changes in hepatic type II iodothyronine deiodinase. Fish Physiol. Biochem. 19: 135–143.

Vera, L.M., A. Montoya, I.M. Pujante, J. Pérez-Sánchez, J.A. Calduch-Giner, J.M. Mancera et al. 2014. Acute stress response in gilthead sea bream (*Sparus aurata* L.) is time-of-day dependent: Physiological and oxidative stress indicators. Chronobiol. Int. 31: 1051–1061.

Vijayan, M.M., P.A. Flett and J.F. Leatherland. 1988. Effect of cortisol on the *in vitro* hepatic conversion of thyroxine to triiodothyronine in brook charr (*Salvelinus fontinalis* mitchill). Gen. Comp. Endocrinol. 70: 312–318.

Walpita, C.N., S.V.H. Grommen, V.M. Darras and S. van der Geyten. 2007. The influence of stress on thyroid hormone production and peripheral deiodination in the Nile tilapia (*Oreochromis niloticus*). Gen. Comp. Endocrinol. 150: 18–25.

Wendelaar Bonga, S.E. 1997. The stress response in fish. Physiol. Rev. 77: 591–625.

Weyts, F.A., G. Flik, J.H. Rombout and B.M. Verburg-van Kemenade. 1998a. Cortisol induces apoptosis in activated B cells, not in other lymphoid cells of the common carp, *Cyprinus carpio* L. Dev. Comp. Immunol. 22: 551–562.

Weyts, F.A., G. Flik and B.M. Verburg-van Kemenade. 1998b. Cortisol inhibits apoptosis in carp neutrophilic granulocytes. Dev. Comp. Immunol. 22: 563–572.

Wunderink, Y.S. 2011. Interaction between the thyroid system and the stress system in the Senegalese sole (*Solea senegalensis*): implications for its cultivation. Doctoral Thesis. University of Cádiz.

Wunderink, Y.S., S. Engels, S. Halm, M. Yúfera, G. Martínez-Rodríguez, G. Flik et al. 2011. Chronic and acute stress responses in Senegalese sole (*Solea senegalensis*): The involvement of cortisol, CRH and CRH-BP. Gen. Comp. Endocrinol. 171: 203–210.

Wunderink, Y.S., E. de Vrieze, J.R. Meltz, S. Halm, G. Martínez-Rodríguez, G. Flik et al. 2012a. Subfunctionalization of POMC paralogues in Senegalese sole (*Solea senegalensis*). Gen. Comp. Endocrinol. 175: 407–415.

Wunderink, Y.S., G. Martínez-Rodríguez, M. Yúfera, I. Martín Montero, G. Flik, J.M. Mancera et al. 2012b. Food deprivation induces chronic stress and affects thyroid hormone metabolism in Senegalese sole (*Solea senegalensis*) post-larvae. Comp. Biochem. Physiol. 162A: 317–322.

B-4.2
Ecotoxicology

Montserrat Solé

1. Introduction

Ecotoxicology studies embrace different scientific disciplines (e.g., chemistry, toxicology, biochemistry, physiology, endocrinology and ecology), the ultimate goal of which is to relate exposure to toxic substances in particular individuals of a given species to ecological consequences for its population. These responses can be measured at several levels of biological organisation (i.e., molecules, cells, tissues, organs, whole organisms). However, this chapter will focus mainly on biochemical responses (mostly enzyme activities) at the individual level that can then be translated to upper levels of biological organisation and in this way act as early warning signals of ecological disturbances and risk assessment (Mearns et al. 2011, Walker et al. 2012). From now on, these biochemical responses shall be termed *biomarkers* which have been defined in a comprehensive fish review (van der Oost et al. 2003). This chapter will also address these appointed biomarkers in *Solea* spp. (including *Solea solea* and *Solea senegalensis*). Responses at other levels of biological hierarchy will only be discussed briefly when referring to ecotoxicology studies using these same species. Other biological aspects will be discussed in other chapters: immunology (B-4.1), histology (B-4.3), physiology (B-4.4) and molecular/genetic (B-5.1).

The choice of a good model species is of extreme importance in ecotoxicology and benthic fish are proposed as ideal sentinels in pollution monitoring programs for assessing local sediment-bound chemical exposures (Lehtonen et al. 2006, Lerebours et al. 2014). Some of the requirements for good sentinels include reduced mobility, responsiveness to pollutants and good adaptability to laboratory conditions. *Solea* spp. meets all these requirements and, therefore, has been the sentinel fish of choice in many recent field and laboratory surveys. In addition, because of their high economic interest in fisheries and aquaculture, sole fish are ideal candidates to study. Therefore, much research has focused on improving knowledge on their biology and optimal cultivation conditions (Imsland et

Institut de Ciencies del Mar (ICM-CSIC). Passeig Marítim de la Barceloneta 37-49 08003 Barcelona, Spain.
Email: msole@icm.csic.es

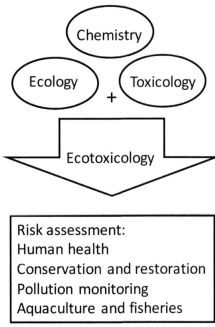

Fig. 1. Some disciplines integrating ecotoxicology studies. Their final goal in risk assessment: implications on issues of human concern from health to economic aspects.

al. 2003). A pre-requisite before using *Solea* spp. in biological monitoring is to characterise the influence of environmental and biological variables in the modulation of the parameters currently selected as biomarkers. This has been the topic of several studies from larvae to juveniles and adults of these species (Fernandez-Diaz et al. 2006, Manchado et al. 2008, Sánchez-Nogué et al. 2013, Solé et al. 2004, 2012, 2014).

2. Bioaccumulation studies

To relate chemical exposures to biological effects is an essential goal in ecotoxicology. Past evidence has shown that chemical analysis alone was unable to predict the consequences of exposures to either single or mixtures of pollutants, as they usually occur in the environment. Thus, nowadays, the inclusion of a battery of biomarkers is encouraged as it complements and integrates the consequences of these exposures to the organism's integrity (Cajaraville et al. 2000, Kerambrun et al. 2012). Nonetheless, due to the high consumption of *Solea* spp. and therefore a human health concern, numerous studies carried out with this species have been based mostly on chemical analysis. That is, studies on the presence of organic pollutants and metals from the Gulf of Lyons and the Catalan Coast (NW Mediterranean) have been conducted on *Solea* spp. (Dierking et al. 2009, Siscar et al. 2013). On the Adriatic coast of Italy (Central Mediterranean), *S. solea* has been surveyed for accumulation of alkylphenols (Ferrara et al. 2005) and metals (Tramati et al. 2011). In the Eastern Mediterranean, studies on bioaccumulated metals in *S. vulgaris/ S. solea* correspond to the Izmir (Kucuksezgin et al. 2006) and Iskenderun (Cogun et al. 2005) bays of Turkey. The Tunisian coast, North Africa (Barhoumi et al. 2009, Ben Ameur et al. 2013b), and a coastal lagoon in Egypt (Mohamed and Gad 2008) have been screened for bioaccumulation of organic pollutants and metals in *S. solea*. In the Southern Iberian Peninsula, *S. senegalensis* have been selected for metals in the bay of Cadiz (Galindo et al. 2012, Usero et al. 2004). Nearby, in the same region, *S. vulgaris/S. solea* was used as sentinel in the mouth of the Tinto and Odiel rivers for presence of organochlorine compounds (Bordajandi et al. 2006) whereas *S. senegalensis* was used for presence in metals (Vicente-Martorell et al. 2009). Also in the Spanish Iberian Peninsula, the Galician coast of Spain, NE Atlantic, has been surveyed for bioaccumulated metals (Mhadhbi et al. 2012) and the coast of Portugal for PCBs alone in Ria de Aveiro and Mondego estuary (Antunes and Gil 2002, Baptista et al. 2013), the Tagus estuary for metals (Franca et al. 2005), all using *S. senegalensis* as sentinel. By contrast, *S. solea/S. vulgaris* was the sentinel for assessing bioaccumulated metals and PCBs in the Gironde estuary of France (Bodin et al. 2014, Durrieu et al. 2005), polychlorinated biphenyls (PCBs), polybrominated diphenyl ethers (PBDEs) and organochlorinated pesticides in the Scheldt estuary (Van Ael et al. 2012, 2014), and hexabromocyclododecane (HBCD) (Janak et al. 2005) and Hg (Baeyens et al. 2003) in the North Sea. Table 1 reports on chemistry studies carried out with *Solea* spp. and a map (Fig. 2) illustrates the geographical location of these studies.

Recent and comprehensive ecotoxicological studies on the Portuguese and Spanish coasts have been carried out combining chemical analysis with several biochemical markers of exposure and effect (mostly histological responses) in *S. senegalensis* and/or *S. solea* (Table 2). They all reveal a capacity to bioaccumulate environmental contaminants via water, sediment and food transfer. The BAF has been revealed for some chemicals but, in most cases, the levels of contaminants reached in fish are lower than those predicted to exhort toxicological consequences or to be of human health concern in the case of organochlorinated compounds (Ben Ameur et al. 2013, Bodin et al. 2014), metals (Galindo et al. 2012), PAHs (Solé et al. 2013) or PAHs and metals (Fonseca et al. 2015), following approved guideline indices.

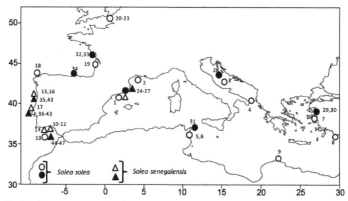

Fig. 2. Map from the Mediterranean and Atlantic regions where *S. solea* and *S. senegalensis* coexist. Filled round symbols indicate field biomarker studies on *S. solea* combining biomarkers and chemical exposures (either measured in water, sediment, and biota or estimated by types of discharges). Filled triangle symbols indicate same type of studies on *S. senegalensis*. Blank symbols indicate bioaccumulation studies alone in *S. solea* (round) and *S. senegalensis* (triangle). Codes: 1–23 correspond to chemistry data only (Table 1). Codes: 24–47 to studies combining biomarker and chemical analysis (Table 3).

Table 1. Field studies based on bioaccumulation of metals and organic pollutants in *Solea* spp. and its map code location in Fig. 2.

Region	Area	Species	Targeted compounds	Reference	Map Code
Western Mediterranean	Catalan Coast, Spain	*S. solea & S. senegalensis*	Metals	Siscar et al. 2013	1
	Gulf of Lyons, France	*S. solea*	PCBs, DDE, Hg	Dierking et al. 2009	2
Central Mediterranean	Adriatic Coast, Italy	*S. solea*	Alkylphenols	Ferrara et al. 2005	3
	Acquatina, Puglia, Italy	*S. solea*	Metals	Tramati et al. 2011	4
	Gulf of Gabes, Tunisian Coast	*S. solea (=vulgaris)*	Cd	Barhoumi et al. 2009	5
		S. solea	PCBs, DDTs, PBDEs, pesticides	Ben Ameur et al. 2013	6
Eastern Mediterranean	Izmir bay, Turkey	*S. solea (=vulgaris)*	metals	Kucuksezgin et al. 2006	7
	Iskenderun bay, Turkey	*S. solea*	Metals	Cogun et al. 2005	8
	Egypt	*S. solea*	Metals and organic pollutants	Mohamed and Gad 2008	9
Atlantic- South Iberian Peninsula	Bay of Cadiz, Spain	*S. senegalensis*	metals	Riba et al. 2004, Usero et al. 2004, Galindo et al. 2012	10,11,12
	Tinto and Odiel estuary, Spain	*S. solea (=vulgaris)*	Organochlorine compounds	Bordajandi et al. 2006	13
	Tinto and Odiel estuary, Spain	*S. senegalensis*	Metals	Vicente-Martorell et al. 2009	14
Atlantic- Central Iberian Peninsula	Ria Aveiro and Mondego estuary, Portugal	*S. senegalensis*	PCBs	Antunes and Gil 2002, Baptista et al. 2013	15 16
	Tagus estuary, Portugal	*S. senegalensis*	metals	Francà et al. 2005	17
Atlantic-North Iberian Peninsula	Galicia Coast, Spain	*S. solea (=vulgaris)*	Metals	Mhadhbi et al. 2012	18
Atlantic-North Sea	Gironde estuary, France	*S. solea*	Metals	Durrieu et al. 2005	19
	Scheldt estuary, Belgium	*S. solea*	PCBs, PBDEs, OCl pesticides	Van Ael et al. 2012, 2014	20,21
	Scheldt estuary, Belgium	*S. solea*	Hg	Baeyens et al. 2003	22
	Scheldt estuary, Belgium	*S. solea*	HBCD	Janak et al. 2005	23

Table 2. Descriptive analysis of enzymatic activities in the S9 fraction of different tissues according to species and sex of adults collected in several low to moderately polluted sites of the NW Mediterranean. The number of individuals of each species and sex used is indicated in brackets. Units: [1]nmol/min/mg prot, [2]pmol/min/mg prot, [3]fmol/min/mg prot, [4]μmol/min/mg prot and [5] nmol MDA/g.w.w.

Parameter	Tissue	*Solea solea*		*Solea senegalensis*	
		Male (18)	Female (30)	Male (16)	Female (7)
Size (cm)		31.7 ± 0.7	38.9 ± 0.7	36.5 ± 0.8	39.0 ± 1.5
Weight (g)		270.4 ± 18.2	617.2 ± 35.4	447.5 ± 26.1	556.4 ± 75.2
AChE[1]	Brain	41.2 ± 3.0	44.9 ± 2.8	36.8 ± 2.5	34.8 ± 4.1
	Muscle	8.7 ± 0.9	7.2 ± 0.4	4.6 ± 0.4	4.7 ± 0.5
	Gills	3.8 ± 0.4	3.2 ± 0.2	2.3 ± 0.2	2.7 ± 0.5
	Gonad	9.8 ± 3.5	1.0 ± 0.1	12.3 ± 1.2	1.3 ± 0.3
CbE[1]	Liver	99.7 ± 7.2	97.7 ± 7.1	54.9 ± 5.2	48.2 ± 8.8
	Gills	6.2 ± 0.3	6.4 ± 0.3	8.3 ± 0.4	8.3 ± 1.0
	Gonad	7.0 ± 0.5	1.9 ± 0.1	10.9 ± 0.4	4.3 ± 0.5
EROD	Liver[2]	2.7 ± 0.5	1.1 ± 0.3	1.1 ± 0.2	0.3 ± 0.1
	Gills[3]	103.0 ± 20.1	180.1 ± 26.5	95.1 ± 15.7	47.2 ± 15.5
GST[1]	Liver	355.4 ± 45.0	399.0 ± 33.7	361.7 ± 19.7	180.4 ± 28.4
	Gills	61.4 ± 6.2	102.0 ± 6.9	54.4 ± 2.5	55.4 ± 6.1
	Gonad	25.8 ± 2.0	33.5 ± 2.6	32.3 ± 1.2	23.2 ± 2.6
CAT[4]	liver	247.2 ± 10.9	237.7 ± 14.4	261.9 ± 18.9	209.3 ± 34
	gills	19.8 ± 2.2	17.4 ± 1.7	15.8 ± 2.3	6.5 ± 0.7
t-GPX[1]	liver	113.1 ± 9.0	112.8 ± 5.2	82.9 ± 6.0	84.5 ± 13.3
	gills	107.2 ± 4.9	100.3 ± 3.4	94.3 ± 4.0	105.7 ± 12.8
GR[1]	liver	18.1 ± 2.6	17.9 ± 1.6	12.3 ± 1.0	14.1 ± 1.3
	gills	68.9 ± 7.2	86.4 ± 7.0	47.1 ± 1.3	47.9 ± 2.7
LP[5]	gills	58.5 ± 9.1	46.3 ± 5.6	71.7 ± 8.4	68.7 ± 11.1
	muscle	22.8 ± 2.0	22.5 ± 1.7	9.7 ± 2.2	4.8 ± 1.0

3. Toxicity assessment

The toxic consequences of chemical exposures can be measured in a number of ways and using several endpoints at different levels of biological organisation: from cell to organism and population. As indicated above, this chapter will focus mainly on sublethal biochemical responses, mostly enzymatic but also molecular, termed from now on as biomarkers, in the frame of environmental toxicology and using the flatfish sole, *Solea* spp. as model organism.

3.1 Biomarkers

Biomarkers are defined as parameters that can be measured and show a quantitative direct relationship with the magnitude of chemical/toxicant exposure. Nonetheless, under natural conditions, organisms are exposed to complex mixtures and the resulting integrated response can be synergistic, antagonist or simply additive (agonist). They can also be classed as (1) biomarkers of exposure when they specifically respond to a chemical class, (2) biomarkers of effect when they indicate a consequence or damage resulting from an exposure and (3) biomarkers of susceptibility, defined as genetic factors, nutritional status, health and life stage that determine a variable degree of vulnerability to chemical exposures. Nowadays, as the number of studies progress, there is an increasing body of evidence to suggest that many chemicals can exert a specific action, as well as a non-specific one, depending on the threshold of the exposures, the tissues targeted and the physiological/developmental status of the organism.

Organic pollutants, and particularly those more hydrophobic, once inside the organism, need to be biotransformed to make them more hydrophilic and, therefore, more readily excretable. This is mostly undertaken by the action of hepatic enzymes involved in phase I metabolism such as cytochrome P450 (CYPs) and carboxylesterases (CbEs), representing oxidation and hydrolysis reactions, respectively. Phase II reactions comprise the addition of an endogenous moiety (e.g., UDP or GSH) over the resulting phase I metabolite or directly over the parental xenobiotic. The main phase II reactions in fish are microsomal uridine diphosphate glucuronosyltransferase (UDPGT) and cytosolic glutathione *S*-transferase (GST). Also, as a result of phase I reactions (oxidation) and aerobic processes, un-stable oxygen molecules or reactive oxygen species (ROS) can be formed. They can be neutralized by the action of antioxidant enzymes to prevent an oxidative stress situation. If the antioxidant system is insufficient to prevent damage, the oxidation of vital molecules (e.g., DNA, proteins, lipids) can take place, causing DNA damage, protein carbonyl formation and lipid peroxidation, all currently used as effect biomarkers. In fact, as a sign of genotoxicity, DNA adducts (Amat et al. 2006, Wessel et al. 2010) and micronuclei formation (Arslan et al. 2015) are a frequent topic of research as effect markers in studies with sole fish. Moreover, in *Solea* spp., as a consequence of hepatic metabolism, biliary metabolites can be formed and used as unequivocal signs of chemical exposure to either polycyclic aromatic hydrocarbons (PAHs) (Hillenweck et al. 2008, Solé et al. 2013, Wessel et al. 2010), PBDEs (Munschy et al. 2010), alcohol polyethoxilates (AEO) (Alvarez-Munoz et al. 2014) or pharmaceuticals (Aceña et al. 2017).

Another frequently used biomarker of neurotoxic chemical exposures (mostly organophosphorus pesticides and carbamates) is the inhibition of the enzyme acetylcholinesterase (AChE) as a sign of neural transmission disruption (Fulton and Key 2001). In recent years, many more chemicals are suggested to have a neurotoxic action also in *Solea* species, such as surfactants, metals and other chemicals (Lopez-Galindo et al. 2010a,b, Oliva et al. 2012a). Moreover, other esterases, such as hepatic CbEs are also susceptible to inhibition by OP pesticides and other contaminants, in some cases by binding to the chemical in a more specific manner than AChE (Wheelock et al. 2008). Thus, the action of CbE has been regarded as a protective mechanism to prevent AChE inhibition.

Specifically, metallothionein (MT) are low molecular weight cytosolic proteins involved in metal transport and homeostasis within an organism (Viarengo et al. 1999). They are also selected as biomarkers of metal exposure and, more importantly, inactivation of toxic metals and therefore are also an efficient way of metal detoxification in *Solea* spp. (Fonseca et al. 2009, Riba et al. 2004, Siscar et al. 2013).

Presence of the egg lipoprotein vitellogenin (VTG) in male plasma has been selected as a reliable sign of endocrine disruption in gonochoristic fish (Sumpter and Jobling 1995). Due to the straight forward consequences of endocrine alterations in fish reproduction, this biomarker is regarded as having high ecological significance and its use is encouraged. Other aspects dealing with sex hormones, VTG and sole reproduction will be the topic of other Chapters (B-1.2 and B-1.3). In addition, the histological examination of key tissues (liver, gills and gonads) for alterations in their normal structure is a frequently used marker of cell damage/endocrine disruption in *Solea* spp. and will also be detailed in other chapters (B-4.3).

Stress proteins are low molecular weight proteins involved in protein folding and transportation within the cell and, in fish; they have been used in several applications as a stress response to multiple factors (Iwama et al. 1999). They are also known as heat shock proteins (HSPs) as they respond promptly to temperature variations although, due to their unspecific nature, they react to many chemical and physical variables in *Solea* spp. (Chairi et al. 2010, Solé et al. 2015). Several stress proteins (HSPs) have been characterised in different tissues and throughout development in *S. senegalensis* from a gene expression perspective (Manchado et al. 2008) and applied to assessing aquaculture rearing conditions of this same species (Salas-Leiton et al. 2010). HSPs have also been applied in *S. solea*, in relation to exposure to the xenoestrogen nonylphenol (Palermo et al. 2012a,b, Cocci et al. 2013).

Other unspecific biomarkers are haematological and immunological parameters which will be dealt with in another chapter (B-4.1) but they have also been considered in an ecotoxicological Mediterranean field study with *Solea* spp. (Solé et al. 2013) and in laboratory exposures to temperature variations as well as temperature and drug combinations (Solé et al. 2015, González-Mira et al. 2016).

3.2 Bioassays

Another way of performing toxicity assessment is through bioassays that link chemical content in a matrix (e.g., sediment) to biological effects using a multi-step and decision making progress together with a multivariate analysis approach (Choueri et al. 2010, DelValls et al. 2002). Acute and chronic tests determining sediment toxicity have frequently used sole as fish model with the final goal of implementing regulations (e.g., disposing of dredged material). Metal toxicity, due to the *Aznalcollar* mining spill, was assessed using *S. senegalensis* exposed for 7 and 14 days to polluted sediment and, in addition to metal analysis, MT was the biomarker targeted (Riba et al. 2004). Complementary chronic assays that include multiple sensible sublethal endpoints are also desirable for environmental quality assessment and regulatory purposes (Jimenez-Tenorio et al. 2007). The latter study comprises field and laboratory aspects using sediment from polluted Spanish harbours and exposing juveniles of *S. senegalensis* under controlled laboratory conditions and measuring several biomarkers of exposure/effect in two tissues: liver and gills (Jimenez-Tenorio et al. 2007). Toxicity assessment of PAHs and metals present in fuel from the *Prestige* spill was also conducted using this approach in *S. senegalensis* (Salamanca et al. 2008). A comprehensive study of Costa et al. (2012b) integrates a multibiomarker scheme, including "omic", transcription and biomarker responses for toxicity assessment due to complex mixtures of chemicals from low to moderately polluted sediments either from field situations or under laboratory exposures. The authors conclude that in these conditions, effect biomarkers (histopathology and DNA damage) were more field site discriminatory and that lab exposures tended to overestimate toxicity in respect of a more realistic scenario provided by field surveys. More recently, Fonseca et al. (2015) also used an integrative approach including several biomarkers, chemical characterisation and other anthropogenic indices, in juveniles of *S. senegalensis* as fish model and with the goal of assessing habitat quality of estuarine breeding areas.

4. Adaptability and susceptibility of *Solea* spp. to stressors

The two selected sole species, *S. solea* and *S. senegalensis* that coexist in Atlantic and Mediterranean waters, also differ in genetic and evolutionary factors that might determine a discrepancy in terms of susceptibility to chemical exposures. To find out which parameters/characteristics render one species more adaptable than another is a key issue in species distribution and adaptability in a fluctuating environment characterised by increasing and more complex chemical exposures and in a climate change scenario (Wittmann and Portner 2013). Predicted increases in temperature, decreases in pH and more extreme variations in salinity may have an impact on this adaptable fish group. Juveniles of *Solea* spp. inhabit estuaries and, therefore, are adapted to modifications in these environmental variables. Their adults dwell in deeper waters; however, they will also have to face the predicted environmental changes in water masses (IPCC 2014). Species segregation, growth rates and habitat suitability of *S. solea* and *S. senegalensis* cohorts have been topics of research in the Tejo estuary in Portugal (Vinagre et al. 2006, 2008, 2013). Adaptability to realistic temperature variations under controlled laboratory conditions has been assessed in juveniles of *S. senegalensis* (Arjona et al. 2010, Castro et al. 2012, Siscar et al. 2014, Solé et al. 2015) and *S. solea* (Schram et al. 2013, Sureau et al. 1989), or in a particular pollutant group (e.g., PCBs) in terms of hypoxia tolerance in juveniles of *S. solea* (Cannas et al. 2013). Diet and the environmental conditions of rearing temperature experienced during early developmental stages were crucial for determining common sole adaptability to hypoxia in juveniles (Zambonino-Infante et al. 2013). Another aspect addressing the use of *S. solea* as model is the maternal transfer of persistent organic pollutants (POPs) to the eggs and its influence of larval survival using an early life stage test (Foekema et al. 2008, 2014). However, a simultaneous exposure study comparing the responses of both species under the same chemical/physical stressors has not yet been carried out. Nonetheless, a comparative approach evaluating species responses under the same field situation *in vivo* and *in vitro* was reported for juveniles (Sánchez-Nogué et al. 2013) and adults (Koenig et al. 2013, Siscar et al. 2015) of these two sympatric species. In general, it was observed that juveniles of *S. senegalensis* expressed higher IC50 values in muscular AChE after *in vitro* exposures to pesticides (less sensitive to pesticide inhibition than *S. solea*). However, *S. solea* also expressed lower IC50s (higher substrate affinity) in CbE activity in several tissues which supports a compensation role to prevent pesticide toxicity. In the case of adults, brain/muscle AChE was better protected in *S. solea* (higher IC50) also with high affinity in liver CbE for these compounds, although with some pesticide dependent peculiarities (Koenig et al. 2013, Siscar et al. 2015). Other protective mechanisms such as phase II GST were equally efficient in both species (Koenig et al. 2013) but phase I EROD activity was usually higher in *S. solea* (Koenig et al. 2013, Sánchez-Nogué et al. 2013, Siscar et al. 2015).

Baseline enzymatic defences could also "naturally" differ in both species as a function of species and sex as the main driving factors. In a recent study with adults from both species and sexes, differences in enzymatic defences and other biomarkers measured in the S10 fraction of several fish tissues could be seen. Derived from this field study, in Table 2 activities/levels for some biomarkers are described in adults collected at low and moderately polluted sites in the NW Mediterranean coast. In general, enzymatic activities were higher in *S. solea* than in *S. senegalensis*, especially when measured in gills. Gonad esterase activities were clearly sex-dependent with activities in males higher than in females. These species, differences were supported in terms of hepatic markers in a more detailed comparative study using cytosol and microsomal subcellular fractions of adult fish (Koenig et al. 2013) but also in the same S10 fraction of juveniles (Sánchez-Nogué et al. 2013).

5. The use of *Solea* spp. as model flatfish species

Flatfish are benthic species frequently used in pollution monitoring studies using an ecotoxicology approach (Akcha et al. 2003, Dupuy et al. 2015, Kopecka and Pempkowiak 2008, Lehtonen et al. 2006, Lerebours et al. 2014, Napierska et al. 2009). The particular use of sole in the Mediterranean and Atlantic region has the added value of being a species of high economic interest reared in hatchery facilities ensuring their availability for research purposes. It is also thanks to their high interest in aquaculture that there are many studies assessing the effects of rearing conditions (e.g., temperature, density, diets) on antioxidant defences, stress and other development parameters in life stages from larvae to adults (Boglino et al. 2014, Castro et al. 2012, Costas et al. 2012, Manchado et al. 2008). Nonetheless, this aquaculture aspect will not be addressed in this chapter.

Since this species is extensively employed in aquaculture production, there is a larger body of literature on the use of *S. senegalensis* as sentinel. Most of the work with this particular species has been carried out in the Southern and Western Iberian Peninsula whereas studies with *S. solea*/*S. vulgaris* mostly correspond to Northern Europe, North Africa and Eastern Mediterranean surveys (Fig. 2; Table 3). Laboratory studies using *Solea* spp. have been conducted to assess exposures to varied chemical groups such as PAHs, organochlorinated compounds, detergents, pesticides and emerging contaminants (Table 4).

Monitoring the consequences of oil spills has been assessed using *S. solea* in the Bay of Biscay (France) after the *Erika* incident (Amat et al. 2006, Claireaux et al. 2004) with DNA adduct formation being one of the most accepted biomarkers of PAH exposure. Laboratory exposure to PAHs also increased DNA adducts and it was mirrored in the biliary PAHs formed, although induction of EROD activity, usually related to PAH exposure, was not evidenced (Wessel et al. 2010). Nonetheless, EROD, other CYP related activities and conjugation enzymes were increased in specimens of *S. solea* sampled in the Adriatic Sea nearby an oil refinery with respect to others from a more distant population (Trisciani et al. 2011). Pollution due to multiple chemical sources: urban, agriculture, shipping and industry was revealed in specific hepatic EROD and CYP1A induction in *S. solea* from the Izmir bay in the Aegean Sea, Eastern Mediterranean (Arinc et al. 2001). A broader multibiomarker approach, also using multiple target tissues (liver, gills, kidney and brain) in the same sole species, was successfully applied to assessing mostly urban and industrial pollution on the Tunisia coastline (Jebali et al. 2013). This study also evidenced which tissue was more adequate for each type of biomarker, the suitability of AChE for site discrimination and that *S. solea* was an adequate sentinel. Several biomarkers in liver and gills of *S. solea* were considered in a field approach aiming to assess the effects of extractive works in the Ebre River over the marine fauna (Crespo and Solé 2016). On the other hand, histological alterations were the unique biomarker selected in a field study also with *S. solea*, conducted alongside the SE Bay of Biscay in the Basque continental shelf (Northern coast of Spain). Sole were exposed, in the field situation, to metals, PCBs and PAHs pollution gradients and the goal was to implement the European Marine Strategy Framework Directive (Cuevas et al.

Table 3. Field studies using *Solea solea* and *Solea senegalensis* as sentinel. Biomarkers evaluated and targeted chemicals.

Sentinel species	Biomarkers measured	Targeted chemicals	Reference	Location	Map code
S. solea & *S. senegalensis*	EROD, antioxidant enzymes, LPO, GST, CbE, histology, biliary PAHs	Metals, organic pollutants	Sanchez-Nogue et al. 2013, Solé et al. 2013, Siscar et al. 2013, 2015	Catalan coast, Spain	24, 25, 26, 27
S. solea	CYP-related enzymes, UDPGT, GST, biliary PAHs, gene induction	PAHs (Oil refinery)	Trisciani et al. 2011	Adriatic coast, Italy	28
	EROD, CYP1A1	Urban, agricultural and industrial pollutants	Arinç and Sen 1999	Izmır bay, Turkey	29
	Micronuclei (genetic damage)	Domestic and industrial pollutants	Arslan et al. 2015	Aliağa Bay, Turkey	30
	AChE, CAT, GST, MT, LPO	Urban and industrial pollutants	Jebali et al. 2013	Tunisia coast	31
	EROD, DNA adducts	PAHs (Erika oil spill)	Amat et al. 2006, Claireaux et al. 2004	Atlantic coast, France	32,33
	Histology	Metals, PCBs, PAHs	Cuevas et al. 2015	Basc country coast, Spain	34
S. senegalensis	Antioxidant enzymes, EROD, GST, LPO	Metals and PAHs	Fonseca et al. 2011a,b	ria Aveiro and Tejo estuaries, Portugal Tejo estuary, Portugal	35 36
	CYP1A, MT, LPO, GST, CAT, histology, VTG	Metals, PCBs, pesticides and PAHs	Costa et al. 2009a, b, 2010, Gonçalves et al. 2013, 2014, Martins et al. 2015	Sado estuary, Portugal	37, 38, 39, 40, 41, 42
	Antioxidant enzymes, EROD, GST, MT, LPO	Metals and PAHs	Fonseca et al. 2015	ria Aveiro, Tejo and Sado estuaries, Portugal	43
	Antioxidant enzymes, GST, LPO, AChE, IDH, LDH, Histology	Metals, PAHs	Oliva et al. 2010, 2012a,b	Huelva estuary, Spain	44,45,46
	Histology	Metals	Galindo-Riano et al. 2015	Huelva estuary, Spain	47

Table 4. Laboratory studies with *Solea* spp. as sentinel species.

Species	Biomarkers measured	Targeted chemicals	Reference
S. solea	EROD, biliary PAHs, genotoxicity	PAHs	Wessel et al. 2010
	VTG, CYPs, HSP, antioxidant capacity gene expression, plasmatic parameters	NP	Palermo et al. 2012a,b, Cocci et al. 2013
	gene induction (CYPs, HSP70, nuclear receptors).	Polluted sediment (PAHs, metals, pesticides, PCBs)	Ribecco et al. 2012
	EROD, DNA adducts, gill histology, Na, K-ATPase	PAHs (fuel No 2)	Claireaux et al. 2004
S. senegalensis	Histopatology, antioxidant enzymes, phosphatases	Alkylbenzene sulfonates (LAS)	Alvarez-Muñoz et al. 2007, 2009
	Metabolomics	Alcohol polyethoxylates (AEO)	Alvarez-Muñoz et al. 2015
	EROD, antioxidant enzymes, LPO, GST, CbE, Na,K-ATPase, histology	Mexel, HCl	Lopez-Galindo et al. 2010a,b
	CYP-related enzymes, CbE, UDPGT, GST, antioxidant enzymes, VTG	NP, EE_2, gemfibrozil, ketoconazole, αNF	Solé et al. 2014
	CYP-related enzymes, CbE, UDPGT, GST, antioxidant enzymes, plasmatic metabolites, osmoregulation, VTG	Ibuprofen, carbamazepine	González-Mira et al. 2016
	AChE, CbE, GST, antioxidant enzymes, LPO	*Prestige* fuel oil	Solé et al. 2008

2015). Furthermore, an exclusively genomic approach was applied to toxicity assessment in a field survey with *S. solea* (Ribecco et al. 2012), and under experimental conditions to waterborne exposures of the endocrine disruptor nonylphenol (NP) and the sex hormone estradiol (Cocci et al. 2013). An effect marker (micronuclei formation) as a sign of DNA damage was the marker adopted in a field study in Aliağa Bay, Turkey, embracing several species with *S. solea* among them (Arslan et al. 2015)

A larger body of evidence on *S. senegalensis* suitability as sentinel is based on studies assessing a larger spectrum of chemicals: metals, PAHs, PCBs and pesticides in estuaries from the Iberian Peninsula. In some surveys, metal pollution due to the historic traditional mining activities and/or metal spills was evaluated using juveniles of *S. senegalensis*. These studies complemented chemical analysis with the use of multiple biomarkers of exposure and effect (Oliva et al. 2012a,b, Riba et al. 2004, 2005). Other surveys evaluated a more diffuse pollution due to river discharges either in the Huelva bay in Spain (Oliva et al. 2010) or in Portugal in the ria de Aveiro and/or Tejo (Fonseca et al. 2011a,b) and/or Sado estuaries (Costa et al. 2008, 2009a,b, 2011, Fonseca et al. 2015, Goncalves et al. 2013). In all of the above studies, chemical analysis of biota and/or local sediment was complemented with multiple biochemical biomarker of exposure (xenobiotic metabolising enzymes, antioxidant defences) and effect (lipid peroxidation, genotoxicity, histological alterations) and also considering metabolomic (Costa et al. 2012b) and transcriptomic (Costa et al. 2011, Osuna-Jimenez et al. 2009) approaches. Histological alterations were also the biomarker of choice in *S. senegalensis* exposed to a particular metal (Costa et al. 2013) or a metal pollution gradient in the SE Iberian Peninsula (Galindo-Riano et al. 2015). Histopathological indices in *S. senegalensis* have been proposed as reliable markers for estuarine ecological risk assessment based on field and laboratory findings (Costa et al. 2009b, 2012a).

Despite an ample use of *Solea* spp. as sentinel in field situations, its suitability under individual chemical exposures has been less documented, with the exception of previously mentioned studies carried out at gene expression level. More recently, the ability of *S. senegalensis* to respond *in vivo* to several chemical classes was tested after intraperitoneal (IP) injections to model CYP modulators (α-naphtoflavone, ketoconazole), to the endocrine disruptor ethinylestradiol (EE$_2$), the lipid regulator gemfibrozil and the surfactant NP (Solé et al. 2014). These single injections revealed a range of responses, in pathways involved in xenobiotic metabolism (CYPs, conjugation reactions and antioxidant defences) as well as alterations in reproductive parameters and sex hormone levels. CYP1A1-related EROD activity was induced after β-naphtoflavone injection and CYP3A4-related BFCOD activity was enhanced after EE$_2$ injection confirming the activation of AhR and PXR receptors, respectively, as in other fish species (Celander 2011). Juveniles of *S. senegalensis* were used to assess sublethal water exposure to the antifoulants sodium hypochlorite (NaClO) and Mexel® (Lopez-Galindo et al. 2010a,b). The parameters measured included stress metabolites in plasma, AChE activity, detoxification and antioxidant enzymes in liver and gills as well as lipid peroxidation (LPO) levels and gill damage. Interesting results were the lack of oxidative stress measured as LPO levels and the response of the antioxidant machinery faced with antifoulants exposures, although NaClO at the highest doses was toxic to fish. Juveniles of the same species were used to assess, under lab exposures, the effects of several dilutions of the water accommodated fraction (WAF) of the *Prestige* oil and multiple biomarker responses, including oxidative stress, were considered (Solé et al. 2008). The assessment of lineal alkylbenzene sulfonates (LAS) exposures in lab facilities has also been conducted on *S. senegalensis* using multiple enzymatic responses and histological examination of selected organs (Alvarez-Munoz et al. 2007, 2009) and more recently on a non-ionic surfactant, alcohol polyethoxylate, using a metabolomic approach (Alvarez-Munoz et al. 2014).

Ecotoxicology of pharmaceuticals and personal care products (PPCPs) in *S. solea* in comparison to other fish has been assessed *in vitro* using the microsomal fraction of the liver (Ribalta et al. 2015, Ribalta and Solé 2014). Overall, *S. solea* was somewhat insensitive to these chemicals (if compared to deep-sea fish) in terms of EROD (CYP1A1) and BFCOD (CYP3A4) interactions but CbE was responsive to lipid regulators such as simvastatin and fenofibrate (Solé and Sanchez-Hernandez 2015). *In vitro* studies are suitable as prior screening tools for conducting further *in vivo* studies (Thibaut et al. 2006). Exposure by IP injection to pharmaceuticals of environmental concern such as ibuprofen and carbamazepine has been conducted in juveniles of *S. senegalensis* and the effects on several plasmatic, osmoregulation and xenobiotic metabolism enzymes evaluated in relation to these exposures but also with respect to two different rearing temperatures (González-Mira et al. 2016). In addition, osmoregulation measured as Na+/K+-ATPase activities was evaluated in several organs/tissues: intestine, kidney and gills and the induction of genes involved in this osmoregulatory system was further measured in gills after drug injections at two acclimation temperatures (González-Mira et al. 2018). As far as we are aware, no studies on nanoparticle effects have been conducted using this species, although a nanoparticle administration method of vitamin C has been carried out on *S. senegalensis* larvae as a potential tool in aquaculture larvae feeding and the effects on their antioxidant capacity was evaluated (Jimenez-Fernandez et al. 2014).

Acknowledgements

The Spanish Ministry of Science and Innovation provided financial support through the DEPURAMAR project (Ref. CTM2010-16611). A. Torreblanca and I. Varó are acknowledged for their contribution to the project.

References

Aceña, J., S. Pérez, P. Eichhorn, M. Solé, C. Porte and D. Barceló. 2017. Metabolite profiling of carbamazepine and ibuprofen in *Solea senegalensis* bile using high-resolution mass spectrometry. Anal. Bioanal. Chem. 409: 5441–5450.

Akcha, F., F.V. Hubert and A. Pfhol-Leszkowicz. 2003. Potential value of the comet assay and DNA adduct measurement in dab (*Limanda limanda*) for assessment of in situ exposure to genotoxic compounds. Mutat. Res-Gen. Tox. 534: 21–32.

Alvarez-Munoz, D., P.A. Lara-Martin, J. Blasco, A. Gomez-Parra and E. Gonzalez-Mazo. 2007. Presence, biotransformation and effects of sulfophenylcarboxylic acids in the benthic fish *Solea senegalensis*. Environ. Int. 33: 565–570.

Alvarez-Munoz, D., A. Gomez-Parra, J. Blasco, C. Sarasquete and E. Gonzalez-Mazo. 2009. Oxidative stress and histopathology damage related to the metabolism of dodecylbenzene sulfonate in Senegalese sole. Chemosphere 74: 1216–1223.

Alvarez-Munoz, D., R. Al-Salhi, A. Abdul-Sada, E. Gonzalez-Mazo and E.M. Hill. 2014. Global Metabolite Profiling Reveals Transformation Pathways and Novel Metabolomic Responses in *Solea senegalensis* after Exposure to a Non-ionic Surfactant. Environ. Sci. Technol. 48: 5203–5210.

Amat, A., T. Burgeot, M. Castegnaro and A. Pfohl-Leszkowicz. 2006. DNA adducts in fish following an oil spill exposure. Environ. Chem. Lett. 4: 93–99.

Antunes, P. and O. Gil. 2002. Accumulation of PCBs in three fish species from Ria de Aveiro, Portugal. Organohalog. Compd. 58: 33–36.

Arinc, E., S. Kocabiyik, A. Sen and E. Su. 2001. Biomonitoring of toxic, carcinogenic pollutants by molecular and biochemical responses of fish cytochrome P4501A1 (CYP1A1) along the Izmir Bay on the Mediterranean Sea. Rapport du Congress de la CIESM 36: 179–180.

Arjona, F.J., I. Ruiz-Jarabo, L. Vargas-Chacoff, M.P.M. del Rio, G. Flik, J.M. Mancera et al. 2010. Acclimation of *Solea senegalensis* to different ambient temperatures: implications for thyroidal status and osmoregulation. Mar. Biol. 157: 1325–1335.

Arslan, O.C., M. Boyacioglu, H. Parlak, S. Katalay and M.A. Karaaslan. 2015. Assessment of micronuclei induction in peripheral blood and gill cells of some fish species from Aliaga Bay Turkey. Mar. Pollut. Bull. 94: 48–54.

Baeyens, W., M. Leermakers, T. Papina, A. Saprykin, N. Brion, J. Noyen et al. 2003. Bioconcentration and biomagnification of mercury and methylmercury in North Sea and Scheldt estuary fish. Arch. Environ. Contam. Toxicol. 45: 498–508.

Baptista, J., P. Pato, E. Pereira, A.C. Duarte and M.A. Pardal. 2013. PCBs in the fish assemblage of a southern European estuary. J. Sea Res. 76: 22–30.

Barhoumi, S., I. Messaoudi, T. Deli, K. Said and A. Kerkeni. 2009. Cadmium bioaccumulation in three benthic fish species, *Salaria basilisca*, *Zosterisessor ophiocephalus* and *Solea vulgaris* collected from the Gulf of Gabes in Tunisia. J. Environ. Sci-China 21: 980–984.

Ben Ameur, W., Y. El Megdiche, E. Eljarrat, S. Ben Hassine, B. Badreddine, T. Souad et al. 2013. Organochlorine and organobromine compounds in a benthic fish (*Solea solea*) from Bizerte Lagoon (northern Tunisia): Implications for human exposure. Ecotox. Environ. Safe. 88: 55–64.

Benzekri, H., P. Armesto, X. Cousin, M. Rovira, D. Crespo, M. Alejandro Merlo et al. 2014. De novo assembly, characterization and functional annotation of Senegalese sole (*Solea senegalensis*) and common sole (*Solea solea*) transcriptomes: integration in a database and design of a microarray. Bmc Genomics 15.

Bodin, N., N. Tapie, K. Le Menach, E. Chassot, P. Elie, E. Rochard et al. 2014. PCB contamination in fish community from the Gironde Estuary (France): Blast from the past. Chemosphere 98: 66–72.

Boglino, A., M.J. Darias, A. Estevez, K.B. Andree, C. Sarasquete, J.B. Ortiz-Delgado et al. 2014. The effect of dietary oxidized lipid levels on growth performance, antioxidant enzyme activities, intestinal lipid deposition and skeletogenesis in Senegalese sole (*Solea senegalensis*) larvae. Aquacult. Nutr. 20: 692–711.

Bordajandi, L.R., I. Martin, E. Abad, J. Rivera and M.J. Gonzalez. 2006. Organochlorine compounds (PCBs, PCDDs and PCDFs) in seafish and seafood from the Spanish Atlantic southwest coast. Chemosphere 64: 1450–1457.

Cajaraville, M.P., M.J. Bebianno, J. Blasco, C. Porte, C. Sarasquete and A. Viarengo. 2000. The use of biomarkers to assess the impact of pollution in coastal environments of the Iberian Peninsula: a practical approach. Sci. Total Environ. 247: 295–311.

Cannas, M., F. Atzori, F. Rupsard, P. Bustamante, V. Loizeau and C. Lefrancois. 2013. PCBs contamination does not alter aerobic metabolism and tolerance to hypoxia of juvenile sole (*Solea solea*, L. 1758). Aquat. Toxicol. 127: 54–60.

Castro, C., A. Perez-Jimenez, I. Guerreiro, H. Peres, M. Castro-Cunha and A. Oliva-Teles. 2012. Effects of temperature and dietary protein level on hepatic oxidative status of Senegalese sole juveniles (*Solea senegalensis*). Comp. Biochem. Physiol. A 163: 372–378.

Celander, M.C. 2011. Cocktail effects on biomarker responses in fish. Aquat. Toxicol. 105: 72–77.

Claireaux, G., Y. Desaunay, F. Akcha, B. Auperin, G. Bocquene, F.N. Budzinski et al. 2004. Influence of oil exposure on the physiology and ecology of the common sole *Solea Solea*: Experimental and field approaches. Aquat. Living Resour. 17: 335–351.

Cocci, P., G. Mosconi and F.A. Palermo. 2013. Effects of 4-nonylphenol on hepatic gene expression of peroxisome proliferator-activated receptors and cytochrome P450 isoforms (CYP1A1 and CYP3A4) in juvenile sole (*Solea solea*). Chemosphere 93: 1176–1181.

Cogun, H., T.A. Yuzereroglu, F. Kargin and O. Firat. 2005. Seasonal variation and tissue distribution of heavy metals in shrimp and fish species from the Yumurtalik coast of Iskenderun Gulf, Mediterranean. Bull. Environ. Contam. Toxicol. 75: 707–715.

Costa, P.M., J. Lobo, S. Caeiro, M. Martins, A.M. Ferreira, M. Caetano et al. 2008. Genotoxic damage in *Solea senegalensis* exposed to sediments from the Sado Estuary (Portugal): Effects of metallic and organic contaminants. Mutat. Res-Gen. Tox 654: 29–37.

Costa, P.M., S. Caeiro, M.S. Diniz, J. Lobo, M. Martins, A.M. Ferreira et al. 2009a. Biochemical endpoints on juvenile *Solea senegalensis* exposed to estuarine sediments: the effect of contaminant mixtures on metallothionein and CYP1A induction. Ecotoxicology 18: 988–1000.

Costa, P.M., M.S. Diniz, S. Caeiro, J. Lobo, M. Martins, A.M. Ferreira et al. 2009b. Histological biomarkers in liver and gills of juvenile *Solea senegalensis* exposed to contaminated estuarine sediments: A weighted indices approach. Aquat. Toxicol. 92: 202–212.

Costa, P.M., C. Miguel, S. Caeiro, J. Lobo, M. Martins, A.M. Ferreira et al. 2011. Transcriptomic analyses in a benthic fish exposed to contaminated estuarine sediments through laboratory and in situ bioassays. Ecotoxicology 20: 1749–1764.

Costa, P.M., S. Caeiro, C. Vale, T.A. DelValls and M.H. Costa. 2012a. Can the integration of multiple biomarkers and sediment geochemistry aid solving the complexity of sediment risk assessment? A case study with a benthic fish. Environ. Pollut. 161: 107–120.

Costa, P.M., E. Chicano-Galvez, S. Caeiro, J. Lobo, M. Martins, A.M. Ferreira et al. 2012b. Hepatic proteome changes in *Solea senegalensis* exposed to contaminated estuarine sediments: a laboratory and *in situ* survey. Ecotoxicology 21: 1194–1207.

Costa, P.M., S. Caeiro and M.H. Costa. 2013. Multi-organ histological observations on juvenile Senegalese soles exposed to low concentrations of waterborne cadmium. Fish Physiol. Biochem. 39: 143–158.

Costas, B., C. Aragao, I. Ruiz-Jarabo, L. Vargas-Chacoff, F.J. Arjona, J.M. Mancera et al. 2012. Different environmental temperatures affect amino acid metabolism in the eurytherm teleost Senegalese sole (*Solea senegalensis* Kaup, 1858) as indicated by changes in plasma metabolites. Amino Acids 43: 327–335.

Crespo, M. and M. Solé. 2016. The use of juvenile *Solea solea* as sentinel in the marine platform of the Ebre Delta: in vitro interaction of emerging contaminants with the liver detoxification system. Environ. Sci. Pollut. Res. 23: 19229–19236.

Cuevas, N., I. Zorita, P.M. Costa, I. Quincoces, J. Larreta and J. Franco. 2015. Histopathological indices in sole (*Solea solea*) and hake (*Merluccius merluccius*) for implementation of the European Marine Strategy Framework Directive along the Basque continental shelf (SE Bay of Biscay). Mar. Pollut. Bull. 94: 185–198.

Chairi, H., C. Fernandez-Diaz, J.I. Navas, M. Manchado, L. Rebordinos and J. Blasco. 2010. In vivo genotoxicity and stress defences in three flatfish species exposed to CuSO4. Ecotox. Environ. Safe. 73: 1279–1285.

Choueri, R.B., A. Cesar, D.M.S. Abessa, R.J. Torres, I. Riba, C.D.S. Pereira et al. 2010. Harmonised framework for ecological risk assessment of sediments from ports and estuarine zones of North and South Atlantic. Ecotoxicology 19: 678–696.

DelValls, T.A., J.M. Forja and A. Gomez-Parra. 2002. Seasonality of contamination, toxicity, and quality values in sediments from littoral ecosystems in the Gulf of Cadiz (SW Spain). Chemosphere 46: 1033–1043.

Dierking, J., E. Wafo, T. Schembri, V. Lagadec, C. Nicolas, Y. Letourneur et al. 2009. Spatial patterns in PCBs, pesticides, mercury and cadmium in the common sole in the NW Mediterranean Sea, and a novel use of contaminants as biomarkers. Mar. Pollut. Bull. 58: 1605–1614.

Dupuy, C., C. Galland, V. Pichereau, W. Sanchez, R. Riso, M. Labonne et al. 2015. Assessment of the European flounder responses to chemical stress in the English Channel, considering biomarkers and life history traits. Mar. Pollut. Bull. 95: 634–645.

Durrieu, G., R. Maury-Brachet, M. Girardin, E. Rochard and A. Boudou. 2005. Contamination by heavy metals (Cd, Zn, Cu, and Hg) of eight fish species in the Gironde estuary (France). Estuaries 28: 581–591.

Fernandez-Diaz, C., J. Kopecka, J.P. Canavate, C. Sarasquete and M. Sole. 2006. Variations on development and stress defences in *Solea senegalensis* larvae fed on live and microencapsulated diets. Aquaculture 251: 573–584.

Ferrara, F., F. Fabietti, M. Delise and E. Funari. 2005. Alkylphenols and alkylphenol ethoxylates contamination of crustaceans and fishes from the Adriatic Sea (Italy). Chemosphere 59: 1145–1150.

Foekema, E.M., C.M. Deerenberg and A.J. Murk. 2008. Prolonged ELS test with the marine flatfish sole (*Solea solea*) shows delayed toxic effects of previous exposure to PCB 126. Aquat. Toxicol. 90: 197–203.

Foekema, E.M., M.L. Parron, M.T. Mergia, E.R.M. Carolus, J.H.J. vd Berg, C. Kwadijk et al. 2014. Internal effect concentrations of organic substances for early life development of egg-exposed fish. Ecotox. Environ. Safe. 101: 14–22.

Fonseca, V., A. Serafim, R. Company, M.J. Bebianno and H. Cabral. 2009. Effect of copper exposure on growth, condition indices and biomarker response in juvenile sole *Solea senegalensis*. Sci. Mar. 73: 51–58.

Fonseca, V.F., S. Franca, A. Serafim, R. Company, B. Lopes, M.J. Bebianno et al. 2011a. Multi-biomarker responses to estuarine habitat contamination in three fish species: *Dicentrarchus labrax*, *Solea senegalensis* and *Pomatoschistus microps*. Aquat. Toxicol. 102: 216–227.

Fonseca, V.F., S. Franca, R.P. Vasconcelos, A. Serafim, R. Company, B. Lopes et al. 2011b. Short-term variability of multiple biomarker response in fish from estuaries: Influence of environmental dynamics. Mar. Environ. Res. 72: 172–178.

Fonseca, V.F., R.P. Vasconcelos, S.E. Tanner, S. Franca, A. Serafim, B. Lopes et al. 2015. Habitat quality of estuarine nursery grounds: Integrating non-biological indicators and multilevel biological responses in *Solea senegalensis*. Ecol. Indic. 58: 335–345.

Franca, S., C. Vinagre, I. Cacador and H.N. Cabral. 2005. Heavy metal concentrations in sediment, benthic invertebrates and fish in three salt marsh areas subjected to different pollution loads in the Tagus Estuary (Portugal). Mar. Pollut. Bull. 50: 998–1003.

Fulton, M.H. and P.B. Key. 2001. Acetylcholinesterase inhibition in estuarine fish and invertebrates as an indicator of organophosphorus insecticide exposure and effects. Environ. Toxicol. Chem. 20: 37–45.

Galindo-Riano, M.D., M. Oliva, J.A. Jurado, D. Sales, M.D. Granado-Castro and F. Lopez-Aguayo. 2015. Comparative Baseline Levels of Heavy Metals and Histopathological Notes in Fish From two Coastal Ecosystems of South-West of Spain. Int. J. Environ. Res. 9: 163–178.

Galindo, M.D., J.A. Jurado, M. Garcia, M.L.G. de Canales, M. Oliva, F. Lopez et al. 2012. Trace metal accumulation in tissues of sole (*Solea senegalensis*) and the relationships with the abiotic environment. Int. J. Anal. Chem. 92: 1072–1092.

Goncalves, C., M. Martins, M.H. Costa, S. Caeiro and P.M. Costa. 2013. Ecological risk assessment of impacted estuarine areas: Integrating histological and biochemical endpoints in wild Senegalese sole. Ecotox. Environ. Safe. 95: 202–211.

Goncalves, C., M. Martins, M.S. Diniz, M.H. Costa, S. Caeiro and P.M. Costa. 2014. May sediment contamination be xenoestrogenic to benthic fish? A case study with *Solea senegalensis*. Mar. Environ. Res. 99: 170–178.

González-Mira, A., I. Varó, M. Solé and A. Torreblanca. 2016. Drugs of environmental concern modify *Solea senegalensis* physiology and biochemistry in a temperature-dependent manner. Environ. Sci. Pollut. Res. 23: 20937–20951.

Gonzalez-Mira, A., A. Torreblanca, F. Hontoria, J.C. Navarro, E. Mananos and I. Varo. 2018. Effects of ibuprofen and carbamazepine on the ion transport system and fatty acid metabolism of temperature conditioned juveniles of *Solea senegalensis*. Ecotox. Environ. Safe. 148: 693–701.

Hillenweck, A., C. Canlet, A. Mauffret, L. Debrauwer, G. Claireaux and J.P. Cravedi. 2008. Characterization of biliary metabolites of fluoranthene in the common sole (*Solea solea*). Environ. Toxicol. Chem. 27: 2575–2581.

Imsland, A.K., A. Foss, L.E.C. Conceicao, M.T. Dinis, D. Delbare, E. Schram et al. 2003. A review of the culture potential of *Solea solea* and *S. senegalensis*. Rev. Fish Biol. Fisher. 13: 379–407.

IPCC. 2014. https://www.ipcc.ch/pdf/assessment-report/ar5/wg2/ar5_wgII_spm_es.pdf.

Iwama, G.K., M.M. Vijayan, R.B. Forsyth and P.A. Ackerman. 1999. Heat shock proteins and physiological stress in fish. Am. Zool. 39: 901–909.

Janak, K., A. Covaci, S. Voorspoels and G. Becher. 2005. Hexabromocyclododecane in marine species from the Western Scheldt Estuary: Diastereoisomer- and enantiomer-specific accumulation. Environ. Sci. Technol. 39: 1987–1994.

Jebali, J., M. Sabbagh, M. Banni, N. Kamel, S. Ben-Khedher, N. M'Hamdi et al. 2013. Multiple biomarkers of pollution effects in *Solea Solea* fish on the Tunisia coastline. Environ. Sci. Pollut. Res. 20: 3812–3821.

Jimenez-Fernandez, E., A. Ruyra, N. Roher, E. Zuasti, C. Infante and C. Fernandez-Diaz. 2014. Nanoparticles as a novel delivery system for vitamin C administration in aquaculture. Aquaculture 432: 426–433.

Jimenez-Tenorio, N., C. Morales-Caselles, J. Kalman, M.J. Salamanca, M. Luisa Gonzalez de Canales, C. Sarasquete et al. 2007. Determining sediment quality for regulatory proposes using fish chronic bioassays. Environ. Int. 33: 474–480.

Kerambrun, E., F. Henry, A. Marechal, W. Sanchez, C. Minier, I. Filipuci et al. 2012. A multibiomarker approach in juvenile turbot, *Scophthalmus maximus*, exposed to contaminated sediments. Ecotox. Environ. Safe. 80: 45–53.

Koenig, S., K. Guillen and M. Sole. 2013. Comparative xenobiotic metabolism capacities and pesticide sensitivity in adults of *Solea solea* and *Solea senegalensis*. Comp. Biochem. Physiol. C. 157: 329–336.

Kopecka, J. and J. Pempkowiak. 2008. Temporal and spatial variations of selected biomarker activities in flounder (*Platichthys flesus*) collected in the Baltic proper. Ecotox. Environ. Safe. 70: 379–391.

Kucuksezgin, F., A. Kontas, O. Altay, E. Uluturhan and E. Darilmaz. 2006. Assessment of marine pollution in Izmir Bay: Nutrient, heavy metal and total hydrocarbon concentrations. Environ. Int. 32: 41–51.

Lehtonen, K.K., D. Schiedek, A. Kohler, T. Lang, P.J. Vuorinen, L. Forlin et al. 2006. The BEEP project in the Baltic Sea: Overview of results and outline for a regional biological effects monitoring strategy. Mar. Pollut. Bull. 53: 523–537.

Lerebours, A., G.D. Stentiford, B.P. Lyons, J.P. Bignell, S.A.P. Derocles and J.M. Rotchell. 2014. Genetic Alterations and Cancer Formation in a European Flatfish at Sites of Different Contaminant Burdens. Environ. Sci. Technol. 48: 10448–10455.

Lopez-Galindo, C., L. Vargas-Chacoff, E. Nebot, J.F. Casanueva, D. Rubio, M. Solé et al. 2010a. Biomarker responses in *Solea senegalensis* exposed to sodium hypochlorite used as antifouling. Chemosphere 78: 885–893.

Lopez-Galindo, C., L. Vargas-Chacoff, E. Nebot, J.F. Casanueva, D. Rubio, M. Solé et al. 2010b. Sublethal effects of the organic antifoulant Mexel (R) 432 on osmoregulation and xenobiotic detoxification in the flatfish *Solea senegalensis*. Chemosphere 79: 78–85.

Manchado, M., E. Salas-Leiton, C. Infante, M. Ponce, E. Asensio, A. Crespo et al. 2008. Molecular characterization, gene expression and transcriptional regulation of cytosolic HSP90 genes in the flatfish Senegalese sole (*Solea senegalensis* Kaup). Gene 416: 77–84.

Martins, C., A.P.A. de Matos, M.H. Costa and P.M. Costa. 2015. Alterations in juvenile flatfish gill epithelia induced by sediment-bound toxicants: A comparative *in situ* and *ex situ* study. Mar. Environ. Res. 112: 122–130.

Mearns, A.J., D.J. Reish, P.S. Oshida, T. Ginn and M.A. Rempel-Hester. 2011. Effects of Pollution on Marine Organisms. Water Environ. Res. 83: 1789–1852.

Mhadhbi, L., A. Palanca, T. Gharred and M. Boumaiza. 2012. Bioaccumulation of Metals in Tissues of *Solea vulgaris* from the outer Coast and Ria de Vigo, NE Atlantic (Spain). Int. J. Environ. Res. 6: 19–24.

Munschy, C., K. Heas-Moisan, C. Tixier, G. Pacepavicius and M. Alaee. 2010. Dietary exposure of juvenile common sole (*Solea solea* L.) to polybrominated diphenyl ethers (PBDEs): Part 2. Formation, bioaccumulation and elimination of hydroxylated metabolites. Environ. Pollut. 158: 3527–3533.

Napierska, D., J. Barsiene, E. Mulkiewicz, M. Podolska and A. Rybakovas. 2009. Biomarker responses in flounder *Platichthys flesus* from the Polish coastal area of the Baltic Sea and applications in biomonitoring. Ecotoxicology 18: 846–859.

Oliva, M., M.L.G. de Canales, C. Gravato, L. Guilhermino and J.A. Perales. 2010. Biochemical effects and polycyclic aromatic hydrocarbons (PAHs) in senegal sole (*Solea senegalensis*) from a Huelva estuary (SW Spain). Ecotox. Environ. Safe. 73: 1842–1851.

Oliva, M., J.A. Perales, C. Gravato, L. Guilhermino and M.D. Galindo-Riano. 2012a. Biomarkers responses in muscle of Senegal sole (*Solea senegalensis*) from a heavy metals and PAHs polluted estuary. Mar. Pollut. Bull. 64: 2097–2108.

Oliva, M., J.J. Vicente, C. Gravato, L. Guilhermino and M.D. Galindo-Riano. 2012b. Oxidative stress biomarkers in Senegal sole, *Solea senegalensis*, to assess the impact of heavy metal pollution in a Huelva estuary (SW Spain): Seasonal and spatial variation. Ecotox. Environ. Safe. 75: 151–162.

Osuna-Jimenez, I., T.D. Williams, M.-J. Prieto-Alamo, N. Abril, J.K. Chipman and C. Pueyo. 2009. Immune- and stress-related transcriptomic responses of *Solea senegalensis* stimulated with lipopolysaccharide and copper sulphate using heterologous cDNA microarrays. Fish Shellfish Immun. 26: 699–706.

Palermo, F.A., P. Cocci, M. Angeletti, A. Polzonetti-Magni and G. Mosconi. 2012a. PCR-ELISA detection of estrogen receptor beta mRNA expression and plasma vitellogenin induction in juvenile sole (*Solea solea*) exposed to waterborne 4-nonylphenol. Chemosphere 86: 919–925.

Palermo, F.A., P. Cocci, M. Nabissi, A. Polzonetti-Magni and G. Mosconi. 2012b. Cortisol response to waterborne 4-nonylphenol exposure leads to increased brain POMC and HSP70 mRNA expressions and reduced total antioxidant capacity in juvenile sole (*Solea solea*). Comp. Biochem. Physiol. C. 156: 135–139.

Riba, I., M.C. Casado-Martinez, J. Blasco and T.A. DelValls. 2004. Bioavailability of heavy metals bound to sediments affected by a mining spill using *Solea senegalensis* and *Scrobicularia plana*. Mar. Environ. Res. 58: 395–399.

Riba, I., J. Blasco, N. Jimenez-Tenorio, M.L.G. de Canales and T.A. DelValls. 2005. Heavy metal bioavailability and effects: II. Histopathology-bioaccumulation relationships caused by mining activities in the Gulf of Cadiz (SW, Spain). Chemosphere 58: 671–682.

Ribalta, C. and M. Solé. 2014. *In vitro* interaction of emerging contaminants with the cytochrome p450 system of mediterranean deep-sea fish. Environ. Sci. Technol. 48: 12327–12335.

Ribalta, C., J.C. Sanchez-Hernandez and M. Solé. 2015. Hepatic biotransformation and antioxidant enzyme activities in Mediterranean fish from different habitat depths. Sci. Total Environ. 532: 176–183.

Ribecco, C., G. Hardiman, R. Sasik, S. Vittori and O. Carnevali. 2012. Teleost fish (*Solea solea*): A novel model for ecotoxicological assay of contaminated sediments. Aquat. Toxicol. 109: 133–142.

Salamanca, M.J., N. Jimenez-Tenorio, M.L.G. de Canales and T.A. Del Valls. 2008. Evaluation of the toxicity of an oil spill conducted through bioassays using the fish *Solea senegalensis*. Cienc. Mar. 34: 339–348.

Salas-Leiton, E., V. Anguis, B. Martin-Antonio, D. Crespo, J.V. Planas, C. Infante et al. 2010. Effects of stocking density and feed ration on growth and gene expression in the Senegalese sole (*Solea senegalensis*): Potential effects on the immune response. Fish Shellfish Immun. 28: 296–302.

Sánchez-Nogué, B., I. Varó and M. Solé. 2013. Comparative analysis of selected biomarkers and pesticide sensitivity in juveniles of *Solea solea* and *Solea senegalensis*. Environ. Sci. Pollut. Res. 20: 3480–3488.

Schram, E., S. Bierman, L.R. Teal, O. Haenen, H. van de Vis and A.D. Rijnsdorp. 2013. Thermal Preference of Juvenile Dover Sole (*Solea solea*) in Relation to Thermal Acclimation and Optimal Growth Temperature. Plos One 8.

Siscar, R., A. Torreblanca, A. Palanques and M. Solé. 2013. Metal concentrations and detoxification mechanisms in *Solea solea* and *Solea senegalensis* from NW Mediterranean fishing grounds. Mar. Pollut. Bull. 77: 90–99.

Siscar, R., A. Torreblanca, J. del Ramo and M. Solé. 2014. Modulation of metallothionein and metal partitioning in liver and kidney of *Solea senegalensis* after long-term acclimation to two environmental temperatures. Environ. Res. 132: 197–205.

Siscar, R., I. Varó and M. Solé. 2015. Hepatic and branchial xenobiotic biomarker responses in *Solea* spp. from several NW Mediterranean fishing grounds. Mar. Environ. Res. 112: 35–43.

Solé, M., J. Potrykus, C. Fernandez-Diaz and J. Blasco. 2004. Variations on stress defences and metallothionein levels in the Senegal sole, *Solea senegalensis*, during early larval stages. Fish Physiol. Biochem. 30: 57–66.

Solé, M., D. Lima, M.A. Reis-Henriques and M.M. Santos. 2008. Stress biomarkers in juvenile senegal sole, *Solea senegalensis*, exposed to the water-accommodated fraction of the "Prestige" fuel oil. Bull. Environ. Contam. Toxicol. 80: 19–23.

Solé, M., S. Vega and I. Varo. 2012. Characterization of type "B" esterases and hepatic CYP450 isoenzimes in Senegalese sole for their further application in monitoring studies. Ecotox. Environ. Safe. 78: 72–79.

Solé, M., M. Manzanera, A. Bartolomé, L. Tort and J. Caixach. 2013. Persistent organic pollutants (POPs) in sediments from fishing grounds in the NW Mediterranean: Ecotoxicological implications for the benthic fish *Solea* sp. Mar. Pollut. Bull. 67: 158–165.

Solé, M., A. Fortuny and E. Mananos. 2014. Effects of selected xenobiotics on hepatic and plasmatic biomarkers in juveniles of *Solea senegalensis*. Environ. Res. 135: 227–235.

Solé, M., I. Varó, A. Gonzalez-Mira and A. Torreblanca. 2015. Xenobiotic metabolism modulation after long-term temperature acclimation in juveniles of *Solea senegalensis*. Mar. Biol. 162: 401–412.

Solé, M. and J.C. Sanchez-Hernandez. 2015. An *in vitro* screening with emerging contaminants reveals inhibition of carboxylesterase activity in aquatic organisms. Aquat. Toxicol. 169: 215–222.

Sumpter, J.P. and S. Jobling. 1995. Vitellogenesis as a biomarker for estrogenic contamination of the aquatic environment. Environ. Health Perspect. 103: 173–178.

Sureau, D., J.P. Lagardere and J.P. Pennec. 1989. Heart-rate and its cholinergic control in the sole (*solea vulgaris*), acclimatized to different temperatures. Comp. Biochem. Physiol. A. 92: 49–51.

Thibaut, R., S. Schnell and C. Porte. 2006. The interference of pharmaceuticals with endogenous and xenobiotic metabolizing enzymes in carp liver: An *in-vitro* study. Environ. Sci. Technol. 40: 5154–5160.

Tramati, C., S. Vizzini, S. Maci, A. Basset and A. Mazzola. 2011. Trace metal contamination in a Mediterranean coastal pond (Acquatina, Puglia). Transit. Water Bull. 5: 124–137.

Trisciani, A., I. Corsi, C. Della Torre, G. Perra and S. Focardi. 2011. Hepatic biotransformation genes and enzymes and PAH metabolites in bile of common sole (*Solea solea*, Linnaeus, 1758) from an oil-contaminated site in the Mediterranean Sea: A field study. Mar. Pollut. Bull. 62: 806–814.

Usero, J., C. Izquierdo, J. Morillo and I. Gracia. 2004. Heavy metals in fish (*Solea vulgaris*, *Anguilla anguilla* and *Liza aurata*) from salt marshes on the southern Atlantic coast of Spain. Environ. Int. 29: 949–956.

Van Ael, E., A. Covaci, R. Blust and L. Bervoets. 2012. Persistent organic pollutants in the Scheldt estuary: Environmental distribution and bioaccumulation. Environ. Int. 48: 17–27.

Van Ael, E., A. Covaci, R. Blust and L. Bervoets. 2014. Persistent organic pollutants in the Scheldt estuary: Environmental distribution and bioaccumulation (vol 48, pg 17, 2012). Environ. Int. 63: 246–251.

Viarengo, A., B. Burlando, F. Dondero, A. Marro and R. Fabbri. 1999. Metallothionein as a tool in biomonitoring programmes. Biomarkers 4: 455–466.

Vicente-Martorell, J.J., M.D. Galindo-Riano, M. Garcia-Vargas and M.D. Granado-Castro. 2009. Bioavailability of heavy metals monitoring water, sediments and fish species from a polluted estuary. J. Hazard. Mater. 162: 823–836.

Vinagre, C., V. Fonseca, H. Cabral and M.J. Costa. 2006. Habitat suitability index models for the juvenile soles, *Solea solea* and *Solea senegalensis*, in the Tagus estuary: Defining variables for species management. Fish. Res. 82: 140–149.

Vinagre, C., V. Fonseca, A. Maia, R. Amara and H. Cabral. 2008. Habitat specific growth rates and condition indices for the sympatric soles *Solea solea* (Linnaeus, 1758) and *Solea senegalensis* Kaup 1858, in the Tagus estuary, Portugal, based on otolith daily increments and RNA-DNA ratio. J. Appl. Ichthyol. 24: 163–169.

Vinagre, C., L. Narciso, M. Pimentel, H.N. Cabral, M.J. Costa and R. Rosa. 2013. Contrasting impacts of climate change across seasons: effects on flatfish cohorts. Reg. Environ. Change 13: 853–859.

Walker, C.H., R.M. Sibly, S.P. Hopkin and D.B. Peakall. 2012. Principles of Ecotoxicology. CRC Press Book. 4th Ed.

Wessel, N., R. Santos, D. Menard, K. Le Menach, V. Buchet, N. Lebayon et al. 2010. Relationship between PAH biotransformation as measured by biliary metabolites and EROD activity, and genotoxicity in juveniles of sole (*Solea solea*). Mar. Environ. Res. 69: S71–S73.

Wheelock, C.E., B.M. Phillips, B.S. Anderson, J.L. Miller, M.J. Miller and B.D. Hammock. 2008. Applications of carboxylesterase activity in environmental monitoring and toxicity identification evaluations (TIEs). Rev. Environ. Contam. Toxicol. 195: 117–178.

Wittmann, A.C. and H.O. Portner. 2013. Sensitivities of extant animal taxa to ocean acidification. Nat. Clim. Change 3: 995–1001.

Zambonino-Infante, J.L., G. Claireaux, B. Ernande, A. Jolivet, P. Quazuguel, A. Severe et al. 2013. Hypoxia tolerance of common sole juveniles depends on dietary regime and temperature at the larval stage: evidence for environmental conditioning. P. Roy. Soc. B Biol. Sci 280: 1–9.

B-4.3

Pathology and Diseases Control

F. Padrós,[1,*] *M. Constenla*[1] and *C. Zarza*[2]

1. Introduction

Health can be described as the way organisms can cope with their environment or particular conditions and live with good levels of metabolic and functional efficiency. Though eminently pelagic and bottom-dwelling species, Soleidae species can live in a relatively wide range of habitats, indicating a remarkable adaptive capacity in the environment (Munroe 2001). Some Soleidae species also present specific trends that make them suitable for fish farming practices. In both situations, fish are susceptible to many different conditions and/or pathogens that can significantly endanger their health status. Biological agents (parasites, bacteria, mesomycetocea, fungi, virus), altered environment (pollution, fisheries, global warming) and unsuitable management and conditions during fish culture are considered the main risks that these species should confront. The way these pathogens and conditions can or cannot overwhelm the resistance capacity of the fish will determine the development of pathologies and how these pathologies impact the health status of these animals.

In the wild, Soleidae species are bottom-dwelling species and have a very close relationship with the benthic boundary level and the upper sediment layer of the river, lake or sea floor. This fact has several important implications for the physiological characteristics such as behaviour, feeding habitats or reproduction. Under rearing conditions, the fact that juveniles and adults of these species nearly always lie on the bottom of the tanks and rarely swim to upper water layers, different from other flatfishes such as turbot, induces

[1] Fish Diseases Diagnostic Service. Departament de Biologia Animal, de Biologia Vegetal i d'Ecologia Facultat de Veterinària. Universitat Autònoma de Barcelona. Bellaterra. Barcelona. Catalonia. Spain.
[2] Skretting ARC. Stavanger. Norway.
* Corresponding author: Francesc.Padros@uab.cat

relevant characteristics in this species in terms of their health condition and susceptibility to diseases. Permanent fish-to-fish contact, accumulation of fish in a narrow area near the bottom with specific water dynamics and water chemical and microbiological quality with strong implications in feeding activity and production of fecal residues are very important specific factors that can affect disease progression and intensity. In fact, cultured *Solea* species are considered as extremely susceptible to several diseases that commonly affect other cultured flatfish and finfish species (Colen et al. 2014).

Soleidae also present some anatomical, morphological and physiological characteristics and adaptations that have important implications for the health conditions and for the development of diseases. Skin and gill mucus production is particularly important in flatfish species and in soleidae as adaptation to their particular way of life. Skin in soleids presents several differences, mainly compared with other fish species living in the water column. Asymmetric skin structure in the ocular and blind sides is easily observable by the different pigmentation patterns (pigmented/white). Pigmentary cell development and distribution are totally different in the two sides and can represent a relevant problem in case of pigment alteration such as albinism and pseudoalbinism in Senegalese sole (Villalta et al. 2005, Darias et al. 2013). Some Soleidae species such as *S. senegalensis* present very small scales with particularly large teeth along their outer edge, very easy to remove event with a gentle skin scraping, making the skin particularly vulnerable to mechanic lesions during management. Dover and Senegalese sole, like other flatfish species, are characterized by a relevant mucus production. Abundance and distribution of mucous cells is also variable *S. senegalensis*, being more frequent in tail and fins than in the body (Sarasquete et al. 1998). Like in other flatfish species, mucus also plays a relevant role in the skin functionality such as respiration, osmoregulation, protection against diseases and the particular relationships established with sand and sediments and also with the potential presence of contaminants in this environment. The particular morphological characteristics of the skin, together with changes in the structure, composition and integrity of the skin associated to stressors or to the set-up of burying activity extensively described by Ellis et al. (1994), mainly in juveniles, have been usually associated to microlesions, skin damage and caudal fin erosion as predisposing factors to some relevant infectious diseases for these species such as Tenacibaculosis (Flexibacteriosis), other bacterial external infections or entrance of other pathogens. Fin and skin alterations have also been described associated to wild Dover sole (Bucke et al. 1983), associated to trawling activity.

It is well known that Soleidae, and particularly *Solea* species, are more susceptible to stress under culture conditions than other reared species including flatfish. The relevance of the stress in these species is extensively developed in Section B-4.1 of this book and it is well known how stress takes a relevant role in the sensitivity to a wide range of primary but also opportunistic diseases.

The particular characteristics in behavior and also in feeding and nutritional requirements are also extensively developed here in other chapters in section B. The relatively low domestication level of these species together with their specific characteristics sometimes leads to consider them as troublesome species compared with other usually reared marine species. Pathologies can be always modulated by underlying suboptimal nutritional status due to inadequate feeding, behavioral problems related to hierarchical organization or presence of low performance fish that are difficult to identify during routine examination of the stocks. Finally, the effects of the specific engineering issues like adequate design of the tanks, water flows, tank bottom hygienic conditions, and water quality in open or recirculation systems on the physiology of the reared fish in the different culture phases also should not be overlooked.

2.　Diseases in cultured and wild soleidae

When conditions are adequate, Senegalese sole usually behaves quite well in on-growing phase, with fast growing and good feed conversion rate; however, one of the main factors that has hampered the production of Senegalese sole has been the increase in the number and intensity of diseases (Padrós et al. 2003, Toranzo et al. 2003). Amongst these problems, most of them are diseases that are already described as affecting other species and also sole seems quite prone to become infected by cohabitation from other fish species in the same facilities. In the early 90s, it was usually the presence of several species such as Gilthead Seabream, *Sparus aurata*, European sea bass, *Dicentrachus labrax*, or Turbot *Scophthalmus maximus* together with Senegalese sole. This was a determining factor for the development of diseases in *S. solea* and *S. senegalensis* and even the economic viability of the farms. However, nowadays sole are reared in specific facilities with controlled ambience, highly reducing the risk of transmission. Nonetheless, some diseases are still common in *S. senegalensis* production systems, possibly associated to the lack of appropriate standardized rearing techniques and sometimes due to poor husbandry or hygienic conditions of the tanks, particularly when using sand at the bottom, or when the temperature exceeds 22°C (Cañavate 2005). The most frequent problems at present are infectious problems, mainly bacterial.

2.1　Pathologies due to virus

2.1.1　Viral nervous necrosis

Nervous necrosis virus (NNV, genus Betanodaviruses) is the aetiological agent of the viral nervous necrosis (VNN) or viral encephalopathy and retinopathy (VER), a devastating neuropathological condition that affects marine fish worldwide (Munday et al. 2002, Shetty et al. 2012). The virus affects the central nervous system; therefore, diseased fish show abnormal behavior (swimming erratically, spinning, fish unable to stay at the bottom of the tank) and can reach mortalities up to 100%, mainly in larvae. Serious and devastating episodes of mortality associated with the presence of VER were detected in juvenile *S. solea*, (Borghesan et al. 2003) (Figs. 1A and 1B) and also from *S. senegalensis* (Starkey et al. 2001, Thiéry et al. 2004, Cutrín et al. 2007, Olveira et al. 2009, Hodneland et al. 2011), although clear clinical signs and abnormal behavior are sometimes not so common. Amongst the different genotypes of the virus, RGNNV and SJNNV play a major role in the areas where *Solea* species are reared in Europe (Mediterranean and South-Atlantic areas). The virulence of SJNNV and RGNNV has been demonstrated in experimental infections (Souto et al. 2015a) and some natural outbreaks with RGNNV genotype (Panzarin et al. 2012) and reassortants (Olveira et al. 2009) have been described. Recently, the emergence of VER reassortants has induced relevant changes in virulence and species susceptibility. In these cases, rearing temperature seems to be very important in the virulence, pathogenicity and survival of the virus in the fish (Souto et al. 2015b). Although this virus can be transmitted horizontally by contact between diseased and healthy fish, the main transmission route in hatcheries is vertical (Barja 2004). Diagnostic in clinical cases can be routinely performed after typical neurological symptoms and confirmation with histology (presence of typical neuronal vacuolization and gliosis in the central nervous system and retina) and confirmation by immunohistochemistry. Classic virus culture in susceptible cell lines and molecular detection and diagnostic techniques based on PCR and, more recently, qPCR using whole larvae and fish from different fish tissues have also been developed. Currently there are no efficient treatments and the recommendation is that affected stocks

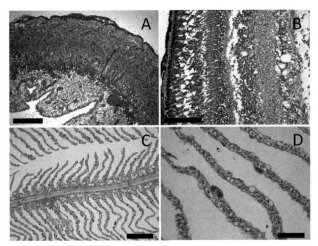

Fig. 1. Viral diseases in Soleidae. (A–B) Betanodavirus (RGNNV genotype) infection in one year-old common sole *Solea solea*. (A) Encephalitis in the hypothalamus with intense gliosis and vascular congestion. Bar scale = 200 µm (B) retinal lesions with vacuolation in different layers (pigment epithelium, inner nuclear layer and ganglion cell layer). Bar scale = 100 µm. (C–D). Hervesvirus in Senegalese sole, *Solea senegalensis* juveniles. (C). Large infected epithelial cells (dark spots) in the epithelium of the gill lamellae. Bar scale = 200 µm. (D). Infected epithelial cells with clear hypertrophic nuclei at higher magnification. Bar scale = 50 µm. (A–D) H/E stain. Figures A and B courtesy of OIE Reference Laboratory for viral encephalopathy and retinopathy of marine fish. Istituto Zooprofilattico Sperimentale delle Venezie.

should be immediately removed and sacrificed when the disease or the virus are detected. Prophylactic measures such as avoiding infected stocks entering in the farm should be taken. Fish should be kept in quarantine and diagnosis techniques should be done in order to select only healthy fish. Broodstock's direct or indirect evaluation for nodavirus presence and, if possible, posterior isolation is paramount to maintain a free stock. Special caution with potentially contaminated food (fresh or frozen) should also be taken. No reports on specific vaccination trials on soleids have been against VER. In contrast, experimental vaccines have been tested for other species. Promising results from experimental oral, bath and injection vaccination in other fish species such as the grouper or Asian sea bass and other species (Lin et al. 2007, Kai and Chi 2008, Nishizawa et al. 2009, Yamashita et al. 2009, Vimal et al. 2016) have been recently published. In addition, some field trials with experimental and pre-commercial vaccines are also under study in different species. These developments may provide a relevant background for a specific and effective vaccine in soleids in the next future.

2.1.2 Herpesvirus-like

Recently, routine health controls indicate that infections associated to herpesvirus are occasionally detected in *S. senegalensis* (Constenla et al. 2015). Affected fish are mainly noticed by behavioral changes and mucus hypersecretion. Histologically, basophilic hypertrophic cells with a hypertrophic nucleus are detected in gills (Figs. 1C and 1D). These cells are usually smaller than those affected by *Herpesvirus scophthalmi* type. These cells are usually accompanied by mild to moderate inflammation and often associated with the presence of proteinaceous exudate and clusters of lymphocytes in the subepithelial space. In addition, similar inflammatory lesions were often observed in kidney and spleen. Transmission electronic microscopy confirmed the presence of polyhedral to round viral capsids with an electron dense core morphologically resembling herpesvirus.

2.1.3 Other virus (associated to diseases, isolated in soleidae or used in viral challenges)

2.1.3.1 Infectious pancreatic necrosis virus (IPNV)

In the 90s, a birnavirus was described as the agent causing almost 100% mortality in wild Senegalese sole broodstock introduced into a culture facility in Southwest Spain (Rodríguez et al. 1997). Affected fish showed dark coloration, hyperactivity and uncoordinated swimming behavior as external signs. The characterization of this virus indicated its similarity to IPNV, reacting positively with the VP2 protein in western blot assays using the anti-IPNV monoclonal antibody AS1 (Rodríguez et al. 1997). However, as birnavirus and IPNV are so widely distributed and affect many different aquatic organisms, its role as a primary pathogen in sole species is not so clear.

2.1.3.2 Lymphocystis disease virus

Lymphocystis disease is caused by an iridovirus with a worldwide geographical distribution that involves a chronic disease characterised by papilloma-like lesions typically on the skin, fins and tail (Walker and Hill 1980). Lymphocystis was detected in wild Dover sole (Bucke 1986) and lymphocystis disease viruses (LCDVs) have been also isolated from cultured Senegalese sole in Southeastern Spain (Alonso et al. 2005, Cano et al. 2010) but without causing clear clinical cases. All these data suggests that soleidae may not be a LCD susceptible disease but a potential carrier.

2.1.3.3 Viral haemorrhagic septicaemia (VHS)

VHS is a well-known disease, especially in farmed salmonids caused by a Novirhabdovirus (Rhabdoviridae). Recently, its detection in wild marine fish becoming more evident led to the conclusions that VHSV is enzootic in marine environment (Skall et al. 2005) but at the same time, the description of different genogroups (Snow et al. 2004) with different susceptibilities of different fish species (including turbot) increases the complexity of the epidemiology of this group. Although no natural VHS outbreaks have been reported for the time being in sole, López-Vazquez et al. (2011) demonstrated the susceptibility of *S. senegalensis* to a VHSV strain isolated from wild Greenland halibut, *Reinhardtius hippoglossoides*, and farmed turbot. In that study, Senegalese sole were infected by an intraperitoneal injection and by immersion, and affected fish showed haemorrhaging of the ventral area and ascitic fluid in the body cavity. In addition, the horizontal transmission of VHSV was confirmed by cohabitation with infected soles, which indicates the potential risk of wild VHSV marine isolates for sole aquaculture (López-Vazquez et al. 2011).

2.1.3.4 Other virus diseases affecting sole

Presence of intracytoplasmic inclusion bodies in erythrocytes similar to those observed in VEN (Viral erythrocytic necrosis) in other species was detected causing low but continuous mortality in a stock of *S. senegalensis* from the wild (Padrós et al. 2003). However, the presence of viral particles could not be demonstrated.

2.2 Pathologies due to bacteria

Nowadays, the main pathological problems for the culture of soles associated to bacteria are Tenacibaculosis (formerly Flexibacteriosis, fin rot or black patch necrosis), Photobacteriosis (formerly Pasteurellosis) and Vibriosis.

2.2.1 Tenacibaculosis/Cutaneous Ulcerative Problems (Black Patch Necrosis)

Tenacibaculosis, which is mainly caused by *Tenacibaculum maritimum* (formerly *Flexibacter maritimum*), can cause significant morbidity and mortality on fish farms in many countries, limiting the culture of economically important marine fish species (Santos et al. 1999). The presence of this pathogen in Europe was first described in Scotland from a *S. solea* suffering from the so-called "black patch necrosis" (BPN) (Bernardet et al. 1990), probably the most important problem in early rearing attempts of these species because of its high incidence (McVicar and White 1979, 1982). The lesions associated with this disease begin as slight blistering of the skin surface or as dark areas between caudal and marginal fins (Figs. 2E and 2F), which rapidly expands, and develop into necrotic ulcers due to the invasion by saprophytic organisms (McVivar and White 1979). Some years later, Cepeda and Santos (2002) isolated for the first time *T. maritimum* from Senegalese sole causing an almost 100% mortality on the affected stocks in a hatchery from southwest Spain. *T. maritimum* was also described affecting other soleidae species such as cultured Wedge sole *Dicologlossa cuneata* (López et al. 2009a). This problem is still quite common in pre-ongrowing and ongrowing of sole culture and it usually appears after stressful situations such as fish handling or grading, with an acute or subacute course and frequently high mortality. Affected sole usually displays several progressive external signs including eroded mouth, rotten fins and skin lesions, typically in dorsal side, superficial erosions that will transform lately in wide deep ulcers. Internally, affected specimens showed paleness of internal organs. *T. maritimum* are Gram-negative thin filamentous bacteria with gliding motility. Typical colonies, flat and pale-yellow with uneven edges and strongly adherent to the medium, grow onto specific *Flexibacter maritimus* Medium (FMM) agar, the most appropriate medium for the successful isolation of this species from fish tissue, after incubation at 20°C to 25°C for 48 hours to 72 hours (Pazos et al. 1996). The species is phenotypically homogeneous, but at least 3 major O-serogroups have been described seemingly related to the host species (Avendaño-Herrera et al. 2004a, 2005): serotype O1, from sole in the northwest of Spain and gilthead sea bream; serotype O2, from turbot; and serotype O3, all strains isolated from sole in Portugal and southern Spain.

Other *Tenacibaculum* species such as *T. discolor* and *T. soleae* were also isolated from diseased Senegalese sole (Piñeiro-Vidal et al. 2008a,b), showing the typical signs observed in fish affected by *T. maritimum*. *T.soleae* was also recently isolated and described from diseased cultured Wedge sole *Dicologlossa cuneata* (López et al. 2010).

Recently, Vilar et al. (2012) described particularly severe ulcerative disease outbreaks in cultured Senegalese sole associated with *T. maritimum*. Grossly, the affected fish showed total loss of epidermis and dermis and extensive necrosis of the muscle layers. Mild to moderate inflammatory response characterised by macrophages and hyaline degenerated muscle cells were described histologically, associated to the lesions.

Although this disease was reported to be highly infectious, it seemed to be both prevented and controlled by providing a sand substrate in the rearing tanks (McVicar and White 1982). This is paradoxical given that the use of sand as a substrate in sole farming has been considered a serious obstacle since a deficient management and hygiene may favor the emergence of bacterial infections (Howell 1997). However, culturing these fish in smooth hard-bottomed tanks without sand or smooth textures have been associated to a series of harmful effects—high incidence of BPN and other skin lesions in flatfish species (Ottesen and Strand 1996, Ottesen et al. 2007), which are problems to be considered not only from an economical point of view, due to reduced survival and growth and the added cost of antibiotics, but also from a consumer standpoint. Many other environmental

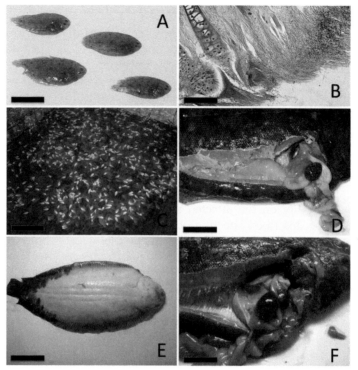

Fig. 2. Bacterial diseases in *Solea senegalensis*. (A–B) Tenacibaculosis in juveniles. (A) Damage in the caudal fin area associated to *Tenacibaculum maritimum* Bar scale = 15 mm. (B) Histopathological aspect of the lesions. Notice the presence of large and thin rods between the fin rays remnants. Bar scale = 25 μm. H/E. (C–D). Pasteurelosis/ Photobacteriosis by *Photobacterium damseale* subs *piscicida*. (C). Severe acute mortality. A high number of the affected or dead fish are turned upsidedown showing the blind side (white). Bar scale = 50 cm. (D) Typical macroscopic lesions : pale liver and enlarged spleen with multiple whitish nodules. Bar scale = 3 cm. (E–F). Vibriosis by *Vibrio anguillarum*. (E) External haemorrhages in the skin of the fish head and body. Bar scale = 5 cm. (F). Enlarged spleen, enlarged gallbladder and ascites in a sole affected by acute vibriosis. Bar scale = 3 cm.

conditions such as higher temperatures, salinity, low water quality, excess of UV light and also management factors (i.e., high density and unbalanced feed) and host-related factors (stress, skin surface condition) have also been described associated to this disease (Avendaño-Herrera et al. 2006).

Presumptive diagnosis is made by anamnesis, presence of typical lesions and observation of bacterial agents in imprints from external lesions and internal organs. Confirmation of diagnosis will be made by isolating bacteria in adequate culture medium and identification by traditional (API), serologic and/or molecular DNA-based methods, such as PCR protocols using the 16S rRNA gene as target (Toyama et al. 1996, Avendaño-Herrera et al. 2004b).

Being a disease that may cause both skin lesions and systemic problems, combined treatments with external disinfectants (such as formalin or hydrogen peroxide) and oral treatments with antibiotics are usually used to control them. Florfenicol, oxitetracycline, flumequine, potentiated sulfonamides and especially enrofloxacin were used for controlling *T. maritimum* outbreaks, although the rapid appearance of resistant strains has already been described (Avendaño-Herrera et al. 2006). Regarding vaccination, there is nowadays a commercially available bacterin to prevent the disease caused by *T. maritimum* in turbot

(Icthiovac TM®), for fish from 20 to 30 g, and also autovaccines are frequently used in sole farms, by bath or IP injection.

2.2.2 Photobacteriosis (formerly pasteurellosis)

Photobacteriosis, caused by *Photobacterium damsela* subsp. *piscicida* (Pdp), is a non-motile, Gram-negative rod bacterium that exhibits bipolar staining (Romalde 2002). It is responsible for high losses in the aquaculture industry since it provokes massive mortalities in cultures of several marine fish species such as Gilthead seabream (Toranzo et al. 1991), Sea bass (Balebona et al. 1992), flatfish Japanese flounder (Fukuda et al. 1996) and many other fish species. All strains of this pathogen are biochemically and serologically homogeneous regardless of the geographic origin and source of isolation; however, two clear separate clonal lineages within Pdp have been detected: a clonal lineage comprising all European isolates and a second one including the Japanese and USA isolates (Toranzo et al. 2005). Since it was recorded in farmed Senegalese sole of southwest Spain (Zorrilla et al. 1999), several sole farms mainly in the south area of Spain have suffered mortalities caused by this disease (Magariños et al. 2003). In most cases, peracute mortalities without apparent lesions (Fig. 2A) are the most typical expression found mainly in juveniles. However, in subacute and chronic cases, external lesions of infected fish only included unspecific symptoms such as erratic swimming, going to surface and then falling to bottom, dark skin coloration, exophthalmia and swelling of the abdominal cavity. Internally, in peracute courses, fish do not show any lesion in internal organs except for a light congestion of organs. However, affected adult fishes showed paleness of liver and kidney and enlargement and development of whitish granulomas of 1–2 mm diameter in the spleen (Fig. 2B), related to the presence of the bacteria (Zorrilla et al. 1999). Severe mortalities occur usually when water temperatures are above 18–20°C. Below this temperature, fish can harbor the pathogen as subclinical infection for long time periods (Magariños et al. 2001).

Diagnosis must be quick and, if possible, when first symptoms are detected because fish stock suddenly stops feeding after these first symptoms of the disease, being impossible to control the problem by conventional medicated feed. With anamnesis, symptoms and lesions observation and detection of the bacteria in internal organs imprints, a preliminary diagnosis can be done and will be enough to establish an emergency treatment. Confirmation will be done lately with isolation of the bacteria and posterior identification by conventional techniques or by molecular and other techniques (Romalde 2002, Andreoni and Magnani 2014).

Choice treatment is medicated feed with appropriate dose of antibiotic during 10–15 days. The most frequently used antibiotics are florfenicol, flumequine and potentiated sulfonamides, although resistances may appear. In addition to antibiotherapy, in some cases, a reduction of the temperature to 15°C for 4 or 5 days will help to decelerate the progression of the disease and, maybe more important, avoid the spread of the disease to other batches. Probiotic use has also been indicated to reduce the infection risk in this disease but also in other bacterial diseases in *S. senegalensis* (García de la Banda et al. 2010).

2.2.3 Vibriosis

Within the Vibrionaceae, the species causing the most economically serious diseases in marine culture are *Listonella* (*Vibrio*) *anguillarum*, *Vibrio ordalii*, *V. salmonicida* and *V. vulnificus* biotype 2. *L. anguillarum* is the aetiological agent of classical Vibriosis and possesses a wide distribution causing a typical haemorrhagic septicaemia in a wide variety of warm and cold water fish species of economic importance (Frans et al. 2011). Of the different known

O serotypes (O1–O23, European serotype designation) among *L. anguillarum* isolates, only serotype O1, O2 and, to a lesser extent, serotype O3 have been associated with mortalities in farmed and feral fish throughout the world (Pedersen et al. 1999). Genetic studies have confirmed the existence of two separate clonal lineages within the major pathogenic serotypes, corresponding to the North European and South European isolates, respectively (Toranzo et al. 2005). Vibrioses affecting common sole and Senegalese sole are usually detected as co-infections or secondary infections associated with an initial Tenacibaculosis (Padrós et al. 2003) but in some cases they can also be primary infections (Manfrin et al. 2003, Paolini et al. 2010). Classical Vibriosis outbreaks in sole juveniles, both O1 and O2 serotypes, have been diagnosed in Spain and Italy. Serotype 01 has been mainly diagnosed in farms located in south of Spain, where sole was in cohabitation with sea bass, whereas serotype 02 in the north, where sole was kept with turbot. Vibriosis in soles usually courses as an acute haemorrhagic septicaemia (Fig. 2C). Preliminary diagnosis is made by anamnesis, lesions and observation of bacteria in imprints from enlarged spleen (Fig. 2D) and kidney and confirmation of the bacteria can be easily achieved by isolation in routine media and identification of the bacteria by traditional or serologic techniques. Molecular techniques can also be applied but traditional isolation and identification is considered as a reliable method for routine diagnostic.

Other Vibrionaceae such as *Vibrio harveyi* and *V. parahaemolyticus* are also pathogenic bacteria which were described in an outbreak of farmed *S. senegalensis* by Zorrilla et al. (2003) causing moderate mortalities in the south of Spain. Main external signs of the disease were skin ulcers and haemorrhagic areas near the fins and mouth (Zorrilla et al. 2003). Similar lesions have been detected in *V. harveyi* infected Wedge sole *Dicoglossa cuneata* (López et al. 2009b). Rico et al. (2008) also characterized strains of *V. harveyi* from diseased-farmed Senegalese sole in Spain from 2000 to 2004, which seems to be causing an increasing number of disease outbreaks. *Vibrio tapetis* was also recently described in cultured and wild (and subsequently captive-held) Wedge sole, *Dicologlossa cuneata* (López et al. 2011) and Dover sole, *Solea solea* (Declerq et al. 2015), respectively. In both cases, affected fish displayed multiple skin lesions and vesicles. Isolation and identification by serotyping and molecular techniques identified closely related strains very similar to originally isolated strains in clam (*Venerupis decussata*). These findings may suggest that *Vibrio tapetis* is a clearly emerging new pathogen for Soleidae species. Gomez-Gil et al. (2012) also describe a new Vibrio species, *Vibrio alfacsensis,* from cultured Senegale sole in two regions of Spain.

There are several studies on vaccinations against these diseases in different fish species (Romalde et al. 2005); more specifically, a divalent vaccine against *Ph. damselae* subsp. *piscicida* and *V. harveyi* that provides short-time protection is being studied in *S.senegalensis* (Arijo et al. 2005). Although many licensed Vibrio vaccines (mostly against *V. anguillarum* infection) area available in the market, no specific vaccines for sole have been registered at present. In contrast, bacterial vaccines registered for other species and also autovaccines against *P. damsela* and *Vibrio* species have been used in some farms with variable results. Recent studies on probiotics to control Photobacteriosis and different *Vibrio* species have given encouraging results (Garcia de la Banda et al. 2012, Tapia-Paniagua et al. 2012, Batista et al. 2013).

2.2.4 Furunculosis

Aeromonas salmonicida subsp. *salmonicida* is the causative agent of the so-called "typical" Furunculosis, which causes economically devastating losses in cultivated salmonids and non-salmonid fish and shows a widespread distribution (Toranzo et al. 2005). In 1992, an outbreak of Furunculosis by *Aeromonas salmonicida* was diagnosed for the first time in turbot reared in floating cages located in northwest of Spain, causing significant economic losses (Toranzo and Barja 1992) and is still a significant problem in some turbot farms. Recently, Magariños et al. (2011) reported A. *salmonicida* subspecies *salmonicida* as the causative agent of a 'typical' Furunculosis outbreak in cultured sole in a marine farm operating in a recirculation system in Galicia. Affected fish showed haemorrhagic areas at the base of the dorsal and ventral fins and, in some cases, ulcerative lesions on the ventral surface. Internally, peritoneal cavities were completely filled with ascitic fluid and livers were extremely pale and showed petechiae. In this particular case, soles were grown in a farm, which also produced turbot, which pointed towards a potential crossed infection of this bacterium from one fish species to another. Diagnostic can be performed using routine microbiological identification techniques or using molecular tools.

2.2.5 *Other bacterial diseases affecting sole*

Similarly, Castro et al. (2012) isolated *Edwarsiella tarda* in Senegalese sole growing in a farm which also produced turbot, which pointed towards a potential crossed infection again. Affected fish showed cutaneous lesions in the dorsal surface, tumefactions around the eyes and haemorrhages in their ventral surface, as well as abundant ascitic fluid, anaemic liver and kidney with petechial haemorrhages for internal lesions.

Lopez et al. (2012) also isolated *Pseudomonas baetica* from adult cultured Wedge sole and demonstrated its pathogenicity for this species.

Regarding lesions in internal organs, granulomas in the kidney and spleen were observed in S. *senegalensis* related to acid-resistant bacteria, Ziehl-Neelsen-positive stain (Padrós and Constenla, personal observations). These bacteria, morphologically resembling *Mycobacterium* spp., may represent a potential new hazard for cultured sole. Transmission of mycobacteria in fishes is poorly understood, but water and associated biofilms are natural habitats for *Mycobacterium* spp. (Pedley et al. 2004) and recirculation systems may play an important role in its transmission in aquaculture.

Although progress has been made on nutrition and feeding, the lack of specific and standardized diets for some early life stages of this species or administration of contaminated live food may also facilitate the development of bacterial infections in soleid larvae. Bacterial enteropathy cases similar to other similar problems in other species such as turbot, seabream and seabass have been observed in larval stages and juveniles of S. *senegalensis* (Padrós et al. 2003).

2.3 *Pathologies due to parasites*

In comparison with other fish groups, a relatively low number of parasitic diseases has been described in cultured Soleidae but in contrast, many parasites have been identified in Soleidae species in the wild. These substantial differences between wild and cultured species lead us to present those associated to wild fish in a specific section (Section 2.3.2 of this chapter).

2.3.1 *Parasites in cultured sole*

In recent years, systemic amoebic disease has become a frequent parasitic problem in cultured Senegalese sole. Although the condition was not associated to high mortalities, reduced growth and high morbidity was noted and fishes showed protuberances on the

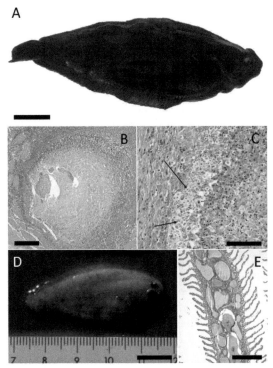

Fig. 3. Parasites and non-infectious diseases in *Solea senegalensis*. (A–C). Amoebic systemic infection by *Endolimax piscium*. (A): macroscopic lesions as body lumps or nodules in the muscle. Bar scale = 15 mm. (B) Granulomatous inflammatory reaction within the muscle. Bar scale = 200 µm. H/E. (C). Granulomatous reaction with small trophozoites in the external layer (arrows). Bar scale= 50 µm. (D) Juvenile with yellow fat necrosis. Notice the yellowish areas near the insertion of the dorsal and anal fins. Bar scale = 10 mm. (E). Multiple thyroid follicles within the gill filaments. Bar scale = 200 µm. H/E.

skin surface (Fig. 3A) in addition to unspecific signs of diseases (lethargy with sporadic and erratic swimming) (Constenla and Padrós 2010). Furthermore, the muscular lesions developed can later prevent the commercialisation of the fish. *Endolimax piscium* (Archamoeba) is the causative agent of this amoebiasis (Constenla et al. 2014) causing a granulomatous inflammatory reaction mainly in muscle but also in different internal organs of the host (Figs. 3B and 3C). This parasite was also detected within the intestinal epithelium and submucosa in both fishes with visible lesions and apparently healthy fishes (Constenla and Padrós 2010). This location seems to be an initial stage in the development of the disease and, consequently, early detection of the parasite in the farm should be considered a priority for the management of this disease in sole culture, since there is no known effective treatment against these parasites. Presumptive diagnosis can be made from external signs and definitive diagnostic includes histology (for symptomatic fish) and molecular methods (PCR and ISH) (Constenla et al. 2016). As with other parasitic diseases, no treatment is available. Management is based on early diagnosis and removal of affected stocks.

Amoeboid organisms similar to *Neoparamoeba* sp. have also been observed in cultured Senegalese sole causing a chronic proliferative mucoid inflammation in gills similar to amoebic gill disease in Atlantic salmon and turbot produced by *Neoparamoeba perurans*, with epithelial hyperplasia and fusion of lamellae in the apical region of some filaments

(Padrós and Constenla, personal observations). At present, the most effective treatment against these gill amoebae in other species seems to be freshwater or hydrogen peroxide baths.

Some sporadic infections by protist parasites like flagellates or ciliates (*Amyloodinium*, *Cryptobia* and *Cryptocaryon*) have also been described (Padrós et al. 2003), usually in cases where sole were reared in ponds. Though infrequently, in cases of massive parasitosis, they can eventually lead to high mortalities, especially in sole grown at high temperatures and in recirculation systems. Prevention and treatment of these diseases are the same as described in other fish species (e.g., marine ornamental fish).

Scuticociliatosis is a parasitic infection caused by ciliated protozoa belonging to the genera *Uronema*, *Miamiensis* and *Philasterides*. Systemic infections by these parasites now represent one of the main challenges for growing marine fish in the world, but although they represent a significant problem in other cultured flatfish species such as turbot (Iglesias et al. 2001) or flounder (Kim et al. 2004), no relevant problems seem to be described or published in *S. solea* or *S. senegalensis* even if they are reared in the same areas as these other flatfish species. However, due to the characteristics of this group of parasites and the high risk and effects in other flatfish species, its importance should not be underestimated for the future.

Another relevant parasitic disease in flatfish species is Enteromyxosis. This disease is mainly found in turbot cultured in Europe and Palenzuela et al. (2007) described an infection by *Enteromyxum scophthalmi* in sole cohabiting with infected turbot. It seems that another important enteric myxozoan *Enteromyxum leei* can also infect sole but only occasionally and after cohabitation or direct contact with other *E. leei* infected fish species such as sea bream (Palenzuela, pers. comm.). Nevertheless, no relevant problems or mortalities have been recorded in *S. solea* or *S. senegalensis* farms. This observation could be interpreted as a lower susceptibility of the soles to the parasite or because of the higher level of biosecurity and control measures in sole farms. In any case, and as in Scuticulociliatosis, the potential risk of enteric myxozoan infection in the future should be taken into account, mainly if the production increases and reaches a critical mass. Other Myxozoa such *Ceratomyxa* sp. have also been observed in sole. Other internal parasites have occasionally been observed in cultured sole *S. senegalensis*, such as myxosporidia found in renal tubules, xenomas of the microsporidian *Tetramicra* sp. in muscle and digenean metacercariae also encysted in the musculature (Padrós et al. 2003). Moreover, ectoparasites such as the leech Hemibdella solea (Hirudinae) have been identified in broodstock of Senegalese sole in the Virginia Institute of Marine Science (Gloucester Point, USA) but it does not seem to adversely affect the fish and can be controlled by reducing water salinity (Dinis et al. 1999).

2.3.2 *Parasites in wild Soleidae species*

Although parasites have been reported to cause disease under cultured conditions, in natural fish population, there is a balance between the host and the pathogen, usually living together in equilibrium. As Soleidae are also relevant species for fisheries with high economic value, oceanographic cruises have become an interesting source of material for the research, including material for parasitological studies. Some of these studies were also encouraged to use the parasites as indicators of their hosts' biology and ecology. Most of the research in this field is focused on macroparasites (Acanthocephala, Monogenea, Digenea Cestoda, Nematoda and Crustacea). Due to the large amount of published data, we decided to compile and summarize them in the Table 1.

Table 1. Parasites from Soleidae species in the wild.

Parasite group	Parasite	Host	Area	References
Ciliophora	Dipartiella simplex	Solea solea	Lake Qarun, Egypt	Ali 2009
	Trichodina gobii	Solea aegyptiaca, Solea senegalensis, Solea solea	Ghar El Melh Lagoon, Bizerte lagoon, Tunis lake	Yemmen et al. 2012a
	Trichodinidae gen. sp.	Solea aegyptiaca, Solea senegalensis, Solea solea	Ghar El Melh Lagoon, Mediterranean Sea	Yemmen et al. 2012a, Constenla el al. 2014
Dinoflagellate	Amyloodinium ocellatum	Solea aegyptiaca, Solea senegalensis, Solea solea	Ghar El Melh Lagoon, Bizerte lagoon	Yemmen et al. 2012a
Myxozoa	Ceratomyxa aegyptiaca	Solea aegyptiaca	Ghar El Melh Lagoon	Yemmen et al. 2012b
	Ceratomyxa sp.	Solea impar, Solea solea	Ghar El Melh Lagoon	Yemmen et al. 2012a
	Zschokkella soleae	Solea solea	Ghar El Melh Lagoon	Yemmen et al. 2013
	Zschokkella sp.	Solea aegyptiaca, Solea senegalensis, Solea solea	Ghar El Melh Lagoon, Bizerte lagoon	Yemmen et al. 2012a
Acanthocephala	Acanthocephaloides cyrusi	Pegusa nasuta	Lake St. Lucia, South Africa	Bray et al. 1988
	Acanthocephaloides distinctus	Microchirus wittei, Monochirus hispidus	Northeast Atlantic Ocean	Vassiliades 1985
	Acanthocephaloides geneticus	Pegusa lascaris, Solea senegalensis	Northeast Atlantic Ocean	Marques et al. 2006a,b
	Acanthocephaloides propinquus	Dicologlossa cuneata, Microchirus azevia, Microchirus variegatus, Monochirus hispidus, Pegusa lascaris, Pegusa nasuta, Solea senegalensis, Solea solea	Northeast Atlantic Ocean, Sea of Marmara, Black Sea	Gaevskaya and Solonchenko 1997, Álvarez et al. 2002, Marques et al. 2006a,b, Oguz and Kvach 2006, Marques et al. 2009
	Acanthocephalus incrassatus	Pegusa lascaris	Northeast Atlantic Ocean	Marques et al. 2006a,b
	Circinatechinorhynchus pseudorhombi	Solea sp., Synaptura sp.	Indian Ocean	Bhattacharya 2007
	Echinorhynchus clavula	Solea senegalensis, Solea solea	Northeast Atlantic Ocean	Carvalho-Varela and Cunha-Ferreira 1987
	Echinorhynchus velli	Synaptura orientalis	Indian Ocean	Bhattacharya 2007
	Echinorhynchus sp.	Euryglossa orientalis	Indian Ocean	Bijukumar 1997a
	Longicollum chabanaudi	Barnardichthys fulvomarginata	no area specified	Golvan and de Buron 1988
	Pseudorhadinorhynchus sp.	Solea senegalensis, Solea solea	Mediterranean Sea	Monje-Ruiz et al. 2015, Solé et al. 2016

Serransentis nadakali	*Zebrias altipinnis*	Indian Ocean	Bijukumar 1997a
Serrasentis sagittifer	*Aesopia cornuta, Dicologlossa cuneata, Dicologlossa hexophthalma, Synaptura orientalis*	Indian Ocean, Northeast Atlantic Ocean	Vassiliades 1985, Amin and Sey 1996, Bhattacharya 2007
Solearhynchus kostylewi	*Pegusa nasuta, Solea solea*	Black Sea, Sea of Marmara	Kostylew 1926, Oguz and Kvach 2006
Pomphorhynchus laevis	*Solea senegalensis, Solea solea*	Northeast Atlantic Ocean	Carvalho-Varela and Cunha-Ferreira 1987
Rhadinorhynchus capensis	*Pegusa nasuta*	no area specified	Golvan and de Buron 1988
Radinorhynchichus sp.	*Microchirus azevia, Pegusa lascaris*	Northeast Atlantic Ocean	Marques et al. 2006a,b
Solearhynchus soleae	*Pegusa impar, Solea solea*	Mediterranean Sea, Sea of Marmara	Golvan and de Buron 1988, Oguz and Kvach 2006
Solearhynchus rhytidotes	*Pegusa impar, Pegusa nasuta*	Mediterranean Sea, Black Sea	Monticelli 1905, Belofastova 2006
Telosentis exiguus	*Pegusa nasuta*	Black Sea	Belofastova and Korniychuk 2000
Yamagutisentis sp.	*Buglossisium luteum*	Mediterranean Sea	Loubes et al. 1988
Digenea			
Allocreadium sp.	*Brachirus orientalis*	Indian Ocean	Bagherpour et al. 2011, Hosseini et al. 2013
Allopodocotyle tunisiensis	*Solea aegyptica*	Mediterranean Sea	Derbel and Neifar 2009
Apocreadium galaicus	*Microchirus variegatus*	Northeast Atlantic Ocean	Sanmartin et al. 1995
Apocreadium sp.	*Microchirus variegatus, Pegusa lascaris*	Northeast Atlantic Ocean	Quintero-Alonso et al. 1988
Aponurus tschugunovi	*Pegusa nasuta*	Black Sea	Gaevskaya and Solonchenko 1997
Derogenes varicus	*Buglossidium luteum, Dicologlossa cuneata, Microchirus variegatus, Pegusa lascaris, Solea senegalensis, Solea solea, Synapturichthys kleinii*	Northeast Atlantic Ocean	Carvalho-Varela and Cunha-Ferreira 1987, Palm et al. 1999, Álvarez et al. 2002, Klimpel et al. 2003, Marques et al. 2006a,b, Marques et al. 2009
Didymozoidae gen. sp.	*Brachyrus sorsogonensis*	North Pacific Ocean	Arthur and Lumanlan-Mayo 1997

Table 1 contd....

...Table 1 contd.

Parasite group	Parasite	Host	Area	References
	Ectenurus lepidus	*Solea solea*	Northeast Atlantic Ocean	Carvalho-Varela and Cunha-Ferreira 1987
	Galactosomum timondavidi	*Solea solea*	Santa Gilla Lagoon, Sardinia, Italy	Culurgioni et al. 2015
	Galactosomum sp.	*Solea senegalensis, Solea solea*	Mediterranean Sea	Constenla et al. 2014, Monje et al. 2015, Solé et al. 2016
	Helicometra fasciata	*Microchirus ocellatus, Microchirus variegatus, Monochirus hispidus, Pegusa nasuta*	Mediterranean Sea, Black Sea	Reversat et al. 1991, Korniychuk and Gaevskaya 1999, Ortis et al. 2002
	Hemipera sp.	*Microchirus azevia, Microchirus variegatus, Pegusa lascaris, Solea senegalensis*	Northeast Atlantic Ocean	Marques et al. 2006b
	Hemiurus appendiculatus	*Pegusa nasuta*	Black Sea	Gaevskaya and Solonchenko 1997
	Hemiuridae metacercaria	*Solea solea*	Northwest Pacific Ocean	Keser et al. 2007
	Hemiuridae gen. sp.	*Solea senegalensis, Solea solea*	Mediterranean Sea	Keser et al. 2007, Monje et al. 2015
	Hemiurus communis	*Solea solea*	Northeast Atlantic Ocean	Carvalho-Varela and Cunha-Ferreira 1987
	Homalometron galaicus	*Dicologlossa cuneata, Microchirus azevia, Microchirus variegatus, Pegusa lascaris, Solea senegalensis*	Northeast Atlantic Ocean	Sanmartin et al. 1995, Álvarez et al. 2002, Marques et al. 2006a,b
	Homalometron senegalese	*Dicologlossa hexophthalma, Synaptura kleinii*	Northeast Atlantic Ocean, Mediterranean Sea	Fischthal and Thomas 1972, Bartoli et al. 2001
	Homalometron wrightae	*Brachirus niger*	Southwest Pacific Ocean	Cribb and Bray 1999
	Lecithochirium musculus	*Buglossidium luteum, Pegusa lascaris, Solea solea, Solea sp.*	Northeast Atlantic Ocean, Mediterranean Sea	Nikolaeva 1966, Lozano et al. 2001, Álvarez et al. 2002
	Lecithochirium rufoviride	*Microchirus azevia, Pegusa lascaris, Solea solea*	Northeast Atlantic Ocean	Álvarez et al. 2002, Marques et al. 2006a,b
	Lepidapedon elongatum	*Solea solea*	Mediterranean Sea	Sey 1970
	Lepocreadioides indicum	*Aesopia cornuta, Zebrias altipinnis, Zebrias synapturoides*	Indian Ocean	Bijukumar 1997a

Lepocreadioides orientalis	*Aesopia cornuta, Zebrias altipinnis, Zebrias synapturoides*	Indian Ocean	Bijukumar 1997b
Lepocreadioides zebrini	*Brachirus orientalis, Zebrias zebra*	Indian Ocean, Northwest Pacific Ocean	Shen and Qiu 1995, Bagherpour et al. 2011, Hosseini et al. 2013
Lintonium vibex	*Buglossidium luteum*	Mediterranean Sea	Hristovski and Jardas 1991
Lomasoma stephanskii	*Microchirus variegatus*	Northeast Atlantic Ocean	Álvarez et al. 2002, Marques et al. 2006b
Lomasoma wardi	*Microchirus variegatus*	Northeast Atlantic Ocean	Quintero-Alonso et al. 1988
Macvicaria cynoglossi	*Aesopia cornuta, Zebrias altipinnis, Zebrias synapturoides*	Indian Ocean	Bijukumar 1997a
Macvicaria jagannathi	*Aesopia cornuta, Zebrias altipinnis, Zebrias synapturoides*	Indian Ocean	Bijukumar 1997a
Macvicaria longicaudus	*Synaptura commersoniana*	Indian Ocean	Bijukumar 1997a
Macvicaria pardachiri	*Pardachirus pavoninus*	Southwest Pacific Ocean	Bray and Justine 2012
Macvicaria soleae	*Dicologlossa cuneata, Microchirus azevia, Microchirus variegatus, Pegusa lascaris, Solea senegalensis, Solea solea*	Northeast Atlantic Ocean	Álvarez et al. 2002, Marques et al. 2006a,b
Otodistomum sp.	*Synapturichthys kleinii*	Northeast Atlantic Ocean	Marques et al. 2009
Podocotyle angulata	*Pegusa lascaris*	Northeast Atlantic Ocean	Álvarez et al. 2002
Podocotyle atomon	*Buglossidium luteum, Pegusa lascaris, Solea solea*	Northeast Atlantic Ocean	Carvalho-Varela and Cunha-Ferreira 1987, Álvarez et al. 2002, Klimpel et al. 2003
Podocotyle sp.	*Solea solea*	Northeast Atlantic Ocean	Durieux et al. 2007
Prodistomum polonii	*Pegusa nasuta*	Black Sea	Bray and Gibson 1990
Proctoeces maculatus	*Solea solea, Zebrias zebra*	Indian Ocean, Northeast Atlantic Ocean	Shen and Qiu 1995, Álvarez et al. 2002
Proctrematoides kuwaiti	*Brachirus orientalis*	Indian Ocean	Sey and Nahhas 1997
Proctrematoides synapturae	*Brachirus orientalis, Pardachirus pavoninus, Synaptura marginata*	Northwest Pacific Ocean, Southwest Pacific Ocean	Machida 2005, Bray and Justine 2012
Prosorhynchus aculeatus	*Solea solea*	Mediterranean Sea	Culurgioni et al. 2015
Prosorhynchus crucibulum	*Dicologlossa cuneata, Microchirus azevia, Microchirus variegatus, Pegusa lascaris, Solea senegalensis, Solea solea, Synaptura lusitanica*	Northeast Atlantic Ocean	Marques et al. 2006a,b, Durieux et al. 2007

Table 1 contd. ...

...Table 1 contd.

Parasite group	Parasite	Host	Area	References
	Prosorhynchus sp.	*Aesopia cornuta, Solea solea, Zebrias altipinnis, Zebrias synaptuaroides*	Indian Ocean, Northeast Atlantic Ocean	Bijukumar 1997a, Durieux et al. 2007
	Stephanostomum lophii	*Microchirus variegatus*	Northeast Atlantic Ocean	Quinteiro et al. 1993
	Stephanostomum sp.	*Microchirus variegatus*	Northeast Atlantic Ocean	Quinteiro-Alonso et al. 1988
	Timoniella spp.	*Solea solea*	Northeast Atlantic Ocean	Durieux et al. 2007
	Timoniella imbutiforme	*Solea solea*	Mediterranean Sea	Culurgioni et al. 2015
	Zoogonoides viviparus	*Microchirus variegatus, Solea solea*	Northeast Atlantic Ocean	Quintero-Alonso et al. 1988, Palm et al. 1999;, Álvarez et al. 2002
	Zoogonus rubellus	*Dicologlossa cuneata, Solea senegalensis*	Northeast Atlantic Ocean	Marques et al. 2006a,b
Monogenean	*Entobdella soleae*	*Dicologlossa cuneata, Solea senegalensis, Solea solea*	Northeast Atlantic Ocean	Little 1929, Kearn and Vasconcelos 1979, Carvalho-Varela and Cunha-Ferreira 1987, Marques et al. 2006a,b
	Gyrodactylus sp.	*Pegusa lascaris, Solea solea*	Northeast Atlantic Ocean	Carvalho-Varela and Cunha-Ferreira 1987, Marques et al. 2006a,b
	Gyrodactylus elegans	*Solea solea*	Northeast Atlantic Ocean	Carvalho-Varela et al. 1981
	Pseudodiplectanum gibsoni	*Microchirus variegatus*	Mediterranean Sea and Atlantic Ocean	Oliver et al. 1987
	Pseudodiplectanum kearni	*Solea solea*	Mediterranean Sea	Vala et al. 1980
	Pseudodiplectanum syrticum	*Synapturichthys kleinii*	Mediterranean Sea	Derbel et al. 2007
Cestoda	*Didymobothrium rudolphii*	*Pegusa impar, Pegusa lascaris, Solea solea*	Mediterranean Sea, Sea of Marmara	Monticelli 1890, Oguz and Bray 2008
	Diphylobothrium sp.	*Synaptura lusitanica*	Northeast Atlantic Ocean	Marques et al. 2006b
	Floriceps sp.	*Aesopia cornuta*	Indian Ocean	El-Naffar et al. 1992
	Grillotia erinaceus	*Solea solea*	Mediterranean Sea	Papoutsoglou and Papaparaskeva-Papoutsoglou 1997

Grillotia minuta	*Pegusa nasuta*	Black Sea	Konyushin and Solonchenko 1978
Grillotia sp.	*Microchirus variegatus, Pegusa lascaris, Solea solea*	Mediterranean Sea, Northeast Atlantic Ocean	Marques et al. 2006b, Álvarez et al. 2002, Keser et al. 2007
Nybelinia lingualis	*Dicologlossa cuneata, Microchirus variegatus, Pegusa lascaris, Solea senegalensis, Solea solea*	Northeast Atlantic Ocean	Álvarez et al. 2002, Marques et al. 2006a,b
Nybelinia jayapaulazariahi	*Aseraggodes macleayanus, Brachirus niger*	South Pacific Ocean	Palm and Beveridge 2002
Phyllobothrium sp.	*Aesopia cornuta, Zebrias altipinnis, Zebrias synapturoides*	Indian Ocean	Bijukumar 1997a
Progrillotia dasyatidis	*Dicologlossa cuneata, Microchirus azevia, Microchirus variegatus, Pegusa lascaris, Solea senegalensis*	Northeast Atlantic Ocean	Marques et al. 2006a,b
Protocephalus sp.	*Solea solea*	Indian Ocean	Bagherpour et al. 2011
Pterobothrium heteracanthum	*Aesopia cornuta*	Indian Ocean	Kardousha 1999
Scolex pleuronectis	*Aesopia cornuta, Dicologlossa cuneata, Microchirus azevia, Microchirus variegatus, Pegusa lascaris, Solea ovata, Solea senegalensis, Solea solea, Synaptura commersonnii, Zebrias altipinnis, Zebrias synapturoides*	Northeast Atlantic Ocean, Mediterranean Sea, Sea of Marmara, Indian Ocean	Carvalho-Varela and Cunha-Ferreira 1987, Bijukumar 1997a, Álvarez et al. 2002, Marques et al. 2006a,b, Keser et al. 2007, Oguz and Bray 2008, Constenla et al. 2014, Monje-Ruiz et al. 2015, Solé et al. 2016.
Trypanorhyncha gen. sp.	*Solea solea*	Mediterranean Sea	Solé et al. 2016
Nematoda *Anisakis simplex sl*	*Dicologlossa cuneata, Microchirus variegatus, Pegusa lascaris*	Northeast Atlantic Ocean	Álvarez et al. 2002, Marques et al. 2006a,b
Anisakis simplex ss	*Pegusa lascaris, Solea senegalensis, Synapturichthys kleinii*	Northeast Atlantic Ocean	Marques et al. 2006c, Marques et al. 2009
Anisakis pegreffii	*Dicologlossa cuneata*	Northeast Atlantic Ocean	Marques et al. 2006c
Anisakidae gen. sp.	*Brachirus sorsogonensis*	Northwest Pacific Ocean	Arthur and Lumanlan-Mayo 1997
Ascaris soleae	*Solea solea*	no area specified	Bruce et al. 1994
Camallanus dollfusi	*Solea elongata*	Indian Ocean	Bashirullah 1973
Capillaria sp.	*Solea elongata*	Indian Ocean	Bashirullah 1973
Capillaridae gen. sp.	*Dicologlossa cuneata, Synaptura lusitanica*	Northeast Atlantic Ocean	Marques et al. 2006b

Table 1 contd. ....

...Table 1 contd.

Parasite group	Parasite	Host	Area	References
	Clavinema mariae	*Zebria zebrinus*	Northwest Pacific Ocean	Yamaguti 1961, Moravec 2006
	Contracaecum rudolphii	*Solea senegalensis, Solea solea*	Santa Gilla Lagoon, Sardinia, Italy, Mediterranean Sea	Culurgioni et al. 2015, Monje et al. 2015
	Cosmocephalus obvelatus	*Solea solea*	Northeast Atlantic Ocean	Palm et al. 1999
	Cucullanus campanae	*Dicologlossa cuneata, Microchirus azevia, Pegusa lascaris, Solea senegalensis, Solea solea*	Mediterranean Sea, Northeast Atlantic Ocean	Petter and Radujkovic 1989, Marques et al. 2006a,b
	Cucullanus heterochrous	*Buglossidium luteum, Microchirus azevia, Microchirus variegatus, Pegusa nasuta, Solea solea*	Black Sea, Northeast Atlantic Ocean, Mediterranean Sea	Gaevskaya and Solonchenko 1997, Palm et al. 1999, Álvarez et al. 2002, Klimpel et al. 2003, Marques et al. 2006a,b, Solé et al. 2016
	Cucullanus minutus	*Pegusa nasuta, Solea solea*	Black Sea, Northeast Atlantic Ocean	Carvalho-Varela and Cunha-Ferreira 1987, Gaevskaya and Solonchenko 1997
	Cucullanus sp.	*Solea solea, Synaptura commersonnii*	Indian Ocean, Northeast Atlantic Ocean	Akram 1994, Palm et al. 1999
	Dichelyne minutus	*Solea solea*	Northeast Atlantic Ocean	Carvalho-Varela et al. 1981, Palm et al. 1999
	Dichelyne sp.	*Solea solea*	Mediterranean Sea	Solé et al. 2016
	Echinocephalus sinensis	*Synaptura cadenati*	Northeast Atlantic Ocean	Obiekezei et al. 1992
	Goezia sp.	*Solea solea*	Northeast Atlantic Ocean	Palm et al. 1999
	Hysterothylacium aduncum	*Buglossidium luteum, Microchirus variegatus, Pegusa lascaris, Pegusa nasuta, Solea solea*	Black Sea, Indian Ocean, Mediterranean Sea, Northeast Atlantic Ocean	Carvalho-Varela and Cunha-Ferreira 1987, Petter and Maillard 1987, Bruce et al. 1994, Gaevskaya and Solonchenko 1997, Álvarez et al. 2002, Klimpel et al. 2003, Marques et al. 2006a,b, Keser et al. 2007, Bagherpour et al. 2011, Tepe and Oguz 2013, Abdel-Ghaffar et al. 2015, Solé et al. 2016

Hysterothylacium fabri	*Buglossidium luteum, Solea solea*	Mediterranean Sea	Hristovski and Jardas 1991, Constenla et al. 2014, Solé et al. 2016
Hysterothylacium reliquens	*Brachirus orientalis, Microchirus azevia, Microchirus boscanion, Microchirus variegatus*	Indian Ocean, Northeast Atlantic Ocean	Petter and Sey 1997, Marques et al. 2006a,b, Marques et al. 2009
Hysterothylacium rhacodes	*Solea aegyptiaca*	Mediterranean Sea	Bruce et al. 1994
Hysterothylacium sp.	*Aesopia cornuta, Microchirus azevia, Pegusa lascaris, Solea ovata, Solea senegalensis, Solea solea, Synaptura commersonnii, Zebrias altipinnis, Zebrias synapturoides*	Indian Ocean, Northeast Atlantic Ocean	Carvalho-Varela and Cunha-Ferreira 1987, Bijukumar 1997a, Marques et al. 2006a,b
Philometra brachiri	*Brachirus orientalis*	Indian Ocean	Moravec and Ali 2014
Procamallanus sp.	*Euryglossa orientalis*	Indian Ocean	Bijukumar 1997a
Raphidascaroides nipponensis	*Zebrias zebra*	Pacific Ocean	Bruce et al. 1994
Spirocamallanus istiblenni	*Soleichthys heterorhinos*	Northwest Pacific Ocean	Hasegawa et al. 1991
Hirudinea			
Caliobdella sp.	*Solea senegalensis*	Northeast Atlantic Ocean	Marques et al. 2006a,b
Hemibdella soleae	*Dicologlossa cuneata, Microchirus azevia, Pegusa lascaris, Solea senegalensis, Solea solea, Synaptura lusitanica, Synapturichthys kleinii*	Northeast Atlantic Ocean	Marques et al. 2006a,b, Marques et al. 2009
Copepoda			
Acanthochondria soleae	*Pegusa lascaris, Solea solea*	Northeast Atlantic Ocean	Marques et al. 2006a,b
Acanthochondria sp.	*Zebrias synapturoides*	Indian Ocean	Bijukumar 1997a
Bomolochus soleae	*Dicologlossa cuneata, Pegusa lascaris, Solea senegalensis, Solea solea, Synapturichthys kleinii*	Northeast Atlantic Ocean	Marques et al. 2006a,b
Bomolochus sp.	*Buglossidium luteum, Solea senegalensis, Solea solea*	Northeast Atlantic Ocean, Mediterranean Sea	Klimpel et al. 2003, Constenla et al. 2014, Monje-Ruiz et al. 2015, Solé et al. 2016
Caligus apodus	*Solea solea*	Mediterranean Sea	Özak et al. 2013
Caligus brevicaudatus	*Pegusa lascaris, Solea senegalensis, Solea solea*	Mediterranean Sea, Northeast Atlantic Ocean, North Pacific Ocean	Choi et al. 1995, Marques et al. 2006a,b, Özak et al. 2013
Caligus elongatus	*Microchirus azevia, Solea solea*	Northeast Atlantic Ocean	Marques et al. 2006a,b
Caligus solea	*Solea solea*	Mediterranean Sea	Demirkale et al. 2014
Ergasilus rostralis	*Euryglossa orientalis*	Indian Ocean	Bijukumar 1997a

Table 1 contd. ...

...Table 1 contd.

Parasite group	Parasite	Host	Area	References
	Ergasilus sp.	*Dicologlossa cuneata*	Northeast Atlantic Ocean	Marques et al. 2006b
	Heterochondria orientalis	*Zebrias fasciatus*	North Pacific Ocean	Moon and Soh 2013
	Heterochondria zebriae	*Zebrias synapturoides*	Indian Ocean	Ho et al. 2000
	Lepeophtheirus europaensis	*Dicologlossa cuneata*	Northeast Atlantic Ocean	Marques et al. 2006b
	Lepeophtheirus pectoralis	*Solea senegalensis*	Northeast Atlantic Ocean	Marques et al. 2006a,b
	Lernaeocera branchialis	*Buglossidium luteum*	Northeast Atlantic Ocean	Klimpel et al. 2003
	Lernaeocera sp.	*Microchirus azevia, Solea solea*	Northeast Atlantic Ocean	Slim 1970, Marques et al. 2006a,b
Isopoda	*Gnathia* sp.	*Microchirus azevia, Pegusa lascaris, Solea solea*	Northeast Atlantic Ocean	Marques et al. 2006a, b
	Nerocila orbignyi	*Solea solea*	Black Sea, Mediterranean Sea	Charfi-Cheikhrouha et al. 2000, Kayis and Ceylan 2011
	Rocinella sp.	*Microchirus azevia, Pegusa lascaris, Solea senegalensis, Synaptura lusitanica*	Northeast Atlantic Ocean	Marques et al. 2006a,b
Pentastomida	*Pentastomida*	*Dicologlossa cuneata, Synapturichthys kleinii*	Northeast Atlantic Ocean	Marques et al. 2006b, Marques et al. 2009

2.5 Non-infectious Diseases

Finally, pathologies of apparently non-infectious origin have also been described by Padrós et al. (2003). Yellow fat necrosis is characterised by the presence of yellowish areas at the base of dorsal and anal fins which correspond to a necrosis of the subdermal adipose tissue (Fig. 3D). It is related to lipid peroxidation and/or excessive exposure to sunlight, and can be associated to secondary bacterial infections. Different levels of kidney damage due to the deposition of minerals in the tubular lumen, similar to the processes of nephrocalcinosis can be sporadically seen in *S. senegalensis*. In addition, "gas bubble disease" expressed as chronic exophthalmia due to the presence of retrobulbar and periocular bubbles were found in Senegalese soles from ponds under hyperoxic conditions (Salas-Leiton et al. 2009). Affected fish also showed bubbles in gills, causing lamellar obstruction, and under the skin all over the body, and deaths were attributed specially to asphyxia and tissue destruction. Proliferation of thyroid follicle (Fig. 3E) has been occasionally observed in Senegalese sole associated to inadequate levels of iodine (Padrós and Constenla, pers observ.).

Pigmentation abnormalities as well as some morphological malformations associated with the migration of the eye are common problems associated with the cultivation of flatfishes. These abnormalities determine the external morphology of the fish, its growth and survival rate, which result in lower market value (Takeuchi et al. 1998, Gavaia et al. 2002). Hypomelanosis or pseudo-albinism, characterized by white patches or areas devoid of normal pigmentation on the ocular surface of the skin, is common in both wild and hatchery reared flatfish. The blind side may display hypermelanosis in the form of dark spots, known as ambicoloration of the skin. Pigmentary disorders and malformations are extensively explained in Chapter B-2.2: Embryonic and larval ontogeny of the Senegalese sole, *Solea senegalensis*: Normal patterns and pathological alterations.

3. Fish disease prevention: general considerations

Most of the pathological problems we have seen in the above section can cause high mortalities in a sole farm in a short period of time. This is mainly due to the apparently high vulnerability to diseases of these species together with the poor response to symptomatic treatments. In sole farming, disease prevention becomes indispensable for the long-term viability of the projects. Besides all these recommendations, it is necessary to bear in mind the relevance of specific physiological and genetic characteristics of the fish, welfare needs, toxicological risks and diet and nutritional requirements of soleidae. All these issues have major direct and indirect influences on the health status or the stocks. We strongly recommend the address to the corresponding sections of this book for further and detailed information.

A health management plan (HMP) covering all aspects of health and welfare have to be in place in the farm. Moreover, in-house fish health professionals and resources, such as a pathology laboratory, have to be available for overseeing the effective application of the HMP and perform routine surveillance and diagnostic work.

The following key aspects about disease prevention should be considered in sole farming:

3.1 Monoculture

Cohabitation with other fish species in the same farming site has to be avoided. Sole has to be farmed in dedicated facilities with no other fish species present. Previous experiences of cohabitation led, in most of the cases, to disease outbreaks due to horizontal transmission of pathogens. Turbot, sea bream and sea bass can be asymptomatic carriers of infectious diseases, and can easily cause epizootic episodes in sole, such as, Viral Nervous Necrosis, Photobacteriosis and Vibriosis.

3.2 Environmental stability

With the high sensitivity of sole to diseases, minimizing stress by controlling water quality in farming conditions is a critical factor. Related to temperature, we not only have to avoid the susceptibility temperature range for infectious agents (low and high), but also any sudden changes that can cause acute stress.

In traditional open systems, such as tanks and land based farms, the environment is very difficult to control. This is one of the reasons why Recirculation Aquaculture Systems (RAS) are becoming the standard for sole farms. RAS are excellent not only for providing the necessary water stability but also from high biosecurity point of view.

Routine monitoring of water parameters, specially temperature, oxygen and salinity, has to be implemented. In RAS systems, other parameters can be considered, such as ammonia, pH or bacterial load.

3.3 Biosecurity

The aim of the biosecurity measures in the farm is to avoid the entry, spread and release of pathogens.

Before entering the farm, fry and brood-stock have to be analyzed at origin and should be sanitary certified. Due to the moderate reliability of some diagnostic techniques when carrier stages of infectious diseases are present, quarantine of the fish inputs is a critical step in sole farms. The quarantine area has to be in a separate building with restriction of personal movement and its own filtration systems. Before going into the general system, new fish inputs have to be kept in quarantine for additional pathological analysis until we are sure they do not carry any disease.

RAS farms, in general, are excellent for external isolation when in high risk areas but, if a pathogen enters the system, the control can be very difficult. This is the reason why in RAS sole farms the quarantine is unavoidable. If we want to farm sole in an open flow system, water filtration before entering the farm and release into the environment has to be considered.

General biosecurity principles of stock management, such as year class separation, all-in-all-out production, restriction of fish and personal movements, establishment of sanitary barriers and fallowing must be applied in sole farming. Special attention has to be paid to the contamination and effluent from the processing plant if present, as in general, this area is closed to the farming area.

Standard Operating Procedures (SOPs) should be in place stating the cleaning and disinfection protocols, and mortality collection and disposal procedures.

3.4 Surveillance and early diagnosis

Due to the fast progression of diseases in sole, an early diagnosis of the pathological problems is essential for the success of any management actions.

The HMP of the farm has to include a disease monitoring and surveillance program. Fish has to be daily monitored for behavior alterations, visible lesions or disease signs. Samples for diagnostic procedures will be taken for routine examinations when a disease is suspected.

There should be dedicated fish health personnel in the farm responsible for application of the surveillance program. An in-house farm laboratory with all necessary resources and equipment for routine pathological diagnostic work is highly recommended. Also, the laboratory should have capacity for preparing and submitting samples for external specialized analysis, such us, bacteriology, histology or virology.

All farm staff has to be trained in diagnostic protocols, including signal alarms' detection and differentiation between normal and sick fish status. Behavior and feed intake alterations generally are an early sign that a disease problem is starting.

Mortality has to be collected daily and examined for disease lesions. After that, mortality is classified and recorded as to suspected causes of death. Any high and/ or unspecified mortality has to be notified to the health responsible/farm manager for further investigation. The farm also should keep records of mortalities, disease outbreak's investigations and treatments applied, in order to have historical data for future analysis when necessary. Dead fish should be removed from the tanks as soon as possible and disposed appropriately.

3.5 Vaccination

Vaccination is the most important single tool for prevention of diseases in salmonid farming. Salmonid farming countries have reduced the use of antibiotics to a minimum due to a country-wide systematic vaccination with licensed vaccines. Unfortunately, volume of production of sole is not high enough to justify the investment of pharmaceutical companies to license and commercialize vaccines for these species. This is the reason why most of the vaccines available in the market for sole are autologous vaccines using pathogens directly isolated from the problematic farm. Which pathogens to include in the vaccine will be based on the farm disease risk assessment. Standard vaccination protocols used commercially for sole are similar to other flatfish species, like turbot. First dip/short-bath vaccination, when fry is followed by a second booster, together with bath vaccination or intraperitoneal injection are the most frequently used protocols.

Aknowledgements

We would like to offer our special thanks to our colleagues Dolores Castro, Carlos Dopazo, Isabel Bandín, José Manuel Leiro, Oswaldo Palenzuela and Anna Toffan for their comments and priceless support in the production of this chapter.

References

Alonso, M.C., I. Cano, E. García-Rosado, D. Castro, J. Lamas, J.L. Barja et al. 2005. Isolation of lymphocystis disease virus from sole, *Solea senegalensis* Kaup, and blackspot sea bream, *Pagellus bogaraveo* (Brünnich). J. Fish Dis. 28: 221–228.

Andreoni, F. and M. Magnani. 2014. Photobacteriosis: Prevention and Diagnosis. J. Immunol. Res. Article ID 793817, 7 pages.

Arijo, S., R. Rico, M. Chabrillon, P. Diaz-Rosales, E. Martínez-Manzanares, M.C. Balelona et al. 2005. Effectiveness of a divalent vaccine for sole, *Solea senegalensis* (Kaup), against *Vibrio harveyi* and *Photobacterium damselae subsp. piscicida*. J. Fish Dis. 28: 33–38.

Avendaño-Herrera, R.E., B. Magariños, S. López-Romalde, J.L. Romalde and A.E. Toranzo. 2004a. Phenotyphic characterization and description of two major O-serotypes in *Tenacibaculum maritimum* strains from marine fishes. Dis. Aquat. Organ. 58: 1–8.

Avendaño-Herrera, R., B. Magariños, A.E. Toranzo, R. Beaz and J.L. Romalde. 2004b. Species-specific polymerase chain reaction primer sets for the diagnosis of *Tenacibaculum maritimum* infection. Dis. Aquat. Organ. 62: 75–83.

Avendaño-Herrera, R.E., B. Magariños, M.A. Moriñigo, J.L. Romalde and A.E. Toranzo. 2005. A novel O-serotype in Tenacibaculum maritimum strains isolated from cultured sole (*Solea senegalensis*). Bull. Eur. Assoc. Fish Pathol. 25: 70–74.

Avendaño-Herrera R.E., A.E. Toranzo and B. Magariños. 2006. Tenacibaculosis infection in marine fish caused by *Tenacibaculum maritimum*: a review. Dis. Aquat. Organ. 71: 255–266.

Balebona, M.C., M.A. Moriñigo, J. Sedano, E. Martinez-Manzanares, A. Vidaurreta, J.J. Borrego et al. 1992. Isolation of *Pasteurella piscicida* from Seabass in southwestern Spain. Bull. Eur. Assoc. Fish Pathol. 12 (1): 168–170.

Barja J.L. 2004. Report about fish viral diseases. pp. 91–102. *In*: Alvarez-Pellitero, P., J.L. Barja, B. Basurco, F. Berthe, A.E. Toranzo (eds.). Mediterranean Aquaculture Diagnostic Laboratories, CIHEAM, Zaragoza. http://om.ciheam.org/om/pdf/b49/04600221.pdf.

Batista, S., S.T. Tapia-Paniagua, M.A. Moriñigo, J.A. Nuñez-Díaz, J.F.M. Gonçalves, R. Barros et al. 2013. Expression of immune response genes in sole (*Solea senegalensis*, Kaup 1858) induced by dietary probiotic supplementation following exposure to *Photobacterium damselae* subsp. *piscicida*. Fish Shellfish Immunol. 34 (6): 1638–1639.

Bernardet, J.F., A.C. Campbell and J.A. Buswell. 1990. *Flexibacter maritimus* is the agent of "black patch necrosis" in Dover sole in Scotland. Dis. Aquat. Organ. 8: 233–237.

Borghesan, F., R. Palazzi, L. Zanella, M. Vascellari, F. Mutinelli, F. Montesi et al. 2003. Viral encephalopathy and retinopathy infection in reared common sole (*Solea solea*). 11th International Conference "Diseases of fish and shellfish" of the European Association of Fish Pathologists (EAFP), 21st–26th September 2003, St Julians (Malta).

Bucke, D., S.W. Feist, M.G. Norton and M.S. Rolfe. 1983. A histopathological report of some epidermal anomalies of Dover sole, *Solea solea* L., and other flatfish species in coastal waters off south-east England. J. Fish Biol. 23: 565–578.

Bucke, D. 1986. A short note on a rare occurrence of lymphocystis in Dover sole, *Solea solea* (L). Bull. Eur. Ass. Fish Pathol. 6: 1.

Cano, I., E.J. Valverde, B. López-Jimena, M.C. Alonso, E. García-Rosado, C. Sarasquete et al. 2010. A new genotype of Lymphocystivirus isolated from cultured gilthead seabream, *Sparus aurata* L., and Senegalese sole, *Solea senegalensis* (Kaup). J. Fish Dis. 33 (8): 695–700.

Cañavate, J.P. 2005. Opciones del lenguado senegalés Solea senegalensis Kaup, 1858 para diversificar la acuicultura marina. Bol. Inst. Esp. Oceanogr. 21 (1-4): 147–154.

Castro, N., A.E. Toranzo, S. Devesa, A. González, S. Nuñez and B. Magariños. 2012. First description of *Edwardsiella tarda* in Senegalese sole, *Solea senegalensis* (Kaup). J. Fish Dis. 35(1): 79–82.

Cepeda, C. and Y. Santos. 2002. First isolation of Flexibacter maritimus from farmed Senegalese sole (*Solea senegalensis*, Kaup) in Spain. Bull. Eur. Ass. Fish Pathol. 22: 388–392.

Colen, R., A. Ramalho, F. Rocha and M.T. Dinis. Cultured Aquatic Species Information Programme. *Solea solea*. Cultured Aquatic Species Information Programme. *In*: FAO Fisheries and Aquaculture Department. Rome. Updated 18 February 2014. http://www.fao.org/fishery/culturedspecies/Solea_spp/en.

Constenla, M. and F. Padrós. 2010. Histopathological and ultrastructural studies on a novel pathological condition in *Solea senegalensis*. Dis. Aquat. Organ. 90: 191–196.

Constenla, M., F. Padrós and O. Palenzuela. 2014. Endolimax piscium sp. nov. (Amoebozoa), causative agent of systemic granulomatous disease of cultured sole, *Solea senegalensis* Kaup. J. Fish Dis. 37(3): 229–40.

Constenla, M., A. Riaza, R. Silva, B. Alonso and F. Padrós. 2015. Comparative histopathology of Herpesvirus infection in *Scophthalmus maximus* and *Solea senegalensis*. 17th EAFP International Conference on diseases of fish and shellfish, Gran Canarias, Spain.

Constenla, M., F. Padrós, R. del Pozo and O. Palenzuela. 2016. Development of different diagnostic techniques for Endolimax piscium (archamoebae) and their applicability in So*lea senegalensis* clinical samples. J. Fish Dis. 39(12): 1433–1443.

Cutrín, J.M., C.P. Dopazo, R. Thiéry, P. Leao, J.G. Olveira, J.L. Barja et al. 2007. Emergence of pathogenic betanodavirus belonging to the SJNNV genogroup in farmed fish species from the Iberian Peninsula. J. Fish Dis. 30: 225–232.

Darias, M.J., K.B Andree, A. Boglino, J. Rotllant, J.M. Cerdà-Reverter, A. Estévez et al. 2013. Morphological and molecular characterization of dietary-induced pseudo-albinism during post-embryonic development of *Solea senegalensis* (Kaup, 1858). PLoS ONE 8: e68844.

Declercq, A.M., K. Chiers, M. Soetaert, A. Lasa, J.L. Romalde, H. Polet et al. 2015. *Vibrio tapetis* isolated from vesicular skin lesions in Dover sole *Solea solea*. Dis. Aquat. Organ. 115(1): 81–6.

Dinis, M.T., L. Ribeiro, F. Soares and C. Sarasquete. 1999. A review on the cultivation potential of *Solea senegalensis* in Spain and in Portugal. Aquaculture 176: 27–38.

Ellis, T., B.R. Hoowell and R.N. Hughes. 1997. The cryptic responses of hatchery-reared sole to a natural sand substratum. J. Fish Biol. 51: 389–401.

Frans, I., C.W. Michiels, P. Bossier, K.A. Willems, B. Lievens and H. Rediers. 2011. *Vibrio anguillarum* as a fish pathogen: virulence factors, diagnosis and prevention. J. Fish Dis. 34: 643–661.

Fukuda, Y., S. Matsuoka, Y. Mizuno and K. Narita. 1996. *Pasteurella piscicida* infection in cultured juvenile Japanese flounder. Fish Pathol. 31 (1): 33–38.

García de La Banda, I., C. Lobo, J.M. León-Rubio, S. Tapia-Paniagua, M.C. Balebona, M.A. Moriñigo, et al. 2010. Influence of two closely related probiotics on juvenile Senegalese sole (*Solea senegalensis*, Kaup 1858) performance and protection against *Photobacterium damselae* subsp. *piscicida*. Aquaculture 306: 281–288.

Garcia de la Banda, I., C. Lobo, M. Chabrillo, J.M. León-Rubio, S. Arijo, G. Pazos et al. 2012. Influence of dietary administration of a probiotic strain *Shewanella putrefaciens* on senegalese sole (*Solea senegalensis*, Kaup 1858) growth, body composition and resistance to *Photobacterium damselae* subsp *piscicida*. Aquaculture Res. 43: 662–669.

Gavaia, P.J., M.T. Dinis and M.L. Cancela. 2002. Osteological development and abnormalities of the vertebral column and caudal skeleton in larval and juvenile stages of hatchery-reared Senegal sole (*Solea senegalensis*). Aquaculture 211: 305–323.

Gomez-Gil, B., A. Roque, L. Chimetto, A.P.B. Moreira, E. Lang and F. Thompson. 2012. *Vibrio alfacsensis* sp. nov., isolated from marine organisms. Int. J. Sys. Evol. Microbiol. 62: 2955–2961.

Hodneland, K., R. García, J.A. Balbuena, C. Zarza and B. Fouz. 2011. Real-time RT-PCR detection of betanodavirus in naturally and experimental infected fish from Spain. J. Fish Dis. 34: 189–202.

Howell, B.R. 1997. A re-appraisal of the potential of the sole, *Solea solea* (L.), for commercial cultivation. Aquaculture 155: 359–369.

Iglesias, R., A. Paramá, M.F. Álvarez, J. Leiro, J. Fernández and M.L. Sanmartín. 2001. *Philasterides dicentrarchi* (Ciliophora, Scuticociliatida) as the causative agent of scuticociliatosis in farmed turbot *Scophthalmus maximus* in Galicia (NW Spain). Dis. Aquat. Org. 46: 47–55.

Kai, Y.H. and S. C. Chi. 2008. Efficacies of inactivated vaccines against betanodavirus in grouper larvae (*Epinephelus coioides*) by bath immunization. Vaccine 26: 1450–1457.

Kim, S.M., J.B. Cho, S.K. Kim, Y.K. Nam and K.H. Kim. 2004. Occurrence of scuticociliatosis in olive flounder *Paralichthys olivaceus* by *Phiasterides dicentrarchi* (Ciliophora: scuticociliatida). Dis. Aquat. Organ. 62(3): 233–238.

Lin, C.C., J.H.Y. Lin, M.S. Chen and H.L.Yang. 2007. An oral nervous necrosis virus vaccine that induces protective immunity in larvae of grouper (*Epinephelus coioides*). Aquaculture 268: 265–273.

López, J.R., S. Núñez, B. Magariños, N. Castro, J.I. Navas, R. de La Herran and A. Toranzo. 2009a. First isolation of *Tenacibaculum maritimum* from wedge sole, *Dicologoglossa cuneata* (Moreau). J. Fish Dis. 32(7): 603–610.

López, J.R., E. de la Roca, S. Núñez, R. de la Herran, J.I. Navas, M. Machado et al. 2009b. Identification of *Vibrio harveyi* isolated from diseased cultured wedge sole *Dicologoglossa cuneata*. Dis. Aquat. Organ. 84(3): 209–217.

López, J.R., M. Piñeiro-Vidal, N. García-Lamas, R. de La Herran, J.I. Navas, I. Hachero-Cruzado et al. 2010. First isolation of *Tenacibaculum soleae* from diseased cultured wedge sole, *Dicologoglossa cuneata* (Moreau), and brill, *Scophthalmus rhombus* (L.). J. Fish Dis. 33(3): 273–278.

López, J.R., S. Balboa, S. Núñez, E. de la Roca, R. de la Herran, J.I. Navas et al. 2011. Characterization of *Vibrio tapetis* strains isolated from diseased cultured Wedge sole (*Dicologoglossa cuneata* Moreau). Res. Vet. Sci. 90(2): 189–95.

López, J.R., A.L. Diéguez, A. Doce, E. de la Roca, R. de la Herran, J.I. Navas et al. 2012. *Pseudomonas baetica* sp. nov., a fish pathogen isolated from wedge sole, *Dicologlossa cuneata* (Moreau). Int. J. Sys. Evol. Microbiol. 62: 874–882.

López-Vázquez, C., M. Conde, C.P. Dopazo, J.L. Barja and I. Bandín. 2011. Susceptibility of juvenile sole *Solea senegalensis* to marine isolates of viral haemorrhagic septicaemia virus from wild and farmed fish. Dis. Aquat. Organ. 93: 11–116.

Magariños, B., N. Couso, M. Noya, P. Merino, A.E. Toranzo and J. Lamas. 2001. Effect of temperature on the development of pasteurellosis in carrier gilthead seabream (*Sparus aurata*). Aquaculture 195(1-2): 17–21.

Magariños, B., J.L. Romalde, S. López-Romalde, M.A. Moriñigo and A.E. Toranzo. 2003. Pathobiological characterisation of *Photobacterium damselae* subsp. *piscicida* isolated from cultured sole (*Solea senegalensis*). Bull. Eur. Ass. Fish Pathol. 23(4): 183–190.

Magariños, B., S. Devesa, A. González, N. Castro and A.E. Toranzo. 2011. Furunculosis in Senegalese sole (*Solea senegalensis*) cultured in a recirculation system. Vet. Rec. 168(16): 431.

Manfrin, A., M. Doimi, P. Antonetti, L. Delgado Montero, K. Quartieri, E. Ramazzo et al. 2003. Primo isolamento di *Vibrio anguillarum* sierotipo O1 dalla sogliola comune (*Solea solea*, L.) in Italia. Bolletino Società Italiana di Patologia Ittica 37: 13–15.

McVicar, A.H. and P.G. White. 1979. Fin and skin necrosis of cultivated Dover sole *Solea solea* (L.). J. Fish Dis. 2: 557–562.

McVicar, A.H. and P.G. White. 1982. The prevention and cure of an infectious disease in cultivated juvenile dover sole, *Solea solea* (L). Aquaculture 26: 3–4.

Munday, B.L., J. Kwang and N. Moody. 2002. Betanodavirus infections of teleost fish: a review. J. Fish Dis. 25: 127–142.

Munroe, T.A. 2001. Soleidae. Soles. pp. 3381–4218. *In*: Carpenter, K.E. and V.H. Niem (eds.). FAO species Identification Guide for Fishery Purposes. The Living Marine Resources of the Western Central Pacific. Volume 6. Bony fishes part 4 (Labridae to Latimeriidae), estuarine crocodiles, sea turtles, sea snakes and marine mammals. FAO, Rome.

Nishizawa, T., I. Takami, Y. Kokawa and M. Yoshimizu. 2009. Fish immunization using a synthetic double-stranded RNA Poly (I:C), an interferon inducer, offers protection against RGNNV, a fish nodavirus. Dis. Aquat. Organ. 83: 115–122.

Olveira, J.G., S. Souto, C.P. Dopazo, R. Thiéry, J.L. Barja and I. Bandín. 2009. Comparative analysis of both genomic segments of betanodaviruses isolated from epizootic outbreaks in farmed fish species provides evidence for genetic reassortment. J. Gen. Virol. 90: 2940–2951.

Ottesen, O.H. and H.K. Strand. 1996. Growth, development and skin abnormalities of halibut, *Hippoglossus hippoglossus*, L., juveniles kept on different bottom substrates. Aquaculture 146: 17–25.

Ottesen, O.H., E.J. Noga and W. Sandaa. 2007. Effect of substrate on progression and healing of skin erosions and epidermal papillomas of Atlantic halibut, *Hippoglossus hippoglossus* (L.). J. Fish Dis. 30: 43–53.

Padrós, F., C. Zarza, A. Estévez, S. Crespo and M.D. Furones. 2003. La patología como factor limitante para el desarrollo del cultivo del lenguado. IX Congreso nacional de acuicultura, Cádiz, Spain. May 12 to 16, pp. 343–345.

Palenzuela, O., M.J. Redondo, E. López and P. Álvarez-Pellitero. 2007. Cultured sole, *Solea senegalensis* is susceptible to *Enteromyxum scophthalmi*, the myxozoan parasite causing turbot emaciative enteritis. Parassitologia 49: 73.

Panzarin, V., A. Fusaro, I. Monne, E. Cappellozza, P.P. Patarnello, I. Capua et al. 2012. Molecular epidemiology and evolutionary dynamics of betanodavirus in southern Europe. Infect. Genet. Evol. 12: 63–70.

Paolini, A., G.E. Magi, L. Gennari, C. Egidi, M. Torresi, M. Vallerani et al. 2010. Severe mortality in common sole (*Solea solea*) at post-weaning stage following environmental stress. Bull. Eur. Ass. Fish Pathol. 30(5): 160.

Pazos, F., Y. Santos, A.R. Macías, S. Núñez and A.E. Toranzo. 1996. Evaluation of media for the succesful culture of *Flexibacter maritimus*. J. Fish. Dis. 19: 193–197.

Pedersen, K., L. Grisez, R. Van Houdt, T. Tianinem, F. Ollevier and J.L. Larsen. 1999. Extended serotyping scheme for *Vibrio anguillarum* with the definition and characterization of seven provisional O-serogroups. Curr. Microbiol. 38: 183–189.

Pedley, S., J. Bartram, G. Rees, A. Dufour and J. Cotruvo. 2004. Pathogenic Mycobacteria in Water: A Guide to Public Health Consequences, Monitoring, and Management. IWA Publishing, London, UK.

Piñeiro-Vidal, M., A. Riaza and Y. Santos. 2008a. *Tenacibaculum discolor* sp. nov. and *Tenacibaculum gallaicum* sp. nov., isolated from sole (*Solea senegalensis*) and turbot (*Psetta maxima*) culture systems. Int. J. Sys. Evol. Microbiol. 58: 21–25.

Piñeiro-Vidal, M., C.G. Carballas, O. Gómez-Barreiro, A. Riaza and Y. Santos. 2008b. *Tenacibaculum soleae* sp. nov., isolated from diseased sole (*Solea senegalensis* Kaup). Int. J. Sys. Evol. Microbiol. 5: 881–885.

Rico, M., S. Tapia-Paniagua, E. Martínez-Manzanares, M.C. Balebona and M.A. Moriñigo. 2008. Characterization of *Vibrio harveyi* strains recovered from diseased farmed Senegalese sole (*Solea senegalensis*). J. Appl. Microbiol. 105(3): 752–760.

Rodríguez, S., M.P. Vilas, M.C. Gutierrez, I. Pérez-Prieto, M.C. Sarasquete and B. Rodríguez. 1997. Isolation and preliminary characterization of a Birnavirus from the sole *Solea senegalensis* in southwest Spain. J. Aquat. Anim. Health 9: 295–300.

Romalde, J.L. 2002. *Photobacterium damselae* subsp. *piscicida*: an integrated view of a bacterial fish pathogen. J. Int. Microbiol. 5(1): 3–9.

Romalde J.L., C. Raveo, S. Lopez-Romalde, R. Avendano-Herrera, B. Magariños and A.E. Toranzo. 2005. Vaccination strategies to prevent emerging diseases for Spanish aquaculture. *In*: Progress in Fish Vaccinology. Dev. Biol. Basel, Karger 121: 85–95.

Sarasquete, C., M.L. González de Canales, J.M. Arellano, J.A. Muñoz-Cueto, L. Ribeiro and M.T. Dinis. 1998. Histochemical study of skin and gills of Senegalese sole, *Solea senegalensis* larvae and adults. Histol. Histopathol. 13: 727–735.

Salas-Leiton, E., B. Cánovas-Conesa, R. Zerolo, J. López-Barea, J.P. Cañavate and J. Alhama. 2009. Proteomics of juvenile Senegal sole (*Solea senegalensis*) affected by gas bubble disease in hyperoxygenated ponds. Mar. Biotechnol. 11: 473–487.

Santos, Y., F. Pazos and J.L. Barja. 1999. *Flexibacter maritimus*, causal agent of flexibacteriosis in marine fish. pp. 1–6. *In*: Olivier, G. [ed.]. ICES Identification Leaflets for Diseases and Parasites of Fish and Shellfish. No. 55. International Council for the Exploration of the Sea. Copenhagen, Denmark.

Shetty, M., B. Maiti, K. Shivakumar Santhosh, M.N. Venugopal and I. Karunasagar. 2012. Betanodavirus of Marine and Freshwater Fish: Distribution, Genomic Organization, Diagnosis and Control Measures. Indian J. Virol. 23(2): 114–123.

Skall, H.F., N.J. Olesen and S. Mellergaard. 2005. Viral haemorrhagic septicaemia virus in marine fish and its implications for fish farming:a review. J. Fish Dis. 28: 509–529.

Snow, M., N. Bain, J. Black, V. Taupin, C.O. Cunningham, J.A. King, H.F. Skall and R.S Raynard. 2004. Genetic population structure of marine viral haemorrhagic septicaemia virus (VHSV). Dis. Aquat. Org. 4; 61(1–2): 11–21.

Souto, S., B. Lopez-Jimena, M.C. Alonso, E. García-Rosado and I. Bandín. 2015a. Experimental susceptibility of European sea bass and Senegaleses sole to different betanodavirus isolates. Vet. Microbiol. 177: 53–61.

Souto, S., J.G. Olveira and I. Bandín. 2015b. Influence of temperature on betanodavirus infection in Senegalese sole (*Solea senegalensis*). Vet. Microbiol. 179(3–4): 162–167.

Starkey, W.G., J.H. Ireland, K.F. Muir, M.E. Jenkins, W.J. Roy, R.H. Richards et al. 2001. Nodavirus infection in Atlantic cod and Dover sole in the UK. Vet. Record. 149(6): 179–181.

Takeuchi, T., J. Dedi, Y. Haga, T. Seikai and T. Watanabe. 1998. Effect of vitamin A compounds on bone deformity in larval Japanese flounder *(Paralichthys olivaceus)*. Aquaculture 169: 155–165.

Tapia-Paniagua, S.T., P. Díaz-Rosales, J.M. León-Rubio, I. García de La Banda, C. Lobo, J.F. Alarcón et al. 2012. Use of the probiotic *Shewanella putrefaciens* Pdp11 on the culture of Senegalese sole (*Solea senegalensis*, Kaup 1858) and gilthead seabream (*Sparus aurata* L.). Aquacult. Int. 20: 1025–1039.

Toranzo, A.E and J.L. Barja. 1992. First report of furunculosis in turbot (*Scophthalmus maximus*) reared in floating cages in Northwest of Spain. Bull. Eur. Ass. Fish Pathol. 12(5): 147–149.

Toranzo, A.E., S. Barreiro, J.F. Casal, A. Figueras, B. Magariños and J.L. Barja. 1991. Pasteurellosis in cultured gilthead seabream (*Sparus aurata*): first report in Spain. Aquaculture 99: 1–15.

Toranzo, A.E., R. Avendaño, C. López-Vázquez, B. Magariños, C.P. Dopazo and J.L. Romalde. 2003. Principales patologías bacterianas y víricas en lenguado cultivado: caracterización de los agentes etiológicos. IX Congreso nacional de acuicultura, Cádiz, May 12 to 16, pp. 354–356.

Toranzo, A.E., M. Beatriz and L.R. Jesús. 2005. A Review of the Main Bacterial Fish Diseases in Mariculture Systems. Aquaculture 246(1): 37–61.

Toyama, T., K. Kita-Tsukamoto and H. Wakabayashi. 1996. Identification of *Flexibacter maritimus*, *Flavobacterium branchiophilum* and *Cytophaga columnaris* by PCR targeted 16S ribosomal DNA. Fish Pathol. 31: 25–31.

Thiéry, R., J. Cozien, C. de Boisseson, S. Kerbart-Boscher and L. Nevarez. 2004. Genomic classification of new betanodavirus isolates by phylogenetic analysis of the coat protein gene suggests a low host-fish species specificity. J. Gen. Virol. 85: 3079–3087.

Vilar, P., L.D. Faílde, R. Bermúdez, F. Vigliano, A. Riaza, R. Silva et al. 2012. Morphopathological features of a severe ulcerative disease outbreak associated with *Tenacibaculum maritimum* in cultivated sole, *Solea senegalensis* (L.). J. Fish Dis. 35: 437–445.

Villalta, M., A. Estévez and M.P. Bransden. 2005. Arachidonic acid enriched live prey induces albinism in Senegalese sole (*Solea senegalensis*) larvae. Aquaculture 245: 193–209.

Vimal, S., M.A. Farook, N. Madan, S. Abdul Majeed, K.S.N. Nambi, G. Taju et al. 2016. Development, distribution and expression of a DNA vaccine against nodavirus in Asian Seabass, *Lates calcarifier* (Bloch, 1790). Aquacult. Res. 47: 41–12.

Walker, D.P. and B.J. Hill. 1980. Studies on the culture assay of infectivity and some *in vitro* properties of Lymphocystis virus. J. Gen. Virol. 51: 385–395.

Yamashita, H., K. Mori and T. Nakai. 2009. Protection conferred against viral nervous necrosis by simultaneous inoculation of aquabirnavirus and inactivated betanodavirus in the seven band grouper, *Epinephelus septemfasciatus* (Thunberg). J. Fish Dis. 32: 201–210.

Zorrilla, I., M.C. Balebona, M.A. Moriñigo, C. Sarasquete and J.J. Borrego. 1999. Isolation and characterization of the causative agent of pasteurellosis, *Photobacterium damsela* spp. *piscicida*, from sole, *Solea senegalensis*. J. Fish Dis. 22: 167–172.

Zorrilla, I., S. Arijo, M. Chabrillon, P. Diaz, E. Martinez-Manzanares, M.C. Balebona et al. 2003. *Vibrio* species isolated from diseased farmed sole, *Solea senegalensis* (Kaup), and evaluation of the potential virulence role of their extracellular products. J. Fish Dis. 26: 103–108.

B-5.1

Osmoregulation

Ignacio Ruiz-Jarabo,[1,2] *Juan Fuentes*[1] and
Juan Miguel Mancera[2,*]

1. Osmoregulatory System

Teleostean fishes are well known for their osmoregulatory capacities. They are able to maintain their "internal milieu" within controlled and (almost) stable levels of ions and osmotic pressure. The maintenance of this homeostasis requires allostatic modifications to adjust to both predictable and unpredictable events (McEwen and Wingfield 2003). Teleosts maintain ionic internal body fluid levels to approximately one-third of those found in seawater (SW), with osmolality values close to 10–12 g L^{-1} salinity (~ 300 mOsm kg^{-1}). However, osmoregulation entails at least two problems dependent on the environmental salinity: (i) teleosts living in a hypo-osmotic environment (e.g., fresh water, FW, ~ 140 mOsm kg^{-1}) must counteract passive loss of ions and gain of water by active, energy dependent uptake of ions (primarily through the gills), and remove excess water by formation of dilute urine; (ii) teleosts in hyper-osmotic environments (e.g., seawater, SW, ~ 1050 mOsm kg^{-1}) must avoid gain of ions, or expel them, and counteract loss of water by drinking. In this case, water is taken up via the intestine into the body while the excess of divalent ions is secreted through the gut and kidney, and monovalent ions (mainly sodium and chloride) through the gills (Evans 2008).

Euryhaline teleosts are able to adapt their osmoregulatory tissues, namely gills, intestine and kidney, to cope with changes in environmental salinity and to maintain a constant osmolality (280–360 mOsm kg^{-1}) in their body fluids (Fiol and Kultz 2007). Therefore, they have the ability to adjust and adapt their osmoregulatory and ion transport strategies to the demands imposed by the surrounding environment. This acclimation capacity, however, depends on the developmental stage of the species (reviewed by Ruiz-Jarabo et al. 2015).

Euryhaline fish are able to live within a wide range of environmental salinities without major consequences. This plasticity requires changes in osmoregulatory organs,

[1] Centre of Marine Sciences (CCMar), Universidade do Algarve, Campus de Gambelas, 8005-139 Faro, Portugal.
[2] Department of Biology, Faculty of Marine and Environmental Sciences, Campus de Excelencia Internacional del Mar (CEI-MAR) University of Cadiz, 11510 Puerto Real, Cádiz, Spain.
* Corresponding author: juanmiguel.mancera@uca.es

but also a metabolic reorganization to meet altered energetic demands associated to the new environment. This adaptation is usually accompanied by modifications in oxygen consumption, which suggests variations in the energy demands for osmoregulation (Soengas et al. 2008).

This review will integrate our knowledge on osmoregulatory system in the Senegalese sole (*Solea senegalensis*, Kaup 1858), a marine teleost that inhabits coastal waters and riverine estuaries, capable of adapting to substantial changes in environmental salinity and temperature (Arjona et al. 2009, 2010), and that it is important for aquaculture (Imsland et al. 2004). In addition, metabolic changes related to osmotic acclimation will be reviewed, focusing on osmoregulatory (gill, kidney, intestine) and non-osmoregulatory organs. We will also briefly describe the endocrine control of the osmoregulatory processes involved in this species.

2. Intestine

Water metabolism is of vital importance for most vertebrates including fish. The mechanisms regulating water ingestion (drinking) seem to vary in the degree of complication between terrestrial and aquatic vertebrates. In terrestrial animals the "drinking" process involves not only physiological but also behavioral mechanisms. Thus, drinking involves a thirst signal, water seeking behavior, water ingestion, swallowing, intestinal processing and finally satiation for drinking termination. This phenomenal integrated process involves a quite complex route for endocrine and neuroendocrine systems in the control of water metabolism. At first glance, the control of water drinking in fish appears somehow less complicated by the fact fish swim and breathe in the water they drink. Therefore, some of the integrated complex processes became less relevant (i.e., water seeking). However, the mechanisms regulating the acquisition, retention and excretion of water in fish remain currently inadequately understood. In addition, a fundamental and common problem to all fish, either from seawater or freshwater, is how to maintain the osmotic disequilibrium of their body fluids with that of the external environment.

Since the Senegalese sole is essentially a seawater inhabiting species this section will focus on the contribution of the intestinal tract to ion regulation in marine teleosts. The intestine of marine fish plays a key role in ion regulation. The ionic disequilibrium of marine fish with their surrounding environment requires high rates of drinking, as part of the osmoregulatory process to compensate the dehydrating effect of seawater in the gills (Evans et al. 2005). One of the main features of the contribution of the intestine to osmoregulation in seawater teleosts relates to water ingestion (drinking) and the processing of the ingested seawater to achieve net absorption of water at intestinal levels. In this context, ion assimilation from the ingested fluid is required to drive water absorption, making the role of the intestine vital to maintain extracellular homeostasis. Previous classical work (Hirano and Mayer-Gostan 1976, Parmelee and Renfro 1983) has shown that the esophagus plays a key role in the initial process of monovalent ions (not water) assimilation. As an example, our work in the sea bream (Marquez and Fuentes 2014) shows a substantial absorption of monovalent ions in the esophagus. The imbibed fluid (seawater) with an osmolality of 1100 mOsm kg^{-1}, reaches the stomach with a substantial decrease in osmolality to around 400 mOsm kg^{-1}. This fluid is further processed in the intestine (Gregorio et al. 2013) where osmolality is further decreased to match (or closely match) plasma osmolality at about 350 mOsm kg^{-1}. This osmolality reduction is the result of several processing mechanisms that prevail in different sections of the gastrointestinal system, which are too extensive and complex to report here in detail. Therefore, we will focus this section in a few selected aspects of intestinal function, which in light of the little information on the intestine of sole in terms of ion regulation should establish future guidelines to expand our knowledge.

In fish the water taken up for physiological purposes involves at least two mechanisms; first: a drinking process and second, in the case of seawater fish, the desalination and the subsequent absorption of water in the intestine. The rates of drinking play a fundamental role in the balance/equilibrium between water ingestion, fluid processing and plasma osmolality (Fuentes and Eddy 1997a). This process is under the regulation of the endocrine system, mainly of the renin-angiotensin system (Fuentes and Eddy 1996). However, model species for study of the drinking process and the endocrine control of drinking have been mainly Salmonids, sharks and eels, few other species have been analyzed.

Drinking responds quickly to changes in salinity, for example rainbow trout maintained in seawater, completely cease drinking when transferred to freshwater within 30 minutes (Fuentes, unpublished). There are several examples of the impact of salinity on drinking than come from Salmonids (Fuentes and Eddy 1997b) and non-Salmonids (Guerreiro et al. 2004). In general, the rates of drinking vary between 0.1–0.4 mL kg^{-1} h^{-1} in freshwater fish depending on the species and fish size, smaller fish have higher drinking rates on a weight basis than juveniles. In contrast, seawater fish juveniles drink in the range of 3–7 mL kg^{-1} h^{-1}, while larvae have phenomenal drinking rates, such as sea bream larvae kept in full strength seawater that drink in the range of 25 mL kg^{-1} h^{-1} (Guerreiro et al. 2001), a process also observed in tilapia (*Oreochromis mossambicus*) larvae, that when hatched and reared in seawater drink in the range of 30–35 mL kg^{-1} h^{-1} (Fuentes, unpublished). In the sole, drinking rates have not been yet analyzed at either stage, and there is not a clear notion of how drinking takes place in this species. However, considering our published results on intestinal fluid characterization (Ruiz-Jarabo et al. 2017), there is an indication of differential drinking rates derived from the changing amount of magnesium accumulation in the intestinal fluid. At a given particular time the intestinal fluid in unfed sole reaches values as high as 170 mM of magnesium in fish kept at 55 ppt and in the order of 130 mM when fish are kept at 38 ppt (full-strength seawater). These values are well in the range of those reported in full-strength seawater for sea bream of circa 130 mM (Fuentes et al. 2006) which drink at a rate of around 5.5 mL kg^{-1} h^{-1} (Guerreiro et al. 2002) or the toadfish (*Opsanus beta*) which accumulate magnesium in the range of 140 mM (Genz et al. 2008), and have drinking rates in the range of 3.3 mL kg^{-1} h^{-1}.

Drinking supplies the crude fluid for potential water absorption, but in order to fuel intestinal net water absorption the fluid ingested has to be processed. The main evidence of selective intestinal processing comes from the particular composition of intestinal fluid in marine fish. In addition to the monovalent ion uptake that takes place in the esophagus, the fluid in the intestine is nearly isosmotic to the plasma of most of the species thus far analyzed. Since it accumulates in the lumen of the intestine, the intestinal fluid magnesium, often 50 to 70 fold higher that in the plasma concentration, has been considered a good proxy for dinking rates. In contrast, monovalent ions, sodium and chloride are often lower than their corresponding plasma levels. In the sea bream fluid sodium is in the range of 90 mM (Fuentes et al. 2006), about half the concentration plasma sodium. In the toadfish, intestinal fluid sodium levels reach values as low as 50 mM (Grosell 2006, 2011). A similar feature of fluid composition was reported for chloride, with intestinal levels in the range of 50% to 70% of plasma chloride levels of matching fish. Interestingly intestinal fluid calcium concentration is at least 2 to 4 times higher that the corresponding plasma levels of the matching fish. It appears that while divalent ions seem to accumulate in the intestinal lumen, monovalent ions are absorbed to drive net water absorption.

The driving force for monovalent ion transport in the intestine in fish is the basolateral Na^+/K^+-ATPase (NKA), which has been demonstrated to be higher at higher salinities in most species including the trout (Fuentes et al. 1997) and the eel (Kim et al. 2008) and this is apparently the case in sole according to our published results. At this stage is relevant to point out the risk of oversimplification of these reports, given the anatomical heterogeneity of the intestine as such and the size of the gastrointestinal tract in most fish models. Most studies have used intestinal regions referred to as anterior intestine or simply intestine excluding the rectum. Recently Grosell (2006, 2011) has provided solid evidence for the involvement of an additional electrogenic mechanism in the intestine of fish a V-type proton pump. Apparently, its biological action on fluid processing is related to the intestinal region under study, an issue also demonstrated in the sea bream (Gregorio et al. 2013). We are currently pursuing this issue in salinity-challenged Senegalese sole.

Driven by the potential generated in the basolateral membrane by the NKA, water absorption seems to rely on Cl^- uptake, a process believed to be mediated exclusively by an apical $Na^+/K^+/2Cl^-$ co-transporter (Musch et al. 1982). However, recent reports have established an important role for apical Cl^-/HCO_3^- anion exchangers (Grosell and Genz 2006, Grosell 2011). The action of both mechanisms seems to be the foundation of low monovalent ion concentration in the intestinal fluid of marine fish. Currently we have some published data to support a high relative importance of apical $Na^+/K^+/2Cl^-$ co-transporters in the intestine of sole, and we are exploring the impact of apical anion transporters in fluid processing. The latter, seems to play a key role in intestinal fluid processing, by enabling the formation of luminal carbonate aggregates, which is an important driving force to facilitate water absorption. However, two conditions are essential for intestinal carbonate aggregate formation: high calcium availability (and/or magnesium) as substrates and high pH to drive the precipitation. While both substrates are in high concentration in the intestinal lumen, i.e., ingested seawater, the alkaline condition is driven by epithelial bicarbonate secretion. Thus, secreted bicarbonate immobilizes calcium in the form of carbonate aggregates, reduces fluid osmolality and, consequently enhances osmotic water absorption, but, only if enough bicarbonate is secreted to the intestinal lumen. The process of intestinal bicarbonate secretion relies on molecular mechanisms that regulate the availability of secreted bicarbonate and indirectly generate carbonate aggregates in the intestinal fluid. The process is far from fully understood and only some elements have been fully characterized (Grosell and Genz 2006, Grosell 2011). We have recently added an important piece to this puzzle demonstrating that the process of bicarbonate secretion is under the regulation of the calciotropic factors PTHrP and stanniocalcin *in vitro* (Fuentes et al. 2010) and *in vivo* (Gregorio et al. 2014). Interestingly the freshwater-adapting hormone prolactin regulates the process of bicarbonate secretion in the intestine of sea bream (Ferlazzo et al. 2012). In addition, our recent work (Carvalho et al. 2012) demonstrates regulatory actions of trans-membrane and soluble adenylyl cyclases with consistent and predictable effects in water absorption in gut preparations and epithelial bicarbonate secretion in the intestine of the sea bream. This demonstrates a functional connection between both processes and highlights the relative importance of anion transporters in the processing of imbibed fluid. In the Senegalese sole these features of intestinal function remain little documented, and we are currently focusing in this species to provide new data on intestinal physiology of sole.

3. Gills

Gills in fish are not just respiratory epithelia (Keys and Willmer 1932) as they are also involved in metabolic waste excretion, ion regulation and blood acid-base balance (Goss et al. 1998). Ionic transport in fish gill epithelium has been deeply studied (Evans et al. 1999, 2005, Hwang et al. 2011), being not our purpose to extensively review the literature available but to highlight the most relevant topics related to Senegalese sole osmoregulation.

The two major cell types in the fish gill epithelium are the mitochondria-rich (MR) chloride cells (CC) (or ionocytes) and the pavement cells (PVC) (Perry 1998). In general, the ionocytes are located primarily on the filament epithelium in most fish with few examples of CC present on the lamellar epithelium. These ionocytes are the major branchial contributors to ionic regulation, acid-base balance, and gas transfer in teleost fish (Perry 1998).

Two types of ionocytes have been identified in euryhaline fish: SW- and FW-type (Hwang et al. 2011). Long-term differentiation of the ionocytes during adaptation to different salinities appears to be regulated by endocrine factors (see below in this chapter). Ultrastructural modifications of the branchial MR cells are thus of physiological significance when environmental salinity is modified (Hwang 1989). Those alterations, described in other flatfish species (*Paralichthys lethostigma*), are associated to changes in branchial expression of ion transport and putative tight junction claudin proteins (Tipsmark et al. 2008), while the number and size of the ionocytes are also modified in Senegalese sole (see Fig. 1) and other euryhaline species (Laiz-Carrión et al. 2005). Moreover, it has been described that changes in the expression of NKA α subunit isoforms are of paramount relevance when fish are acclimated to different environmental salinities. Hence, Atlantic salmon NKAα1a abundance is higher in FW while the isoform NKAα1b increased in SW (McCormick et al. 2009). Similarly, recent studies in Senegalese sole confirmed this differentiated NKA subunits isoforms expression. Thus *atp1a1a*, *atp1b1a* and *atp1b1b* transcripts levels increased significantly when transferred to high SW, while *atp1a3a*, *atp1b3a* and *atp1b3b* transcripts increased at low salinity (Armesto et al. 2014, 2015b). However, if we go a step further, rapid responses related to the activation of the NKA proteins occurred in fish gills when transferred to different environmental salinities, as was described for killifish (Mancera et al. 2000) or Senegalese sole (Arjona et al. 2007, Herrera et al. 2012).

SW-type of marine teleost fish mainly secrete Na^+, Cl^-, K^+ and NH_4^+ and absorb Ca^{2+} (Flik et al. 1983, Marshall and Bryson 1998), but a more complete diagram may be seen in Fig. 2. In SW, teleost fish gills rely on the electrochemical gradient provided by the basolateral Na^+/K^+-ATPase (NKA) to actively secrete ions. In this sense, NKA moves K^+ into the cell while pumps Na^+ ions into the paracellular space, while basolateral recycling of K^+ is achieved through an inwardly rectifying K^+ channel (eKir) (Suzuki et al. 1999). The basolateral $Na^+/K^+/2Cl^-$ co-transporter (NKCC1) mediates the entry of Na^+ and Cl^- into the cellular compartment, followed by passive exit of Cl^- through the apical cystic fibrosis transmembrane receptor (CFTR) (reviewed by Marshall and Bryson 1998, Evans et al. 1999). The increased chloride ions render the apical crypts with a negative electrical charge, enabling paracellular Na^+ (and also NH_4^+) to passively cross through the tight-junction proteins, occludins and claudins (Wilkie 1997). Some authors have described K^+ channels also in the apical membrane, thus allowing its passive exit (Marshall and Bryson 1998, Evans et al. 1999). Ca^{2+} uptake is via apical epithelial Ca^{2+} channels (ECaC) and basolateral Na^+/Ca^{2+} exchange (NCX) (Hwang and Chou 2013) and plasma membrane Ca^{2+}-ATPase (PMCA) (Flik et al. 1983). Metabolic CO_2 crosses the basolateral membrane and intracellular carbonic anhydrases (CA) produce HCO_3^- and H^+. This bicarbonate may exit the cell through the CFTR-type channels (Wilson et al. 2000b). Excretion of metabolic nitrogen compounds such as ammonia (NH_3) involved Rhesus (Rh) glycoproteins. Thus, basolateral Rhbg, but also passive diffusion through the membranes increased intracellular NH_3 concentrations, leading apical Rhcg glycoproteins to excrete this metabolite (Wu

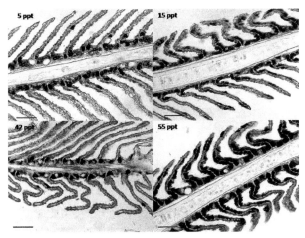

Fig. 1. Ionocytes in *S. senegalensis* acclimated to different environmental salinities. MR cells were detected by immunohistochemistry using a polyclonal anti NKA antibody (as explained before in Laiz-Carrion et al. 2005). Scale bar 50 mm.

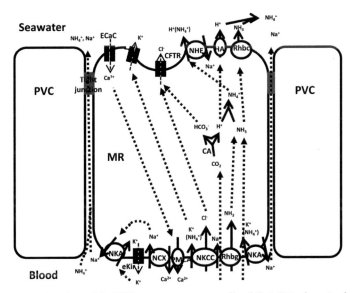

Fig. 2. Seawater ionocytes in teleost fish. PVC means pavement cell, while MR is the mitochondria-rich cell or chloride cell. Other acronyms are described in the main text. Solid lines indicate active transport; dashed lines show diffusion or exchange through membrane channels or across the cell. Figure modified from Wilkie (1997), Marshall and Bryson (1998), Wilson et al. (2000b) and Hwang et al. (2011). We have intentionally skipped the osmoregulatory actions of pavement and accessory cells in order to get a clearer outline of ion movement across the gills. Further information regarding these processes is available in the published work of Sullivan et al. (1995), McCormick (2001), O'Donnell et al. (2001) and Tse et al. (2007).

et al. 2010, Hwang et al. 2011). Ammonia reacts with the protons derived from CA activity producing the cation ammonium (NH_4^+) (Wilson et al. 2000b). This ammonium may mimic Na^+ and/or H^+ ions, also entering the cell via the basolateral NKA, being excreted by a Na^+/H^+ exchanger (NHE) (Hwang et al. 2011). The excess of H^+ is secreted by an apical V-type H^+-ATPase (HA) (Wilson et al. 2000b), which supports the conversion of the NH_3 into the less toxic compound NH_4^+ (Wilkie 1997).

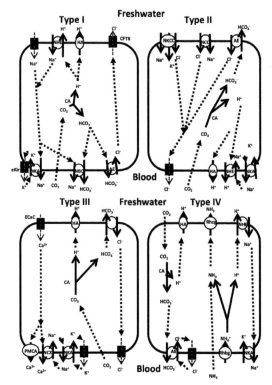

Fig. 3. Freshwater ionocyte types in teleost fish. Present information must be carefully considered as the presence/absence of ion transporters/channels shown herein, and their relationships, should change depending on the species studied. Further details as in legend of Fig. 2.

In FW, NaCl uptake by the gill was estimated to account for about the 1.8% of whole animal O_2 consumption, being this O_2 consumption higher in FW than in SW (Morgan and Iwama 1998). Furthermore, the general account for ammonia excretion through the gills is almost 90% in FW, while goes from 50 to 85% in SW acclimated fish, being the reminder excreted by the kidney (Wilkie 1997).

It is noteworthy to mention that, while ionocyte models in SW do not substantially vary between fish species, the population of these cells in FW shows great modifications (Hwang et al. 2011). At least four different types have been described in FW (Hiroi and McCormick 2012) depending on the expression of ion transporters, morphologies and functions. Some authors (Hwang and Chou 2013) have described them depending on their net functioning as: (1) transepithelial H^+ secretion, Na^+ uptake and NH_4^+ excretion; (2) Ca^{2+} uptake; (3) Na^+/Cl^- uptake; and (4) K^+ secretion. For better understanding these modifications, and according to the studies of Evans et al. (1999), Wilson et al. (2000a), Evans (2008) and Hwang et al. (2011), we have consolidated what is known so far and the resulting information has been condensed in Fig. 3. Thus, we have reduced the vast amount of models available into four types of FW-ionocytes. Types I to IV FW-ionocytes refer to what we have considered calling "Na⁺ uptake cell" (type I), "Cl⁻ uptake cell" (type II), "Ca²⁺ uptake cell" (type III) and "NH₃ secretory cell" (type IV).

Type I ionocyte basically captures Na^+ while secreting H^+ and Cl^- (Fenwick et al. 1999). It has been established that a proton pump (HA) co-localizes in the apical membrane with an ENaC-like sodium pump channel (Wilson et al. 2000a) allowing the entrance of Na$^+$ into the cell, being also postulated the presence of Na$^+$/H$^+$ exchangers (NHE) in this side (revised by Evans 2008). The basolateral NKA forced the movement of Na$^+$ ions into the body fluids, while an eKir channel recycles the K$^+$ in this part of the membrane (Evans et al. 1999). There is some evidence that a basolateral Na$^+$/HCO$_3^-$ co-transporter (NBCe1b) may also be involved in the basolateral transport of Na$^+$ (Evans 2008). HCO$_3^-$ may be provided by intracellular CA, which also produces H$^+$. CFTR proteins have been described in the apical membrane of these cells (Evans et al. 1999, Hwang et al. 2011), which secrete Cl$^-$ that have entered the cell through basolateral Cl$^-$/HCO$_3^-$ exchangers (also called anion exchangers or AE) (Evans et al. 1999).

Type II ionocytes are also called NCC cells by some authors as they show Na$^+$/Cl$^-$ co-transporters (NCC) but also NKCC in the apical membrane (Evans 2008, Hwang et al. 2011, Hiroi and McCormick 2012). The NKA α1 subunit atp1a1a.2 is specifically localized in NCC cells of zebrafish (Liao et al. 2009). There are evidences for an apical AE (Slc4 in tilapia; Slc26 in stingray, Slc4a1b in zebrafish) that also co-localize with a basolateral HA (Wilson et al. 2000a, Piermarini et al. 2002, Hwang and Chou 2013), inverting their positions respecting to the described type I ionocytes. Chloride uptake includes a basolateral chloride-channel (ClC) (Tang and Lee 2011) while HCO$_3^-$ and H$^+$ are supplied by the action of an intracellular CA. Finally, those protons may also evidence some Na$^+$/H$^+$ exchange (NHE3b) in the basolateral membrane (reviewed by Evans et al. 1999, Evans 2008).

Type III MR cells have a HA in the apical membrane that generates a favorable electrochemical gradient for the passive diffusion (uptake) of Ca^{2+} through an epithelial Ca^{2+} channel (ECaC) and acidifies the boundary layer. Basolateral plasma membrane Ca^{2+}-ATPase (PMCA2) and Na$^+$/Ca^{2+} exchanger (NCX1b) introduce Ca^{2+} into the body fluids (Flik et al. 1983, Hwang et al. 2011). In the apical membrane an AE, driven by low HCO$_3^-$ in the boundary layer, drives Cl$^-$ into the cell and Cl$^-$ exits via an anion channel, likely CFTR channel (Evans et al. 2005). It is remarkable that, over a total of six NKA α1 subunit genes in zebrafish, these cells expressed only the atp1a1a.1 (Liao et al. 2009).

Type IV cells are well described in the review of Hwang and Chou (2013) and are also called H$^+$-ATPase-rich cells (HR). They were first identified as the major cell type responsible for acid secretion function of the gills. At the apical membrane there is a NHE (NHE3b), while the basolateral membrane shows an AE (AE1b) and a NKA constituted by the α1 subunit atp1a1a.5. Chloride anions necessary for AE activity may be recycled by basolateral chloride channels. Intracellular CA (CA2-like a) produces the H$^+$ necessary for apical HA and the NHE function, and HCO$_3^-$ is used up by the basolateral AE. Metabolic nitrogenous residues may enter into the cells by passive diffusion of NH$_3$, or by means of Rhbg NH$_4^+$ transporters (Wu et al. 2010), while an apical Rhcg1 excretes NH$_3$ to the environment. Thus, it is demonstrated that the epithelial transport of these cells are related to not only to H$^+$ secretion but also to Na$^+$ and HCO$_3^-$ uptake and NH$_3$ excretion.

After the analysis of the putative ionocyte models in SW and FW fish, we can postulate that the NKA and the HA pumps are of vital importance, as they provide the electrochemical potential necessary to take up/secrete ions. Moreover, those enzymes are present in any of the assumed models (Figs. 2 and 3) regardless of environmental salinity. Thus, it would be coherent to analyze the behavior of both ATPases in response to changes in environmental salinity in order to get a clear view of the osmoregulatory processes that take place in the gills. In Senegalese sole, NKA α and β subunits isoforms has been studied through their mRNA expression, *in situ* hybridization or immunohistochemical techniques (Armesto et al. 2014, 2015b), while the biochemical activities of the NKA and HA enzymes in fish maintained in a wide range of environmental salinities were also analyzed (Arjona et al. 2007, 2009, Ruiz-Jarabo et al. 2017). To the best of our knowledge, the presence of different isoforms of the NKA α and β subunit subunits in the gills (*atp1a1b*, *atp1a3a*, *atp1a3b*, *atp1b1a*, *atp1b1b*, *atp1b2*, *atp1b2b*, *atp1b3a*, *atp1b3b* and *atp1b4*) is salinity dependent. This supports the idea that differentiated populations (mostly due to their morphology and function features) of ionocytes appear in accordance to osmoregulatory external stimuli (McCormick 2001) such as salinity changes in the water. Moreover, the increase in ionocyte number and size in HSW-acclimated sole (Fig. 1), as well as the increase in the NKA activity in this environment (Arjona et al. 2009), indicates that fish acclimated to hyper-osmotic media should invest great amounts of energy in order to extrude ions through the gills. These results are in agreement with those previously described for other Pleuronectiformes such as the wedge sole (Herrera et al. 2009), the turbot (Imsland et al. 2003) or the brill (Ruiz-Jarabo and Mancera, unpublished). Noticeably, when we analyzed HA activity referred to protein concentration of the gill homogenate, the HA activity in 5 ppt-exposed sole did not vary compared to control SW-acclimated fish (Ruiz-Jarabo et al. 2017). After the analysis of Fig. 1 it results evident that HA activity is higher in low salinity waters as there are few ionocytes in this environment, thus increasing the ratio "activity : ionocytes" substantially. It may be hypothesized that in low salinity water, the HA is necessary to restore plasma osmolality; while the NKA is the main ionic pump in high salinity water.

The study of fish branchial cells results of paramount significance in ion transport regulation, as they serve as relevant models (acting as a pseudo kidney in fish) that could be aligned, for instance, with biomedical science (Arjona et al. 2014). In this sense, the zebrafish has been employed as a useful model to study renal diseases in humans (Arjona et al. 2013). Thus, cultured respiratory cells of gill provide a model for studying ion and water transport (Avella and Ehrenfeld 1997), but other alternatives are available. In this sense, the opercular skin resembles the branchial filament epithelium (Burgess et al. 1998) but its flat shape facilitates putative *ex vivo* assays related to ion transport (Martos-Sitcha et al. 2015). Due to the euryhaline capacity of the Senegalese sole, this species results in phenomenal biological model that may be employed in osmoregulatory studies.

4. Kidney

Euryhaline fish capable of inhabit low salinity water and SW, like the Senegalese sole, possess the full complement of the vertebrate nephron: glomerulus, proximal tubule, distal tubule and collecting duct (Beyenbach 2004). The importance of the kidney in fish is evident, as it acts as an important osmoregulatory tissue (reviewed by Beyenbach 2004, McDonald 2007). Hence, it is well established that fish in hypo-osmotic environments produce abundant and diluted urine, while in SW urine is scanty and concentrated (Brown et al. 1980) (Fig. 4).

The glomerular ion and water transport mechanisms are similar in FW and SW fish. In the glomerulus, the Bowman's capsule filtrates the blood, leaving small proteins and small molecules such as glucose, Na^+, Cl^-, amino acids, etc., pass freely into Bowman's space and into the proximal tubule. The filtrate is iso-osmotic to the blood plasma in teleost fish and presents a similar concentration of ions. The glomerular filtration rate depends on the environmental salinity in which the fish is evolving, being high in FW-fish and low in SW-acclimated individuals (Graham et al. 2003).

The major osmoregulatory differences in fish occur in the proximal tubules as SW and FW fish mainly secrete and reabsorb ions, respectively, in this region. Proximal tubules

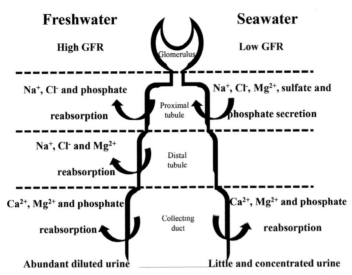

Fig. 4. General model of a nephron in FW and SW-acclimated fish. GFR means glomerular filtration rate. Arrows indicate active transport of ions. This figure summarizes the available information obtained from Renfro (1997), Elger et al. (1998), Dantzler (2003), Beyenbach (2004) and Arjona et al. (2014).

of SW fish accomplish net secretion of Na^+, Cl^-, Mg^{2+} and sulfates (Beyenbach 2004), being the concentrations of Na^+ and Cl^- in the secreted fluid nearly isomolar to the blood plasma (Beyenbach et al. 1986). These secretory processes actively depend on the activity of basolateral NKA pumps. Moreover, reabsorption in hypo-osmotic environments also involves active transport of Na^+ by basolateral NKA and passive reabsorption of Cl^- (Dantzler 2003). The proximal tubule epithelium is also related to absorption/secretion of phosphate, with strong evidences confirming endocrine modulation by a plethora of hormones. In *Pleuronectes americanus*, another flatfish species, phosphate reabsorption is stimulated by stanniocalcin, somatolactin, prolactin and growth hormone (Renfro 1997). To date, there is a lack of knowledge regarding the control of renal filtration by hormones in *S. senegalensis*.

The distal convolute tubule of FW-species such as the zebrafish presents an increased Mg^{2+} uptake through specific channels (TRPM6) (Arjona et al. 2014), but also a chloride channel and a Na^+/Cl^- symporter (NCC) are involved in the reabsorption of Na^+ and Cl^- ions in FW-tilapia (Yan et al. 2013). To the best of our knowledge, ion transport in the distal tubule in SW-species is scarce, with no information available for the Senegalese sole.

No clear information is available regarding the importance of the collecting ducts in fish osmoregulation beyond some studies reporting phosphate transport processes in this region (Beyenbach 2004, Graham et al. 2003). Furthermore, the Senegalese sole has been never used as a biological model for this purpose despite its osmoregulatory advantages and plasticity.

It is remarkable that the NKA appeared in the epithelial cells of proximal tubules, distal tubules, and collecting tubules in fish acclimated at various salinities, as was described before in *Tetraodon nigroviridis* (Lin et al. 2004). The importance of this enzyme is thus evident in fish nephrons regardless the external salinity medium. In hypo-osmotic environments, a higher reabsorption rate of salts is demanded after the glomerular filtrate. In this sense, it has been described that renal NKA activity of euryhaline species is higher in FW-living specimens than in SW (Venturini et al. 1992, Lin et al. 2004). However, Senegalese sole did not show any differences in renal net NKA activity after being acclimated to a wide range of environmental salinities (Arjona et al. 2007, 2009), as was described before in rainbow trout (Fuentes et al. 1997). This could be due to the impossibility of distinguishing between different areas of the nephron, or due to differentiated cells within the nephron sensitive to changes in environmental salinity (as was the case for branchial ionocytes). In this regard, we have described above that the NKA is necessary for ion secretion but also for reabsorption, and the unaltered levels of this enzyme could be sustained by differentiated populations of other ion transporters depending on the external media. Furthermore, in a close-related flatfish species, the wedge sole (*Dicologoglossa cuneata*), renal NKA activity increased with environmental salinity (Herrera et al. 2009). This enhancement could be attributed to a reduction in urine production and/or to an enhancement in ion transport in a hyper-osmotic environment. In this regard, it has been hypothesized that up to 50% of total ammonia is excreted through the kidney in SW-fish, thus increasing ion-transport mechanisms when compared to FW-fish (Wilkie 1997).

Despite the importance of kidney in ion and water balance on teleost fish, scarce information is available for Senegalese sole. Further studies are thus necessary to fully understand the osmoregulatory pathways in this species. The aquaculture industry will be benefited in this way as a result of improved knowledge regarding the acclimation processes.

5. Integrated osmoregulatory response in Senegalese sole

All the above osmoregulatory tissues present an integrated activity in the control of plasmatic ions concentrations (Venturini et al. 1992), as we have recently seen in *S. senegalensis* (Ruiz-Jarabo et al. 2017). To provide a holistic view of this system, Fig. 5 shows what physiological changes occur in the sole after long-term acclimation to different environmental salinities.

Thus, fish maintained at 5 ppt for 30 days drastically decreased their plasma osmolality and main ions concentration, evidencing haemodilution. In this sense, anterior intestine and rectum increased their DIDS-sensitive (associated to Cl^-/HCO_3^- exchangers) as well as HA and H^+/K^+-ATPase (HKA) activities as a consequence of an active chloride and potassium uptake from the ingested water. In this hypo-osmotic salinity gills are passively gaining water, so that an active ion uptake takes place. The kidney, in glomerular-nephron species such as the sole, serves as a water pump which produces abundant diluted urine, while the proximal and distal tubules of the nephron increase their ion uptake processes. In hyperosmotic environments such as the SW, or salinities as high as 55 ppt, the Senegalese sole drinks enormous amounts of water. As these animals are being passively dehydrated (note the huge difference between plasma and water osmolalities herein), the anterior intestine must increase its active Na^+/Cl^- uptake processes as demonstrated by the high NKA, DIDS-sensitive, NKCC and HKA activities in this tissue, so that water may be absorbed due to the sharp decrease in the intestinal fluid osmolality. The rectum appears as a place where sole capture water before being completely removed from the guts. Hence, DIDS-sensitive and HA activities increased in the rectum in order to secrete bicarbonate to the lumen, favoring carbonate precipitates formation and subsequent absorption of the released water. The excess of monovalent ions is secreted through the gills, thus enhancing the NKA activity in this tissue. In hyper-osmotic environments, kidney produced concentrated urine as a consequence of nephron water uptake and active ion secretion. Although the NKA is the enzyme that promotes ion uptake or secretion in the nephron, no differences are described in *S. senegalensis* maintained at different environmental salinities for 10 weeks (Arjona et al. 2009). The reason is that current biochemical techniques for

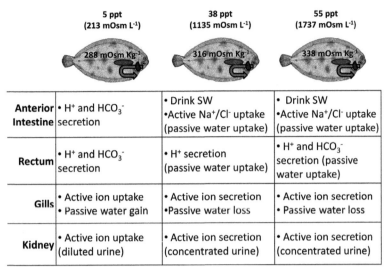

	5 ppt (213 mOsm L⁻¹) 288 mOsm Kg⁻¹	38 ppt (1135 mOsm L⁻¹) 316 mOsm Kg⁻¹	55 ppt (1737 mOsm L⁻¹) 338 mOsm Kg⁻¹
Anterior Intestine	• H⁺ and HCO₃⁻ secretion	• Drink SW •Active Na⁺/Cl⁻ uptake (passive water uptake)	• Drink SW •Active Na⁺/Cl⁻ uptake (passive water uptake)
Rectum	• H⁺ and HCO₃⁻ secretion	• H⁺ secretion (passive water uptake)	• H⁺ and HCO₃⁻ secretion (passive water uptake)
Gills	• Active ion uptake • Passive water gain	• Active ion secretion •Passive water loss	• Active ion secretion • Passive water loss
Kidney	• Active ion uptake (diluted urine)	• Active ion secretion (concentrated urine)	• Active ion secretion (concentrated urine)

Fig. 5. Osmoregulatory mechanisms of *Solea senegalensis* submitted to different environmental salinities. Senegalese sole acclimated to a wide range of environmental salinities may assemble their osmoregulatory tissues (viz. anterior and posterior intestine, gills and kidney) coordinately (Ruiz-Jarabo et al. 2017).

the analysis of NKA activity in this species did not differentiate between regions of the nephron, requiring further approaches to clearly understand renal functioning in the sole.

6. Hormonal control

Endocrine control of osmoregulatory system involves a panel of hypophyseal and extrahypophyseal hormones, through their direct or indirect action in target organs. Several and excellent reviews have been published on this topic in teleost (see Takei and McCormick 2013). Notwithstanding the importance of hormones in the osmoregulatory process, there is scarce information on endocrine control of osmoregulatory system in *S. senegalensis* (Arjona et al. 2008, 2010, Armesto et al. 2015a). In this review we will focus on cortisol, growth hormone/insulin-like growth factor-I axis, and thyroid hormones, as well as on renin–angiotensin system. However, the role of other endocrine factors on the osmoregulatory process in this species is still unknown, and deserves further studies.

6.1 Cortisol

Cortisol is the main steroid produced and released by interrenal tissue in teleost fish. It shows both glucocorticoid and mineralocorticoid activities, as fish express both mineralocorticoid and glucocorticoid receptors with physiological affinity for cortisol (Mommsen et al. 1999). Accordingly, a role for cortisol in the control of several processes such as intermediary metabolism, growth, stress, and immune function has been repeatedly demonstrated in teleost fish (Wendelaar Bonga 1997, Mommsen et al. 1999). The osmoregulatory role of cortisol in fish has been reviewed before (McCormick 2001, Takei and McCormick 2013).

Cortisol plays an important function in the adaptation to hyper-osmotic environments, and is classically considered a 'SW promoting hormone' in euryhaline fish. Cortisol induces salinity tolerance in fish that are migrating to, or experimentally transferred to, higher salinities. The osmoregulatory role of this hormone involves several osmoregulatory organs (see above). In gills, cortisol-treatment increases development and proliferation of gill chloride cells, NKA activity, expression of NKAα-subunit mRNA, and expression and abundance of $Na^+/K^+/2Cl^-$ cotransporter, both in Salmonid and non-Salmonid species (reviews McCormick 1995, 2001, Takei and McCormick 2013). Cortisol also stimulates NKA activity, together with ion and water absorption in the intestine, and it improves the drinking response after transfer to SW (see Fuentes et al. 1996, Veillette et al. 1995).

In addition to its classical hypo-osmoregulatory capacity, there is ample evidence for hyper-osmoregulatory actions of cortisol in fish. For instance, cortisol has been shown to increase both NKA α1a and NKA α1b mRNA expression in the gills of juvenile Atlantic salmon held in FW (McCormick 2001, Tipsmark and Madsen 2009), while it also stimulates ionocyte proliferation and increases in ion uptake in FW rainbow trout (Laurent and Perry 1990). However the synthesis and release of the main FW-adapting hormone, prolactin (PRL), is directly inhibited by cortisol (reviewed by McCormick 2013). Little is known of the interaction between cortisol and PRL in FW fish. Conspicuously, it is well established that cortisol, as a glucocorticoid hormone in fish, mediates in the reallocation of energy stores (Soengas et al. 2008) as it was described in *S. senegalensis* (Arjona et al. 2011, Wunderink et al. 2011). Hence, the mediation of cortisol herein may indicate an allostatic state in *S. senegalensis* that allows homeostasis in low salinities by the consumption of metabolic reserves (Arjona et al. 2009). In addition, cortisol treatment (i) restores plasma osmolality and ion levels in hypophysectomized fish, (ii) increases gill chloride cells surface and sodium/chloride influx, (iii) enhances whole-body calcium uptake and branchial calcium pump, and (siv) stimulates gill H^+-ATPase activity (related to sodium uptake in hypo-osmotic environments) (see: Mancera et al. 1994, 2002, McCormick 2001).

Acclimation of juvenile *S. senegalensis* to different environmental salinities (5, 15, SW, and 55 ppt) enhanced plasma cortisol values in all salinities different from SW during the first 18 days post-transfer (Arjona et al. 2007). This rise in cortisol levels could be responsible for branchial NKA activity stimulation observed after transfer to 55 ppt (Arjona et al. 2007), but to date there is no enough information to substantiate a direct relationship of cortisol and ATPase activity or mRNA expression in sole acclimated to low salinity environments.

Similarly, to hyper-osmotic transfer, long-term acclimation to decreased environmental salinity (5 and 15 ppt) also enhanced plasma cortisol values (Arjona et al. 2009). According to the proposed role of cortisol in the acclimation to low salinity environments, this could originate the adaptation into ionic uptake systems in gills (see McCormick 2001, Takei and McCormick 2013). However, cortisol also presents a glucocorticoid role and could contribute to increase the energy metabolites availability for ion regulation in the adaptive and chronic regulatory period. Therefore, plasma cortisol increases observed in those environments with salinity ≤ 15 ppt and ≥ 55 ppt could be reflection of chronic stress (Arjona et al. 2009).

Treatment of *S. senegalensis* with intra-peritoneal slow-release coconut oil implants containing cortisol up-regulated branchial and renal NKA activities (Arjona et al. 2008). This action could be related to a direct effect on ionocytes. However, the same study indicated that hepatic and renal outer-ring deiodination (ORD) activities were also up-regulated in cortisol-injected fish, suggesting that cortisol regulates local T3 bioavailability in *S. senegalensis* via ORD in an organ-specific manner (see thyroid section), with indirect consequences on the use of energy metabolites.

6.2 Growth hormone and insulin-like growth factor I

The 'somatomedin hypothesis' (Holly and Wass 1989) implies that growth hormone (GH) has direct actions on target tissues, as well as exerts localized actions (in different organs, e.g., gills, cartilage, liver) via production of insulin-like growth factors I or II (IGF-I, IGF-

II). By doing so, IGF-I could for instance carry out many of the physiological actions of GH. Therefore, instead of isolated effects of GH, the physiological actions of GH are often referred to as those of the GH/IGF-axis.

The GH/IGF-axis plays a clear role in promotion of fish growth, differentiation and energy metabolism (McLean and Donaldson 1993). In addition, this axis seems to play a clear role in SW acclimation in Salmonids, whereas in non-salmonid teleosts the osmoregulatory role is less evident (McCormick 1995, Mancera and McCormick 1998).

In a time course experiment juvenile specimens of *S. senegalensis* were submitted to hypo-osmotic (from SW to 5 ppt) and hyper-osmotic (from SW to 55 ppt) acute transfer and GH and IGF-I mRNA expression assessed at different times (1, 3, and 10 days) (Mohammed-Geba et al. 2013). GH pituitary mRNA expression enhanced after acclimation to hyperosmotic environments, while hepatic IGF-I expression was reduced in fish at extreme salinities, suggesting a role for this hormone in osmoregulation rather than just in growth. The enhanced pituitary GH expression observed in high SW (HSW)-acclimated specimens could originate high plasma levels of this hormone, supporting, together with cortisol, the increase in gill NKA activity and *atp1a1a* expression observed in HSW-transferred fish (Arjona et al. 2007, Armesto et al. 2014). Also, hepatic IGF-I expression decreased in 5 and 55 ppt-acclimated fish, coinciding with high plasma cortisol levels. This observation may suggest the existence of a negative feedback effect on main growth related genes in *S. senegalensis* juveniles, during the adaptation to environments of extremely high or low salinity. This is also in agreement with previously reported results for this species, where environmental salinities different from SW induced low growth rates (Arjona et al. 2007, 2009). However, more comprehensive studies of the factors involved in the GH/IGF system (hypothalamic factors, GH, IGF's, IGFBP, receptors, etc.) are necessary in order to clarify the osmoregulatory role of GH/IGF axis in *S. senegalensis*.

6.3 *Thyroid hormones*

Thyroid hormones (THs, thyroxine, T4, and 3,5,3'-triiodo-L-thyronine, T3) influence a plethora of physiological processes (i.e., growth, metamorphosis, etc.). However, the osmoregulatory role of these hormones in teleost is rather inconclusive (Grau 1988, Takei and McCormick 2013).

Thyroid hormone receptors have been found in gills, kidney and intestine of different teleosts (see Power et al. 2001). In those organs promoter regions of NKA subunits have positive thyroid response elements and this could suggest a stimulatory action of T3 on NKA expression/activity, and a subsequent role of this hormone in fish osmoregulation. T3 or T4 treatment increase gill NKA activity and chloride cell density in some Salmonids but not in others (Madsen 1990, McCormick 1995). Remarkably, T3, GH and cortisol, positively interact in SW acclimation of Salmonids, with an important role for T3 in the up regulation of the number of cortisol receptors in the gills (Shrimpton and McCormick 1998).

Also in non-salmonid teleosts, a role for thyroid hormones in osmoregulation is disputed. Branchial NKA was stimulated or unaffected by treatment with T3 o T4 (tilapia: Dange 1986, killifish: Mancera and McCormick 1999, gilthead sea bream: Vargas-Chacoff et al. 2016). In tilapia, cooperation between T4 and cortisol increasing gill NKA activity has been reported (Dange 1986). This observation leads to the hypothesis that, similarly to Salmonids, the influence of thyroid hormones may increase gill cortisol receptors number in this osmoregulatory tissue.

In the sole, *S. senegalensis* treatment with intraperitoneal slow-release implants containing T3 increased significantly plasma-free T3 levels, indicating the effectiveness of hormonal treatment. Branchial and renal NKA activities decreased after two weeks post-treatment, but the ORD activities were unaffected (Arjona et al. 2011). No clear relationships between THs and osmoregulation could be inferred from the latter research. Further studies could be necessary in order to expose a possible interaction between T3 and GH and/or cortisol in SW acclimation, as it has been demonstrated in other teleosts (Dange 1986, Shrimpton and McCormick 1998).

6.4 *Renin-angiotensin system*

The renin-angiotensin system (RAS) plays a key role in controlling blood volume and pressure, drinking rate, and electrolyte homeostasis in vertebrates (Takei et al. 2004). In teleosts, RAS is activated in response to high salinities and this agree with the hypo-osmoregulatory role of angiotensin II: (i) stimulation of drinking, (ii) elevation of blood pressure, (iii) regulation of renal filtration, (iv) modulation of NKA activity in gills and kidney, (v) stimulation of cortisol secretion (Takei and McCormick 2013).

Recently, cDNA sequences of different compounds of RAS system have been obtained in *S. senegalensis*: angiotensinogen (*agt*), renin (*ren*), angiotensin-converting enzyme (*ace*), angiotensin-converting enzyme 2 (*ace2*), as well as angiotensin II receptor type I (*agtr1*) and type II (*agtr2*) (Armesto et al. 2015a). In juvenile fish tissue, these transcripts were detected mainly in osmoregulatory (kidney: *ren*, *agtr1* and *agtr2*; intestine: *ace* and *ace2*) and non-osmoregulatory (liver: *agt* and *agtr2*; brain: *agtr1*) tissues. Similarly to other teleost the highest *agt* transcript levels were found in liver, while for *ren* was found in the kidney. Interestingly, the highest *ace* transcript levels were observed in the intestine suggesting this osmoregulatory organ as the major site for Ang II production through ACE activity. Transfer of juveniles from SW (35 ppt) to different environmental salinities (10 and 60 ppt) during 4 days induced changes in RAS system expression in osmoregulatory and non-osmoregulatory organs, putatively related to rapid adaptation to environmental salinity modifications (Armesto et al. 2015a), as it has been shown in other teleosts (Takei and McCormick 2013).

7. Effects of salinity on growth

The Senegalese sole, during acclimation to different environmental salinities, has to cope with both ionic, volume regulatory, respiratory and acid-base disturbances which initially may limit the euryhaline capacity, as was described before in other teleosts (Madsen et al. 1996). To cope with those disturbances, the sole consumes a certain amount of energy to meet the metabolic cost of ionic and osmotic regulation, being established that 10 to more than 50% of the total fish energy budget is devoted to osmoregulation (Boeuf and Payan 2001). It has been hypothesized that if the external environment is iso-osmotic to the internal body fluids, the costs of ionic regulation was lowered, improving the growth and food conversion efficiency of the fish (Imsland et al. 2003). But this hypothesis depends not only on the flatfish species but also on their life stage (Ruiz-Jarabo et al. 2015).

Changes in environmental salinity trigger a series of physiological responses in *S. senegalensis* that are grouped as primary, secondary and tertiary (Barton 2002). Primary responses were studied by Arjona and coworkers (2007), who established the prompt release of cortisol into the circulation in salinities (5, 15 and 55 ppt) different from SW (38 ppt). The increases of this glucocorticoid produce immediate actions at the tissue level (Mommsen et al. 1999). Thus, cortisol elicits mobilization of energy metabolites such as glucose and lactate to fuel the osmoregulatory organs (viz. gills, kidney and intestine).

Also grouped as secondary responses to changes in environmental salinity, the activity of the main ionic transporters is modified in those tissues (Arjona et al. 2007), forcing the animals to increase their oxygen consumption (Herrera et al. 2012). One week after the transfer to different environmental salinities, Senegalese sole reach a new allostatic state that allows the homeostasis at salinities different from SW. Tertiary stress responses extend to the level of the organism and refer to aspects of whole-animal performance such as changes in growth (Wedemeyer et al. 1990). Besides environmental salinity, some other variables that are able to modify the osmoregulatory system and therefore growth in the Senegalese sole are temperature (Arjona et al. 2010) and crowding density (Wunderink et al. 2011). The energy metabolic stores play a vital role herein as they sustain the energy input for ionic and osmotic regulation, being the liver described as the major source supplying carbohydrate metabolites to osmoregulatory organs (Tseng and Hwang 2008, Arjona et al. 2009). In Senegalese sole other tissues such as white muscle has been also studied as important places for amino acid turnover to be used in oxidation/gluconeogenesis in the liver (Arjona et al. 2009), as this metabolites seem to play an important role in the osmoregulatory adjustment in this species (Aragão et al. 2010). Juveniles of Senegalese sole, maintained at different environmental salinities for three months, present an allostatic state at hypo- and iso-osmotic media with altered and sustained activity of primary mediators of the stress response such as cortisol (McEwen and Wingfield 2003, Arjona et al. 2009). During this period, homeostatic imbalance is tolerated for limited periods only if feed intake and/or energy stores can fuel homeostatic mechanisms. This is at the cost of the energy budget that is destined for growth, as was reflected in the lower body weights in those animals maintained at 15 ppt. Better growth rates for this species are achieved in a range of salinities that goes from 25 to 39 ppt salinity (Arjona et al. 2009). This improved growth and food conversion efficiency at intermediate salinities are in agreement with other flatfish species, like *Scophthalmus maximus* (Imsland et al. 2001) or *Paralichthys orbignyanus* (Sampaio and Bianchini 2002). These authors linked the lower growth rates at hypo-osmotic salinities to a higher expenditure associated to increased branchial NKA activity in this environment.

Thus, *S. senegalensis* juveniles and adults performed better in salinities close to that of the sea. Notwithstanding with the importance of the osmoregulatory organs and their metabolism in fish growth, there is a relevant endocrine component. Hence, the euryhaline capacity is controlled by hormones such as GH (Fuentes and Eddy 1997c), promoting acclimation to SW, PRL that promotes acclimation to FW, cortisol that interacts with both hormones thus having a dual osmoregulatory action, and thyroid hormones that support the action of GH and cortisol in promoting SW acclimation (McCormick 2001). The role of these hormones in growth is well described in fish. In the chronic regulatory period of Senegalese sole, cortisol increased in salinities different from SW, while free plasma T4 do not return to basal levels, indicating the involvement of both hormones as primary mediators that sustain allostatic state (Arjona et al. 2008). Their interactions and effects result in better growth performance at SW environments for *S. senegalensis*.

Acknowledgments

JMM appreciates the continued support of the Spanish Ministry of Economic Affairs and Competitiveness (MINECO) through different projects in the last years. JF current research is funded by FCT (Portugal) through project PTDC/MAR-BIO/3034/2014. CCMar is supported by national funds from FCT through project UID/Multi/04326/2013. The authors thank to Dra. Carolina Balmaceda by Fig. 1.

References

Aragao, C., B. Costas, L. Vargas-Chacoff, I. Ruiz-Jarabo, M.T. Dinis, J.M. Mancera and L.E. Conceicao. 2010. Changes in plasma amino acid levels in a euryhaline fish exposed to different environmental salinities. Amino Acids 38: 311–7.

Arjona, F.J., L. Vargas-Chacoff, I. Ruiz-Jarabo, M.P. Martin del Rio and J.M. Mancera. 2007. Osmoregulatory response of Senegalese sole (*Solea senegalensis*) to changes in environmental salinity. Comp. Biochem. Physiol. A. Mol. Integr. Physiol. 148: 413–21.

Arjona, F.J., L. Vargas-Chacoff, M.P. Martin del Rio, G. Flik, J.M. Mancera and P.H. Klaren. 2008. The involvement of thyroid hormones and cortisol in the osmotic acclimation of *Solea senegalensis*. Gen. Comp. Endocrinol. 155: 796–803.

Arjona, F.J., L. Vargas-Chacoff, I. Ruiz-Jarabo, O. Gonçalves, I. Pâscoa, M.P. Martín del Río and J.M. Mancera. 2009. Tertiary stress responses in Senegalese sole (*Solea senegalensis* Kaup, 1858) to osmotic challenge: Implications for osmoregulation, energy metabolism and growth. Aquaculture 287: 419–426.

Arjona, F.J., I. Ruiz-Jarabo, L. Vargas-Chacoff, M.P. Martín del Río, G. Flik, J.M. Mancera and P.H. Klaren. 2010. Acclimation of *Solea senegalensis* to different ambient temperatures: implications for thyroidal status and osmoregulation. Marine Biology 157: 1325–1335.

Arjona, F.J., Y.X. Chen, G. Flik, R.J Bindels and J.G. Hoenderop. 2013. Tissue-specific expression and in vivo regulation of zebrafish orthologues of mammalian genes related to symptomatic hypomagnesemia. Pflugers Arch.

Arjona, F.J., J.H. de Baaij, K.P. Schlingmann, A.L., Lameris, E. van Wijk, G. Flik et al. 2014. CNNM2 mutations cause impaired brain development and seizures in patients with hypomagnesemia. PLoS Genet 10, e1004267.

Armesto, P., M.A. Campinho, A. Rodriguez-Rua, X. Cousin, D.M. Power, M. Manchado and C. Infante. 2014. Molecular characterization and transcriptional regulation of the Na +/K+ ATPase alpha subunit isoforms during development and salinity challenge in a teleost fish, the Senegalese sole (*Solea senegalensis*). Comp. Biochem. Physiol. B. Biochem. Mol. Biol. 175: 23–38.

Armesto, P., X. Cousin, E. Salas-Leiton, E. Asensio, M. Manchado and C. Infante. 2015a. Molecular characterization and transcriptional regulation of the renin-angiotensin system genes in Senegalese sole (*Solea senegalensis* Kaup, 1858): differential gene regulation by salinity. Comp. Biochem. Physiol. A. Mol. Integr. Physiol. 184: 6–19.

Armesto, P., C. Infante, X. Cousin, M. Ponce and M. Manchado. 2015b. Molecular and functional characterization of seven Na+/K+-ATPase beta subunit paralogs in Senegalese sole (*Solea senegalensis* Kaup, 1858). Comp. Biochem. Physiol. B. Biochem. Mol. Biol. 182: 14–26.

Avella, M. and J. Ehrenfeld.1997. Fish Gill Respiratory Cells in Culture: A New Model for Cl--secreting Epithelia. Journal of Membrane Biology 156: 87–97.

Barton, B.A. 2002. Stress in fishes: a diversity of responses with particular reference to changes in circulating corticosteroids. Integr. Comp. Biol. 42: 517–25.

Beyenbach, K.W., D.H. Petzel and W.H. Cliff. 1986. Renal proximal tubule of flounder. physiological-properties. American Journal of Physiology 250: R608–R615.

Beyenbach, K.W. 2004. Kidneys sans glomeruli. Am. J. Physiol. Renal. Physiol. 286: F811–27.

Boeuf, G. and P. Payan. 2001. How should salinity influence fish growth? Comparative Biochemistry and Physiology C-Toxicology & Pharmacology 130: 411–423.

Brown, J.A., J.A. Oliver, I.W. Henderson and B.A. Jackson. 1980. Angiotensin and single nephron glomerular function in the trout *Salmo gairdneri*. Am. J. Physiol. 239: R509–14.

Burgess, D.W., W.S. Marshall and C.M. Wood. 1998. Ionic transport by the opercular epithelia of freshwater acclimated tilapia (*Oreochromis niloticus*) and killifish (*Fundulus heteroclitus*). Comparative Biochemistry and Physiology A-Molecular and Integrative Physiology 121: 155–164.

Carvalho, E.S., S.F. Gregorio, D.M. Power, A.V. Canario and J. Fuentes. 2012. Water absorption and bicarbonate secretion in the intestine of the sea bream are regulated by transmembrane and soluble adenylyl cyclase stimulation. J. Comp. Physiol. B. 182: 1069–1080.

Dange, A.D. 1986. Branchial Na⁺-K⁺-ATPase Activity in Fresh-Water or Saltwater Acclimated Tilapia, *Oreochromis* (Sarotherodon) *Mossambicus*—Effects of Cortisol and Thyroxine. General and Comparative Endocrinology 62: 341–343.

Dantzler, W.H. 2003. Regulation of renal proximal and distal tubule transport: sodium, chloride and organic anions. Comparative Biochemistry and Physiology Part A 136: 453–478.

Elger, M., A. Werner, P. Herter, B. Kohl, R.K. Kinne and H. Hentschel. 1998. Na-P(i) cotransport sites in proximal tubule and collecting tubule of winter flounder (*Pleuronectes americanus*). Am. J. Physiol. 274: F374–83.

Evans, D.H., P.M. Piermarini and W.T.W. Potts. 1999. Ionic transport in the fish gill epithelium. J. Exp. Zool. 283: 641–652.

Evans, D.H., P.M. Piermarini and K.P. Choe. 2005. The multifunctional fish gill: dominant site of gas exchange, osmoregulation, acid-base regulation, and excretion of nitrogenous waste. Physiol. Rev. 85: 97–177.

Evans, D.H. 2008. Teleost fish osmoregulation: what have we learned since August Krogh, Homer Smith, and Ancel Keys. Am. J. Physiol. Regul. Integr. Comp. Physiol. 295: R704–13.

Fenwick, J.C., S.E. Wendelaar Bonga and G. Flik. 1999. *In vivo* bafilomycin-sensitive Na(+) uptake in young freshwater fish. J. Exp. Biol. 202 Pt 24: 3659–66.

Ferlazzo, A., E.S. Carvalho, S.F. Gregorio, D.M. Power, A.V. Canario, F. Trischitta and J. Fuentes. 2012. Prolactin regulates luminal bicarbonate secretion in the intestine of the sea bream (Sparus aurata L.). J. Exp. Biol. 215 3836–3844.

Fiol, D.F. and D. Kultz. 2007. Osmotic stress sensing and signaling in fishes. FEBS J 274: 5790–8.

Flik, G., S.E. Wendelaar Bonga and J.C. Fenwick. 1983. Ca^{2+}-dependent phosphatase and ATPase activities in eel gill plasma membranes I. Identification of Ca^{2+}-activated ATPase activities with non-specific phosphatase activities. Comp. Biochem. Physiol. B. 76: 745–54.

Fuentes, J., J.C. McGeer and F.B. Eddy. 1996. Drinking rate in juvenile Atlantic salmon, *Salmo salar* L fry in response to a nitric oxide donor, sodium nitroprusside and an inhibitor of angiotensin converting enzyme, enalapril. Fish Physiol. Biochem. 15: 65–9.

Fuentes, J. and F.B. Eddy. 1997a. Drinking in freshwater, euryhaline and marine teleosts. pp. 135–149. *In*: Hazon, N., F.B. Eddy and G. Flik [eds.]. Ionic Regulation in Animals. New York: Heidelberg: Springer-Verlag.

Fuentes, J. and F.B. Eddy. 1997b. Effect of manipulation of the renin-angiotensin system in control of drinking in juvenile Atlantic salmon (*Salmo salar* L.) in fresh water and after transfer to sea water. J. Comp. Physiol. B 167: 438–443.

Fuentes, J. and F.B. Eddy. 1997c. Drinking in Atlantic salmon presmolts and smolts in response to growth hormone and salinity. Comp. Biochem. Physiol. A. Physiol. 117: 487–91.

Fuentes, J., J.L. Soengas, P. Rey and E. Rebolledo. 1997. Progressive transfer to seawater enhances intestinal and branchial Na^+-K^+-ATPase activity in non-anadromous rainbow trout. Aquaculture International 5: 217–227.

Fuentes, J., J. Figueiredo, D.M. Power and A.V. Canario. 2006. Parathyroid hormone-related protein regulates intestinal calcium transport in sea bream (*Sparus auratus*). Am. J. Physiol. Regul. Integr. Comp. Physiol. 291 R1: 499–1506.

Fuentes, J., D.M. Power and A.V. Canario. 2010. Parathyroid hormone-related protein-stanniocalcin antagonism in regulation of bicarbonate secretion and calcium precipitation in a marine fish intestine. Am. J. Physiol. Regul. Integr. Comp. Physiol. 299: R150–158.

Genz, J., J.R. Taylor and M. Grosell. 2008. Effects of salinity on intestinal bicarbonate secretion and compensatory regulation of acid-base balance in *Opsanus beta*. J. Exp. Biol. 211: 2327–2335.

Graham, C., P. Nalbant, B. Scholermann, H. Hentschel, R.K. Kinne and A. Werner. 2003. Characterization of a type IIb sodium-phosphate cotransporter from zebrafish (*Danio rerio*) kidney. Am. J. Physiol. Renal. Physiol. 284: F727–36.

Grau, E.G. 1988. Environmental influences on thyroid function in teleost fish. American Zoologist 28: 329–335.

Gregorio, S.F., E.S. Carvalho, S. Encarnaçao, J.M. Wilson, D.M. Power, A.V. Canario and J. Fuentes. 2013. Adaptation to different salinities exposes functional specialization in the intestine of the sea bream (*Sparus aurata* L.). J. Exp. Biol. 216: 470–479.

Grosell, M. 2006. Intestinal anion exchange in marine fish osmoregulation. J. Exp. Biol. 209: 2813–2827.

Grosell, M. and J. Genz. 2006. Ouabain-sensitive bicarbonate secretion and acid absorption by the marine teleost fish intestine play a role in osmoregulation. Am. J. Physiol. Regul. Integr. Comp. Physiol. 291: R1145–1156.

Grosell, M. 2011. Intestinal anion exchange in marine teleosts is involved in osmoregulation and contributes to the oceanic inorganic carbon cycle. Acta Physiol. (Oxf) 202: 421–434.

Guerreiro, P.M., J. Fuentes, D.M. Power, P.M. Ingleton, G. Flik and A.V. Canario. 2001. Parathyroid hormone-related protein: a calcium regulatory factor in sea bream (*Sparus aurata* L.) larvae. Am. J. Physiol. Regul. Integr. Comp. Physiol. 281: R855–860.

Guerreiro, P.M., J. Fuentes, A.V. Canario and D.M. Power. 2002. Calcium balance in sea bream (*Sparus aurata*): the effect of oestradiol-17beta. J. Endocrinol. 173: 377–385.

Guerreiro, P.M., J. Fuentes, G. Flik, J. Rotllant, D.M. Power and A.V. Canario. 2004. Water calcium concentration modifies whole-body calcium uptake in sea bream larvae during short-term adaptation to altered salinities. J. Exp. Biol. 207: 645–653.

Herrera, M., L. Vargas-Chacoff, I. Hachero, I. Ruiz-Jarabo, A. Rodiles, J.I. Navas and J.M. Mancera. 2009. Osmoregulatory changes in wedge sole (*Dicologoglossa cuneata* Moreau, 1881) after acclimation to different environmental salinities. Aquaculture Research 40: 762–771.

Herrera, M., C. Aragao, I. Hachero, I. Ruiz-Jarabo, L. Vargas-Chacoff, J.M. Mancera and L. Conceiçao. 2012. Physiological short-term response to sudden salinity change in the Senegalese sole (*Solea senegalensis*). Fish Physiol. Biochem. 38: 1741–1751.

Hirano, T. and N. Mayer-Gostan. 1976. Eel esophagus as an osmoregulatory organ. Proc. Natl. Acad. Sci. USA 73: 1348–1350.

Hiroi, J. and S.D. McCormick. 2012. New insights into gill ionocyte and ion transporter function in euryhaline and diadromous fish. Respiratory Physiology & Neurobiology 184: 257–268.

Holly, J.M. and J.A. Wass. 1989. Insulin-like growth factors; autocrine, paracrine or endocrine? New perspectives of the somatomedin hypothesis in the light of recent developments. J. Endocrinol. 122: 611–8.

Hwang, P.P., T.H. Lee and L.Y. Lin. 2011. Ion regulation in fish gills: recent progress in the cellular and molecular mechanisms. Am. J. Physiol. Regul. Integr. Comp. Physiol. 301: R28–47.

Hwang, P.P. and M.Y. Chou. 2013. Zebrafish as an animal model to study ion homeostasis. Pflugers Arch 465: 1233–47.

Hwang, P.P., C.M. Sun and S.M. Wu. 1989. Changes of Plasma Osmolality, Chloride Concentration and Gill Na-K-Atpase Activity in Tilapia *Oreochromis mossambicus* during Seawater Acclimation. Marine Biology 100: 295–299.

Imsland, A.K., A. Foss, S. Gunnarsson, M.H.G. Berntssen, R. FitzGerald, S.W. Bonga, E.v. Ham, G. Naevdal and S.O. Stefansson. 2001. The interaction of temperature and salinity on growth and food conversion in juvenile turbot (*Scophthalmus maximus*). Aquaculture 198: 353–367.

Imsland, A.K., S. Gunnarsson, A. Foss and S.O. Stefansson. 2003. Gill Na+,K+-ATPase activity, plasma chloride and osmolality in juvenile turbot (*Scophthalmus maximus*) reared at different temperatures and salinities. Aquaculture 218: 671–683.

Imsland, A.K., A. Foss, L.E.C. Conceiçao, M.T. Dinis, D. Delbare, E. Schram, A. Kamstra, P. Rema and P. White 2004. A review of the culture potential of *Solea solea* and *S. senegalensis*. Reviews in Fish Biology and Fisheries 13: 379–407.

Kim, Y.K., H. Ideuchi, S. Watanabe, S.I. Park, M. Huh and T. Kaneko. 2008. Rectal water absorption in seawater-adapted Japanese eel *Anguilla japonica*. Comp. Biochem. Physiol. A. Mol. Integr. Physiol. 151: 533–541.

Keys, A. and E.N. Willmer. 1932. "Chloride secreting cells" in the gills of fishes, with special reference to the common eel. Journal of Physiology-London 76: 368–378.

Laiz-Carrion, R., S. Sangiao-Alvarellos, J.M. Guzman, M.P. Martin del Rio, J.M. Miguez, J.L. Soengas and J.M. Mancera. 2002. Energy metabolism in fish tissues related to osmoregulation and cortisol action. Fish Physiology and Biochemistry 27: 179–188.

Laiz-Carrion, R., P.M. Guerreiro, J. Fuentes, A.V.M. Canario, M.P. Martin del Rio and J.M. Mancera. 2005. Branchial osmoregulatory response to salinity in the gilthead sea bream, *Sparus auratus*. Journal of Experimental Zoology Part A Comparative Experimental Biology 303A: 563–576.

Laurent, P. and S.F. Perry. 1990. Effects of Cortisol on Gill Chloride Cell Morphology and Ionic Uptake in the Fresh-Water Trout, *Salmo gairdneri*. Cell and Tissue Research 259: 429–442.

Liao, B.K., R.D. Chen and P.P. Hwang. 2009. Expression regulation of Na+-K+-ATPase alpha1-subunit subtypes in zebrafish gill ionocytes. Am. J. Physiol. Regul. Integr. Comp. Physiol. 296: R1897–906.

Lin, C.H., C.L. Huang, C.H. Yang, T.H. Lee and P.P. Hwang. 2004. Time-course changes in the expression of Na, K-ATPase and the morphometry of mitochondrion-rich cells in gills of euryhaline tilapia (*Oreochromis mossambicus*) during freshwater acclimation. J. Exp. Zool. A. Comp. Exp. Biol. 301: 85–96.

Madsen, S.S. 1990. Effect of Repetitive Cortisol and Thyroxine Injections on Chloride Cell Number and Na+/K+-ATPase Activity in Gills of Fresh-Water Acclimated Rainbow-Trout, *Salmo gairdneri*. Comparative Biochemistry and Physiology a-Physiology 95: 171–175.

Madsen, S.S., B.K. Larsen and F.B. Jensen. 1996. Effects of freshwater to seawater transfer on osmoregulation, acid base balance and respiration in river migrating whitefish (*Coregonus lavaretus*). Journal of Comparative Physiology B-Biochemical Systemic and Environmental Physiology 166: 101–109.

Mancera, J.M., J.M. Perez-Figares and P. Fernandez-Llebrez. 1994. Effect of cortisol on brackish water adaptation in the euryhaline gilthead sea bream (*Sparus aurata* L.). Comparative Biochemistry and Physiology A 107: 397–402.

Mancera, J.M. and S.D. McCormick. 1998. Osmoregulatory actions of the GH:IGF axis in non-salmonid teleosts. Comp. Biochem. Physiol. B. 121: 43–48.

Mancera, J.M. and S.D. McCormick. 1999. Influence of cortisol, growth hormone, insulin-like growth factor I and 3,3´,5-triiodo-L-thyronine on hypoosmoregulatory ability in the euryhaline teleost *Fundulus heteroclitus*. Fish Physiology and Biochemistry 21: 25–33.

Mancera, J.M. and S.D. McCormick. 2000. Rapid activation of gill Na(+),K(+)-ATPase in the euryhaline teleost *Fundulus heteroclitus*. J. Exp. Zool. 287: 263–74.

Marquez, L. and J. Fuentes. 2014. *In vitro* characterization of acid secretion in the gilthead sea bream (*Sparus aurata*) stomach. Comp. Biochem. Physiol. A. Mol. Integr. Physiol. 167: 52–58.

Marshall, W.S. and S.E. Bryson. 1998. Transport mechanisms of seawater teleost chloride cells: an inclusive model of a multifunctional cell. Comp. Biochem. Physiol. A. Mol. Integr. Physiol. 119: 97–106.

Martos-Sitcha, J.A., G. MartinezRodriguez, J.M. Mancera and J. Fuentes (2015). AVT and IT regulate ion transport across the opercular epithelium of killifish (*Fundulus heteroclitus*) and gilthead sea bream (*Sparus aurata*). Comp. Biochem. Physiol. A. Mol. Integr Physiol. 182: 93–101.

McCormick, S.D. 1995. Hormonal control of gill Na$^+$/K$^+$-ATPase and chloride cell function. pp. 285–315. *In*: Wood, C.M. and T. J. Shuttleworth [eds.]. Fish Physiology, vol. 14, Cellular and Molecular Approaches to Fish Ionic Regulation. New York: Academic Press.

McCormick, S.D. 2001. Endocrine control of osmoregulation in teleost fish. American Zoologist 41: 781–794.

McCormick, S.D., A.M. Regish and A.K. Christensen. 2009. Distinct freshwater and seawater isoforms of Na$^+$/K$^+$-ATPase in gill chloride cells of Atlantic salmon. J. Exp. Biol. 212: 3994–4001.

McDonald, M.D. 2007. The renal contribution to salt and water balance. pp. 309–332. *In*: Baldiserotto, B., J.M. Mancera and B.G. Kapoor [eds.]. Fish Osmoregulation. New Delhi, India: Sciencie Publishers Enfield (NH) & IBH Publishing.

McEwen, B.S. and J.C. Wingfield. 2003. The concept of allostasis in biology and biomedicine. Horm. Behav. 43: 2–15.

McLean, E. and E.M. Donaldson. 1993. The role of growth hormone in growth of poikilotherms. pp. 43–71. *In*: Schreibman, M.P., C.G. Scanes and P. K. T. Pang [eds.]. The Endocrinology of Growth, Development and Metabolism in Vertebrates. New York: Academic Press.

Mohammed Geba, K., L. Vargas-Chacoff, I. Ruiz-Jarabo, J.M. Mancera and G. Martinez-Rodriguez. 2011. Time course responses of GH/IGF-I axis to abrupt changes in environmental salinity in the senegalese sole *Solea senegalensis*. pp. 54–57. *In*: Wilson, J.M. A. Damasceno-Oliveira and J. Coimbra [eds.]. Avanços em endocrinologia Comparativa, vol. V. Oporto: Centro Interdisciplinar de Investigaçao Marinha e Ambiental, Universidade do Porto.

Mommsen, T.P., M.M. Vijayan and T.W. Moon. 1999. Cortisol in teleosts: dynamics, mechanisms of action, and metabolic regulation. Reviews in Fish Biology and Fisheries 9: 211–268.

Morgan, J.D. and G.K. Iwama. 1998. Salinity effects on oxygen consumption, gill Na$^+$,K$^+$-ATPase and ion regulation in juvenile coho salmon. Journal of Fish Biology 53: 1110–1119.

Musch, M.W., S.A. Orellana, L.S. Kimberg, M. Field, D.R. Halm, E.J. Krasny and R.A. Frizzell. 1982. Na$^+$-K$^+$-Cl$^-$ co-transport in the intestine of a marine teleost. Nature 300: 351–353.

O'Donnell, M.J., S.P. Kelly, C.A. Nurse and C.M. Wood. 2001. A maxi Cl$^-$ channel in cultured pavement cells from the gills of the freshwater rainbow trout *Oncorhynchus mykiss*. Journal of Experimental Biology 204: 1783–1794.

Parmelee, J.T. and J.L. Renfro. 1983. Esophageal desalination of seawater in flounder: role of active sodium transport. Am. J. Physiol. 245: R888–893.

Perry, S.F. 1998. Relationships between branchial chloride cells and gas transfer in freshwater fish. Comp. Biochem. Physiol. A. Mol. Integr. Physiol. 119: 9–16.

Piermarini, P.M., J.W. Verlander, I.E. Royaux and D.H. Evans. 2002. Pendrin immunoreactivity in the gill epithelium of a euryhaline elasmobranch. Am. J. Physiol. Regul. Integr. Comp. Physiol. 283: R983–92.

Power, D.M., L. Llewellyn, M. Faustino, M.A. Nowell, B.T. Bjornsson, I.E. Einarsdottir, A.V. Canario and G.E. Sweeney. 2001. Thyroid hormones in growth and development of fish. Comparative Biochemistry and Physiology C-Toxicology & Pharmacology 130: 447–459.

Renfro, J.L. 1997. Hormonal regulation of renal inorganic phosphate transport in the winter flounder, *Pleuronectes americanus*. Fish Physiology and Biochemistry 17: 377–383.

Ruiz-Jarabo, I., M. Herrera, I. Hachero-Cruzado, L. Vargas-Chacoff, J.M. Mancera and F.J. Arjona. 2015. Environmental salinity and osmoregulatory processes in cultured flatfish. Aquaculture Research 46: 10–29.

Ruiz-Jarabo, I., A. Barany, I. Jerez-Cepa, J.M. Mancera and J. Fuentes. 2017. Intestinal response to salinity challenge in the Senegalese sole (*Solea senegalensis*). Comp. Biochem. Physiol. A 204: 57–64.

Sampaio, L.A. and A. Bianchini. 2002. Salinity effects on osmoregulation and growth of the euryhaline flounder *Paralichthys orbignyanus*. Journal of Experimental Marine Biology and Ecology 269: 187–196.

Shrimpton, J.M. and S.D. McCormick. 1998. Regulation of gill cytosolic corticosteroid receptors in juvenile Atlantic salmon: interaction effects of growth hormone with prolactin and triiodothyronine. Gen. Comp. Endocrinol. 112: 262–74.

Soengas, J.L., S. Sangiao-Alvarellos, R. Laiz-Carrion and J.M. Mancera. 2008. Energy metabolism and osmotic acclimation in teleost fish. pp. 277–307. *In*: Baldiserotto, B., J.M. Mancera and B.G. Kapoor [eds.]. Fish Osmoregulation. Enfield, NH: Science Publishers.

Sullivan, G., J. Fryer and S. Perry. 1995. Immunolocalization of proton pumps (H⁺-ATPase) in pavement cells of rainbow trout gill. J. Exp. Biol. 198: 2619–29.

Suzuki, Y., M. Itakura, M. Kashiwagi, N. Nakamura, T. Matsuki et al. 1999. Identification by differential display of a hypertonicity-inducible inward rectifier potassium channel highly expressed in chloride cells. J. Biol. Chem. 274: 11376–82.

Takei, Y., J.M. Joss, W. Kloas and J.C. Rankin. 2004. Identification of angiotensin I in several vertebrate species: its structural and functional evolution. Gen. Comp. Endocrinol. 135: 286–92.

Takei, Y. and S.D. McCormick. 2013. Hormonal control of fish euryhalinity. pp. 69–124. *In*: McCormick, S.D., C. J. Brauner and A. P. Farrell [eds.]. Fish Physiology, vol. 32, Amsterdam: Academic Press.

Tang, C.H. and T.H. Lee. 2011. Ion-deficient environment induces the expression of basolateral chloride channel, ClC-3-like protein, in gill mitochondrion-rich cells for chloride uptake of the tilapia *Oreochromis mossambicus*. Physiol. Biochem. Zool. 84: 54–67.

Tipsmark, C.K., J.A. Luckenbach, S.S. Madsen, P. Kiilerich and R.J. Borski. 2008. Osmoregulation and expression of ion transport proteins and putative claudins in the gill of Southern Flounder (*Paralichthys lethostigma*). Comparative Biochemistry and Physiology a-Molecular & Integrative Physiology 150: 265–273.

Tipsmark, C.K. and S.S. Madsen. 2009. Distinct hormonal regulation of Na(+),K(+)-atpase genes in the gill of Atlantic salmon (*Salmo salar* L.). J. Endocrinol. 203: 301–10.

Tse, W.K., D.W. Au and C.K. Wong. 2007. Effect of osmotic shrinkage and hormones on the expression of Na⁺/H⁺ exchanger-1, Na⁺/K⁺/2Cl⁻ cotransporter and Na⁺/K⁺-ATPase in gill pavement cells of freshwater adapted Japanese eel, *Anguilla japonica*. J. Exp. Biol. 210: 2113–20.

Tseng, Y.C. and P.P. Hwang. 2008. Some insights into energy metabolism for osmoregulation in fish. Comparative Biochemistry and Physiology C-Toxicology & Pharmacology 148: 419–429.

Veillette, P.A., K. Sundell and J.L. Specker. 1995. Cortisol mediates the increase in intestinal fluid absorption in Atlantic salmon during parr-smolt transformation. Gen. Comp. Endocrinol. 97: 250–8.

Venturini, G., E. Cataldi, G. Marino, P. Pucci, L. Garibaldi, P. Bronzi and S. Cataudella. 1992. Serum Ions Concentration and Atpase Activity in Gills, Kidney and Esophagus of European Sea Bass (*Dicentrarchus labrax*, Pisces, Perciformes) during Acclimation Trials to Fresh-Water. Comparative Biochemistry and Physiology a-Physiology 103: 451–454.

Wedemeyer, G.A., B.A. Barton and D.J. McLeay. 1990. Stress and acclimation. pp. 451–489. *In*: Schreck, C.B. and P.B. Moyle [eds.]. Methods of Fish Biology. Bethesda MD: American Fisheries Society.

Wendelaar Bonga, S.E. 1997. The stress response in fish. Physiological Reviews 77: 591–625.

Wilkie, M.P. 1997. Mechanisms of ammonia excretion across fish gills. Comparative Biochemistry and Physiology a-Molecular & Integrative Physiology 118: 39–50.

Wilson, J.M., P. Laurent, B.L. Tufts, D.J. Benos, M. Donowitz, A.W. Vogl and D.J. Randall. 2000a. NaCl uptake by the branchial epithelium in freshwater teleost fish: An immunological approach to ion-transport protein localization. Journal of Experimental Biology 203: 2279–2296.

Wilson, J.M., D.J. Randall, M. Donowitz, A.W. Vogl and A.K. Ip. 2000b. Immunolocalization of ion-transport proteins to branchial epithelium mitochondria-rich cells in the mudskipper (*Periophthalmodon schlosseri*). J. Exp. Biol. 203: 2297–310.

Wu, S.C., J.L. Horng, S.T. Liu, P.P. Hwang, Z.H. Wen, C.S. Lin and L.Y. Lin. 2010. Ammonium-dependent sodium uptake in mitochondrion-rich cells of medaka (*Oryzias latipes*) larvae. Am. J. Physiol. Cell Physiol. 298: C237–50.

Wunderink, Y.S., S. Engels, S. Halm, M. Yufera, G. Martinez-Rodriguez, G. Flik, P.H. Klaren and J.M. Mancera. 2011. Chronic and acute stress responses in Senegalese sole (*Solea senegalensis*): the involvement of cortisol, CRH and CRH-BP. Gen. Comp. Endocrinol. 171: 203–10.

Yan, B.Q., Z.H. Wang and J.L. Zhao. 2013. Mechanism of osmoregulatory adaptation in tilapia. Molecular Biology Reports 40: 925–931.

B-6.1

Genetic and Genomic Characterization of Soles

Manuel Manchado,[1,*] *Josep V. Planas,*[2] *Xavier Cousin,*[3] *Laureana Rebordinos*[4] and *M. Gonzalo Claros*[5]

1. Introduction

Genome architecture of soles has been intensively explored in the last years, particularly with the arrival of Next Generation Sequencing (NGS) techniques. Hence, a large amount of transcriptomic and genomic data has been generated facilitating the identification of molecular markers and the development of expression tools required to decipher the molecular mechanisms that govern important traits such as growth, development or reproduction and the effects of diets, micronutrients or immunostimulants. In this chapter, the main features of the transcriptome, genome and epigenome of *S. senegalensis* and *S. solea*, when available, are described. Moreover, the main uses and applications of these resources to design novel expression tools in Senegalese sole, with some examples will be reviewed. Finally, due to the importance of these species in aquaculture, main advances in molecular markers and the implementation of genetic breeding programs will be discussed emphasising in the current bottlenecks, tools, methodologies and strategies designed for an optimal management of genetic resources in the industry.

[1] IFAPA Centro El Toruño, IFAPA, Consejeria de Agricultura, Pesca y Medio Rural. 11500 El Puerto de Santa María (Cádiz), Spain.

[2] Departament de Fisiologia i Immunologia, Facultat de Biologia, Universitat de Barcelona and Institut de Biomedicina de la Universitat de Barcelona (IBUB), 08028 Barcelona, Spain

[3] IFREMER, Laboratoire Adaptation et Adaptabilités des Animaux et des Systèmes, F-34250 Palavas-les-Flots, France; Inra, UMR GABI, F-78350 Jouy-en-Josas, France.

[4] Laboratorio de Genética. Facultad de Ciencias del Mar y Ambientales, Universidad de Cádiz, Polígono del Río San Pedro, 11510, Puerto Real, Cádiz, Spain.

[5] Departamento de Biología Molecular y Bioquímica, Facultad de Ciencias, Campus de Teatinos s/n, Universidad de Málaga, 29071 Málaga, Spain

* Corresponding author: manuel.manchado@juntadeandalucia.es

2. Genome and transcriptome characterization in sole

2.1 Transcriptome assembly, annotation and characterization

Transcriptomes of soles were studied in different projects as a primary tool to identify expressed sequence tags (ESTs) for the discovery of new molecular markers, coding genes and regulatory transcripts. The first large genomic studies in Senegalese sole were carried out using the Sanger technology in the framework of the Pleurogene project. This consortium produced several normalized libraries from adult tissues (brain, stomach, intestine, liver, ovary, and testis), larval stages (pre-metamorphosis, metamorphosis), juvenile stages (post-metamorphosis, abnormal fish) and undifferentiated gonads that yielded a total of 10,185 ESTs (Cerda et al. 2008). Later studies based on NGS technologies increased significantly the number of genomic resources available in both soles. A first study in *S. solea* using the Roche/454 technology succeeded in assembling 22,252 transcripts in a combined analysis of larvae and muscle samples, of which 16,731 where annotated (Ferraresso et al. 2013). A later study in both *S. senegalensis* and *S. solea* produced more than 1,800 million reads in each species using the Roche/454 and Illumina NGS technologies with a transcriptome coverage ranging from 1,384× to 2,543× (Benzekri et al. 2014). In this way, global transcriptomes accounting for 697,125 transcripts in *S. senegalensis* and 523,637 transcripts in *S. solea* were assembled. These transcriptomes were generated from embryos and larvae at different developmental stages in *S. solea* (Illumina) and a mix of adult tissues and larval stages in *S. senegalensis* (454 and Illumina). Moreover, specific transcriptomes for hypothalamus, pituitary, immune-related organs (head kidney, spleen, brain, thymus and gills), osmoregulatory organs (kidney, intestine, gill, and brain), testis and ovary using the Roche/454 technology were generated in *S. senegalensis*. The number of transcripts assembled for each organ ranged between 55,469 and 117,058 in hypothalamic and immune-related organ libraries, respectively (Benzekri et al. 2014).

A full analysis of the transcriptome in the two sole species showed a total of 45,063 and 38,402 transcripts with different orthologous identifiers when compared with zebrafish in *S. senegalensis* and *S. solea*, respectively, of which 18,738 and 22,683 were complete ORFs, respectively. The deduced number of protein-coding genes is closer to the number estimated in the sequenced genomes of fish species including the closely related flatfish, the tongue sole (*Cynoglossus semilaevis*), coelacanth and elephant shark (Prachumwat et al. 2008, Amemiya et al. 2013, Chen et al. 2014, Venkatesh et al. 2014), with the exception of zebrafish that possesses a slightly higher number of protein-coding genes, 26,206 (Kettleborough et al. 2013). This indicates that an almost complete transcriptome has now been obtained.

A further characterization of transcriptomes using gene ontology showed curiously that category representation in tissue-specific transcriptomes (Fig. 1A) was slightly different from global transcriptomes (Fig. 1B). Although the categories "Cellular process", "Single-organism process" and "Metabolic process" were the most represented in all transcriptomes, their percentages dropped in global transcriptomes (i.e., Cellular process from 25 to 17%). In contrast, other categories such as "Developmental process" and "Growth" were quite better represented in the latter (1.6 vs. 6.0 and 0.2 vs. 6.4%, respectively) and this can be explained by the fact that development and growth are regulated at multi-tissue level and also because global transcriptomes included developing and growing stages (embryos and larvae). Interestingly, the category percentages between *S. senegalensis* and *S. solea* were quite similar confirming an excellent and equally representation of both transcriptomes.

A

B

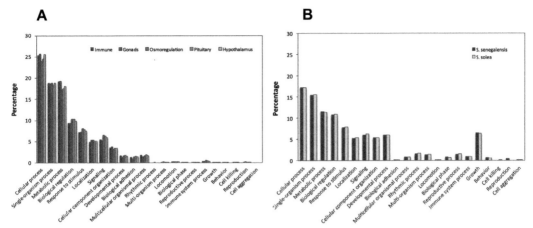

Fig. 1. GO representation (%) for biological process. (A): Transcriptomes for immune- and osmoregulation-related organs, gonads, pituitary and hypothalamus (Roche/454); (B): Global transcriptomes for *S. senegalensis* and *S. solea*.

Besides offering access to genomic resources and paving the way for the development of novel tools as will be discussed below, all this transcriptomic information has provided new clues about evolution of flatfish. The comparison between the two sole transcriptomes, with respect to tongue sole (*C. semilaevis*) and other more distantly teleosts allowed for the identification of sole-, flatfish- and fish-specific genes based on orthology and blast analysis. The former evidenced a high degree of similarity between both soles with ~ 65% of the transcripts sharing the same zebrafish ortholog. In the latter, a reciprocal blast identified 11,953 different transcripts as true sole Blast-based orthologs (BBO). From this set, the vast majority were annotated (98.2%) with a zebrafish ortholog (93.8%). However, a small subset of annotated BBO transcripts (351 transcripts shared by both *S. senegalensis* and *S. solea*) lacked zebrafish orthology (mainly immune system genes) and may represent lineage-specific genes that have appeared and thereafter subfunctionalized or neofunctionalized during evolution. Moreover, a total of 210 orthologs remained as unannotated transcripts, of which 137 had a testcode ≥ 0.94, suggesting that they were actually coding sequences (Benzekri et al. 2014). When considering together those 137 transcripts and the 351 annotated transcripts without zebrafish orthologs (total number of 488 transcripts), 386 (79%) had orthologs in *C. semilaevis* and 317 (65%) in other teleosts, indicating that those genes had been retained or lost during *Acanthomorpha* evolution. Interestingly, 118 transcripts (24%) showed orthology only to *C. semilaevis*, possibly defining a set of flatfish- specific transcripts and 53 (~ 11%) failed to show any identity with any other teleost possibly defining a set of sole-specific transcripts (Benzekri et al. 2014).

2.2 Unraveling the genome

2.2.1 Karyotype, cytogenetic and genetic maps in sole

The karyotype of soles (Fig. 2A) is composed of 2 n = 42 (Libertini et al. 2002, Vega et al. 2002). In Senegalese sole, it comprises three pairs of metacentric, two pairs of submeta-subtelocentric, four pairs of subtelocentric and twelve pairs of acrocentric chromosomes that are difficult to distinguish according to shape and size. Moreover, no sexual chromosomes could be identified at morphological level (Vega et al. 2002). Some

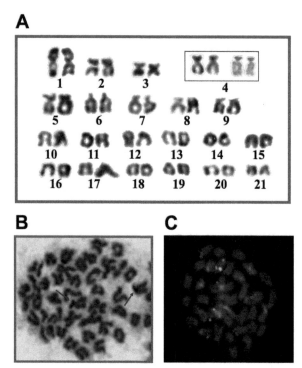

Fig. 2. Karyotype characteristics in *S. senegalensis*. (A), 21 chromosome pairs and the pair harboring the nucleolus organizer regions (NOR), visualized using the silver stain method; (B). *S. senegalensis* metaphases showing Ag-NORs (arrows); (C), Multiple Fluorescence *in situ* hybridization (mFISH) showing location of *sox3* (blue), *sox6* (green), *sox8* (pink) and *sox9* (*red*) genes in different chromosome pairs.

cytogenetic studies characterized molecular markers specific of chromosomes or their arms. The hybridization with the telomeric $(TTAGGG)_n$ repeat displayed small signals in all chromosomal telomeres. The Ag-NORs (Fig. 2B) and the 45S rDNA were mapped to a medium-sized submetacentric chromosomal pair and the 5S rDNA showed a major signal on the short arm of a medium-sized submetacentric pair (Libertini et al. 2002, Cross et al. 2006). Another study demonstrated that *S. senegalensis* possesses two different classes of 5S rDNA corresponding to the major locus previously described co-localized with the 45S rDNA and one minor near the centromeric region of an acrocentric pair (Cross et al. 2006, Manchado et al. 2006).

These cytogenetic techniques were also applied in the Senegalese sole to the development of a BAC-FISH map (Garcia-Cegarra et al. 2013) using clones from a BAC library (Ponce et al. 2011) as probes and the two-color FISH to discriminate chromosome position. In this study, 10 clones were hybridized obtaining clear signals for nine of them in one chromosome pair. The two-color FISH also confirmed that 4 BACs were co-localized in two chromosome pairs. These BAC clones were sequenced and provided the gene structure of 23 genes related to the endocrine and immune systems, allowing for the building of the first draft for a cytogenetic map in this species. Further studies are currently ongoing to complete the karyotype characterization of this species focusing on BACs that contain genes associated with sex determination and sex-differentiation using a Multiple-FISH technique. The combination of all these cytogenetic results provides markers anchored in 15 pairs of chromosomes out the total of 21 in the karyotype of *S. senegalensis* (Portela-Bens et al. 2016).

Moreover, genome architecture in soles has been studied by using molecular markers such as microsatellites (SSR) and single nucleotide polymorphism (SNPs) to build genetic maps. In the Senegalese sole, two haploid families were used to locate the 129 SSR markers in linkage groups (LGs) providing a consensus map that consisted of 27 LGs with an average density of 4.7 markers per LG spanning 1,004 cM (Molina-Luzon et al. 2015a). In the common sole, only a linkage map that included 423 SNPs derived from ESTs and 8 neutral SSRs has been reported. The total map length was 1,233.8 cM organized in 38 LGs with a size varying between 0 to 92.1 cM in each group (Diopere et al. 2014).

2.2.2 Physical map and genome assembly in the Senegalese sole

In addition to cytogenetic and genetic maps, a preliminary physical map for the Senegalese sole genome has been obtained. This preliminary draft genome corresponds to a female and it has been built from $2,983 \times 10^6$ Illumina paired-end reads and 8.3×10^6 long paired-end reads from the Roche/454 platform. After a complex bioinformatics protocol (Benzekri 2016) based on SeqTrimNext, RAY, SOAPdenovo, NUCMER, GAM-NGS, GapCloser and SSPACE software for pre-processing, assembly, scaffolding and gap-closing, a draft of the *S. senegalensis* genome was achieved that consisted of 34,176 scaffolds. The longest scaffold was 638.2 kbp in length. The N50 was 85.6 kbp, the mean length of scaffolds was of 14.5 kbp with an overall indetermination of 14,3 kbp. Although all scaffolds provide a putative genome size of 600.3 Mbp, the most reliable estimation was obtained with the KmerGenie software (Chikhi and Medvedev 2014), based on the *k*-mer frequency resulting in 612.3 Mbp, in agreement with the cytogenetical size estimation of 713 Mbp haploid genome in *S. solea* (Libertini et al. 2002). This current draft status indicates that the female *S. senegalensis* genome is smaller than Nile tilapia (*Oreochromis niloticus*) genome [927 Mbp (Brawand et al. 2014)] or the zebrafish (*Danio rerio*) genome [1,371 Mbp (Howe et al. 2013)] and slightly larger than that of the tongue sole (*Cynoglossus semilaevis*) genome (470 Mbp (Chen et al. 2014)].

2.2.3 Synteny in the Senegalese sole

Synteny analysis using the scaffolds of the draft genome for the apolipoprotein gene family revealed a high degree of conservation for gene arrangement across taxa (Fig. 3) (Roman-Padilla et al. 2016). This gene family was structured in two gene clusters: cluster A containing the duplicated genes *apoA-IVAa1* and *apoA-IVAa2* inverted in head-to-head orientation with the latter closer to the *apoEa* in Acanthopterygii and cluster B containing an array of 6 apolipoproteins (*apoA-IVBa4*, *apoA-IVBa3*, *apoEb*, *apoC-I*, *apo14* and *apoC-II*) in the same transcriptional orientation except the most distant head-to-head inverted *apoA-IVBa4*. Interestingly, these gene arrangements were highly conserved across teleosts, particularly with *C. semilaevis* (tongue sole) and *Gasterosteus aculeatus* (three-spined stickleback), and less conserved in the model zebrafish.

Therefore, the extent of similarity between Senegalese sole and tongue sole genomes was also investigated using the nine longest scaffolds of the sole draft genome and the last version of the tongue sole genome [GCF_000523035.1 (Chen et al. 2014)]. The results based on tongue sole proteins or Senegalese sole transcripts indicated that most genes contained in Senegalese sole scaffolds were syntenic with tongue sole chromosomes. This is particularly clear when the comparison is carried out using transcripts (Manchado et al. 2016). A further synteny analysis using GEvo, from the CoGe (Comparative Genomics) suite (Lyons et al. 2008) also confirmed (Fig. 4) the high conservation of gene arrangements

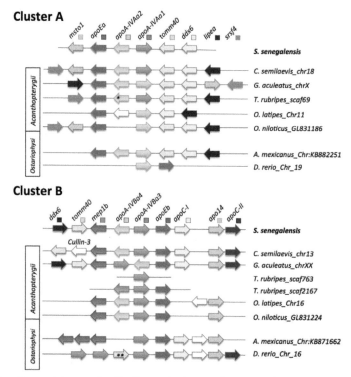

Fig. 3. Synteny of apolipoprotein gene family. Cluster A, Synteny for *apoA-IVAa1*, *apoA-IVAa2* and *apoEa*. Cluster B, Synteny for *apoA-IVBa3*, *apoA-IVBa4*, *apoEb*, *apoC-I*, *apoC14* and *apoC-II*. The species names and the chromosome or scaffold location are indicated on the right. Each gene is represented by a color within each cluster. The coding direction is indicated by the pointed end. *Source:* Reprinted from Comp. Biochem. Physiol. B. Biochem. Mol. Biol., 191. Roman-Padilla et al., Genomic characterization and expression analysis of four apolipoprotein A-IV paralogs in Senegalese sole (Solea senegalensis Kaup), pages 84–98, Copyright (2016), with permission from Elsevier.

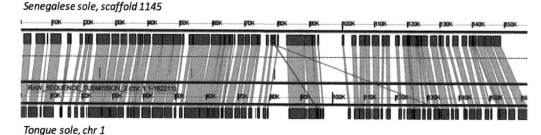

Fig. 4. Synteny analysis of scaffold 1145 with the corresponding fragment of chromosome 1 in tongue sole indicating small local variations.

between both species, exemplified by the alignment of scaffold 1145 in Senegalese sole with the corresponding fragment of chromosome 1 in tongue sole.

Currently, additional efforts are ongoing to use the scaffolds to construct the super-scaffold corresponding to the putative 21 chromosomes in sole (Fig. 5). Interestingly, only 7,971 scaffolds (most of them [7,899 scaffolds] longer than 500 bp) were required to cover the tongue sole chromosomes, providing a reduced genome size for the Senegalese sole of

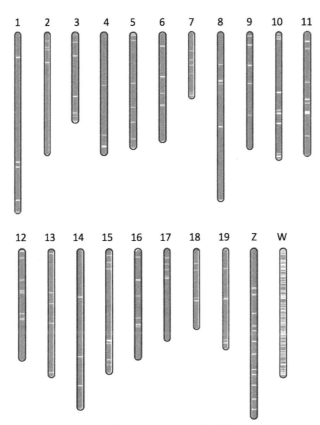

Fig. 5. Super-scaffold coverage of tongue sole chromosomes. Shaded zones correspond to regions covered by Senegalese sole scaffolds.

464,5 Mbp. In spite of the size similarity between the Senegalese sole «reduced genome» and the tongue sole genome (470 Mbp, see above), the chromosome coverage ranged from 99.6% for chromosome 1 to 115.1% for chromosome 9 (Benzekri 2016, Manchado et al. 2016) with an average coverage of 109.0% ± 4.5%. The fact that chromosome coverage was > 100% can be explained due to the higher size of the Senegalese sole genome (Fig. 5). The reliability of these super-scaffolds has been also tested by the co-localization of Senegalese sole transcripts from SoleaDB (Benzekri et al. 2014) on tongue sole chromosomes and Senegalese sole super-scaffolds. These data demonstrate that 32.0–45.0% of scaffolds within each super-scaffold contained transcripts, most of them homologous genes in the same location that the tongue sole chromosome. The exceptions to this rule were percentages for chromosomes 20 and Z with values of 62.1% and the chromosome W with no scaffold encoding for any protein. Therefore, it seems that Senegalese sole super-scaffolds provide a reasonable sequence arrangement. Moreover, a comparison between the position of genetic map and superscaffolds has been done. *In silico* analysis located 113 out of the 129 SSR markers (based on the primers when full-length sequences were not available) onto Senegalese sole scaffolds, while only 10 SSRs presented an uncertain localization, and 6 could not be located. These mapping results lead to the conclusion that (i) most of chromosomes were assigned directly to one or two LGs (Table 1), that (ii) chromosomes 3, 4 and 19 did no contain any SSR, and hence, no LG could be assigned, that (iii) LG1 was split between chromosome 9 and 17 (linked to LG14) and that (iv) LG20, LG22 and LG24 were

Table 1. Main features of super-scaffolds (SSC) assembled in Senegalese sole. The homolog chromosome (chr) in tongue sole, the number of scaffolds in each SSC and those mapping transcripts at SoleaDB, transcripts with matching position between Senegalese sole and tongue sole and the correspondence with linkage groups (LG) of genetic map are indicated.

SSC	Tongue sole	Scaffolds with transcripts (%)	Confirmed position (%)	LG
SSC1	chr_1	260/658 (39.5)	230/260 (88.5)	LG15
SSC2	chr_2	145/356 (40.7)	130/145 (89.7)	LG8
SSC3	chr_3	116/248 (46.8)	89/116 (76.7)	-
SSC4	chr_4	137/332 (14.3)	105/137 (76.6)	-
SSC5	chr_5	149/392 (38.0)	123/149 (82.6)	LG3 + LG26
SSC6	chr_6	128/310 (41.3)	110/128 (85.9)	LG6
SSC7	chr_7	108/260 (41.5)	94/108 (87.0)	LG11
SSC8	chr_8	222/571 (38.9)	189/222 (85.1)	LG7 + LG19
SSC9	chr_9	142/409 (34.7)	131/142 (92.3)	LG1*
SSC10	chr_10	139/354 (39.2)	126/139 (90.6)	LG10
SSC11	chr_11	150/413 (36.3)	119/150 (79.3)	LG2
SSC12	chr_12	114/349 (32.7)	106/114 (93.0)	LG12
SSC13	chr_13	149/411 (36.2)	140/149 (94.0)	LG16
SSC14	chr_14	192/601 (32.0)	173/192 (90.1)	LG9 + LG18
SSC15	chr_15	145/395 (36.7)	130/145 (89.7)	LG13 + LG23
SSC16	chr_16	118/341 (34.6)	106/118 (89.8)	LG17 + LG25
SSC17	chr_17	136/302 (45.0)	127/136 (93.4)	LG1* + LG14
SSC18	chr_18	91/278 (32.7)	80/91 (87.9)	LG5
SSC19	chr_19	136/314 (43.3)	119/136 (87.5)	–
SSC20	chr_20	133/214 (62.1)	96/133 (72.2)	LG21 + LG27
SSCzx	chr_z	285/459 (62.1)	116/285 (40.7)	LG4

LG1 contains SSRs mapping on two different super-scaffolds.

not mapped on super-scaffolds (Table 1). The high consistency between the super-scaffolds and the distribution of SSRs on coherent LGs confirmed that the assignment of scaffolds to super-scaffolds is reliable. Nevertheless, it would be necessary in a near future to develop high density maps to validate the correct position of the scaffolds and extend the synteny analyses among flatfishes to accurately assign the 38 (Diopere et al. 2014) or 27 (Molina-Luzon et al. 2015a) LGs of these soles.

2.2.4 Characterization of putative sexual chromosomes in Senegalese sole

Karyotype analysis, as commented above, cannot distinguish between autosomes and sexual chromosomes. Nevertheless, *C. semilaevis* represents an exception to this rule with clear differences in size between chromosomes Z and W (Chen et al. 2014). *In silico* analysis mapping our Senegalese sole genome draft clearly identified scaffolds positioned on tongue sole chromosome Z covering 99.8% of its length (with a N50 of 93.8 kbp) but not chromosome W (only a 40.7% coverage was obtained when matching criteria were relaxed, what lead to locate only repetitive sequences), as depicted in Fig. 5. About 62.1% of SSCzx scaffolds encoded for polypeptides suggesting that sexual SSCzx super-scaffold in Senegalese sole could be related with female sex determination. Multicolor fluorescence in situ hybridization (mFISH) showed co-location of *dmrt1-dmrt2-drmt3* in the largest metacentric chromosome of *S. senegalensis* coinciding with the Z chromosome of *C. semilaevis*, a finding that would make this potentially a proto-sexual chromosome (Rebordinos et al. 2015, Portela-Bens et al. 2016). Moreover, since the female genome did not cover chromosome W of tongue sole, we hypothesized a homogametic (XX) sex determination system in Senegalese sole, as previously suggested using gynogenetic animals (Molina-Luzon et al. 2015b).

However, we cannot exclude that the absence of transcript coverage for chromosome W could be due to a reorganization and dispersion of chromosome W genes from tongue sole over the chromosomes of Senegalese sole. In fact, tongue sole chromosomes Z and W have been reported to contain 1,294 and 331 genes, respectively (Chen et al. 2014). After discarding genes with similar sequence (40% identity in at least 40% length of the sequence) in both chromosomes, 738 genes from Z and 46 genes from W were clearly different. These genes allowed the identification of 4,377 and 387 transcripts from SoleaDB that were orthologous to those Z and W genes, respectively. These transcripts have been mapped on female scaffolds in order to determine if they are located on the SSCzx or elsewhere and are a promising source of female-specific sequences.

2.3 *Epigenetics in sole*

In addition to genome and transcriptome, epigenetic regulation represents a third regulatory level that modulates genome activity and gene expression. In this way, animals integrate signals from several environmental factors (e.g., temperature) that in some cases can last a long time after the causing event. This explains how early imprinting (e.g., during embryonic and larval stages) can durably affect later physiology and performance. In aquaculture, the knowledge of epigenome has become a key tool to modulate early events to program specific phenotypes. However, the epigenetic regulation comprises complex mechanisms such as chromatin remodeling (DNA methylation and histone modification) and RNA-mediated modifications (non-coding RNA and microRNA). To date, very few genome-wide methylation profiles have been determined in aquaculture fish while a target-gene approach has been used to reveal environmental and/or sexual imprinting in several species (Navarro-Martin et al. 2011, Chen et al. 2014). Particularly in soles, temperature manipulation during early stages in *S. senegalensis* induces long lasting effects on muscle differentiation and sex ratios. Temperature cycles modify sex population ratio and induce a biased production of males and females (Blanco-Vives et al. 2011). Moreover, during the period from embryonic incubation until hatching, temperature induces a shift in metabolic capacity, growth and muscle cellularity in later stages (Campos et al. 2013a, Campos et al. 2013c). These changes were associated with altered methylation patterns and miRNA regulation. In this way, low temperatures (~ 15°C) significantly increase the methylation of the *myog* promoter in skeletal muscle, a myogenic regulatory factor (Campos et al. 2013b). Also, deep sequencing identified 320 miRNAs, some of them differentially regulated according to temperature during embryonic development with higher incubation temperatures increasing the expression of miRNAs related with growth, myogenesis, lipid metabolism and energy production in some specific developmental stages, confirming the involvement of both epigenetic mechanisms to regulate larval plasticity.

3. Genomic tools in sole

3.1 SoleaDB, a database for genomic resources in Solea spp.

To facilitate the use and management of genomic resources in soles, a specific database named SoleaDB that hosts all transcriptomic information for Senegalese sole and common sole has been released (Fig. 6; http://www.juntadeandalucia.es/agriculturaypesca/ifapa/soleadb_ifapa/). This database currently houses annotated sequences from 454 and Illumina projects developed in both species providing tissue-specific as well as species-specific transcriptomes. SoleaDB design is based on a user-friendly interface that can be browsed anonymously to facilitate researchers to search, identify and download easily all transcriptome information.

The «*home button*» tab contains general information about SoleaDB, released versions and the Assembly pipeline followed for each assembly. On the right panel, the user can check the information about versions of the bioinformatic tools used in the assembly pipeline, the current database release and funding credits. This page can be recalled by means of the «*Home*» button in the navigation bar.

The «*Assemblies button*» button takes the user to the start-up point for transcriptome navigation displaying the data structured as tab panels; note that the term "assemblies" is used here as synonym of "transcriptomes". The «All assemblies» tab appears selected

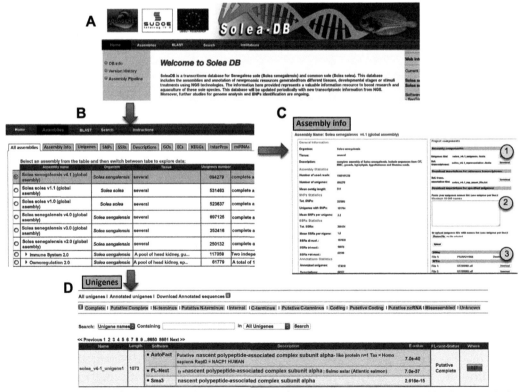

Fig. 6. SoleaDB description. (A), «*home button*» tab for database information; (B), «*Assemblies button*» tab for transcriptome selection; (C), «*Assembly info*» tab for transcriptome information. 1, fasta file and annotations, 2, window to download information for targeted transcripts, 3, raw data; (D), «*Unigenes tab*» for browsing transcripts within in each assembly.

by default showing all available transcriptomes (Fig. 6B). Although the most recent *S. senegalensis* transcriptome (currently *v*4.1) is selected by default, the user can select alternative transcriptomes by clicking on them. At the moment, users can only select and browse one transcriptome at each time. The next tab, «Assembly info» is divided into two sides, the left side containing detailed information about the chosen assembly (Fig. 6C). The right side offers access to the whole set and reference transcripts in Fasta format as well as their corresponding annotations and zebrafish ortholog identifiers. For 454 data, reports for sequence cleaning and the raw sequence data can be downloaded whereas those for Illumina were deposited in the GenBank SRA repository [(Accession numbers in Benzekri et al. (2014)]. Also, a window for downloading information for targeting transcripts using names has been included, , which is useful for RNA-seq analysis to download annotations for differentially expressed transcripts. The users will obtain a plain text file with fields separated by tabs that can be managed in different programs for functional analysis in other platforms as Cluego in Cytoscape or Webstalt (Bindea et al. 2009, Wang et al. 2013).

The «*Unigenes tab*» allows for browsing transcripts within each assembly (transcriptome) displaying the annotation as obtained by AutoFact, Full-LenghterNext and Sma3 annotators, ORF status and if included or not in the reference transcriptome. To highlight the reliability of annotations, common words in the description for each transcript are marked in green. Each transcript is identified by a unique code indicating the assembly version. Also, a link that directs the user to the corresponding transcript included in the reference transcriptome or the ortholog transcript in the other Solea species was added below the identifier.

The default view corresponds to the «Annotated transcripts» button, but all transcripts can be accessed using the «All transcripts» button (Fig. 6D). Moreover, transcripts can be browsed according to their ORF properties. For example, «Complete» button (Fig. 6D) filters transcripts that code for a complete protein; «Coding» button retains transcripts without predicted ORF that should code for a protein based on the testcode test. Also, transcripts corresponding to putative ncRNA precursors can be selected. Moreover, a word-targeted search can also be done.

Each transcript can be browsed by clicking on its identifiers displaying the information about sequence (available to download in Fasta format), the «*Annotations*» block containing the assigned descriptions by the three programs mentioned above, putative gene names as well as GOs, ECs, KEGG pathways associated, all accompanied by the associated *E*-value to assess annotation quality. The «Descriptions», «GO», «EC», «InterPro» and «KEGG» Tabs allow for a navigation following the annotations of transcriptomes. Clicking on one description, the user can access the collection of transcripts with such descriptions. This approach is particularly interesting to explore specific regulatory or metabolic pathways or identification of paralogous genes in the transcriptomes. Moreover, RefSeq and ENSEMBLE orthologues in zebrafish are included to facilitate gene-enrichment and functional analyses. The «ORF prediction» block (Fig. 3) describes the putative ORF, if complete, the position of start and/or stop codons, and the alignment that supports such predictions. The ORF prediction is an extremely useful information that will find direct use in laboratories (e.g., for primer or probes design).

The «SSR» and «SNPs» tabs identify the transcripts containing these molecular markers. The transcripts can be browsed by the type of SSR repeat motif. Moreover, molecular markers and positions in each transcript are shown for all transcripts.

To search information in the database, the Blast or text (keywords) search can be used. A blast-based search engine with customizable *E*-value for nucleotide (blastn) or amino acids (blastx) has been built against the transcriptomes selected by the corresponding checkbox. The user can download the results as an HTML file (the same information shown on screen) or as the direct blast output. Also, the user can search keywords, through all «Assemblies» panel.

3.2 High- and mid-throughput expression tools in the Senegalese sole. Examples

All these genomic resources in soles have facilitated the design of high-throughput gene expression tools such as custom microarrays. In common sole, a microarray containing 14,674 oligonucleotides was developed and tested on larval stages describing the main expression patterns during metamorphosis (Ferraresso et al. 2013). In Senegalese sole, the first oligonucleotide microarray developed contained 5,087 unique sequences spotted on glass slides (Cerda et al. 2008). This microarray platform was initially used to identify changes in gene expression levels among wild Senegalese sole males at different stages during the progression of spermatogenesis and identified approximately 400 differentially expressed transcripts (Forne et al. 2011). Approximately 400 transcripts were differentially expressed during the progression of spermatogenesis in wild male fish. These changes indicated mostly an activation of pathways related with transcriptional and translational activity in the maturing testis, paralleled with changes in transcripts involved in the ubiquitin/proteasome system, suggesting chromatin reorganization events in the final maturational phases of the testis. Furthermore, testicular maturation was accompanied by increased transcript levels of genes involved in sperm maturation and binding to the egg, suggesting a molecular basis for the increase in fertilizable competence in the Senegalese sole. This same microarray platform was used to describe the transcriptomic effects (90 differentially expressed genes) of *in vivo* administration of human chorionic gonadotropin, a placental gonadotropin used to mimic the effects of luteinizing hormone, on the wild Senegalese sole testis (Marin-Juez et al. 2011). They included a number of up-regulated genes reported to act as transcriptional and translational regulators supporting the transcriptomic changes observed during the progression of natural spermatogenesis in this species.

Transcriptomic profiling using microarray was also carried out for ovarian development. In this study, 118 transcripts were differentially expressed from vitellogenesis to final maturation, with increases in the expression of genes involved in energy production, antioxidant protection and zona radiata organization (Tingaud-Sequeira et al. 2009). Interestingly, atresia of ovarian follicles was associated with an up-regulation of genes involved in yolk reabsorption, leukocyte chemoattraction, angiogenesis and apoptosis prevention, suggesting that these events are part of the natural mechanism leading to oocyte degeneration and reabsorption that takes place after spawning.

More recently, a much more comprehensive microarray the Senegalese sole was developed that contained 43,303 probes based on the Agilent 4 × 44 K format (Benzekri et al. 2014). This microarray was designed using the 252,416 unigenes of *S. senegalensis v3* filtered to select the 30,119 complete and longest non-redundant transcripts, which expanded the maximum possible to the 3'-end, the most divergent region in sequence. Moreover, this set included 22,612 putative new coding genes in order increase representation of this kind of genes. Finally, all transcripts with orthologue and a set of 13,881 putative coding transcripts with the highest testcode index were selected as the 44,000 unigenes to synthetize probes for microarray. This tool was applied to the study of expression profiles in embryos incubated at two salinities (10 and 30 ppt) during the first 3 days after hatching. Statistical analysis showed a total of 2,816 differentially expressed transcripts ($P < 0.05$) of which

2,641 were up-regulated and 175 down-regulated at 10 ppt. Most of them were involved in osmoregulation, inflammation and innate immune system, chaperones, antioxidant enzymes, catabolic enzymes, vitamin A and retinoic acid metabolism and scavenging and bone and cartilage metabolism (Benzekri et al. 2014).

In addition to microarrays, the transcriptome information has been used to optimize mid-throughput tools for quantitative real-time PCR (qPCR) to evaluate a set of specific genes related with the innate immune system, lipid metabolism, oxidative stress and osmoregulation (Hachero-Cruzado et al. 2014, Jimenez-Fernandez et al. 2015, Montero et al. 2015). These tools have revealed as useful in larvae to successfully identify the molecular mechanisms associated with diet components in larvae and juvenile fish. A combined strategy based on RNA-seq followed by validation with mid-density qPCR arrays demonstrated that early pelagic larvae were able to establish compensatory mechanisms to manage correctly the lipid contents and store energy to accomplish successfully metamorphosis when supplied diets with different triacylglycerol (TAG) contents. Larvae fed with high levels of TAG activated in a coordinated way those genes related to lipid mobilization and transport such as apolipoproteins. In contrast, when the TAG content was limiting, they activated metabolic pathways related to fatty acid and TAG biosynthesis as well as protein to compensate energy expenditure and sustain growth rate (Hachero-Cruzado et al. 2014). Also, these qPCR arrays have been applied to elucidate the effects of micronutrients in larvae for the improvement of diet formulations and rearing protocols. Diets with different levels of vitamin C modulated differentially 16 transcripts related to antioxidant defence (glutathione peroxidase (1), tissue structure (collagens), stress (heat shock proteins 70 kDa, glucocorticoid receptors), glycolytic pathway, osmoregulation and pigmentation. In this way, the ascorbic acid levels during the rotifer-feeding period appear to be critical affecting growth, metamorphosis progress and malformations with long-term effects (Jimenez-Fernandez et al. 2015).

In juveniles, a qPCR array for immune-related genes was applied to evaluate the effects in the intestine of diets with high replacement of fish-based oils by alternative ingredients of terrestrial plants. Substituting up to 100% of fish oil by a vegetable blend of rapeseed, soybean and linseed oils had no consequences on sole juvenile performance and feed utilization (Montero et al. 2015), however, transcriptomic responses in the gut of animals fed with a 100% replacement of fish by soybean oil induced an over-expression of genes related to inflammation.

3.3 *Reference sole transcriptomes for RNA-seq studies*

Although microarray and qPCR arrays are very applied tools to study targeted gene expression, novel applications of NGS for the discovery of regulatory sequences or pathways are becoming increasingly popular. However, these whole-transcriptome analyses (RNA-seq) require reference transcriptomes that are correctly annotated and curated. In soles, the assembled transcriptomes (see above) included an extremely high number of transcripts that deviated greatly from the expected number of real transcripts in fish. Some obvious explanations for this discrepancy include misidentified alleles as genes, the presence of fragmented transcripts, spliced forms, immature mRNAs and their combinations as reported in Roman-Padilla et al. (2016). These issues, which artificially increase the actual number of transcripts, can therefore complicate expression statistics or functional analyses. To provide a smart solution, an in-house script with a strategy similar to the selection of probes for the microarray was used in order to identify representative transcripts in soles. Briefly, representative transcripts were selected from (i) the longest transcripts with unique and different orthologous ID and (ii) the putative, non-redundant new transcripts. This resulted in the selection of 59,514 transcripts for *S. senegalensis* (*v*4.1; 8.5% of initial set of valid transcripts) and 54,005 transcripts for *S. solea* (*v*1.1; 10.2%). With regards to other vertebrates, the number of protein-coding genes identified in these two species evidences that these reference transcriptomes still include a certain level of redundancy, which would deserve additional refining. The efficiency of this strategy in identifying non-redundant transcripts has been implemented in the software pipeline Full-LengtherNext (Seoane et al. in preparation). *In silico* analysis to confirm the adequacy of these reference transcriptomes was assessed by mapping of useful RNA-seq reads and obtaining up to 82.3–87.5% of reads onto transcripts similarly to whole-transcriptome mapping.

3.4 Molecular markers

Previous to NGS data arrival, some microsatellite markers (SSR) from ESTs and anonymous DNA regions had been characterized in the Senegalese sole (Funes et al. 2004, Porta et al. 2004, Chen et al. 2008, De La Herran et al. 2008, Molina-Luzon et al. 2012) and common sole (Iyengar et al. 2000, Garoia et al. 2006). More recently, high-throughput transcriptome characterization has provided a total of 266,434 and 316,388 SSRs (affecting 266,434 and 316,388 transcripts, respectively) in *S. senegalensis v4* and *S. solea v1* transcriptomes, respectively (Benzekri et al. 2014). SSRs with dinucleotide repeats were the most abundant and were mainly located in the UTRs. Interestingly, a cross-species comparison between Solea species identified 12,418 and 18,486 SSRs in ortholog transcripts. From these, 6,596 and 6,772 SSRs were fully conserved between orthologues in *S. senegalensis* and *S. solea*, respectively while 1,273 and 4,803 SSRs in *S. senegalensis* and *S. solea*, respectively, were considered species-specific as they were only found in the orthologs of only one species. Regarding SNPs, 337,315 SNPs were identified in *S. senegalensisv4* and 381,404 in *S. soleav1* transcriptomes with a significant proportion of SNPs occurring in transcripts containing an ORF (32.4% and 30.3%, respectively). All these sources of putative molecular markers represent an important source of molecular markers that can be applied to several other purposes including tools for pedigree assignment, genetic selection, population genetics and comparative genomics.

4. Genetic breeding in sole

The sole industry is rapidly growing and it is expected to develop significantly in the next few years. Currently, the hatchery and nursery facilities are producing fertilized eggs using wild broodstock. Although this strategy is adequate in the short term, it is not sustainable and assumes significant risks associated with disease control and production of high-quality fish and also involves higher costs due to the continuous extraction of soles from salt marshes and adaptation to captivity. Hence, once the hatchery and nursery routines have been established and standardized in RAS systems, genetic breeding programs have become a priority in order to reduce the high variability of juvenile performance and reduce the time to harvest while maintaining an optimal quality of the final product (Manchado et al. 2015a, Manchado et al. 2016).

Nevertheless, the design of genetic breeding programs, at least for the Senegalese sole, needs to address two major factors: (i) the low number of families produced in spontaneous spawns mainly constrained by the reproductive behaviour of soles and the low volume of sperm produced by males; (ii) the reproduction failure of F1 soles unable to yield fecundated eggs. The courtship behaviour that soles undergo at night prior to egg release and fecundation is required for a synchronization of gametes release (also referred to as a pseudocopula) and represents a key step in the production of fecundated eggs due to the extremely low volume of sperm production in this species (max 100 µl per male). This behaviour constitutes a major constraint to family production. Progeny analysis based on microsatellite assignments revealed that most of the spawns (61.7%) were composed of one family (Martin et al. 2014). Moreover, half-sib progeny in the same spawning rarely occurs and family representation is quite repetitive through different spawns even across yearly spawning seasons due to dominant pairs that reproduce preferably accounting for most of the offspring produced (Porta et al. 2006, Martin et al. 2014, Manchado et al. 2015a). As a consequence, the number of families that can be obtained under the standard reproduction strategies used by the industry is low with a high bias in the offspring and with a rapid loss of genetic variability and effective number of breeders (Porta et al. 2006, Manchado et al. 2015a). A similar situation has been described in *S. solea* (Blonk et al. 2009), demonstrating that genetic management is very important in the sole industry in order to manage the loss of genetic variability due to inbreeding. Therefore, it is necessary to establish models to successfully reproduce these species controlling the parental contributions to the next generation. Although artificial fertilization methodologies based on the use of hormone therapies have been optimized in the next few years (Morais et al. 2014), the variable quality and low quantity of sperm collected makes this technology feasible for research but not easy to apply at an industrial scale even for fish evaluation. As an alternative, new (and complementary) designs devoted to maximize family production are currently being evaluated that combine procedures for spawn synchronization in tanks with a low number of fish for simultaneously production of a higher number of families in an easy manner. This strategy could produce biological material in a high enough number to carry out larval rearing in communal tanks and downstream procedures (weaning, grading, …) that will facilitate animal testing and a real imbrication of the routine procedures for production in the industry.

Another factor that limits genetic breeding programs is the reproductive dysfunction of F1 in *S. senegalensis*. Recent studies have demonstrated that this problem is related to a lack of sexual behaviour of F1 males (Morais et al. 2014) and will be reviewed elsewhere in this book. Previous data from our group indicates that F1 females can successfully produced fertilized spawns when they are in contact with wild males (Manchado et al. 2015a). These data open the possibility to use F1 females as breeders in genetic selection programs establishing a combined strategy for selection that includes family selection for F_n females and progeny testing for wild males (Gjedrem et al. 2009). The situation is different in *S. solea*, which F1 reproduce regularly in captivity (Marie-Laure Bégout, personal communication).

In addition to reproductive features, genetic programs require additional technologies adapted to soles that assist in the identification of individuals and keep pedigree records including physical tagging methodologies and molecular pedigree analysis tools. In general, breeding programs in marine species, including soles, rear larvae from different families in communally in tanks to reduce the environmental component of variation as much as possible (Blonk et al. 2010b, Lee-Montero et al. 2015, Negrin-Baez et al. 2015). After weaning, sole juveniles (~ 0.8–1 g at 150 dph) can be easily tagged using different devices to maintain individual life history records. Although visible elastomeric tags have been successfully tested for individual growth evaluation based on a complex code based in number, color and location of marks (Salas-Leiton et al. 2010), they are not easy to apply in a higher number of individuals and are prone to errors. As an alternative, RFID devices seem to be the most adequate tag for genetic improvement programs. Nowadays, there exist various devices differing in size and cost to be applied intraperitoneally in soles, allowing automatic registration of several variables in a highly precise manner (Aparicio et al. 2015, Carballo et al. 2018).

Also, several molecular tools have been developed to reconstruct the pedigree of individuals. These studies have facilitated the development of parental assignment tools mainly multiplex assays based in a low number of loci (6–8 loci) suitable to be applied for parentage assessment of a reduced number of breeders (5–41 parents). Nevertheless, powerful multiplex tools based on genome analysis and the optimization of a higher number of polymorphic SSR markers are being developed to be routinely applied to large genetic breeding programs over several generations (Manchado et al. 2015b).

Heritability estimates in soles are still limited but some studies in common sole indicate that values for growth and shape are similar to other marine species. Heritability for weight at harvest ranged between 0.06 and 0.25 (Blonk et al. 2010b, 2010a, Blonk et al. 2010c, Mas-Muñoz et al. 2013), for body length was 0.28 (Blonk et al. 2010a) and for specific growth rate within the range of 0.2–0.3 (Mas-Muñoz et al. 2013). For morphometric parameters, heritability for ellipticity shape ranged between 0.34–0.45 depending on the measuring methodology (Blonk et al. 2010c). These values indicate that genetic breeding programs can help to improve these growth and commercialization traits. Interestingly, moderate negative correlations between shape and body weight or body height (but not body length) were identified indicating that a mixed selection index should be used to avoid a gain in shape circularity if fish with a faster growth is selected (Blonk et al. 2010c). Moreover, a genotype by environment interaction was determined in soles reared in ponds and RAS systems (Mas-Muñoz et al. 2013), making necessary specific breeding programs for each production system although at the moment most facilities have moved to produce soles in RAS systems.

5. Concluding remarks

Transcriptomes have proven to be a powerful tool for physiological studies dedicated to the understanding of breeding issues and other biological questions in aquaculture. However, currently the number of unique genes identified in soles far exceeds the actual number found in other species, indicating that further refining is needed to increase its usefulness. Genetic maps, including linkage, cytogenetic and physical maps are now available in *S. senegalensis*, providing complementary information to understand genome organization and evolution in soles. The sequencing of a male genome will help to clarify sexual chromosome organization given that a female genome is now assembled using the genome of tongue sole, with sex chromosomes ZZ for males and ZW for females, as reference. The new sequencing of the *S. solea* genome (male and female) would be a practical tool to generate new knowledge about the sex determination, speciation process and the common features between these two closely-related flatfish species. Particularly, the species comparison would be of high interest to shed light on the causes of the inability of *S. senegalensis*, but not *S. solea*, F1 to successfully reproduce. The epigenome, still in its infancy in soles, could be a complementary useful tool to identify possible mechanisms underlying reproductive problems. Further studies are required in order to characterize environmentally-cued regulation for either avoiding detrimental breeding conditions or to modulate growth potential. Finally present discrepancy between cytogenetic and genomic-based genetic maps, which differ in the number of chromosomes or linkage groups, is likely due to the rather low density of cytogenetic maps. The tremendous increase in putative molecular markers as a result of genome sequencing opens possibilities to achieve such an improvement. Taken together, the present (and future) omics data offer fantastic perspectives to contribute to the resolution of bottlenecks and the improvement of soles breeding including the development of genetic and selection programs.

Acknowledgements

This study has been funded by project AQUAGENET (SOE2/P1/E287) program INTERREG IVB SUDOE and European Regional Development Fund (FEDER/ERDF) and INIA and EU through FEDER 2014–2020 "Programa Operativo de Crecimiento Inteligente" project RTA2013-00023-C02-01, RTA2017-00054-C03 and MICINN (AGL2011- 25596 and AGL2014-51860-C2-1-P).

References

Amemiya, C.T., J. Alfoldi, A.P. Lee, S. Fan, H. Philippe, I. Maccallum et al. 2013. The African coelacanth genome provides insights into tetrapod evolution. Nature 496: 311–316.

Aparicio, M., A. Crespo, C. Carballo, C. Berbel, M.L. Begout, M. Manchado et al. 2015. Evaluación de diferentes sistema electrónicos de marcaje (RFID) para su aplicación en mejora genética de *Solea senegalensis* y *Sparus aurata*. XV Congreso Nacional y I Congreso Ibérico de Acuicultura, Huelva, Spain.

Benzekri, H., P. Armesto, X. Cousin, M. Rovira, D. Crespo, M.A. Merlo et al. 2014. *De novo* assembly, characterization and functional annotation of Senegalese sole (*Solea senegalensis*) and common sole (*Solea solea*) transcriptomes: integration in a database and design of a microarray. BMC Genomics 15: 952.

Benzekri, Hicham. 2016. Aportación bioinformática a la biología de especies marinas. Biochemistry and Molecular Biology, Universidad de Málaga, Málaga.

Bindea, G., B. Mlecnik, H. Hackl, P. Charoentong, M. Tosolini, A. Kirilovsky et al. 2009. ClueGO: a Cytoscape plug-in to decipher functionally grouped gene ontology and pathway annotation networks. Bioinformatics 25: 1091–1093.

Blanco-Vives, B., L.M. Vera, J. Ramos, M.J. Bayarri, E. Mañanos and F.J. Sanchez-Vazquez. 2011. Exposure of larvae to daily thermocycles affects gonad development, sex ratio, and sexual steroids in *Solea senegalensis*, Kaup. J. Exp. Zool. A. Ecol. Genet. Physiol. 315: 162–169.

Blonk, R., J. Komen, A. Kamstra, R. Crooijmans and J.A. van Arendonk. 2009. Levels of inbreeding in group mating captive broodstock populations of common sole (*Solea solea*), inferred from parental relatedness and contribution. Aquaculture 289: 26–31.

Blonk, R.J., H. Komen, A. Kamstra and J.A. van Arendonk. 2010a. Effects of grading on heritability estimates under commercial conditions: A case study with common sole, *Solea solea*. Aquaculture 300: 43–49.

Blonk, R.J., H. Komen, A. Kamstra and J.A. van Arendonk. 2010b. Estimating breeding values with molecular relatedness and reconstructed pedigrees in natural mating populations of common sole, *Solea solea*. Genetics 184: 213–219.

Blonk, R.J., H. Komen, A. Tenghe, A. Kamstra and J.A.M. van Arendonk. 2010c. Heritability of shape in common sole, *Solea solea*, estimated from image analysis data. Aquaculture 307:6-11.

Brawand, D., C.E. Wagner, Y.I. Li, M. Malinsky, I. Keller, S. Fan et al. 2014. The genomic substrate for adaptive radiation in African cichlid fish. Nature 513: 375–381.

Campos, C., M.F. Castanheira, S. Engrola, L.M. Valente, J.M. Fernandes and L.E. Conceicao. 2013a. Rearing temperature affects Senegalese sole (*Solea senegalensis*) larvae protein metabolic capacity. Fish Physiol Biochem. 39: 1485–1496.

Campos, C., L. Valente, L. Conceiçao, S. Engrola and J. Fernandes. 2013b. Temperature affects methylation of the myogenin putative promoter, its expression and muscle cellularity in Senegalese sole larvae. Epigenetics 8.

Campos, C., L.M. Valente, L.E. Conceicao, S. Engrola, V. Sousa, E. Rocha et al. 2013c. Incubation temperature induces changes in muscle cellularity and gene expression in Senegalese sole (*Solea senegalensis*). Gene 516: 209–217.

Carballo, C., C. Berbel, I. Guerrero-Cózar, E. Jiménez-Fernández, X. Cousin, M.L. Bégout and M. Manchado. 2018. Evaluation of different tags on survival, growth and stress response in the flatfish Senegalese sole. Aquaculture 494: 10–18.

Cerda, J., J. Mercade, J.J. Lozano, M. Manchado, A. Tingaud-Sequeira, A. Astola et al. 2008. Genomic resources for a commercial flatfish, the Senegalese sole (Solea senegalensis): EST sequencing, oligo microarray design, and development of the Soleamold bioinformatic platform. BMC Genomics 9: 508.

Chen, S.-L., C.-W. Shao, G.-B. Xu, X.-L. Liao and Y.-S. Tian. 2008. Development of 15 novel dinucleotide microsatellite markers in the Senegalese sole *Solea senegalensis*. Fisheries Sci. 74: 1357–1359.

Chen, S., G. Zhang, C. Shao, Q. Huang, G. Liu, P. Zhang et al. 2014. Whole-genome sequence of a flatfish provides insights into ZW sex chromosome evolution and adaptation to a benthic lifestyle. Nat. Genet 46: 253–260.

Chikhi, R. and P. Medvedev. 2014. Informed and automated k-mer size selection for genome assembly. Bioinformatics 30: 31–37.

Cross, I., A. Merlo, M. Manchado, C. Infante, J.P. Canavate and L. Rebordinos. 2006. Cytogenetic characterization of the sole *Solea senegalensis* (Teleostei: Pleuronectiformes: Soleidae): Ag-NOR, (GATA)n, (TTAGGG)n and ribosomal genes by one-color and two-color FISH. Genetica 128: 253–259.

De La Herran, R., F. Robles, J.I. Navas, A.M. Hamman-Khalifa, M. Herrera, I. Hachero et al. 2008. A highly accurate, single PCR reaction for parentage assignment in Senegal sole based on eight informative microsatellite loci. Aquaculture Res. 39: 1169–1174.

Diopere, E., G.E. Maes, H. Komen, F.A. Volckaert and M.A. Groenen. 2014. A genetic linkage map of sole (*Solea solea*): a tool for evolutionary and comparative analyses of exploited (flat)fishes. PLoS One 9: e115040.

Ferraresso, S., A. Bonaldo, L. Parma, S. Cinotti, P. Massi, L. Bargelloni et al. 2013. Exploring the larval transcriptome of the common sole (*Solea solea* L.). BMC Genomics 14: 315.

Forne, I., B. Castellana, R. Marin-Juez, J. Cerda, J. Abian and J.V. Planas. 2011. Transcriptional and proteomic profiling of flatfish (*Solea senegalensis*) spermatogenesis. Proteomics 11: 2195–2211.

Funes, V., E. Zuasti, G. Catanese, C. Infante and M. Manchado. 2004. Isolation and characterization of ten microsatellite loci for Senegal sole (*Solea senegalensis* Kaup). Mol. Ecol. Notes 4: 339–341.

Garcia-Cegarra, A., M.A. Merlo, M. Ponce, S. Portela-Bens, I. Cross, M. Manchado et al. 2013. A preliminary genetic map in *Solea senegalensis* (Pleuronectiformes, Soleidae) using BAC-FISH and next-generation sequencing. Cytogenet. Genome Res. 141: 227–240.

Garoia, F., S. Marzola, I. Guarniero, M. Trentini and F. Tinti. 2006. Isolation of polymorphic DNA microsatellites in the common sole *Solea vulgaris*. Mol. Ecol. Notes 6: 144–146.

Gjedrem, T. and M. Baranski. 2009. Selective breeding in Aquacutlure: An introduction. Edited by J.L. Nielsen. Vol. 10, Reviews: Methods and Technologies in Fish Biology and Fisheries. Heidelberg: Springer.

Hachero-Cruzado, I., A. Rodriguez-Rua, J. Roman-Padilla, M. Ponce, C. Fernandez-Diaz and M. Manchado. 2014. Characterization of the genomic responses in early Senegalese sole larvae fed diets with different dietary triacylglycerol and total lipids levels. Comp. Biochem. Physiol. Part D Genomics Proteomics 12: 61–73.

Howe, K., M.D. Clark, C.F. Torroja, J. Torrance, C. Berthelot, M. Muffato et al. 2013. The zebrafish reference genome sequence and its relationship to the human genome. Nature 496: 498–503.

Iyengar, A., S. Piyapattanakorn, D.M. Stone, D.A. Heipel, B.R. Howell, S.M. Baynes et al. 2000. Identification of microsatellite repeats in turbot (*Scophthalmus maximus*) and Dover sole (*Solea solea*) using a RAPD-based technique: characterization of microsatellite markers in Dover sole. Mar Biotechnol (NY) 2: 49–56.

Jimenez-Fernandez, E., M. Ponce, A. Rodriguez-Rua, E. Zuasti, M. Manchado and C. Fernandez-Diaz. 2015. Effect of dietary vitamin C level during early larval stages in Senegalese sole (*Solea senegalensis*). Aquaculture 443: 65–76.

Kettleborough, R.N., E.M. Busch-Nentwich, S.A. Harvey, C.M. Dooley, E. de Bruijn, F. van Eeden et al. 2013. A systematic genome-wide analysis of zebrafish protein-coding gene function. Nature 496: 494–497.

Lee-Montero, I., A. Navarro, D. Negrin-Baez, M.J. Zamorano, C. Berbel, J.A. Sanchez et al. 2015. Genetic parameters and genotype-environment interactions for skeleton deformities and growth traits at different ages on gilthead seabream (*Sparus aurata* L.) in four Spanish regions. Anim Genet. 46: 164–174.

Libertini, A., M. Mandrioli, M.S. Colomba, D. Bertotto, A. Francescon and R. Vitturi. 2002. A cytogenetic study of the common sole, Solea solea, from the Northern Adriatic Sea. Chromosome Science 6: 63–66.

Lyons, E. and M. Freeling. 2008. How to usefully compare homologous plant genes and chromosomes as DNA sequences. Plant J 53: 661–673.

Manchado, M., E. Zuasti, I. Cross, A. Merlo, C. Infante and L. Rebordinos. 2006. Molecular characterization and chromosomal mapping of the 5S rRNA gene in *Solea senegalensis*: a new linkage to the U1, U2, and U5 small nuclear RNA genes. Genome 49: 79–86.

Manchado, M., M. Aparicio, A.M. Crespo, C. Berbel and R. Zerolo. 2015a. Caracterización genética y primeras estimas de heredabilidad para crecimiento en lenguado senegalés. XV Congreso Nacional y I Congreso Ibérico de Acuicultura, Huelva, Spain.

Manchado, M., H. Benzekri, P. Seoane, R. Bautista, J.J. Sánchez, X. Cousin et al. 2015b. Development of genomic tools in senegalese sole: transcriptome assembly, annotated database, microarray and genome draft. The International Symposium on Genetics in Aquaculture XII., Santiago de Compostela, Spain.

Manchado, M., J.V. Planas, X. Cousin, L. Rebordinos and G. Claros. 2016. Chapter 9: Current status in other finfish species Description of current genomic resources for the gilthead seabream (*Sparus aurata*) and soles (*Solea senegalensis* and *Solea solea*). pp. 195–221. *In*: MacKenzie, S. and S. Jentoft [eds.]. Genomics in Aquaculture.

Marin-Juez, R., B. Castellana, M. Manchado and J.V. Planas. 2011. Molecular identification of genes involved in testicular steroid synthesis and characterization of the response to gonadotropic stimulation in the Senegalese sole (*Solea senegalensis*) testis. Gen. Comp. Endocrinol. 172: 130–139.

Martin, I., I. Rasines, M. Gómez, C. Rodríguez, P. Martinez and O. Chereguini. 2014. Evolution of egg production and parental contribution in Senegalese sole, *Solea senegalensis*, during four consecutive spawning seasons. Aquaculture 424-425: 45–52.

Mas-Muñoz, J., R. Blonk, J.W. Schrama, J.A. van Arendonk and H. Komen. 2013. Genotype by environment interaction for growth of sole (*Solea solea*) reared in an intensive aquaculture system and in a semi-natural environment. Aquaculture 230–235: 410–411.

Molina-Luzon, M.J., J.R. Lopez, R. Navajas-Perez, F. Robles, C. Ruiz-Rejon and R. De La Herran. 2012. Validation and comparison of microsatellite markers derived from Senegalese sole (*Solea senegalensis*, Kaup) genomic and expressed sequence tags libraries. Mol. Ecol. Resour. 12: 956–966.

Molina-Luzon, M.J., M. Hermida, R. Navajas-Perez, F. Robles, J.I. Navas, C. Ruiz-Rejon et al. 2015a. First haploid genetic map based on microsatellite markers in Senegalese sole (*Solea senegalensis*, Kaup 1858). Mar Biotechnol (NY) 17: 8–22.

Molina-Luzon, M.J., J.R. Lopez, F. Robles, R. Navajas-Perez, C. Ruiz-Rejon, R. De la Herran et al. 2015b. Chromosomal manipulation in Senegalese sole (Solea senegalensis Kaup, 1858): induction of triploidy and gynogenesis. J. Appl. Genet. 56: 77–84.

Montero, D., V. Benitez-Dorta, M.J. Caballero, M. Ponce, S. Torrecillas, M. Izquierdo et al. 2015. Dietary vegetable oils: effects on the expression of immune-related genes in Senegalese sole (*Solea senegalensis*) intestine. Fish Shellfish Immunol. 44: 100–108.

Morais, S., C. Aragão, E. Cabrita, L.E.C. Conceição, M. Constenla, B. Costas et al. 2014. New developments and biological insights into the farming of *Solea senegalensis* reinforcing its aquaculture potential. Rev. Aquacult. 6: 1–37.

Navarro-Martin, L., J. Vinas, L. Ribas, N. Diaz, A. Gutierrez, L. Di Croce et al. 2011. DNA methylation of the gonadal aromatase (*cyp19a*) promoter is involved in temperature-dependent sex ratio shifts in the European sea bass. PLoS Genet 7: e1002447.

Negrin-Baez, D., A. Navarro, I. Lee-Montero, M. Soula, J.M. Afonso and M.J. Zamorano. 2015. Inheritance of skeletal deformities in gilthead seabream (Sparus aurata) - lack of operculum, lordosis, vertebral fusion and LSK complex. J. Anim. Sci. 93: 53–61.

Ponce, M., E. Salas-Leiton, A. Garcia-Cegarra, A. Boglino, O. Coste, C. Infante et al. 2011. Genomic characterization, phylogeny and gene regulation of g-type lysozyme in sole (*Solea senegalensis*). Fish Shellfish Immunol. 31: 925–937.

Porta, J. and M.C. Alvarez. 2004. Development and characterization of microsatellites from Senegal sole (*Solea senegalensis*). Mol. Ecol. Notes 4: 277–279.

Porta, J., J.M. Porta, G. Martínez-Rodríguez and M.C. Alvarez. 2006. Development of a microsatellite multiplex PCR for Senegalese sole (*Solea senegalensis*) and its application to broodstock management. Aquaculture 256: 159–166.

Portela-Bens, S., M.A. Merlo, M.E. Rodríguez, I. Cross, M. Manchado, N. Kosyakova, T. Liehr and L. Rebordinos. 2016. Integrated gene mapping and synteny studies give insights into the evolution of a sex proto-chromosome in *Solea senegalensis*. Chromosoma 60: 441–453.

Prachumwat, A. and W.H. Li. 2008. Gene number expansion and contraction in vertebrate genomes with respect to invertebrate genomes. Genome Res 18: 221–232.

Rebordinos, L., A. Merlo, M.E. Rodriguez, I. Cross, M. Manchado, N. Kosyakova et al. 2015. Evidencias de la existencia de un proto-cromosoma sexual en *Solea senegalensis*. XV Congreso Nacional y I Congreso Ibérico de Acuicultura, Huelva, Spain.

Roman-Padilla, J., A. Rodriguez-Rua, M.G. Claros, I. Hachero-Cruzado and M. Manchado. 2016. Genomic characterization and expression analysis of four apolipoprotein A-IV paralogs in Senegalese sole (*Solea senegalensis* Kaup). Comp. Biochem. Physiol. B. Biochem. Mol. Biol. 191: 84–98.

Salas-leiton, E., V. Anguis, A. Rodriguez-Rua and J.P. Cañavate. 2010. Stocking homogeneous size groups does not improve growth performance of Senegalese sole (*Solea senegalensis*, Kaup 1858) juveniles: Individual growth related to fish size. Aquacult Eng. 43: 108–113.

Tingaud-Sequeira, A., F. Chauvigne, J. Lozano, M.J. Agulleiro, E. Asensio and J. Cerda. 2009. New insights into molecular pathways associated with flatfish ovarian development and atresia revealed by transcriptional analysis. BMC Genomics 10: 434.

Vega, L., E. Díaz, I. Cross and L. Rebordinos. 2002. Caracterizacion citogenetica e isoenzimatica del lenguado *Solea senegalensis* Kaup, 1858. Bol. Inst. Esp. Oceanogr. 18: 245–250.

Venkatesh, B., A.P. Lee, V. Ravi, A.K. Maurya, M.M. Lian, J.B. Swann et al. 2014. Elephant shark genome provides unique insights into gnathostome evolution. Nature 505: 174–179.

Wang, J., D. Duncan, Z. Shi and B. Zhang. 2013. WEB-based GEne SeT AnaLysis Toolkit (WebGestalt): update 2013. Nucleic Acids Res 41: W77–83.

Index